ADVANCES IN CHEMICAL PHYSICS

VOLUME 114

EDITORIAL BOARD

Advances in
CHEMICAL PHYSICS

Edited by

I. PRIGOGINE

Center for Studies in Statistical Mechanics and Complex Systems
The University of Texas
Austin, Texas
and
International Solvay Institutes
Université Libre de Bruxelles
Brussels, Belgium

and

STUART A. RICE

Department of Chemistry
and
The James Franck Institute
The University of Chicago
Chicago, Illinois

VOLUME 114

AN INTERSCIENCE® PUBLICATION
JOHN WILEY & SONS, INC.
NEW YORK • CHICHESTER • WEINHEIM • BRISBANE • SINGAPORE • TORONTO

This book is printed on acid-free paper. ∞

An Interscience® Publication

Copyright © 2000 by John Wiley & Sons, Inc. All rights reserved.

Published simultaneously in Canada.

For ordering and customer service, call 1-800-CALL-WILEY.

Library of Congress Catalog Number: 58-9935

ISBN 0-471-39267-7

Printed in the United States of America.

10 9 8 7 6 5 4 3 2 1

CONTRIBUTORS TO VOLUME 114

A. I. BURSHTEIN, The Weizmann Institute of Science, Rehovot 76100, Israel

TUCKER CARRINGTON, JR., Département de chimie, Université de Montréal, C.P. 6128, succursale Centre-ville, Montreal H3C3J7, Quebec, Canada

M. GRUEBELE, Department of Chemistry and Beckman Institute for Advanced Science and Technology, University of Illinois, Urbana, IL 61801

JOHN C. LIGHT, Department of Chemistry and The James Franck Institute, The University of Chicago, Chicago, IL 60637

ARLENE M. LOUGHAN, Department of Applied Mathematics and Theoretical Physics, The Queen's University of Belfast, Belfast, Northern Ireland

MOSHE SHAPIRO, Department of Chemical Physics, The Weizmann Institute of Science, Rehovot 76100, Israel

BRIAN SUTCLIFFE, Department of Chemistry, University of York, York Y01 5DD, England and Laboratoire de Chimie Physique Moléculaire, Université Libre de Bruxelles, B-1050 (present address)

INTRODUCTION

Few of us can any longer keep up with the flood of scientific literature, even in specialized subfields. Any attempt to do more and be broadly educated with respect to a large domain of science has the appearance of tilting at windmills. Yet the synthesis of ideas drawn from different subjects into new, powerful, general concepts is as valuable as ever, and the desire to remain educated persists in all scientists. This series, *Advances in Chemical Physics*, is devoted to helping the reader obtain general information about a wide variety of topics in chemical physics, a field that we interpret very broadly. Our intent is to have experts present comprehensive analyses of subjects of interest and to encourage the expression of individual points of view. We hope that this approach to the presentation of an overview of a subject will both stimulate new research and serve as a personalized learning text for beginners in a field.

I. PRIGOGINE
STUART A. RICE

CONTENTS

ADVANCES IN CHEMICAL PHYSICS

VOLUME 114

THE DECOUPLING OF ELECTRONIC AND NUCLEAR MOTIONS IN THE ISOLATED MOLECULE SCHRÖDINGER HAMILTONIAN

BRIAN SUTCLIFFE

*Department of Chemistry, University of York, York, England and Laboratoire de Chimie Physique Moléculaire, Université Libre de Bruxelles, Belgium**

CONTENTS

*Present address.

Advances in Chemical Physics, Volume 114, Edited by I. Prigogine and Stuart A. Rice.
ISBN 0-471-39267-7 © 2000 John Wiley & Sons, Inc.

I. INTRODUCTION

The idea that the proper way to treat molecules in quantum mechanics is to try to separate the electronic and nuclear motions as far as possible, dates from the very earliest days of the subject. The genesis of the idea is usually attributed to Born and Oppenheimer [1], but it is an idea that was in the air at the time, for the earliest papers in which the idea is used predate the publication of their paper. The physical picture that informs the attempted separation is one well known and widely used even in classical mechanics, namely, division of the problem into a set of rapidly moving particles, here electrons, and a much more slowly moving set, here the nuclei. Experience is that it is wise to attempt to separate such incommensurate motions both to calculate efficiently and to obtain a useful physical picture.

The object of the separation in the molecular case is to formulate an electronic motion problem in which the nuclear positions can be treated as parameters and whose solutions can be used to solve the nuclear motion problem. The insights arising from classical chemistry seem to predicate that, for lowish energies, the nuclear motion function should be strongly peaked at a nuclear geometry that corresponds to the traditional molecular geometry. A function of this kind would allow a good account of the electronic structure of a molecule to be given in terms of a single choice for the nuclear geometry.

The discussion in this chapter attempts to clarify in what circumstances and to what extent quantum mechanics can legitimately be used to support the classical chemical picture of an isolated molecule that retains a reasonably well specified shape while performing small vibrations and undergoing essentially rigid rotations. Carl Eckart [2] was among the first to discuss how this picture might be supported, but he did so in a context that assumed the separation of electronic and nuclear motions. In his approach, the electrons are regarded simply as providing a potential. This potential is invariant under all uniform translations and rigid rotations of the nuclei that form the molecule. It is usually referred to as a *potential-energy surface* (strictly, *hyper*surface) and the nuclei are said to move on this surface. Eckart actually treats the nuclear motions by classical rather than quantum mechanics but it is his approach and developments from it that have dominated the interpretation of molecular spectra since 1936, and we shall return to it later in the chapter.

Schrödinger's Hamiltonian describing the molecule as a system of N charged particles in a coordinate frame fixed in the laboratory is

$$\hat{H}(\mathbf{x}) = -\frac{\hbar^2}{2}\sum_{i=1}^{N}\frac{1}{m_i}\nabla^2(\mathbf{x}_i) + \frac{e^2}{8\pi\epsilon_0}\sum_{i,j=1}^{N}{}'\frac{Z_iZ_j}{x_{ij}} \qquad (1.1)$$

where the separation between particles is defined by

$$x_{ij}^2 = \sum_{\alpha}(x_{\alpha j} - x_{\alpha i})^2 \qquad (1.2)$$

It is convenient to regard $\mathbf{x}_i$ as a column matrix of three Cartesian components $x_{\alpha i}$, $\alpha = x, y, z$ and to regard $\mathbf{x}_i$ collectively as the $3 \times N$ matrix $\mathbf{x}$. Each of the particles has mass m_i and charge $Z_i e$. The charge numbers Z_i are positive for a nucleus and minus one for an electron. In a neutral system the charge numbers sum to zero.

To distinguish between electrons and nuclei, the variables are split into two sets, one set consisting of L variables, $\mathbf{x}_i^e$, describing the electrons and the other set of H variables, $\mathbf{x}_i^n$, describing the nuclei and $N = L + H$. When it is necessary to emphasise this split, Eq. (1.1) will be denoted $\hat{H}(\mathbf{x}^n, \mathbf{x}^e)$.

If the full problem has eigenstates that are square-integrable so that

$$\hat{H}(\mathbf{x}^n, \mathbf{x}^e)\psi(\mathbf{x}^n, \mathbf{x}^e) = E\psi(\mathbf{x}^n, \mathbf{x}^e) \qquad (1.3)$$

then, as Hunter showed [3], an eigenstate can be written rigorously as a product of the required form

$$\psi(\mathbf{x}^n, \mathbf{x}^e) = \Phi(\mathbf{x}^n)\phi(\mathbf{x}^n, \mathbf{x}^e)$$

The function $\Phi(\mathbf{x}^n)$ (chosen to be nontrivial) is determined as a solution of an effective nuclear motion equation, obtained from the full equation [Eq. (1.3)] by multiplying from the left by $\phi^*(\mathbf{x}^n, \mathbf{x}^e)$ and integrating over all $\mathbf{x}_i^e$. The difficulty with this approach is that it seems not to be possible to specify the required electronic function $\phi(\mathbf{x}^n, \mathbf{x}^e)$ except in terms of the full exact solution. But if that solution were known, it would at least be arguable that one might not be interested in separating electronic from nuclear motions. However, it is possible to guess a plausible electronic function, to formulate the required nuclear motion problem and to use its solutions to improve the electronic function and so on, until a satisfactory solution is obtained.

Traditionally a good guess for the electronic wavefunction is supposedly provided by a solution of the clamped nuclei electronic Hamiltonian:

$$\hat{H}^{cn}(\mathbf{a}, \mathbf{x}^e) = -\frac{\hbar^2}{2m} \sum_{i=1}^{L} \nabla^2(\mathbf{x}_i^e) - \frac{e^2}{4\pi\epsilon_0} \sum_{i=1}^{H} \sum_{j=1}^{L} \frac{Z_i}{|\mathbf{x}_j^e - \mathbf{a}_i|}$$

$$+ \frac{e^2}{8\pi\epsilon_0} \sum_{i,j=1}^{N}{}' \frac{1}{|\mathbf{x}_i^e - \mathbf{x}_j^e|} \tag{1.4}$$

This Hamiltonian is obtained from the original one [Eq. (1.1)] by assigning the values $\mathbf{a}_i$ to the nuclear variables $\mathbf{x}_i^n$, hence the designation *clamped nuclei* for this form. Within the electronic problem each nuclear position $\mathbf{a}_i$ is treated as a parameter. For solution of the entire problem, the electronic wavefunction must be available for all values of these parameters. The energy obtained from the solution of this problem depends on the nuclear parameters and is commonly called the *electronic energy*. It is usual to think of the potential-energy surface, used in the Eckart approach, as formed by adding the electronic and the classical nuclear repulsion energy.

One way to improve a guess might be to extend the single product form to a sum of products form using more of the spectrum of the clamped nuclei Hamiltonian. Thus the wavefunction for the full problem is written as

$$\Psi(\mathbf{x}) = \sum_p \Phi_p(\mathbf{x}^n)\psi_p(\mathbf{x}^n, \mathbf{x}^e) \tag{1.5}$$

where $\psi_p(\mathbf{x}^n, \mathbf{x}^e)$ is the pth clamped nuclei solution.

This is essentially the approach advocated by Born* in the early 1950s, which is given in Appendix VIII in the book by Born and Huang [5]. This approach to the problem of separation, we shall call the *standard* approach. It should be noticed that it is a distinctly different method from that originally proposed by Born and Oppenheimer. Thus no discussion of that original work is provided in the present chapter. Discussions of that work can be found, for example, in Combes [6] and Klein et al. [7]. Other methods to separating electronic and nuclear motions have been proposed; for examples, see Hagedorn [8,9] and Essén [10]. An article by Ballhausen and Hansen [11] provides a careful survey of usage in the context of the Born approach and suggests what the various possible methods might best be called.

---

*Actually this approach was first proposed by John Slater [4], but apparently it was not employed by any other worker and he did not develop it further.

But it is not possible to use the Hunter approach or the standard approach to approximate it, because the Hamiltonian (1.1) is invariant under uniform translations in the frame fixed in the laboratory. This means that the center of molecular mass moves through space like a free particle and the states of a free particle are not quantised and eigenfunctions are not square-integrable. The center-of-mass motion must therefore be separated out to disentangle any bound states from the continuum, and this must be done in such a way as to lead to a translationally invariant form for the potential-energy surface if the Eckart approach is to be maintained.

The molecular Hamiltonian is invariant under all orthogonal transformations (rotation reflections) of the particle variables in the frame fixed in the laboratory. The usual potential-energy surface is similarly invariant so it is sensible to separate as far as possible the orientational motions of the system from its purely internal motions because it is in terms of the internal motions that the potential-energy surface is expressed. The internal motions comprise dilations, contractions, and deformations of a specified configuration of particle variables. Expressed colloquially, the potential-energy surface is a function of the molecular geometry only.

The Hamiltonian is also invariant under the permutation of the variable sets of all identical particles, and it is natural to require, if possible, that the potential energy surface be invariant under permutations of the variable sets of the identical nuclei. But in any case it is essential that the permutational properties of the various parts of the decoupled wavefunction be well specified to ensure that they will be properly symmetric or antisymmetric, according to particle type, when spin variables are included. These observations seem perfectly harmless, but attainment of the objectives expressed by them causes some very unpleasant complications.

The way in which translational motion can be removed from the problem is well understood from classical mechanics. However, it involves an essentially arbitrary choice of translationally invariant coordinates, and there is always one less such coordinate than the original number because of the center-of-mass coordinate. After this separation is made, it is clearly a matter of opinion and/or convention how the translationally invariant coordinates should be identified. So the role of the coordinates in specifying either electronic or nuclear motions becomes problematic.

The separation of orientation variables from internal variables is also a well-understood problem, but in order to achieve the separation, the three orientation variables must be specified in terms of a particular way of fixing a coordinate frame in the (nonrigid) assembly of particles. This choice is, like the choice of translationally invariant coordinates, quite arbitrary. Whatever choice is made, however, there will always be a configuration of the particles that causes the definition of the frame to fail. This can be appreciated by imagining a three-particle system and mentally fixing the frame in it so as to place all three particles in a plane. This defines an axis normal to the plane, say the z axis, and the x and y axes can then be chosen at will to form a righthanded orthogonal system of axes. But if the three particles are collinear, then the frame definition fails. In addition to this complication, the definition of three orientation variables removes a further three variables from the translationally invariant ones to leave $3N - 6$ variables to describe the internal motions. The internal coordinates must be invariant to any orthogonal transformation of the translationally invariant coordinates and so must be expressible in terms of scalar products of these coordinates. The choice here is again quite arbitrary in terms of the already arbitrary choice of translationally invariant coordinates.

The internal and orientation coordinates can obviously be expressed directly in terms of the original coordinates in the laboratory frame, and so there is no absolute need to consider the translational motion in a separate step. However, to do so aids clear exposition. Where it is necessary to reexpress the internal and orientation coordinates in the original set, as it is when talking about the Eckart Hamiltonian, we shall do so by a distinct discussion.

A permutation of the variables describing a set of identical particles will naturally leave the center-of-mass coordinate unchanged, but it will induce a linear transformation among the translationally invariant coordinates, and that transformation will certainly not have the form of a standard permutation matrix; nor will it be orthogonal. It is also perfectly plausible for such a permutation to induce changes in the orientation variables and in the internal coordinates, and it might well actually mix the orientation variables and the internal coordinates. If this happens, any separation of orientational from internal motion would be difficult.

Any account of decoupling electronic and nuclear motions has to face up to these difficulties and also to carry conviction in the choices required at each stage to remove the inherent arbitrariness in the process. But it has to be recognized that such choices are matters of judgment, if not quite matters of taste, and hence matters about which reasonable disagreements are possible. It is, however, quite disconcerting to appreciate (supposing that the account that we shall give cannot be replaced with one involving fewer judgments) how many physical pictures that inform much experimental thinking owe at least as much to the insights of approximate solution makers as they do to the underlying quantum mechanics of the problem.

The plan of the review is first to consider the translationally invariant problem, making as few assumptions as possible. The object of this consideration is to outline what is known about the spectrum of the problem and the extent to which discrete bound states might occur. This is not just a formal problem, for physical experience is that not all combinations of atoms can result in stable molecules, and it is therefore a matter of some importance to know whether the system being considered has solutions of the required kind.

Next the translationally invariant coordinates are split into two parts, one of which can be plausibly identified with the electronic motion and the other with the nuclear motion. Further specialisations of these choices are then made to keep standard behavior under permutations of both electrons and of nuclei for the translationally invariant electronic coordinates and to separate as far as possible the perceived electronic and nuclear motions in the Hamiltonian. A guiding principle in the choices made is to try to achieve a form for the electronic part of the problem as close as possible to the clamped nuclei form.

Using this choice of translation-free coordinates, the orientation and internal coordinates are introduced by fixing a coordinate frame in the body in such a way as to continue, as far as possible, the separation of the perceived electronic and nuclear motions in the Hamiltonian. By constructing matrix elements of both the dipole and the Hamiltonian operator between the rotational functions, we may obtain effective operators for the purely internal motions. Since the transformation to orientation and internal coordinates is nonlinear, the Jacobian is not trivial and its construction and properties are also discussed.

An account is given of the full rotation-reflection symmetry of both the Hamiltonian and of the dipole operator. The effect of permutations of identical particles upon the chosen coordinates is considered, and a general discussion of the maximal symmetry of the problem at this stage is attempted.

A portion of the Hamiltonian that can plausibly be designated an electronic Hamiltonian is then identified and its relationship with the usual clamped nuclei Hamiltonian is exhibited. The potential-energy surface is then studied, and the problems that arise on the intersection of two such surfaces are discussed. The general uncoupling problem is then discussed, and the status of the rotating vibrating molecule picture is examined, especially from the standpoint of the Eckart Hamiltonian.

In the light of this, a summary is given of the parts of the problem that can be considered as well posed and hence well understood and what difficulties still remain in the way of a full account that is congruent with our usual approach to describing an isolated molecule.

II. REMOVING TRANSLATIONAL MOTION

To remove the center-of-mass motion from the full molecule Hamiltonian, all that is needed is a coordinate transformation symbolized by

$$(\mathbf{t}\,\mathbf{X}_\mathrm{T}) = \mathbf{x}\,\mathbf{V} \tag{2.1}$$

where $\mathbf{t}$ is a $3 \times (N-1)$ matrix and $\mathbf{X}_T$ is a 3×1 matrix, so that the combined (bracketed) matrix on the left is $3 \times N$. $\mathbf{V}$ is an $N \times N$ matrix that, from the structure of the left side of (2.1), has a special last column whose elements are

$$V_{iN} = M_T^{-1} m_i, \qquad M_T = \sum_{i=1}^{N} m_i. \tag{2.2}$$

Hence $\mathbf{X}_T$ is the standard center-of-mass coordinate,

$$\mathbf{X}_T = M_T^{-1} \sum_{i=1}^{N} m_i \mathbf{x}_i \tag{2.3}$$

As the coordinates $\mathbf{t}_j, j = 1, 2, \ldots, N-1$ are to be translationally invariant, we require on each remaining column of $\mathbf{V}$

$$\sum_{i=1}^{N} V_{ij} = 0, \qquad j = 1, 2, \ldots, N-1 \tag{2.4}$$

and it is easy to see that this last equation forces $\mathbf{t}_j \rightarrow \mathbf{t}_j$ as $\mathbf{x}_i \rightarrow \mathbf{x}_i + \mathbf{a}$, all i. The $\mathbf{t}_i$ are independent if the inverse transformation

$$\mathbf{x} = (\mathbf{t}\, \mathbf{X}_T) \mathbf{V}^{-1} \tag{2.5}$$

exists. The structure of the right side of this equation shows that the bottom row of $\mathbf{V}^{-1}$ is special, and, without loss of generality, we may require its elements to be

$$(\mathbf{V}^{-1})_{Ni} = 1, \qquad i = 1, 2 \ldots, N \tag{2.6}$$

The inverse requirement on the remainder of $\mathbf{V}^{-1}$ implies that

$$\sum_{i=1}^{N} (\mathbf{V}^{-1})_{ji} m_i = 0, \qquad j = 1, 2, \ldots, N-1 \tag{2.7}$$

When we write the column matrix of the Cartesian components of the partial derivative operator as $\partial/\partial \mathbf{x}_i$, the coordinate change (2.1) gives

$$\frac{\partial}{\partial \mathbf{x}_i} = \sum_{j=1}^{N-1} V_{ij} \frac{\partial}{\partial \mathbf{t}_j} + m_i M_T^{-1} \frac{\partial}{\partial \mathbf{X}_T} \tag{2.8}$$

and when it seems more convenient, this column matrix of derivative operators will also be denoted as the vector grad operator $\vec{\nabla}(\mathbf{x}_i)$.

Defining a set of translationally invariant coordinates is often referred to as "specifying a coordinate frame fixed in space."

A. The Translationally Invariant Hamiltonian

The Hamiltonian (1.1) in the new coordinates becomes

$$\hat{H}(\mathbf{t}, \mathbf{X}_\mathrm{T}) = -\frac{\hbar^2}{2} \sum_{i,j=1}^{N-1} \frac{1}{\mu_{ij}} \vec{\nabla}(\mathbf{t}_i) \cdot \vec{\nabla}(\mathbf{t}_j) + \frac{e^2}{8\pi\epsilon_0} \sum_{i,j=1}^{N}{}' \frac{Z_i Z_j}{r_{ij}(\mathbf{t})} - \frac{\hbar^2}{2M_\mathrm{T}} \nabla^2(\mathbf{X}_\mathrm{T}).$$

$$(2.9)$$

Here

$$\frac{1}{\mu_{ij}} = \sum_{k=1}^{N} m_k^{-1} V_{ki} V_{kj} \qquad i,j = 1, 2, \ldots, N-1 \qquad (2.10)$$

The $(N-1)$-dimensional square matrix composed of all the $1/\mu_{ij}$ is denoted as $\boldsymbol{\mu}^{-1}$. As will be seen later, it is inverse to a matrix of reduced masses $\boldsymbol{\mu}$, but in our approach it is straightforward to construct the elements of the inverse directly. The operator r_{ij} is just x_{ij} as given by (2.2) but expressed as a function of $\mathbf{t}_i$. Thus

$$r_{ij}(\mathbf{t}) = \left(\sum_\alpha \left(\sum_{k=1}^{N-1} ((\mathbf{V}^{-1})_{kj} - (\mathbf{V}^{-1})_{ki}) t_{\alpha k} \right)^2 \right)^{1/2} \qquad (2.11)$$

In (2.9) the $\vec{\nabla}(\mathbf{t}_i)$ are grad operators expressed in the Cartesian components of $\mathbf{t}_i$ and the last term represents the center-of-mass kinetic energy. Since the center-of-mass variable does not enter the potential term, the center-of-mass problem may be separated off completely so that the full solution is of the form

$$T(\mathbf{X}_\mathrm{T})\Psi(\mathbf{t}) \qquad (2.12)$$

where $\Psi(\mathbf{t})$ is a solution to the problem specified by the first two terms in (2.9) which will be denoted collectively by $\hat{H}(\mathbf{t})$ and referred to as the *translationally invariant* Hamiltonian.

B. The Translationally Invariant Angular Momentum Operator

We shall need the angular momentum operator and the electric dipole operator in terms of X_T and the $\mathbf{t}_i$; the total angular momentum operator may be written as

$$\hat{\mathbf{L}}(\mathbf{x}) = \frac{\hbar}{i} \sum_{i=1}^{N} \hat{\mathbf{x}}_i \frac{\partial}{\partial \mathbf{x}_i} \qquad (2.13)$$

where $\hat{\mathbf{L}}(\mathbf{x})$ is a column matrix of Cartesian components and the skew-symmetric matrix $\hat{\mathbf{x}}_i$ is

$$\hat{\mathbf{x}}_i = \begin{pmatrix} 0 & -x_{zi} & x_{yi} \\ x_{zi} & 0 & -x_{xi} \\ -x_{yi} & x_{xi} & 0 \end{pmatrix} \qquad (2.14)$$

The matrix $\hat{\mathbf{x}}_i$ can also be written in terms of the infinitesimal rotation generators

$$\mathbf{M}^x = \begin{pmatrix} 0 & 0 & 0 \\ 0 & 0 & 1 \\ 0 & -1 & 0 \end{pmatrix} \quad \mathbf{M}^y = \begin{pmatrix} 0 & 0 & -1 \\ 0 & 0 & 0 \\ 1 & 0 & 0 \end{pmatrix} \quad \mathbf{M}^z = \begin{pmatrix} 0 & 1 & 0 \\ -1 & 0 & 0 \\ 0 & 0 & 0 \end{pmatrix}$$

$$(2.15)$$

so that

$$\hat{\mathbf{x}}_i = \sum_{\alpha} x_{\alpha i} \mathbf{M}^{\alpha T} \qquad (2.16)$$

A variable symbol with a caret over it will, from here on, be used to denote a skew-symmetric matrix as defined by (2.16). The matrices are simply representations of the antisymmetric (Levi–Civita) tensor, such that $\mathbf{M}^{\alpha}_{\beta\gamma}$ is $e_{\alpha\beta\gamma}$.

Although the angular momentum operator is not invariant under a general orthogonal transformation of particle coordinates, it is invariant under coordinate inversion. It is also invariant under the permutation of any of the coordinate sets, not just those of identical particles.

Transforming to coordinates $\mathbf{X}_T, \mathbf{t}_i$ gives

$$\hat{\mathbf{L}}(\mathbf{x}) \rightarrow \frac{\hbar}{i} \sum_{i=1}^{N-1} \hat{\mathbf{t}}_i \frac{\partial}{\partial \mathbf{t}_i} + \frac{\hbar}{i} \hat{\mathbf{X}}_T \frac{\partial}{\partial \mathbf{X}_T} \qquad (2.17)$$

and after this point the first term will be denoted as $\hat{\mathbf{L}}(\mathbf{t})$ and called the *translationally invariant angular momentum*. The square of this operator and its z component commute, and both commute with the translationally invariant Hamiltonian. It is therefore possible to choose the eigenfunctions (if there are any) of the translationally invariant Hamiltonian to be angular momentum eigenfunctions.

C. The Translationally Invariant Dipole Operator

The total dipole operator is

$$\mathbf{d}(\mathbf{x}) = e \sum_{i=1}^{N} Z_i \mathbf{x}_i \tag{2.18}$$

and simple transformation using (2.5) leads to

$$\mathbf{d}(\mathbf{t}, \mathbf{X}_T) = e \sum_{i=1}^{N-1} \tilde{Z}_i \mathbf{t}_i + e Z_T \mathbf{X}_T \tag{2.19}$$

The first term in (2.19) will be denoted $\mathbf{d}(\mathbf{t})$ and the effective charges are given by

$$\tilde{Z}_i = \sum_{j=1}^{N} (\mathbf{V}^{-1})_{ij} Z_j, \qquad Z_T = \sum_{i=1}^{N} Z_i \tag{2.20}$$

As is to be expected, the center-of-mass-dependent term in the dipole vanishes if the system is neutral, that is, if $Z_T = 0$.

The dipole operator is not invariant under a general orthogonal transformation of particle coordinates, but it only changes sign under coordinate inversion. It is invariant under the permutation of the coordinate sets of particles with identical charges.

D. Example: Translationally Invariant Coordinates

Taking the neutral ammonia molecule NH_3 as an example with $N = 14$, let the variables representing the electrons be $\mathbf{x}_i$, $i = 1, 2, \ldots, 10$. Let each of these variables be associated with a charge -1 and a mass m. Let the variables denoting the protons, each with charge $+1$ and mass m_p be $\mathbf{x}_i$, $i = 11, 12, 13$ and that representing the nitrogen nucleus with charge $+7$ and mass m_N be denoted $\mathbf{x}_{14}$.

A perfectly reasonable set of translationally invariant coordinates is the set in which the nitrogen nucleus is taken as the origin of coordinates:

$$\mathbf{t}_i = \mathbf{x}_i - \mathbf{x}_{14}, \qquad i = 1, \ldots, 13$$

Another possibility would be to choose the center of nuclear mass $\mathbf{X}$ as the origin, where

$$\mathbf{X} = M^{-1}(m_p\mathbf{x}_{11} + m_p\mathbf{x}_{12} + m_p\mathbf{x}_{13} + m_N\mathbf{x}_{14}), \qquad M = (m_N + 3m_p)$$

and with such a choice it would be quite arbitrary which of the $\mathbf{x}_i$ were excluded from the definitions of the $\mathbf{t}_i$.

Or if ammonia were considered a collision complex of H and NH_2, one of the protons could be chosen as the origin for one of the electrons and the nitrogen as the origin for the remaining electrons and protons with the vector distance between the proton first chosen and the nitrogen nucleus as the remaining variable. So the translationally invariant coordinates might be

$$\mathbf{t}_1 = \mathbf{x}_1 - \mathbf{x}_{11}$$

$$\mathbf{t}_i = \mathbf{x}_i - \mathbf{x}_{14}, \qquad i = 2,\ldots,13$$

It would be also be possible to choose a single electronic variable alone as the coordinate origin so that

$$\mathbf{t}_{i-1} = \mathbf{x}_i - \mathbf{x}_1, \qquad i = 2,\ldots,14$$

and although perhaps a bit perverse, there could be no technical objections to such a choice.

For definiteness, consider the choice of the nitrogen nucleus as the origin. Then the inverse transformation produces

$$\mathbf{x}_i = \mathbf{t}_i - m_p M^{-1}(\mathbf{t}_{11} + \mathbf{t}_{12} + \mathbf{t}_{13}) + \mathbf{X}, \qquad i = 1,2,\ldots,10$$

$$\mathbf{x}_i = \mathbf{t}_i - m_p M^{-1}(\mathbf{t}_{11} + \mathbf{t}_{12} + \mathbf{t}_{13}) + \mathbf{X}, \qquad i = 11,12,13$$

$$\mathbf{x}_{14} = -m_p M^{-1}(\mathbf{t}_{11} + \mathbf{t}_{12} + \mathbf{t}_{13}) + \mathbf{X}$$

In these expressions the center of nuclear mass has been used simply for compactness. It can be eliminated in favor of the center of mass using

$$\mathbf{X}_T = \mathbf{X} + M_T^{-1}m\sum_{i=1}^{10}\mathbf{t}_i - 10m_p M^{-1}mM_T^{-1}(\mathbf{t}_{11} + \mathbf{t}_{12} + \mathbf{t}_{13})$$

and its presence does not imply any kind of approximation.

The inverse reduced mass matrix in this choice has elements

$$\frac{1}{\mu_{ii}} = \frac{1}{m} + \frac{1}{m_N}, \quad i = 1, 2, \ldots, 10, \qquad \frac{1}{\mu_{ij}} = \frac{1}{m_N}, \quad i \neq j = 1, 2, \ldots, 10$$

$$\frac{1}{\mu_{ii}} = \frac{1}{m_p} + \frac{1}{m_N}, \quad i = 11, 12, 13, \qquad \frac{1}{\mu_{ij}} = \frac{1}{m_N}, \quad i \neq j = 11, 12, 13$$

$$\frac{1}{\mu_{ij}} = \frac{1}{m_p} + \frac{1}{m_N}, \quad i = 11, 12, 13, \qquad j = 1, 2, \ldots, 10$$

and the interparticle distances are

$$x_{ij} = |\mathbf{t}_i - \mathbf{t}_j|, \qquad i \neq j = 1, 2, \ldots, 13$$
$$x_{i14} = |\mathbf{t}_i|, \qquad i = 1, 2, \ldots, 13$$

E.   The Bound States of the Translationally Invariant Problem

Whether $\hat{H}(\mathbf{t})$ has bound states is rather problematic. Bound-state eigenfunctions are square-integrable, have negative energies and lie below the start of the continuous spectrum of the system. The location of the bottom of the continuous spectrum a system can be determined, in principle, with the aid of the Hunziker–van Winter–Zhislin (HVZ) theorem. Accounts of this theorem can be found in Vol. IV of Reed and Simon [12] and in Vol. 3 of Thirring [13], but roughly speaking, it asserts that this limit is the lowest energy at which the system can break up into two noninteracting clusters. For atoms the limit is obviously at the first ionization energy and so is (in hartrees), 0 for the hydrogen atom, nearly -2 for the helium atom, and so on. The tricky part of the problem is to determine whether there are any states below this limit and if so, how many there are.

In the single-nucleus case, that is, for an atom rather than a molecule, if it is electrically positive or neutral, it was shown first by Zhislin [14] and later by Uchiyama [15], that there are an infinite number of bound states. Proofs of this are accessible in Thirring [13] or Simon, [16]. If the system is negative then it has at most a finite number of bound states as, again, was first shown by Zhislin [17]. For example, Nyden Hill [18] showed explicitly that the H$^-$ ion had only one bound state. If the system has more than one nucleus but the nuclei are held clamped then its spectral properties are similar to those for an atom [19]. If the nuclei are allowed to move, however, then general results are few. It is known that if a molecule gets either too positive or too negative, it does not have any bound states at all [20,21]. For a neutral system, Simon [22] argued very persuasively that only if the position of the bottom of the continuum is determined by breakup into a pair of oppositely charged ionic clusters will a neutral molecule have an infinite number of bound states. If the clusters are neutral, then there will be, at most, only a finite number of bound states. Vugal'ter and Zhislin [23], were able to show rigorously that Simon's belief about the neutral clusters was well founded, and Evans et al. [24], were able to demonstrate the validity of his belief regarding the charged clusters. An examination of tables of experimental values of electron affinities and ionization energies seems to negate the possibility that any diatomic molecule has an infinite number of bound states. This observation is not inconsistent with spectroscopic experience. The awkward problem is to know whether a neutral system has any bound states at all. Ordinary chemical experience suggests that at least some atomic combinations do not have any bound states but, so far, there are no rigorous results to support the theory that a particular kind of neutral system has no bound states. As for showing that a molecular system has some bound states, at present the most that has been proved is that the hydrogen molecule has at least one bound state [25]. It is necessary, therefore, in order to make progress, to summon up courage and go ahead as if, for all problems of interest, there were a number of bound states.

Of course, this emphasis on bound states should not obscure the fact that even using the translationally invariant Hamiltonian, there will be, among its solutions, continuum states resulting from the relative motion of two or more fragments. The energies of the continuum states will be above the energies of the bound states (if any) and thus will not give rise to the problems that the translation continuum does. Such states can be treated, formally at least in an expansion, by stipulating that the sum includes integration over the continuum. In future expansions this interpretation will be assumed where appropriate, but in practice, of course, such integration cannot be achieved. These continuum states are of the greatest interest in any discussions of scattering and hence of reactions.

It is now apparent why the separation of translation is problematic for the identification of electrons and nuclei. In the translationally invariant Hamiltonian the inverse effective mass matrix μ^{-1} and the form of the potential functions involving the r_{ij} depend intimately on the choice of V, and this choice is essentially arbitrary. In particular, it should be observed that because there are only $N - 1$ translationally invariant variables, they cannot—except in the most conventional of senses—be regarded as particle coordinates and that the nondiagonal nature of μ^{-1} and the peculiar form of the r_{ij} also militate against any simple particle interpretation of the translationally invariant Hamiltonian. It is thus not an entirely straightforward matter to identify electrons and nuclei once this separation has been made.

F. Distinguishing Electronic and Nuclear Motions

Because the spectrum of the translationally invariant Hamiltonian is independent of the choice of $\mathbf{V}$, the way in which it is chosen would be immaterial if it were possible to construct exact solutions. As it is necessary in practice to use an approximation scheme, it is rational to choose a $\mathbf{V}$ adapted to the scheme. Here the aim is to design the approximate wavefunctions so as to decouple electronic and nuclear motions as far as possible. Ideally, the electronic part of the approximate wavefunction should consist of solutions of an electronic Hamiltonian that is as much like the clamped nuclei Hamiltonian as possible and whose eigenvalues can be identified with electronic energies as functions of the nuclear coordinates. The nuclear part of the approximation should, again ideally, consist of solutions to a problem composed of the nuclear motion kinetic-energy operator expressed in terms of coordinates that arise from the original nuclear coordinates alone together with a potential that consists of a sum of the electronic and the potential energy of nuclear repulsion.

It seems reasonable, therefore, to require that the translationally invariant nuclear coordinates be expressible entirely in terms of the original nuclear coordinates. Thus, analogously to (2.1)

$$(\mathbf{t}^n \mathbf{X}) = \mathbf{x}^n \mathbf{V}^n \qquad (2.21)$$

Here $\mathbf{t}^n$ is a $3 \times (H-1)$ matrix and $\mathbf{X}$ is a 3×1 matrix. $\mathbf{V}^n$ is an H by H matrix whose last column is special, with elements

$$V_{iH}^n = M^{-1} m_i, \qquad M = \sum_{i=1}^{H} m_i \qquad (2.22)$$

so that $\mathbf{X}$ is the coordinate of the centre of nuclear mass. The elements in each of the first $H-1$ columns of $\mathbf{V}^n$ each sum to zero, precisely as in (2.4), to ensure translational invariance.

If the $\mathbf{t}^n$ are independent, then

$$\mathbf{x}^n = (\mathbf{t}^n \mathbf{X})(\mathbf{V}^n)^{-1} \qquad (2.23)$$

just as in (2.5) and, just as in (2.6), the bottom row of $(\mathbf{V}^n)^{-1}$ is special with elements

$$((\mathbf{V}^n)^{-1})_{Hi} = 1, \qquad i = 1, 2, \ldots, H \qquad (2.24)$$

while like (2.7) the inverse requirement on the remaining rows gives

$$\sum_{i=1}^{H}((\mathbf{V}^n)^{-1})_{ji}m_i = 0, \qquad j = 1, 2, \ldots, H-1 \qquad (2.25)$$

The comparable electronic coordinates will have to involve the original nuclear coordinates so that (2.21) becomes generalized to

$$(\mathbf{t}^e\mathbf{t}^n\mathbf{X}) = (\mathbf{x}^e\mathbf{x}^n)\begin{pmatrix} \mathbf{V}^e & \mathbf{0} \\ \mathbf{V}^{ne} & \mathbf{V}^n \end{pmatrix} \qquad (2.26)$$

where $\mathbf{t}^e$ is a $3 \times L$ matrix. It is not possible to choose $\mathbf{V}^{ne}$ to be a null matrix and to satisfy simultaneously the translational invariance requirements as specified by (2.4) while leaving the whole matrix nonsingular. Given that it exists, the inverse of (2.26) may be written as

$$(\mathbf{x}^e\mathbf{x}^n) = (\mathbf{t}^e\mathbf{t}^n\mathbf{X})\begin{pmatrix} (\mathbf{V}^e)^{-1} & \mathbf{0} \\ \mathbf{B} & (\mathbf{V}^n)^{-1} \end{pmatrix} \qquad (2.27)$$

where

$$\mathbf{B} = -(\mathbf{V}^n)^{-1}\mathbf{V}^{ne}(\mathbf{V}^e)^{-1} \qquad (2.28)$$

The bottom row of $\mathbf{B}$ is special in the same way as is the bottom row of $(\mathbf{V}^n)^{-1}$ and consists of the elements

$$B_{Hi} = 1, \qquad i = 1, 2, \ldots, L \qquad (2.29)$$

Using (2.27), we obtain

$$\mathbf{x}_i^e = \mathbf{X} + \sum_{j=1}^{L} \mathbf{t}_j^e((\mathbf{V}^e)^{-1})_{ji} + \sum_{j=1}^{H-1} \mathbf{t}_j^n B_{ji} \qquad (2.30)$$

and

$$\mathbf{x}_i^n = \mathbf{X} + \sum_{j=1}^{H-1} \mathbf{t}_j^n((\mathbf{V}^n)^{-1})_{ji} \qquad (2.31)$$

A change from $\mathbf{X}$ to $\mathbf{X}_T$ has no effect on the expressions for $\mathbf{t}^e$ and $\mathbf{t}^n$, but it does affect the expression for the inverse. Using (2.27) and the definitions of

X and $\mathbf{X}_T$, it follows that

$$\mathbf{X}_T = \mathbf{X} + \sum_{i=1}^{L} s_i^e \mathbf{t}_i^e + \sum_{i=1}^{H-1} s_i^n \mathbf{t}_i^n \qquad (2.32)$$

where

$$s_i^e = M_T^{-1} m \sum_{j=1}^{L} ((\mathbf{V}^e)^{-1})_{ij}, \qquad s_i^n = M_T^{-1} m \sum_{j=1}^{L} B_{ij} \qquad (2.33)$$

so that using (2.32) **X** can be eliminated in favor of $\mathbf{X}_T$ whenever necessary.

The derivative operator (2.8) can now be distinguished as consisting of two parts:

$$\frac{\partial}{\partial \mathbf{x}_i^e} = \sum_{j=1}^{L} V_{ij}^e \frac{\partial}{\partial \mathbf{t}_j^e} + m M_T^{-1} \frac{\partial}{\partial \mathbf{X}_T} \qquad (2.34)$$

$$\frac{\partial}{\partial \mathbf{x}_i^n} = \sum_{j=1}^{L} V_{ij}^{ne} \frac{\partial}{\partial \mathbf{t}_j^e} + \sum_{j=1}^{H-1} V_{ij}^n \frac{\partial}{\partial \mathbf{t}_j^n} + m_i M_T^{-1} \frac{\partial}{\partial \mathbf{X}_T} \qquad (2.35)$$

1.   Example: Translationally Invariant Coordinates Identifying Electrons and Nuclei

It would obviously not be possible to cast the problem in the form presented in (2.26) when using a single electronic variable alone as origin, but that does not seem too large a restriction. All the other choices made earlier do conform to the desired division, but the choice made to describe the collision complex gives rise to some problems. To see these, label the electronic coordinates $\mathbf{x}_i^e, i = 1, 2, \ldots, 10$ and the nuclear coordinates $\mathbf{x}_i^n, i = 1, 2, 3, 4$ with the nitrogen as 4. The translationally invariant coordinates are then obtained by rewriting the earlier result in the notation of this section:

$$\mathbf{t}_1^e = \mathbf{x}_1^e - \mathbf{x}_1^n$$
$$\mathbf{t}_i^e = \mathbf{x}_i^e - \mathbf{x}_4^n, \qquad i = 2, \ldots, 10$$
$$\mathbf{t}_i^n = \mathbf{x}_i^n - \mathbf{x}_4^n, \qquad i = 1, 2, 3$$

The inverse here is

$$\mathbf{x}_i^n = \mathbf{t}_i^n - m_p M^{-1}(\mathbf{t}_1^n + \mathbf{t}_2^n + \mathbf{t}_3^n) + \mathbf{X}, \qquad i = 1, 2, 3$$
$$\mathbf{x}_4^n = -m_p M^{-1}(\mathbf{t}_1^n + \mathbf{t}_2^n + \mathbf{t}_3^n) + \mathbf{X}$$
$$\mathbf{x}_1^e = \mathbf{t}_1^e + \mathbf{t}_1^n - m_p M^{-1}(\mathbf{t}_1^n + \mathbf{t}_2^n + \mathbf{t}_3^n) + \mathbf{X}$$
$$\mathbf{x}_i^e = \mathbf{t}_i^e - m_p M^{-1}(\mathbf{t}_1^n + \mathbf{t}_2^n + \mathbf{t}_3^n) + \mathbf{X}, \qquad i = 2, 3, \ldots, 10$$

$$(2.36)$$

so that

$$\mathbf{x}_1^e - \mathbf{x}_2^e = \mathbf{t}_1^e - \mathbf{t}_2^e + \mathbf{t}_1^n$$

and thus certain of the interelectronic variables involve what are, nominally, the nuclear variables in the transformed set. Furthermore, if the variables $\mathbf{x}_1^n$ and $\mathbf{x}_2^n$ are permuted, then

$$\mathbf{t}_1^e \to \mathbf{t}_1^e + \mathbf{t}_1^n - \mathbf{t}_2^n$$

which is a rather untoward mixing of the nominal nuclear and electronic variables.

G. The Hamiltonian Operator with Electrons Identified

The translationally invariant kinetic-energy operator arising from the first term in (2.9) expands into three parts:

$$\hat{K}(\mathbf{t}) \to \hat{K}^e(\mathbf{t}^e) + \hat{K}^n(\mathbf{t}^n) + \hat{K}^{en}(\mathbf{t}^n, \mathbf{t}^e) \tag{2.37}$$

Here

$$\hat{K}^e(\mathbf{t}^e) = -\frac{\hbar^2}{2} \sum_{i,j=1}^{L} \frac{1}{\mu_{ij}^e} \vec{\nabla}(\mathbf{t}_i^e) \cdot \vec{\nabla}(\mathbf{t}_j^e) \tag{2.38}$$

with

$$\frac{1}{\mu_{ij}^e} = m^{-1} \sum_{k=1}^{L} V_{ki}^e V_{kj}^e + \sum_{k=1}^{H} m_k^{-1} V_{ki}^{ne} V_{kj}^{ne} \tag{2.39}$$

and

$$\hat{K}^n(\mathbf{t}^n) = -\frac{\hbar^2}{2} \sum_{i,j=1}^{H-1} \frac{1}{\mu_{ij}^n} \vec{\nabla}(\mathbf{t}_i^n) \cdot \vec{\nabla}(\mathbf{t}_j^n) \tag{2.40}$$

in which $1/\mu_{ij}^n$ is defined just as in (2.10) but in terms of only the nuclear masses and using $\mathbf{V}^n$. Finally

$$\hat{K}^{en}(\mathbf{t}^n, \mathbf{t}^e) = -\frac{\hbar^2}{2} \sum_{i=1}^{H-1} \sum_{j=1}^{L} \frac{1}{\mu_{ij}^{ne}} \left(\vec{\nabla}(\mathbf{t}_i^n) \cdot \vec{\nabla}(\mathbf{t}_j^e) + \vec{\nabla}(\mathbf{t}_j^e) \cdot \vec{\nabla}(\mathbf{t}_i^n) \right) \tag{2.41}$$

with

$$\frac{1}{\mu_{ij}^{ne}} = \sum_{k=1}^{H} m_k^{-1} V_{ki}^{n} V_{kj}^{ne} \tag{2.42}$$

The rather cumbersome form in which $\hat{K}^{en}$ is written is to anticipate the fact that a change of variables to orientation and internal coordinates, it will no longer generally be the case that the operators $\vec{\nabla}(\mathbf{t}_j^{e})$ and $\vec{\nabla}(\mathbf{t}_i^{n})$ commute.

The interparticle distances needed for the potential-energy operator can now be written distinguishing the particle types. Using (2.30), $\mathbf{x}_j^{e} - \mathbf{x}_i^{e}$ becomes

$$\mathbf{x}_j^{e} - \mathbf{x}_i^{e} = \sum_{k=1}^{L} \mathbf{t}_k^{e}(((\mathbf{V}^{e})^{-1})_{kj} - ((\mathbf{V}^{e})^{-1})_{ki}) + \sum_{k=1}^{H-1} \mathbf{t}_k^{n}(B_{kj} - B_{ki}) \tag{2.43}$$

and so is not generally expressible in terms of the $\mathbf{t}_i^{e}$ alone and furthermore, the form is not invariant under permutation of identical nuclei, nor does it change in the ordinary way under the permutation of electrons. Similar problems arise with $\mathbf{x}_j^{n} - \mathbf{x}_i^{e}$:

$$\mathbf{x}_i^{n} - \mathbf{x}_j^{e} = \sum_{k=1}^{H-1} \mathbf{t}_k^{n}((\mathbf{V}^{n})_{ki}^{-1} - B_{kj}) + \sum_{k=1}^{L} \mathbf{t}_k^{e}(\mathbf{V}^{e})_{kj}^{-1} \tag{2.44}$$

H. The Angular Momentum Operator with Electrons Identified

The angular momentum operator constituting the first term in (2.17) can be realized directly in terms of $\mathbf{t}^{e}$ and $\mathbf{t}^{n}$ to become

$$\hat{\mathbf{L}}(\mathbf{t}^{n}, \mathbf{t}^{e}) = \frac{\hbar}{i} \sum_{i=1}^{H-1} \hat{\mathbf{t}}_i^{n} \frac{\partial}{\partial \mathbf{t}_i^{n}} + \frac{\hbar}{i} \sum_{i=1}^{L} \hat{\mathbf{t}}_i^{e} \frac{\partial}{\partial \mathbf{t}_i^{e}} \tag{2.45}$$

The first part of (2.45) can be regarded as the translationally invariant angular momentum associated with the nuclei and the second part as the translationally invariant angular momentum associated with the electrons.

I. The Dipole Operator with Electrons identified

The dipole operator constitute the first term in (2.19) becomes

$$\mathbf{d}(\mathbf{t}^{n}, \mathbf{t}^{e}) = e \sum_{i=1}^{H-1} \tilde{Z}_i^{n} \mathbf{t}_i^{n} - e \sum_{i=1}^{L} \tilde{Z}_i^{e} \mathbf{t}_i^{e} \tag{2.46}$$

where

$$\tilde{Z}_i^n = \sum_{j=1}^{H}((\mathbf{V}^n)^{-1})_{ij}Z_j - \sum_{j=1}^{L}B_{ij} - Z_T s_i^n \tag{2.47}$$

$$\tilde{Z}_i^e = \sum_{j=1}^{L}((\mathbf{V}^e)^{-1})_{ij} - Z_T s_i^e \tag{2.48}$$

in which $\mathbf{V}^{-1}$ in (2.20) has been realized using (2.27) and (2.32).

J. Permutationally Restricted Translationally Invariant Coordinates

Even though the choices made so far regarding translationally invariant coordinates do identify electrons and nuclei, they do not do so, as has been seen, in a settled way. That is, permutations of the original coordinates of one class of particle can induce changes in the translationally invariant coordinates used to designate the other class. It would therefore seem wise to restrict further the form of $\mathbf{V}^{ne}$ and of $\mathbf{V}^e$ so that permutations of any set of the original coordinates induce changes only in the corresponding set of translationally invariant coordinates.

The general permutation of identical particles can be written as

$$\mathscr{P}(\mathbf{x}^e\mathbf{x}^n) = (\mathbf{x}^e\mathbf{x}^n)\begin{pmatrix} \mathbf{P}^e & \mathbf{0} \\ \mathbf{0} & \mathbf{P}^n \end{pmatrix} \tag{2.49}$$

where $\mathbf{P}^e$ and $\mathbf{P}^n$ are standard permutation matrices.

Using (2.26) and (2.27), it follows that

$$\mathscr{P}(\mathbf{t}^e\mathbf{t}^n\mathbf{X}) = (\mathbf{t}^e\mathbf{t}^n\mathbf{X})\begin{pmatrix} (\mathbf{V}^e)^{-1} & \mathbf{0} \\ \mathbf{B} & (\mathbf{V}^n)^{-1} \end{pmatrix}\begin{pmatrix} \mathbf{P}^e & \mathbf{0} \\ \mathbf{0} & \mathbf{P}^n \end{pmatrix}\begin{pmatrix} \mathbf{V}^e & \mathbf{0} \\ \mathbf{V}^{ne} & \mathbf{V}^n \end{pmatrix} \tag{2.50}$$

To achieve the required invariance, the matrix on the right side of this equation must be block diagonal, and this occurs only if

$$\mathbf{B}\mathbf{P}^e\mathbf{V}^e + (\mathbf{V}^n)^{-1}\mathbf{P}^n\mathbf{V}^{ne} = \mathbf{0}_{H,L} \tag{2.51}$$

The most general way in which this can be achieved is to require the following relations to hold:

$$\mathbf{P}^e\mathbf{V}^e = \mathbf{V}^e\mathbf{P}^e \tag{2.52}$$

$$\mathbf{V}^{ne}\mathbf{P}^e = \mathbf{V}^{ne} \tag{2.53}$$

$$\mathbf{P}^n\mathbf{V}^{ne} = \mathbf{V}^{ne} \tag{2.54}$$

If these relations hold then $\mathbf{P}^e$ can be taken as a common factor to the right and, using (2.28) for $\mathbf{B}$, we may take $(\mathbf{V}^n)^{-1}$ as a common factor to the left in (2.51). The factored matrix then vanishes identically.

The most general form for $\mathbf{V}^e$ that satisfies (2.52) is

$$(\mathbf{V}^e)_{ij} = \delta_{ij} + a \tag{2.55}$$

where a is a constant. The inverse has elements

$$((\mathbf{V}^e)^{-1})_{ij} = \delta_{ij} - \frac{a}{1 + La} \tag{2.56}$$

This shows that a can take any value (including 0) except $-1/L$.

The physical content of the requirement (2.53) is that any electronic variable in the problem should have exactly the same relationship to the nuclear variables as does any other electronic variable. The physical content of the requirement (2.54) is that every member of a set of identical nuclei must enter into the definition of $\mathbf{t}^e$ in the same way. Thus (2.53) is satisfied by requiring all the columns of $\mathbf{V}^{ne}$ to be identical, and from now on a typical column will be denoted $\mathbf{v}$. If the entries in $\mathbf{v}$ are identical for identical nuclei, then (2.54) is also satisfied.

Using these results in (2.58), it follows that the columns of $\mathbf{B}$ are identical to one another, and the typical column will from now on be written as $\mathbf{b}$. Using (2.30) with these restrictions, it follows that $\mathbf{x}_{ij}^e$ becomes $\mathbf{t}_{ij}^e$, as required. Using the results in (2.33) gives

$$s_i^e = \frac{m}{M_T(1 + La)}, \qquad s_i^n = M_T^{-1} L m b_i, \qquad \mathbf{b} = -\frac{(\mathbf{V}^n)^{-1}\mathbf{v}}{(1 + La)} \tag{2.57}$$

With these choices the explicit forms for the translationally invariant coordinates are

$$\mathbf{t}_i^e = \mathbf{x}_i^e + a \sum_{j=1}^{L} \mathbf{x}_j^e + \sum_{j=1}^{H} v_j \mathbf{x}_j^n \tag{2.58}$$

$$\mathbf{t}_i^n = \sum_{j=1}^{H} \mathbf{x}_j^n \mathbf{V}_{ji}^n, \qquad i = 1, 2, \ldots, H - 1 \tag{2.59}$$

and for translational invariance of $\mathbf{t}_i^e$, from (2.58), we obtain

$$(1 + La) = -\sum_{j=1}^{H} v_j \tag{2.60}$$

Their inverses are

$$\mathbf{x}_i^e = \mathbf{X} + \mathbf{t}_i^e - \frac{a}{(1+La)} \sum_{j=1}^{L} \mathbf{t}_j^e + \sum_{j=1}^{H-1} b_j \mathbf{t}_j^n \tag{2.61}$$

$$\mathbf{x}_i^n = \mathbf{X} + \sum_{j=1}^{H-1} \mathbf{t}_j^n ((\mathbf{V}^n)^{-1})_{ji} \tag{2.62}$$

If relations (2.52)–(2.54) are satisfied, it is easy to show that if in (2.50) $\mathbf{X}$ is replaced by $\mathbf{X}_T$, the equation generalizes to

$$\mathscr{P}(\mathbf{t}^e \mathbf{t}^n \mathbf{X}_T) = (\mathbf{t}^e \mathbf{t}^n \mathbf{X}_T) \begin{pmatrix} \mathbf{P}^e & 0 & 0 \\ 0 & \mathbf{H} & 0 \\ 0 & 0 & 1 \end{pmatrix} \tag{2.63}$$

where

$$(\mathbf{H})_{ij} = ((\mathbf{V}^n)^{-1} \mathbf{P}^n \mathbf{V}^n)_{ij}, \qquad i,j = 1, 2, \ldots, H-1 \tag{2.64}$$

The matrix $\mathbf{H}$ is generally not in standard permutational form, nor is it orthogonal even though it has determinant ± 1 according to the sign of $|\mathbf{P}^n|$.

The form of the derivative operators and of the variables remains unchanged from that given in the previous section, but the partition made here does enable a more specific structure to be given to them with parts attributable to the types of particle.

The derivative operator (2.34) can now simplified as

$$\frac{\partial}{\partial \mathbf{x}_i^e} = \sum_{j=1}^{L} (\delta_{ij} + a) \frac{\partial}{\partial \mathbf{t}_j^e} + m M_T^{-1} \frac{\partial}{\partial \mathbf{X}_T} \tag{2.65}$$

and (2.35) as

$$\frac{\partial}{\partial \mathbf{x}_i^n} = v_i \sum_{j=1}^{L} \frac{\partial}{\partial \mathbf{t}_j^e} + \sum_{j=1}^{H-1} V_{ij}^n \frac{\partial}{\partial \mathbf{t}_j^n} + m_i M_T^{-1} \frac{\partial}{\partial \mathbf{X}_T} \tag{2.66}$$

1. Example: Permutationally Restricted Translationally Invariant Coordinates for NH₃ with Electrons and Nuclei Identified

It should now be clear that the coordinates chosen according to (2.26) and (2.27) in which the conditions (2.52)–(2.54) are satisfied, are a restricted choice. The choice of coordinates made earlier to describe ammonia as a collision complex, with (2.36) yielding $(\mathbf{V}^n)^{-1}$ and using (2.58), is not sufficiently restricted. A little algebra shows that with this choice, because not all the columns of $\mathbf{V}^{ne}$ are identical, neither are all the columns of $\mathbf{B}$. The first column of $\mathbf{B}$ is

$$\mathbf{b}_1 = \begin{pmatrix} (M - m_p)M^{-1} \\ -m_pM^{-1} \\ -m_pM^{-1} \\ 1 \end{pmatrix}$$

whereas all the other columns are

$$\mathbf{b} = \begin{pmatrix} -m_pM^{-1} \\ -m_pM^{-1} \\ -m_pM^{-1} \\ 1 \end{pmatrix}$$

but this is sufficient to ensure that (2.51) is not true and hence that (2.50) is not block diagonal.

However, choosing the nitrogen as the origin for *all* the electronic coordinates leads to identical columns $\mathbf{b}$

$$\mathbf{b} = \begin{pmatrix} -m_pM^{-1} \\ -m_pM^{-1} \\ -m_pM^{-1} \\ 1 \end{pmatrix}$$

as required.

Another possible choice for the electronic origin is the center of nuclear mass (scaled if required); this choice leads to a set of identical $\mathbf{b}$ of the form

$$\mathbf{b} = \begin{pmatrix} 0 \\ 0 \\ 0 \\ 1 \end{pmatrix}$$

as anticipated above.

Even with a fixed choice of electronic origin, different choices of $\mathbf{V}^n$ can lead to differing forms for $\mathbf{b}$. Thus suppose that the electronic coordinates are described as above with the nitrogen nucleus as the origin but that translationally invariant coordinates for the nuclei are defined as

$$\mathbf{t}_i^n = \mathbf{x}_i^n - \mathbf{x}_3^n, \quad i = 1, 2$$
$$\mathbf{t}_3^n = \mathbf{x}_4^n - M_3^{-1} m_p(\mathbf{x}_1^n + \mathbf{x}_2^n + \mathbf{x}_3^n) \tag{2.67}$$

where the sum of the proton masses is M_3. The vectors $\mathbf{t}_1^n$ and $\mathbf{t}_2^n$ define vectors from proton 3 to protons 1 and 2, while $\mathbf{t}_3^n$ positions the nitrogen nucleus with respect to the centre of nuclear mass of the three protons. The inverse transformation is

$$\mathbf{x}_1^n = 2m_p M_3^{-1} \mathbf{t}_1^n - m_p M_3^{-1} \mathbf{t}_2^n - m_N M^{-1} \mathbf{t}_3^n + \mathbf{X}$$
$$\mathbf{x}_2^n = -m_p M_3^{-1} \mathbf{t}_1^n + 2m_p M_3^{-1} \mathbf{t}_2^n - m_N M^{-1} \mathbf{t}_3^n + \mathbf{X}$$
$$\mathbf{x}_3^n = -m_p M_3^{-1} \mathbf{t}_1^n - m_p M_3^{-1} \mathbf{t}_2^n - m_N M^{-1} \mathbf{t}_3^n + \mathbf{X} \tag{2.68}$$
$$\mathbf{x}_4^n = M_3 M^{-1} \mathbf{t}_3^n + \mathbf{X}$$
$$\mathbf{x}_i^e = \mathbf{t}_i^e + M_3 M^{-1} \mathbf{t}_3^n + \mathbf{X}, \quad i = 1, 2, 3, \ldots, 10$$

and hence $\mathbf{b}$ is

$$\mathbf{b} = \begin{pmatrix} 0 \\ 0 \\ M_3 M^{-1} \\ 1 \end{pmatrix}$$

Examining the form taken by $\mathbf{H}$ as defined in (2.64) with the choices for the nuclear variables made above. If the nitrogen nucleus is chosen as the origin, so that $\mathbf{t}_i^n = \mathbf{x}_i^n - \mathbf{x}_4^n$, $i = 1, 2, 3$, and the inverse transformation is given by (2.36), then in this case $\mathbf{H}$ does actually have the form of the standard permutation matrix for each proton permutation. This is because the coordinates are nitrogen–proton bond vectors and the permutations simply permute the bond vectors exactly as the protons are permuted. However, if the nuclear variables are chosen as in (2.67) with the inverse given by (2.68), then $\mathbf{H}$ retains a standard form only for the proton permutation (12). Otherwise it has a general linear transformation form.

K. The Permutationally Restricted Translationally Invariant Hamiltonian Identifying Electrons

The electronic kinetic-energy operator (2.38) simplifies, and, adding the electron repulsion term, the electronic Hamiltonian is

$$\hat{H}^e(\mathbf{t}^e) = -\frac{\hbar^2}{2\mu} \sum_{i=1}^{L} \nabla^2(\mathbf{t}_i^e) - \frac{\hbar^2}{2\bar{\mu}} \sum_{i,j=1}^{L}{}' \vec{\nabla}(\mathbf{t}_i^e) \cdot \vec{\nabla}(\mathbf{t}_j^e) + \frac{e^2}{8\pi\epsilon_0} \sum_{i,j=1}^{L}{}' \frac{1}{|\mathbf{t}_j^e - \mathbf{t}_i^e|}$$

$$\tag{2.69}$$

with

$$\frac{1}{\mu} = \frac{1}{m} + \frac{1}{\bar{\mu}} \tag{2.70}$$

$$\frac{1}{\bar{\mu}} = m^{-1}a(2 + La) + \sum_{k=1}^{H} m_k^{-1} v_k^2 \tag{2.71}$$

From (2.40) and the nuclear repulsion, the nuclear Hamiltonian is

$$\hat{H}^{\mathrm{n}}(\mathbf{t}^{\mathrm{n}}) = -\frac{\hbar^2}{2} \sum_{i,j=1}^{H-1} \frac{1}{\mu_{ij}^{\mathrm{n}}} \vec{\nabla}(\mathbf{t}_i^{\mathrm{n}}) \cdot \vec{\nabla}(\mathbf{t}_j^{\mathrm{n}}) + \frac{e^2}{8\pi\epsilon_0} \sum_{i,j=1}^{H}{}' \frac{Z_i Z_j}{r_{ij}(\mathbf{t}^{\mathrm{n}})} \tag{2.72}$$

where $r_{ij}(\mathbf{t}^{\mathrm{n}})$ is defined just as in (2.11) but using only the $\mathbf{t}_i^{\mathrm{n}}$ and $(\mathbf{V}^{\mathrm{n}})^{-1}$ and, using (2.41), the interaction Hamiltonian is

$$\hat{H}^{\mathrm{en}}(\mathbf{t}^{\mathrm{n}}, \mathbf{t}^{\mathrm{e}}) = -\frac{\hbar^2}{2} \sum_{i=1}^{H-1} \frac{1}{\mu_i} \sum_{j=1}^{L} \left(\vec{\nabla}(\mathbf{t}_i^{\mathrm{n}}) \cdot \vec{\nabla}(\mathbf{t}_j^{\mathrm{e}}) + \vec{\nabla}(\mathbf{t}_j^{\mathrm{e}}) \cdot \vec{\nabla}(\mathbf{t}_i^{\mathrm{n}}) \right)$$
$$- \frac{e^2}{4\pi\epsilon_0} \sum_{i=1}^{H} \sum_{j=1}^{L} \frac{Z_i}{r_{ij}'(\mathbf{t}^{\mathrm{n}}, \mathbf{t}^{\mathrm{e}})} \tag{2.73}$$

because

$$\frac{1}{\mu_{ij}^{\mathrm{ne}}} \rightarrow \frac{1}{\mu_i} = \sum_{k=1}^{H} m_k^{-1} v_k V_{ki}^{\mathrm{n}} \tag{2.74}$$

whereas r_{ij}' is the electron–nucleus distance, and so it is the modulus

$$|\mathbf{x}_i^{\mathrm{n}} - \mathbf{x}_j^{\mathrm{e}}| = \left| \sum_{k=1}^{H-1} \mathbf{t}_k^{\mathrm{n}}((\mathbf{V}^{\mathrm{n}})_{ki}^{-1} - b_k) + \frac{a}{1 + La} \sum_{k=1}^{L} \mathbf{t}_k^{\mathrm{e}} - \mathbf{t}_j^{\mathrm{e}} \right| \tag{2.75}$$

The choice of translationally invariant electronic coordinates renders $\hat{H}^{\mathrm{e}}(\mathbf{t}^{\mathrm{e}})$ trivially invariant under permutations of the original electronic coordinates and independent of any particular choice of translationally invariant nuclear coordinates. Similarly $\hat{H}^{\mathrm{n}}(\mathbf{t}^{\mathrm{n}})$ is independent of any particular choice of translationally invariant electronic coordinates and can be shown, after some algebra, to be invariant under any permutation of the original coordinates of identical nuclei. The interaction operator $\hat{H}^{\mathrm{en}}(\mathbf{t}^{\mathrm{n}}, \mathbf{t}^{\mathrm{e}})$ is obviously invariant under a permutation of the original electronic coordinates and, again after a little algebra, can be shown to be invariant under any permutation of the original coordinates of identical nuclei.

The coupling in the kinetic-energy expression between the electronic and and nuclear coordinates as the first term in (2.73) can be removed if the elements of $\mathbf{v}$ are chosen as

$$v_k = -\alpha M^{-1} m_k, \qquad \alpha = (1 + La) \qquad (2.76)$$

where the choice for α is determined by the translational invariance requirements as explained in connection with (2.58). Thus the $\mathbf{t}_i^e$ are the electronic coordinates referred to the center of nuclear mass scaled by α. This choice satisfies the permutation restrictions and with it from (2.25) and (2.57) $b_i = 0, (i \neq H)$, so s_i^n vanishes as well, and substituting for v_k into (2.74), $1/\mu_i$ vanishes. The term is (2.73) then simplifies to

$$\hat{H}^{\text{en}}(\mathbf{t}^n, \mathbf{t}^e) = -\frac{e^2}{4\pi\epsilon_0} \sum_{i=1}^{H} \sum_{j=1}^{L} \frac{Z_i}{r'_{ij}(\mathbf{t}^n, \mathbf{t}^e)} \qquad (2.77)$$

and the electron-nucleus distance expression r'_{ij} becomes

$$|\mathbf{x}_i^n - \mathbf{x}_j^e| = \left| \sum_{k=1}^{H-1} \mathbf{t}_k^n (\mathbf{V}^n)_{ki}^{-1} + \frac{a}{1 + La} \sum_{k=1}^{L} \mathbf{t}_k^e - \mathbf{t}_j^e \right| \qquad (2.78)$$

and (2.71) simplifies to

$$\frac{1}{\mu} = m^{-1} a (2 + La) + \alpha^2 M^{-1} \qquad (2.79)$$

In the special case where the center of nuclear mass is chosen as the origin, then $a = 0$ and thus $\alpha = 1$ and

$$\frac{1}{\mu} = \frac{1}{M}$$

while (2.78) simplifies further to

$$|\mathbf{x}_i^n - \mathbf{x}_j^e| = \left| \sum_{k=1}^{H-1} \mathbf{t}_k^n (\mathbf{V}^n)_{ki}^{-1} - \mathbf{t}_j^e \right|$$

L.   The Permutationally Restricted Translationally Invariant Angular Momentum Operator Identifying Electrons

The angular momentum operator is given by (2.45) regardless of whether the permutation restrictions are satisfied. However with the permutation restrictions enforced on the coordinates the two parts are now separately invariant. The first part of (2.45), regarded as the translationally invariant angular momentum associated with the nuclei, has its square and its z component commute and commute with $\hat{H}^n$ as given in (2.72). Similarly, the second part, regarded as the translationally invariant angular momentum associated with the electrons, has its relevant parts commute with $\hat{H}^e$ in (2.69). Of course, neither operator separately commutes with the interaction part of the Hamiltonian $\hat{H}^{en}$ in (2.73), although their sum does. It is thus possible, if desired, to construct angular momentum eigenfunctions separately for the nuclear motion and for the electronic motion and then to couple them using the standard techniques of angular momentum algebra to a total angular momentum eigenfunction for the whole system.

M.   The Permutationally Restricted Translationally Invariant Dipole Operator Identifying Electrons

The dipole operator in (2.46) simplifies to

$$\mathbf{d}(\mathbf{t}^n, \mathbf{t}^e) = e \sum_{i=1}^{H-1} \tilde{Z}_i \mathbf{t}_i^n - e\tilde{Z} \sum_{i=1}^{L} \mathbf{t}_i^e \qquad (2.80)$$

where, from (2.47) and (2.57)

$$\tilde{Z}_i = \sum_{j=1}^{H} ((\mathbf{V}^n)^{-1})_{ij} Z_j - \frac{L(M_T + mZ_T)b_i}{M_T} \qquad (2.81)$$

and from (2.48), (2.57) and (2.55)

$$\tilde{Z} = \frac{(M_T + mZ_T)}{M_T(1 + La)} \qquad (2.82)$$

The first part of (2.80) can be shown to be invariant under a permutation of the original coordinates of any nuclei with identical charges and is independent of the choice of the $\mathbf{t}^e$. The second part is obviously invariant under the permutation of the original electronic coordinates and independent of any choice of the $\mathbf{t}^n$.

N. Example: Permutationally Restricted Translationally Invariant Operators for NH_3

The angular momentum operator needs no further specification in this context, but explicit forms for the other operators are given below.

1. A Kinetic-Energy Operator for NH_3

Making the choice (2.67) for the translationally invariant nuclear coordinates and taking the origin for the electronic coordinates as the nitrogen nucleus (so that $a = 0$), the nonzero elements of the matrix of inverse reduced masses are

$$\frac{1}{\mu_{11}^n} = \frac{1}{\mu_{22}^n} = \frac{1}{m_p} + \frac{1}{m_N}, \qquad \frac{1}{\mu_{33}^n} = \frac{1}{m_N} + \frac{1}{M_3}$$

$$\frac{1}{\mu_{12}^n} = \frac{1}{m_N}$$

$$\frac{1}{\mu} = \frac{1}{m_N}, \qquad \frac{1}{\mu_i} = \frac{\delta_{i3}}{m_N}$$

and the kinetic-energy operator in these coordinates is

$$-\frac{\hbar^2}{2\mu} \sum_{i=1}^{10} \nabla^2(\mathbf{t}_i^e) - \frac{\hbar^2}{2m_N} \sum_{i,j=1}^{10}{}' \vec{\nabla}(\mathbf{t}_i^e) \cdot \vec{\nabla}(\mathbf{t}_j^e)$$

$$-\frac{\hbar^2}{2} \frac{1}{m_N} \sum_{j=1}^{10} \left(\vec{\nabla}(\mathbf{t}_3^n) \cdot \vec{\nabla}(\mathbf{t}_j^e) + \vec{\nabla}(\mathbf{t}_j^e) \cdot \vec{\nabla}(\mathbf{t}_3^n) \right)$$

$$-\frac{\hbar^2}{2} \sum_{i,j=1}^{2} \frac{1}{\mu_{ij}^n} \vec{\nabla}(\mathbf{t}_i^n) \cdot \vec{\nabla}(\mathbf{t}_j^n) - \frac{\hbar^2}{2\mu_{33}^n} \nabla^2(\mathbf{t}_3^n) \qquad (2.83)$$

2. Some Dipole Moment Operators for NH_3

The proton charges Z_i, $i = 1, 2, 3$ are all $Z_p = 1$ and the charge on the nitrogen nucleus Z_4 is $Z_N = 7$ so that with 10 electrons the total charge Z_T is 0. The electronic origin will be chosen with $a = 0$. It follows that the effective electronic charge for the dipole, $\tilde{Z}$ as given by (2.82), is simply 1 and that the last term in (2.81) is Lb_i.

If we denote by Q_i that part of the effective nuclear charge arising from the first term in (2.81), then in the case where the nitrogen nucleus has been chosen as the nuclear origin:

$$Q_i = \left(1 - \frac{3m_p}{M}\right) Z_p - \left(\frac{m_p}{M}\right) Z_N = Z_p - \frac{m_p Z_{nuc}}{M}, \qquad i = 1, 2, 3$$

while the choice (2.67) has been made

$$Q_i = 0, \quad i = 1,2, \qquad Q_3 = (M_3 Z_N - 3m_N Z_p) = 3(m_p Z_N - m_N Z_p) \quad (2.84)$$

When the center of nuclear mass is chosen as the electronic origin, the b_i vanish and so the preceding expressions are directly those for the effective charges of the nuclear part of the dipole. If the nitrogen nucleus is chosen as the electronic origin, then in the first case

$$\tilde{Z}_i = Q_i + \frac{10 m_p}{M} = Z_p = 1, \qquad i = 1,2,3$$

while in the second case

$$\tilde{Z}_i = -\frac{10 M_3}{M}, \qquad i = 1,2, \qquad \tilde{Z}_3 = -3 \qquad (2.85)$$

If we take m_N to be about $14 m_p$, then a rough idea of the orders of magnitude may be obtained. With this choice, the Q_i in the first case are $\frac{7}{17}$, while Q_3 in the second case is $\frac{-21}{17}$ and $\tilde{Z}_i$ is $\frac{-30}{17}$, $i = 1,2$. It seems physically quite reasonable that the nuclear part of the dipole associated with the NH bond vectors in the first case should each be the same, as should be that associated with the HH bond vectors in the second case. But there seems to be no obvious way to rationalize the signs or the magnitudes of the effective charges.

O. Symmetry in the Translationally Invariant Permutationally Restricted Space

The solutions to the translationally invariant permutationally restricted problem as formulated here can be chosen as

$$\Psi_s^{J,M,r}(\mathbf{t}^n, \mathbf{t}^e) \qquad (2.86)$$

where J is the integer that specifies the eigenvalue of $\hat{L}^2(\mathbf{t})$ and M is an integer that lies in the interval $(-J, J)$ and is the eigenvalue of $\hat{L}_z(\mathbf{t})$ and r is the parity index. Alternatively, J can be regarded as labeling an irreducible representation (irrep) of the rotation group in three dimensions $SO(3)$ and M as labeling its rows (columns). The index r then associates this irrep with the appropriate irrep of the full orthogonal group $O(3)$. All $2J + 1$ components of the function (2.86) are degenerate. The set of numbers s specify the irreducible representations of the symmetric group relevant to each collection of identical particles. Thus, for example, one part of the set will specify the the irrep of $\mathcal{S}_L$, the symmetric group of order $L!$ relevant to the description of the electronic part of the problem. In practice, not all irreps will be accessible. So, for example, for electrons (or any spin $\frac{1}{2}$ particles), the spin functions can belong only to representations of $\mathcal{S}_L$ that can be described by Young diagrams with at most two rows. To make the total space–spin function properly antisymmetric, the allowed space functions can belong only to representations formed from the Young diagrams conjugate to the spin ones. And there are similar, although generally more complicated, sorts of restrictions for the symmetries of the nuclear part of the problem with spins other than $\frac{1}{2}$. We shall simply note these matters here and not develop them further. We would expect the spatial functions (2.86) to consist of degenerate sets according to the value of J and according to the irreps to which they belong, of the symmetric groups of the sets of identical particles. In what follows, to simplify the writing here and because it is difficult to make use of it, the subscript **s** will generally be suppressed. The index r will also be suppressed until it is explicitly required.

III. FIXING A FRAME IN THE BODY

The discussion so far describes a reasonable and reasonably convenient way of partitioning $\mathbf{V}$ and hence its inverse, if the division of the problem into electronic and nuclear parts is to be recognized and made explicit. From here on it will be assumed that a coordinate system has been chosen according to (2.26) and (2.27) but there will be some results for which the permutation restriction conditions (2.52)–(2.54) need not be satisfied, and these will be explicitly noted when appropriate. However, with a suitably restricted coordinate choice, one could attempt a solution $\Psi(\mathbf{t})$ to the problem specified by $\hat{H}(\mathbf{t})$ in the form

$$\Psi(\mathbf{t}) = \sum_{\mathrm{p}} \Phi_{\mathrm{p}}(\mathbf{t}^{\mathrm{n}})\psi_{\mathrm{p}}(\mathbf{t}^{\mathrm{n}}, \mathbf{t}^{\mathrm{e}}) \tag{3.1}$$

The electronic coordinates can be plausibly identified because they remain unchanged under nuclear permutations and transform in the standard manner under permutations of the original electronic coordinates. The form of $\hat{H}^{\mathrm{e}}(\mathbf{t}^{\mathrm{e}})$ together with the electron–nucleus attraction term can reasonably be thought to have some solutions expressible in terms of square-integrable functions, but it is still not equivalent to the required electronic Hamiltonian. The reason is that the operator formed from this sum depends on $3H - 3$ nuclear coordinates rather than the $3H - 6$ on which the required electronic Hamiltonian depends. To circumvent this difficulty, we must, as indicated earlier, account for the invariance of the full problem under rotation reflections in such a way that the remaining nuclear variables "carry" all such motions. This is done by defining a coordinate frame that rotates in a defined way with the system. This is usually called *defining* or *embedding* a frame fixed in the body. In the present case, the definition must involve only the nuclear variables.

For a system with more than two nuclei, one can transform the coordinates t^n such that rotational motion can be expressed in terms of three orientation variables, with the remaining motions expressed in terms of variables (commonly called *internal coordinates*) that are invariant under all orthogonal transformations of the t^n. For $H = 2$, only two orientation variables are required, and this case (the diatomic molecule) is rather special and is excluded from all subsequent discussion. To construct the frame fixed in the body, it is supposed that the three orientation variables are specified by means of an orthogonal matrix C, the elements of which are expressed as functions of three Eulerian angles $\phi_m, m = 1, 2, 3$, which are orientation variables. We require that the the matrix C be specified entirely in terms of the H nuclear variables, and so there will be just $3H - 6$ internal variables for the nuclei.

Thus the nuclear Cartesian coordinates t^n are considered related to a set z^n by

$$t^n = Cz^n \tag{3.2}$$

so the matrix C may be considered as a direction cosine matrix, relating the laboratory frame to the frame fixed in the body. The laboratory frame may always be chosen as a righthanded frame, but it is not always the case that there is the freedom to choose the frame fixed in the body as a righthanded one. Since z^n are fixed in the body, not all their $3H - 3$ components are independent, for there must be three relations between them. Hence components of z_i^n must be writable in terms of $3H - 6$ independent internal coordinates $q_i, i = 1, 2, \ldots, 3H - 6$. Some of the q_i may be components of z_i^n, but generally q_i are expressible in terms of scalar products of the t_i^n (and equally of the z_i^n) since scalar products are the most general constructions that are invariant under orthogonal transformations of their constituent vectors. If only proper orthogonal transformations are considered, the scalar triple products are also invariants, but they change sign under improper operations.

The electronic variables fixed in the body are then defined in terms of the preceding transformation by

$$z_i = C^T t_i^e, \qquad i = 1, 2, \ldots, L \tag{3.3}$$

in which the superscript on the electronic variables fixed in the body has been dropped. Equations (3.2) and (3.3) *define* in the frame fixed in the body by means of $\mathbf{C}$, the Cartesian form of the variables. Thus *any* orthogonal transformation of the translationally invariant coordinates (including inversion) leaves them, by definition, unchanged.

To express the translationally invariant differential operators in terms of the orientation and internal coordinates, we must obtain expressions for the partial derivatives of these coordinates with respect to the translationally invariant ones. Only the derivatives involving the $\mathbf{t}^n$ present any problems. There has been much previous work along these lines, but most of it has been in the context of particular coordinate choices. However, both Chapuisat and his colleagues [26,27] and Handy and his [28,29], among others, for example, Lukka [30], have presented rather general and abstract accounts of the methodology. Additional reviews have been provided by Sutcliffe [31–33]. The results that are quoted below prior to (3.24), do not depend on the permutation restrictions imposed in (2.52)–(2.54) but only on the separated forms (2.26) and (2.27).

To deal with the orientational variables, we note that

$$\frac{\partial}{\partial t^n_{\alpha i}}(\mathbf{C}^{\mathrm{T}}\mathbf{C}) = \mathbf{0}_3 \tag{3.4}$$

because $\mathbf{C}$ is an orthogonal matrix and hence $\mathbf{C}^{\mathrm{T}}\mathbf{C} = \mathbf{E}_3$. Therefore

$$\frac{\partial \mathbf{C}^{\mathrm{T}}}{\partial t^n_{\alpha i}}\mathbf{C} = \hat{\omega}^{\alpha i} \tag{3.5}$$

where $\hat{\omega}^{\alpha i}$ is a skew-symmetric matrix of the same form as (2.14), containing three independent elements $\omega^{\alpha i}_\gamma$. Using the form (2.16), we see that (3.5) becomes

$$\frac{\partial \mathbf{C}}{\partial t^n_{\alpha i}} = \sum_\gamma \omega^{\alpha i}_\gamma \mathbf{C}\mathbf{M}^\gamma \tag{3.6}$$

We introduce the matrix with elements $\Omega^i_{\beta\gamma}$ such that

$$\omega^{\alpha i}_\gamma = \sum_\beta C_{\alpha\beta}\Omega^i_{\beta\gamma} \tag{3.7}$$

so that the elements of the matrix Ω^i are functions of the internal coordinates only. Hence (3.6) becomes

$$\frac{\partial \mathbf{C}}{\partial t^n_{\alpha i}} = \sum_\beta (\mathbf{C}\mathbf{M}^\beta)(\mathbf{C}\Omega^i)_{\alpha\beta} \tag{3.8}$$

Because $\mathbf{C}$ is a function of the ϕ_m only, it follows that

$$\frac{\partial \mathbf{C}}{\partial t_{\alpha i}^n} = \sum_{m=1}^{3} \frac{\partial \mathbf{C}}{\partial \phi_m} \frac{\partial \phi_m}{\partial t_{\alpha i}^n} \tag{3.9}$$

and by analogy with (3.4)–(3.6), it follows that

$$\frac{\partial \mathbf{C}}{\partial \phi_m} = \sum_{\gamma} (\mathbf{D}^{-1})_{m\gamma} \mathbf{C} \mathbf{M}^{\gamma} \tag{3.10}$$

where $(\mathbf{D}^{-1})_{m\gamma}$ is a function of the ϕ_m only and plays the same role in (3.10) that $\omega_{\gamma}^{\alpha i}$ does in (3.6).

Strictly speaking, since the elements of $\mathbf{C}$ in (3.8) are assumed to be functions of the t_i^n but in (3.9) and (3.10) they are assumed to be functions of the ϕ_m, a different symbol for $\mathbf{C}$ should be used in the second case. However, the usage is clear from the context, and so no distinction will be made.

From (3.8)–(3.10) it follows that

$$\sum_{m=1}^{3} (\mathbf{D}^{-1})_{m\gamma} \frac{\partial \phi_m}{\partial t_{\alpha i}} = (\mathbf{C}\Omega^i)_{\alpha\gamma}$$

so finally

$$\frac{\partial \phi_m}{\partial t_{\alpha i}^n} = (\mathbf{C}\Omega^i \mathbf{D})_{\alpha m} \tag{3.11}$$

The process so far is, of course, purely formal since $\mathbf{C}$ has not been specified in terms of the t_i or the ϕ_m.

A similar formal process establishes that

$$\frac{\partial q_k}{\partial t_{\alpha i}^n} = (\mathbf{C}\mathbf{Q}^i)_{\alpha k} \tag{3.12}$$

where the elements of $\mathbf{Q}^i$ are dependent only on internal variables because the q_k are functions only of scalar products of the t_i.

This establishes the form for the only problematic parts of the Jacobian matrix for the transformation from the $(\phi, \mathbf{q}, \mathbf{z})$ to the $(\mathbf{t}^e, \mathbf{t}^n)$. In summary, the Jacobian matrix elements are

$$\frac{\partial \phi_m}{\partial t^n_{\alpha i}} = (C\Omega^i D)_{\alpha m}, \qquad \frac{\partial \phi_m}{\partial t^e_{\alpha i}} = 0 \qquad (3.13)$$

$$\frac{\partial q_k}{\partial t^n_{\alpha i}} = (CQ^i)_{\alpha k}, \qquad \frac{\partial q_k}{\partial t^e_{\alpha i}} = 0 \qquad (3.14)$$

$$\frac{\partial z_{\gamma j}}{\partial t^n_{\alpha i}} = (C\Omega^i \hat{z}_j)_{\alpha \gamma} \qquad \frac{\partial z_{\gamma j}}{\partial t^e_{\alpha i}} = \delta_{ij} C_{\alpha \gamma} \qquad (3.15)$$

where the elements of Q^i and of Ω^i are dependent on internal variables only whereas the elements of C and of D are functions of the Eulerian angles only.

Although not all the $z^n_{\beta i}$ can be linearly independent, they all possess derivatives with respect to the $t^n_{\epsilon j}$, which, from (3.2) and (3.8), have the form

$$\frac{\partial z^n_{\beta i}}{\partial t^n_{\epsilon j}} = (C\Omega^j \hat{z}^n_i)_{\epsilon \beta} + C_{\epsilon \beta} \delta_{ij} \qquad (3.16)$$

It is also sometimes possible to express the constraint conditions on the $z^n_{\beta i}$ in the form $f_m(z^n) = 0$, $m = 1, 2, 3$ and hence as $f_m(C^T t^n) \equiv g_m(t^n) = 0$, $m = 1, 2, 3$. In that case

$$\frac{\partial g_m}{\partial t^n_{\epsilon j}} = (C(-\Omega^j T + S^j))_{\epsilon m} = 0$$

where

$$S^j_{\epsilon m} = \frac{\partial f_m}{\partial z^n_{\epsilon j}}, \qquad T = \sum_{i=1}^{H-1} \hat{z}^{nT}_i S^i$$

The derivative with respect to $z^n_{\epsilon j}$ is perfectly well defined in the usual way even though the $z^n_{\epsilon j}$ are not all independent variables because $f_m(z^n)$ is an explicit function of all of them. If T is nonsingular then one can write

$$\Omega^i = S^i T^{-1} \qquad (3.17)$$

This result was apparently first noticed by Sørensen [34] in the context of the Eckart Hamiltonian.

The derivatives of the translationally invariant coordinates in terms of the orientation and internal coordinates are as follows:

$$\frac{\partial}{\partial t_i^n} = \mathbf{C}\left(\mathbf{\Omega}^i \mathbf{D}\frac{\partial}{\partial \boldsymbol{\phi}} + \mathbf{Q}^i\frac{\partial}{\partial \mathbf{q}} + \mathbf{\Omega}^i \sum_{j=1}^{L} \hat{\mathbf{z}}_j\frac{\partial}{\partial \mathbf{z}_j}\right) \qquad (3.18)$$

and

$$\frac{\partial}{\partial t_i^e} = \mathbf{C}\frac{\partial}{\partial \mathbf{z}_i} \qquad (3.19)$$

where $\partial/\partial\boldsymbol{\phi}$ and $\partial/\partial\mathbf{q}$ are column matrices of 3 and $3N - 6$ partial derivatives, respectively, and $\partial/\partial t_i$ and $\partial/\partial \mathbf{z}_i$ are column matrices of 3 partial derivatives.

There are similar developments in the expressions associated with the inverse transformation and the elements of the inverse Jacobian matrix are

$$\frac{\partial t_{\alpha i}^n}{\partial \phi_m} = (\mathbf{C}\hat{\mathbf{z}}_i^n \mathbf{D}^{-T})_{\alpha m}, \qquad \frac{\partial t_{\alpha i}^e}{\partial \phi_m} = (\mathbf{C}\hat{\mathbf{z}}_i \mathbf{D}^{-T})_{\alpha m} \qquad (3.20)$$

$$\frac{\partial t_{\alpha i}^n}{\partial q_k} = (\mathbf{C}\tilde{\mathbf{Q}}^i)_{\alpha k} \qquad \frac{\partial t_{\alpha i}^e}{\partial q_k} = 0 \qquad (3.21)$$

$$\frac{\partial t_{\alpha i}^n}{\partial z_{\gamma j}} = 0 \qquad \frac{\partial t_{\alpha i}^e}{\partial z_{\gamma j}} = \delta_{ij}C_{\alpha\gamma} \qquad (3.22)$$

where the elements of $\tilde{\mathbf{Q}}^i$ are functions of the internal coordinates alone.

The relationship between the Jacobian matrix and its inverse leads to the following expressions:

$$\sum_{i=1}^{H-1} \hat{\mathbf{z}}_i^{nT}\mathbf{\Omega}^i = \mathbf{E}_3, \qquad \sum_{i=1}^{H-1} \tilde{\mathbf{Q}}^{iT}\mathbf{Q}^i = \mathbf{E}_{3N-6}$$

$$\sum_{i=1}^{H-1} \hat{\mathbf{z}}_i^{nT}\mathbf{Q}^i = \mathbf{0}_{3,3N-6}, \qquad \sum_{i=1}^{H-1} \tilde{\mathbf{Q}}^{iT}\mathbf{\Omega}^i = \mathbf{0}_{3N-6,3} \qquad (3.23)$$

$$\mathbf{\Omega}^i \hat{\mathbf{z}}_j^{nT} + \mathbf{Q}^i \tilde{\mathbf{Q}}^{jT} = \delta_{ij}\mathbf{E}_3 \qquad (3.24)$$

These expressions are helpful in the formal manipulations that lead to expressions for the operators in orientation and internal coordinates. They are also the origin of the "sum rules", which constitute such a part of the manipulation of the Eckart Hamiltonian [35].

The symmetry properties of the internal and angular coordinates are seldom immediately apparent, and they will be discussed separately from operator forms. But it should be noted here that, assuming the translationally invariant coordinates to be permutationally restricted, although permutations of nuclear variables will not affect the electronic coordinates expressed in a frame fixed in the body, such permutations will generally induce changes in both the internal and orientational coordinates. In particular, a change can be induced in an Euler angle that is expressible only in terms of the Euler angles and the internal coordinates.

A. The Permutationally Restricted Angular Momentum Operator in a Frame Fixed in the Body

The translationally invariant angular momentum operator becomes

$$\hat{\mathbf{L}}(\mathbf{t}) = -\frac{\hbar}{i}|\mathbf{C}|\mathbf{C}\mathbf{D}\frac{\partial}{\partial\boldsymbol{\phi}} = -|\mathbf{C}|\mathbf{C}\hat{\mathbf{L}}(\boldsymbol{\phi}) \tag{3.25}$$

where $|\mathbf{C}|$ is either plus or minus one according to whether $\mathbf{C}$ corresponds to a proper or improper rotation. This term arises because on the formal variable change (3.2) or (3.3)

$$\hat{\mathbf{t}}_i = |\mathbf{C}|\mathbf{C}\hat{\mathbf{z}}_i\mathbf{C}^{\mathrm{T}} \tag{3.26}$$

where $\mathbf{t}$ and $\mathbf{z}$ can be the nuclear or the electronic variables.

There is at this stage an element of choice for the definition of the angular momentum in the frame fixed in the body, and in (3.25) it can be seen that we have chosen

$$\hat{\mathbf{L}}(\boldsymbol{\phi}) = \frac{\hbar}{i}\mathbf{D}\frac{\partial}{\partial\boldsymbol{\phi}} \tag{3.27}$$

Usually, the negative of this operator is chosen. However, a little algebra shows that in either case $\hat{L}^2(\boldsymbol{\phi}) \equiv \hat{L}^2(\mathbf{t})$ and that $\hat{\mathbf{L}}_z(\boldsymbol{\phi})$ and $\hat{\mathbf{L}}_z(\mathbf{t})$ commute with $\hat{\mathbf{L}}^2$, so one can find a complete set of angular momentum eigenfunctions $|JMk\rangle$ such that

$$\hat{L}^2(\mathbf{t})|JMk\rangle = \hat{L}^2(\boldsymbol{\phi})|JMk\rangle = \hbar^2 J(J+1)|JMk\rangle$$
$$\hat{L}_z(\mathbf{t})|JMk\rangle = \hbar M|JMk\rangle$$
$$\hat{L}_z(\boldsymbol{\phi})|JMk\rangle = \hbar k|JMk\rangle \tag{3.28}$$

The functions $|JMk\rangle$ are often called *symmetric-top* eigenfunctions.

Choosing the angular momentum operator in the frame fixed in the body according to (3.25) means that the components of this operator obey the *standard* commutation conditions, so that the angular momentum eigenfunctions, $|JMk\rangle$ have the standard properties of those defined in Brink and Satchler [36] or in Biedenharn and Louck [37]. Explicitly, if $\mathbf{C}$ is parameterized by the standard Euler angle choice made in Brink and Satchler [36] or Biedenharn and Louck [37], then

$$|JMk\rangle = \left(\frac{2J+1}{8\pi^2}\right)^{1/2} (-1)^k \mathcal{D}^{J*}_{M-k}(\boldsymbol{\phi})$$
$$\mathcal{D}^{J*}_{M-k}(\boldsymbol{\phi}) = e^{iM\phi_1} d^J_{M-k}(\phi_2) e^{-ik\phi_3} \tag{3.29}$$

where $\mathcal{D}^J$ is the standard Wigner matrix as defined in Brink and Satchler [36] or Biedenharn and Louck [37]. If the more usual choice of the negative of (3.27) is made, its components obey the celebrated anomalous commutation conditions and the relevant symmetric-top functions are proportional to $\mathcal{D}^{J*}_{Mk}$. With the present choice, however, the stepup and stepdown operators may be defined in the usual way as $L_{\pm} = L_x \pm iL_y$, then

$$\hat{L}_{\pm}(\mathbf{t})|JMk\rangle = \hbar C^{\pm}_{JM}|JM \pm 1k\rangle$$
$$\hat{L}_{\pm}(\boldsymbol{\phi})|JMk\rangle = \hbar C^{\pm}_{Jk}|JMk \pm 1\rangle \tag{3.30}$$

where the phase conventions are chosen as the standard Condon–Shortley ones [38] so that

$$C^{\pm}_{Jj} = [J(J+1) - j(j \pm 1)]^{1/2} \tag{3.31}$$

A more extended discussion of these matters can be found in Section 3.8 of Biedenharn and Louck [37] and also in Brown and Howard [39] and Zare [40].

It can be shown [37,41] that whatever the parameterization, for any $\mathbf{C}$ in $SO(3)$, the Wigner $\mathcal{D}^1$ matrix can be written as

$$\mathcal{D}^1 = \mathbf{X}^{\dagger}\mathbf{C}\mathbf{X} \tag{3.32}$$

with

$$\mathbf{X} = \begin{pmatrix} \dfrac{-1}{\sqrt{2}} & 0 & \dfrac{1}{\sqrt{2}} \\ \dfrac{-i}{\sqrt{2}} & 0 & \dfrac{-i}{\sqrt{2}} \\ 0 & 1 & 0 \end{pmatrix} \tag{3.33}$$

provided $C_{\alpha\beta}$ is ordered $\alpha, \beta = x, y, z$ and the indices on $\mathscr{D}^1$ run $+1, 0, -1$ across each row and down each column. For the present choice of operators it may furthermore be shown that

$$|1Mk\rangle = \left(\frac{3}{8\pi^2}\right)(-1)^k \mathscr{D}^{1*}_{M-k} = \left(\frac{3}{8\pi^2}\right)^{1/2}(\mathbf{X}^{\mathrm{T}}\mathbf{CX})_{Mk} = \left(\frac{3}{8\pi^2}\right)^{1/2}\mathbf{D}^1_{Mk}(\mathbf{C})$$

(3.34)

It should be emphasized that for any choice of Eulerian angle definition, it is always possible to construct another matrix $\mathbf{C}$, simply by multiplying the original choice by $-\mathbf{E}_3$. There is no choice of transformed Eulerian angles that will result in this other matrix. Thus, to specify the transformation completely, it is necessary to specify the Eulerian angles and $|\mathbf{C}|$, the parity of the transformation. If the original choice of matrix represents a proper rotation with parity $+1$, the other matrix represents a reflection with parity -1. It follows from (3.10) that both matrices give rise to the same matrix $\mathbf{D}$ so that from (3.25) $\hat{\mathbf{L}}(\mathbf{t})$ is invariant under inversion, as is required. For the time being we shall neglect this possibility and consider simply proper rotations so that attention will be confined to $SO(3)$.

B. The Permutationally Restricted Hamiltonian in a Frame Fixed in the Body

The complete kinetic-energy operator may be written as

$$\hat{K}(\mathbf{z}) + \hat{K}(\mathbf{q}, \mathbf{z}) + \hat{K}(\boldsymbol{\phi}, \mathbf{q}, \mathbf{z})$$

(3.35)

The first term in (3.35) arises trivially from the kinetic-energy part of (2.69) simply by replacing the $\mathbf{t}^e$ by the $\mathbf{z}$ and so is

$$\hat{K}(\mathbf{z}) = -\frac{\hbar^2}{2\mu}\sum_{i=1}^{L}\nabla^2(\mathbf{z}_i) - \frac{\hbar^2}{2\bar{\mu}}\sum_{i,j=1}^{L}{}'\vec{\nabla}(\mathbf{z}_i)\cdot\vec{\nabla}(\mathbf{z}_j)$$

(3.36)

The transformation of the nuclear part of the translationally invariant kinetic-energy operator from (2.72) and (2.73) into the coordinates $\boldsymbol{\phi}$ and $\mathbf{q}$ is long and tedious, but the final result can be stated directly; as the derivation is mechanical, simply involving letting (3.18) operate on itself and on (3.19) and summing over i and j, there is no need to go into details. The resulting operators are

$$\hat{K}(\boldsymbol{\phi}, \mathbf{q}, \mathbf{z}) = \frac{1}{2}\left(\sum_{\alpha\beta}\kappa_{\alpha\beta}\hat{L}_\alpha\hat{L}_\beta + \hbar\sum_\alpha\bar{\lambda}_\alpha\hat{L}_\alpha\right)$$

(3.37)

and

$$\hat{K}(\mathbf{q}, \mathbf{z}) = -\frac{\hbar^2}{2} \left(\sum_{k,l=1}^{3H-6} g_{kl} \frac{\partial^2}{\partial q_k \partial q_l} + \sum_{k=1}^{3H-6} h_k \frac{\partial}{\partial q_k} \right)$$

$$+ \frac{\hbar^2}{2} \left(\sum_{\alpha\beta} \kappa_{\alpha\beta} \hat{l}_\alpha \hat{l}_\beta + \sum_\alpha (\lambda_\alpha + \lambda'_\alpha) \hat{l}_\alpha \right)$$

$$- \hbar^2 \sum_{j=1}^{L} \sum_\alpha \left(\sum_{k=1}^{3H-6} g'_{\alpha k} \frac{\partial^2}{\partial q_k \partial z_{\alpha j}} + h'_\alpha \frac{\partial}{\partial z_{\alpha j}} \right) \tag{3.38}$$

where κ is an inverse generalized inertia tensor defined as the 3×3 matrix

$$\kappa = \sum_{i,j=1}^{H-1} \frac{1}{\mu_{ij}^n} \Omega^{iT} \Omega^j \tag{3.39}$$

and

$$\overline{\lambda}_\alpha = (\lambda_\alpha + \lambda'_\alpha + 2(\kappa \hat{\mathbf{l}})_\alpha) \tag{3.40}$$

with $\hat{\mathbf{l}}$ as a 3×1 column matrix of Cartesian components

$$\hat{\mathbf{l}} = \frac{1}{i} \sum_{i=1}^{L} \hat{\mathbf{z}}_i \frac{\partial}{\partial \mathbf{z}_i} \tag{3.41}$$

The other components of $\overline{\lambda}_\alpha$ are

$$\lambda_\alpha = \frac{1}{i} \left(\nu_\alpha + 2\tau^T \frac{\partial}{\partial \mathbf{q}_\alpha} \right) \tag{3.42}$$

with the $(3H - 6) \times 3$ matrix τ defined as

$$\tau = \sum_{i,j=1}^{H-1} \frac{1}{\mu_{ij}^n} \mathbf{Q}^{iT} \Omega^j \tag{3.43}$$

and

$$\nu_\alpha = \sum_{i,j=1}^{H-1} \frac{1}{\mu_{ij}^n} \left(\sum_\beta (\Omega^{iT} \mathbf{M}^\beta \Omega^j)_{\beta\alpha} + \sum_{l=1}^{3H-6} \left(\mathbf{Q}^{iT} \frac{\partial}{\partial q_l} \Omega^j \right)_{l\alpha} \right) \tag{3.44}$$

also

$$\lambda'_\alpha = \frac{2}{i}\sum_{i=1}^{H-1}\sum_{j=1}^{L}\sum_{\beta}\frac{1}{\mu_i}\Omega^i_{\beta\alpha}\frac{\partial}{\partial z_{\beta j}} \tag{3.45}$$

The terms in (3.42) and (3.45) are associated with the Coriolis coupling, and so no coordinate system can be found in which they will vanish.

The $(3H - 6) \times (3H - 6)$ matrix $\mathbf{g}$ is given by

$$\mathbf{g} = \sum_{i,j=1}^{H-1}\frac{1}{\mu^n_{ij}}\mathbf{Q}^{iT}\mathbf{Q}^j \tag{3.46}$$

while $\mathbf{g}'$ is a $3 \times (3H - 6)$ matrix with elements

$$\mathbf{g}' = \sum_{i=1}^{H-1}\frac{1}{\mu_i}\mathbf{Q}^i \tag{3.47}$$

In the terms linear in the derivatives of the coordinates

$$h_k = \sum_{i,j=1}^{H-1}\frac{1}{\mu^n_{ij}}\left(\sum_{\beta}(\Omega^{iT}\mathbf{M}^\beta\mathbf{Q}^j)_{\beta k} + \sum_{l=1}^{3H-6}\left(\mathbf{Q}^{iT}\frac{\partial}{\partial q_l}\mathbf{Q}^j\right)_{lk}\right) \tag{3.48}$$

and

$$h'_\alpha = \frac{1}{2}\sum_{i=1}^{H-1}\frac{1}{\mu_i}\sum_{\beta}((\Omega^i + \Omega^{iT})\mathbf{M}^\beta)_{\beta\alpha} \tag{3.49}$$

It is possible to choose a coordinate system in which these last two terms vanish, but in general, they do not disappear. In these equations, the terms marked with a prime, such as h'_α, arise from the kinetic-energy term in (2.73) and so are absent when that term vanishes, as it does when the center of nuclear mass is chosen as the electronic origin.

The potential-energy operator is

$$V(\mathbf{q}, \mathbf{z}) = \frac{e^2}{8\pi\epsilon_0}\sum_{i,j=1}^{L}{}'\frac{1}{|\mathbf{z}_j - \mathbf{z}_i|} + \frac{e^2}{8\pi\epsilon_0}\sum_{i,j=1}^{H}{}'\frac{Z_iZ_j}{r_{ij}(\mathbf{z}^n)} - \frac{e^2}{4\pi\epsilon_0}\sum_{i=1}^{H}\sum_{j=1}^{L}\frac{Z_i}{r'_{ij}(\mathbf{z}^n, \mathbf{z}^e)}$$

or

$$V(\mathbf{q}, \mathbf{z}) = V^{\mathrm{e}}(\mathbf{z}) + V^{\mathrm{n}}(\mathbf{q}) - V^{\mathrm{ne}}(\mathbf{q}, \mathbf{z}) \qquad (3.50)$$

where r'_{ij} is the electron-nucleus distance and so [see (2.75)] is the modulus

$$|\mathbf{x}_i^{\mathrm{n}} - \mathbf{x}_j^{\mathrm{e}}| = \left| \sum_{k=1}^{H-1} \mathbf{z}_k^{\mathrm{n}}(\mathbf{q})((\mathbf{V}^{\mathrm{n}})_{ki}^{-1} - b_k) + \frac{a}{1 + La} \sum_{k=1}^{L} \mathbf{z}_k - \mathbf{z}_j \right| \qquad (3.51)$$

while r_{ij} is defined as explained just below (2.70) but with $z_{\alpha k}^{\mathrm{n}}(\mathbf{q})$ replacing $t_{\alpha k}^{\mathrm{n}}$

Although both $\hat{\mathbf{L}}_z(\boldsymbol{\phi})$ and $\hat{\mathbf{L}}_z(\mathbf{t})$ commute with $\hat{\mathbf{L}}^2$, only $\hat{\mathbf{L}}_z(\mathbf{t})$ and $\hat{\mathbf{L}}^2$ commute with the Hamiltonian so that the eigenfunctions $\Psi^{J,M}(\mathbf{t}^{\mathrm{n}}, \mathbf{t}^{\mathrm{e}})$ from (2.86) can be written in the form

$$\Psi^{J,M}(\mathbf{t}^{\mathrm{n}}, \mathbf{t}^{\mathrm{e}}) \rightarrow \Psi^{J,M}(\boldsymbol{\phi}, \mathbf{q}, \mathbf{z}) = \sum_{k=-J}^{+J} \Phi_k^J(\mathbf{q}, \mathbf{z}) |JMk\rangle \qquad (3.52)$$

where the $|JMk\rangle$ are angular momentum eigenfunctions and the internal coordinate function on the right side cannot depend on M because, in the absence of a field, the energy of the system does not depend on M. Eigenfunctions of this kind form a basis for irreducible representations of $SO(3)$, as required.

C. The Permutationally Restricted Dipole Operator in a Frame Fixed in the Body

The dipole operator (2.80) is

$$
\begin{aligned}
\mathbf{d}(\mathbf{t}^{\mathrm{n}}, \mathbf{t}^{\mathrm{e}}) &= e \sum_{i=1}^{H-1} \tilde{Z}_i \mathbf{C} \mathbf{z}_i^{\mathrm{n}} - e\tilde{Z} \sum_{i=1}^{L} \mathbf{C} \mathbf{z}_i \\
&= e\mathbf{C}(\mathbf{d}(\mathbf{q}) - \mathbf{d}(\mathbf{z}))
\end{aligned}
\qquad (3.53)
$$

Confining attention to $SO(3)$, it follows from (3.33) that

$$\mathbf{X}^{\dagger}\mathbf{d}(\mathbf{t}^{\mathrm{n}}, \mathbf{t}^{\mathrm{e}}) = e\mathscr{D}^1 \mathbf{X}^{\dagger}(\mathbf{d}(\mathbf{q}) - \mathbf{d}(\mathbf{z})) \qquad (3.54)$$

It is not, however, the practice to work with the form $\mathbf{X}^{\dagger}\mathbf{d}$, but rather with the "spherical" form

$$\mathbf{d}^{\mathrm{s}} = \mathbf{X}^{\mathrm{T}}\mathbf{d} \qquad (3.55)$$

where, for example

$$\mathbf{X}^{T}\mathbf{z} = \begin{pmatrix} \dfrac{-(z_x + iz_y)}{\sqrt{2}} \\ z_z \\ \dfrac{z_x - iz_y}{\sqrt{2}} \end{pmatrix} \tag{3.56}$$

Taking the complex conjugate of both sides of (3.54) gives

$$\mathbf{d}^{s}(\mathbf{t}^{n}, \mathbf{t}^{e}) = \mathscr{D}^{1^{*}}(\mathbf{d}^{s}(\mathbf{q}) - \mathbf{d}^{s}(\mathbf{z})) = \mathscr{D}^{1^{*}}\mathbf{d}^{s}(\mathbf{q}, \mathbf{z}) \tag{3.57}$$

with the column elements labeled $(d^{s}_{+1}, d^{s}_{0}, d^{s}_{-1})$.

D. The Jacobian for the Transformation to a Frame Fixed in the Body

As the transformation (2.26) is linear, its Jacobian is simply a constant that can be ignored; the transformation from the $\mathbf{t}^{e}_{i}$ to the $\mathbf{z}_{i}$ is essentially a constant orthogonal one with a unit Jacobian. The transformation from the $\mathbf{t}^{n}_{i}$ to the Eulerian angles and the internal coordinates is nonlinear and has a Jacobian $|\mathbf{J}|^{-1}$, where $\mathbf{J}$ is the matrix constructed from the nuclear terms in (3.13) and (3.14). The nonlinearity is a topologic consequence of any transformation that allows rotational motion to be separated [42], and there is always some conformation of the particles that causes the Jacobian to vanish. Clearly, where the Jacobian vanishes, the transformation is undefined. This failure manifests itself in the Hamiltonian by the presence of terms that diverge unless, acting on the wavefunction, they vanish. This can occur either by cancellation or if the wavefunction itself is vanishingly small in the divergent region.

The origin of these divergences is not physical; they arise simply as a consequence of the choice of coordinates. A particular choice can obviously preclude the description of a possible physical state of a system. Thus, suppose that a triatomic is described according to Eckart's approach [2,35], with the reference geometry specified as bent. In this case the Jacobian vanishes when the internal coordinates correspond to a linear geometry. The problem then becomes ill-conditioned for states with large-amplitude angular deformations. Such states are physically reasonable, but they cannot be described according to this formulation.

The important point is that the nonlinear transformation cannot be globally valid. As it has only local validity, one can at most derive a local Hamiltonian that is valid within a particular domain. According to general topologic considerations [42], one can construct a sequence of transformations that have common ranges of validity sufficient for passage from one to another to cover the whole space.

The volume element for integration is

$$dt = |\mathbf{J}|^{-1} d\boldsymbol{\phi}\, d\mathbf{q}\, d\mathbf{z} \qquad (3.58)$$

with the volume element for $\mathbf{z}$ in standard Cartesian form. It is sometimes more convenient to construct the determinant of the metric derived from the nuclear terms (3.13) and (3.14) that will be equal to $|\mathbf{J}|^{-2}$, to within a constant factor. The determinant is

$$|\mathbf{D}^{-1}|^2 \begin{vmatrix} \boldsymbol{\kappa} & \boldsymbol{\tau}^{\mathrm{T}} \\ \boldsymbol{\tau} & \mathbf{g} \end{vmatrix}^{-1} \qquad (3.59)$$

where the matrices in the partitions are given by (3.39), (3.46), and (3.43). To within a constant factor

$$|\mathbf{J}|^{-1} = |\mathbf{D}|^{-1}|\boldsymbol{\kappa}|^{-(1/2)}|\mathbf{g} - \boldsymbol{\tau}\boldsymbol{\kappa}^{-1}\boldsymbol{\tau}^{\mathrm{T}}|^{-(1/2)} \qquad (3.60)$$

It is the factor $|\mathbf{D}|^{-1}$ that is the angular part of the Jacobian and in the standard parameterization $|\mathbf{D}|^{-1} = \sin\phi_2$ as required for the usual interpretation of the matrix elements. The remaining terms in (3.60) are functions of the q_k alone. It is perhaps worthwhile to notice that in the Podolsky approach to the construction of a frame fixed in the body, which is the approach used by Watson [35] (and indeed many others) it is the form of the Jacobian arising from (3.60) that is naturally used.

The choice is sometimes made to incorporate the internal coordinate part of the Jacobian (or some of it) into the definition of the Hamiltonian. This is a fairly familiar process when working in spherical polars, for example, where the radial volume element $r^2 dr$ can be reduced to dr by writing the trial wavefunction $\psi(r)$ as $r^{-1}P(r)$ and modifying the Hamiltonian to refer to $P(r)$. This modification changes the derivative terms in the operator by $\partial/\partial r \rightarrow (\partial/\partial r - 1/r)$ and so on but alters none of the multiplicative or $\partial/\partial\theta$ terms. The resulting Hamiltonian is often said to be in *manifestly Hermitian* form. Particular examples of this kind of construction can be found in Watson [35] and in Louck [43], while a general account is given in Section 35 of Kemble [44]. This process is often extremely useful in practice with specific coordinate choices; however, it does not simplify matters at the level of formal exposition, and in what follows we shall *not* explicitly consider its incorporation. But that it can be incorporated, in whole or in part, should always be borne in mind when identifying operator forms. Thus, if the Jacobian were incorporated into the function on which the operator on the right of (3.18) were working, it would have a changed form appropriate to working on what remains of the function when the Jacobian has been dealt with. Similarly, the forms of the kinetic-energy operators would be modified.

E. Example: NH_3 in a Frame Fixed in the Body with Electrons and Nuclei Identified

It might seem natural to describe the nuclear motions in ammonia by means of internal coordinates arising from the translationally invariant bond-vector coordinates introduced in Section II.F.1 and discussed at the end of Section II.J.1. The natural choice of internal coordinates in this context would seem to be the three bond lengths and the three interbond angles. But the three interbond angles are not independent coordinates. This can be easily appreciated by considering the case where the nitrogen nucleus and the three protons are coplanar and hence their sum must be 2π. It is not at all easy to choose a set of independent internal coordinates with this choice of translationally invariant coordinates. However, if we make the choice of translationally invariant coordinates (2.67), we can avoid these problems. The nuclear coordinates are two interproton vectors and the nitrogen vector positioned with respect to the center of proton mass. All the important points may be illustrated simply and clearly by considering just the nuclear kinetic-energy operator, the last term in (2.83). The relevant expressions in the frame fixed in the body are then (3.37) but with $\bar{\lambda}_\alpha$ replaced by λ_α together with the first term in (3.38). No separation of nuclear and electronic motion is implied in this exposition. The electronic terms are simply not written down to aid the clarity of the exposition. They can be easily, if cumbersomely, included.

Let us choose the embedding matrix $\mathbf{C}$ to place the protons in the x–y plane so that the two translationally invariant coordinates $\mathbf{t}_1^n$ and $\mathbf{t}_2^n$, are transformed to a set $\mathbf{z}_i^n$ fixed in the body by means of an orthogonal matrix $\mathbf{C}$ according to

$$\mathbf{t}^n = \mathbf{C}\,\mathbf{z}^n \tag{3.61}$$

In order to define fully the three Euler angles that specify $\mathbf{C}$, there must be three relations among the components of the $\mathbf{z}_i^n$ and, in consequence, the components of $\mathbf{z}^n$ must be specifiable in terms of the three rotationally invariant internal coordinates. These internal coordinates are chosen here to be r_i, the length of the $\mathbf{t}_i^n$ $i = 1, 2$, and θ the angle between them. The matrix $\mathbf{C}$ is chosen to put the two $\mathbf{z}_i^n$ in the x–y plane. The x axis in the frame fixed in the body is such that r_1 makes with it an angle $a\theta$ and r_2 an angle $(1 - a)\theta$

with a in the interval $(0,1)$. It is further required that $|\mathbf{C}| = +1$ in order to keep a righthanded coordinate frame. Thus, in general

$$
\mathbf{z}^n = \mathbf{C}^T \mathbf{t}^n = \begin{pmatrix} z_{x1}^n & z_{x2}^n \\ z_{y1}^n & z_{y2}^n \\ 0 & 0 \end{pmatrix} \tag{3.62}
$$

or

$$
\mathbf{z}^n = \begin{pmatrix} r_1\cos a\theta & r_2\cos(1-a)\theta \\ -r_1\sin a\theta & r_2\sin(1-a)\theta \\ 0 & 0 \end{pmatrix} \tag{3.63}
$$

It is, of course, of no physical significance that the x–y plane has been chosen as the embedding plane, it is merely a matter of convention and convenience. Relationships between the various possible orientation schemes are discussed shortly.

The nitrogen atom variable in the frame fixed in the body may be chosen in Cartesian form and so is

$$
\mathbf{z}_3^n = \mathbf{C}^T \mathbf{t}_3^n \tag{3.64}
$$

and the the derivatives of the translationally invariant nuclear coordinates become

$$
\frac{\partial}{\partial \mathbf{t}_i^n} = \mathbf{C}\left(\Omega^i \mathbf{D}\frac{\partial}{\partial \phi} + \mathbf{Q}^i \frac{\partial}{\partial \mathbf{q}} + \Omega^i \hat{\mathbf{z}}_3^n \frac{\partial}{\partial \mathbf{z}_3^n} \right), \qquad i = 1,2 \tag{3.65}
$$

and

$$
\frac{\partial}{\partial \mathbf{t}_3^n} = \mathbf{C}\frac{\partial}{\partial \mathbf{z}_3^n} \tag{3.66}
$$

where the three components of $\mathbf{z}_3^n$ have been kept explicitly as internal coordinates.

To realize these two equations, explicit forms must be found for Ω^i and $\mathbf{Q}^i$ and as an example of how to do this, consider the derivative

$$
\frac{\partial z_{zi}^n}{\partial r_{\epsilon j}^n} = (\mathbf{C}\Omega^j \hat{\mathbf{z}}_i^n)_{\epsilon z} + C_{\epsilon z}\delta_{ij} = 0
$$

obtained using (3.16) and (3.63). Eliminating $\mathbf{C}$ yields

$$\Omega^j_{\beta x} z^n_{yi} - \Omega^j_{\beta y} z^n_{xi} + \delta_{ij}\delta_{\alpha\beta} = 0$$

and solving the simultaneous equations implicit here gives

$$\Omega^1_{zx} = \frac{\cos(1-a)\theta}{r_1 \sin\theta}, \qquad \Omega^2_{zx} = \frac{-\cos a\theta}{r_2 \sin\theta}$$

$$\Omega^1_{zy} = \frac{\sin(1-a)\theta}{r_1 \sin\theta}, \qquad \Omega^2_{zy} = \frac{\sin a\theta}{r_2 \sin\theta}$$

Expanding $\cos(1-a)\theta$ and reexpressing it in terms of trigonometric functions of $a\theta$ and θ yields the following, after some algebra:

$$\Omega^1_{xz} = \frac{-(1-a)\sin a\theta}{r_1}, \qquad \Omega^2_{xz} = \frac{a\,\sin(1-a)\theta}{r_2}$$

$$\Omega^1_{yz} = \frac{-(1-a)\cos a\theta}{r_1}, \qquad \Omega^2_{yz} = \frac{-a\,\cos(1-a)\theta}{r_2}$$

All other elements of the Ω^i vanish.

If the internal coordinates q_1, q_2, q_3 are identified as r_1, r_2, θ and these are expressed as functions of the $\mathbf{t}^n_{\epsilon i}$, then differentiation with respect to these Cartesians followed by the elimination of $\mathbf{C}$ gives

$$Q^1_{x1} = \cos a\theta, \qquad Q^1_{y1} = -\sin a\theta$$

$$Q^2_{x2} = \cos(1-a)\theta, \qquad Q^2_{y2} = \sin(1-a)\theta$$

$$Q^1_{x3} = \frac{\cos\theta\cos a\theta - \cos(1-a)\theta}{r_1 \sin\theta}, \qquad Q^1_{y3} = \frac{-(\cos\theta\sin a\theta + \sin(1-a)\theta)}{r_1 \sin\theta}$$

$$Q^2_{x3} = \frac{\cos\theta\cos(1-a)\theta - \cos a\theta}{r_2 \sin\theta}, \qquad Q^2_{y3} = \frac{(\cos\theta\sin(1-a)\theta + \sin a\theta)}{r_2 \sin\theta}$$

All other elements of the $\mathbf{Q}^i$ vanish.

1. The Kinetic-Energy Operator in a Frame Fixed in the Body for NH_3

The nuclear kinetic-energy operator can now be constructed by routine algebra. The last term in (2.83) changes trivially as

$$-\frac{\hbar^2}{2\mu^n_{33}}\nabla^2(\mathbf{t}^n_3) \rightarrow \hat{K}^{atom} = -\frac{\hbar^2}{2\mu^n_{33}}\nabla^2(\mathbf{z}^n_3)$$

and it is convenient to introduce the atomic angular momentum

$$\hat{\mathbf{j}} = \frac{1}{i}\mathbf{z}_3^n \times \vec{\nabla}(\mathbf{z}_3^n)$$

where the vector product $\times$ is to be realized in terms of the Levi–Civita tensor, $\epsilon_{\alpha\beta\gamma}$. This angular momentum is not a constant of the motion.

The first term in the last line of (2.83) expands into

$$\hat{K}_V^{(1)} + \hat{K}_V^{(2)} + \hat{K}_{VA} + \hat{K}_{VR} \tag{3.67}$$

with the first three terms arising from the transformed form (3.38) and the last term arising from (3.37)

$$\hat{K}_V^{(1)} = -\frac{\hbar^2}{2}\left[\frac{1}{\mu_{11}^n r_1^2}\left(\frac{\partial}{\partial r_1}r_1^2\frac{\partial}{\partial r_1} + \frac{1}{\sin\theta}\frac{\partial}{\partial\theta}\sin\theta\frac{\partial}{\partial\theta}\right)\right.$$
$$\left. + \frac{1}{\mu_{22}^n r_2^2}\left(\frac{\partial}{\partial r_2}r_2^2\frac{\partial}{\partial r_2} + \frac{1}{\sin\theta}\frac{\partial}{\partial\theta}\sin\theta\frac{\partial}{\partial\theta}\right)\right]$$

$$\hat{K}_V^{(2)} = +\frac{\hbar^2}{\mu_{12}^n}\left[-\cos\theta\frac{\partial^2}{\partial r_1\partial r_2} + \frac{\cos\theta}{r_1 r_2}\left(\frac{1}{\sin\theta}\frac{\partial}{\partial\theta}\sin\theta\frac{\partial}{\partial\theta}\right)\right.$$
$$\left. + \sin\theta\left(\frac{1}{r_1}\frac{\partial}{\partial r_2} + \frac{1}{r_2}\frac{\partial}{\partial r_1} + \frac{1}{r_1 r_2}\right)\frac{\partial}{\partial\theta}\right]$$

$$\hat{K}_{VA} = \frac{\hbar^2}{2}\left[\kappa_{xx}\hat{j}_x^2 + \kappa_{yy}\hat{j}_y^2 + \kappa_{zz}\hat{j}_z^2 + \kappa_{xy}(\hat{j}_x\hat{j}_y + \hat{j}_y\hat{j}_x)\right]$$
$$+ \frac{\hbar^2}{2}\eta_z\hat{j}_z$$

$$\hat{K}_{VR} = \frac{1}{2}\left[\kappa_{xx}\hat{L}_x^2 + \kappa_{yy}\hat{L}_y^2 + \kappa_{zz}\hat{L}_z^2 + \kappa_{xy}(\hat{L}_x\hat{L}_y + \hat{L}_y\hat{L}_x)\right]$$
$$+ \frac{\hbar}{2}\sum_\alpha\lambda_\alpha\hat{L}_\alpha$$

$$\lambda_\alpha = \eta_\alpha + 2(\boldsymbol{\kappa}\hat{\mathbf{j}})_\alpha$$

$$\eta_\alpha = \delta_{\alpha z}\frac{2}{i}\left[\left(\frac{(1-a)}{\mu_{11}^n r_1^2} - \frac{a}{\mu_{22}^n r_2^2}\right)\left(\frac{\partial}{\partial\theta} + \frac{\cot\theta}{2}\right) + \frac{(2a-1)}{\mu_{12}^n r_1 r_2}\left(\cos\theta\frac{\partial}{\partial\theta} + \frac{1}{2\sin\theta}\right)\right.$$
$$\left. + \frac{\sin\theta}{\mu_{12}^n}\left(\frac{a}{r_2}\frac{\partial}{\partial r_1} - \frac{(1-a)}{r_1}\frac{\partial}{\partial r_2}\right)\right]$$

$$\kappa_{xx} = \frac{1}{\sin^2\theta}\left(\frac{\cos^2(1-a)\theta}{\mu_{11}^n r_1^2} + \frac{\cos^2 a\theta}{\mu_{22}^n r_2^2} - 2\frac{\cos a\theta\cos(1-a)\theta}{\mu_{12}^n r_1 r_2}\right)$$

$$\kappa_{yy} = \frac{1}{\sin^2\theta}\left(\frac{\sin^2(1-a)\theta}{\mu_{11}^n r_1^2} + \frac{\sin^2 a\theta}{\mu_{22}^n r_2^2} + 2\frac{\sin a\theta \sin(1-a)\theta}{\mu_{12}^n r_1 r_2}\right)$$

$$\kappa_{zz} = \frac{(1-a)^2}{\mu_{11}^n r_1^2} + \frac{a^2}{\mu_{22}^n r_2^2} + \frac{2a(1-a)\cos\theta}{\mu_{12}^n r_1 r_2}$$

$$\kappa_{xy} = \frac{1}{\sin^2\theta}\left(\frac{\cos(1-a)\theta\sin(1-a)\theta}{\mu_{11}^n r_1^2} - \frac{\cos a\theta \sin a\theta}{\mu_{22}^n r_2^2}\right.$$
$$\left.+ \frac{(\sin a\theta\cos(1-a)\theta - \cos a\theta\sin(1-a)\theta)}{\mu_{12}^n r_1 r_2}\right)$$

The h_k from (3.48) and the v_α from (3.44) have been incorporated into the expressions above, but for later purposes it is useful to present them explicitly. They are

$$h_1 = \frac{2}{\mu_{11}^n r_1}, \quad h_2 = \frac{2}{\mu_{22}^n r_2}, \quad h_3 = \left(\frac{1}{\mu_{11}^n r_1^2} + \frac{1}{\mu_{22}^n r_2^2}\right)\cot\theta - \frac{2}{\mu_{12}^n r_1 r_2 \sin\theta}$$

$$(3.68)$$

and

$$v_\alpha = 2\delta_{\alpha z}\left(\frac{\cot\theta}{2}\left(\frac{1-a}{\mu_{11}^n r_1^2} - \frac{a}{\mu_{22}^n r_2^2}\right) + \frac{2a-1}{2\mu_{12}^n r_1 r_2 \sin\theta}\right) \qquad (3.69)$$

The internal coordinate part of the Jacobian for the transformation is $r_1^2 r_2^2 \sin\theta$, and that for the Euler angle part is $\sin\phi_2$, where ϕ_2 is the second Euler angle. The range of r_1 and r_2 is $(0,\infty)$ and of θ and ϕ_2 is $(0,\pi)$.

2. The Potential Energy Operator in a Frame Fixed in the Body for NH_3

The potential-energy operators as defined in (3.50) involve the interparticle distances that may be developed from (2.68)

The interelectronic distances are of the usual form and need no further consideration, but the internuclear distances are

$$r_{12} = |\mathbf{z}_1^n - \mathbf{z}_2^n|, r_{13} = |\mathbf{z}_1^n|, \quad r_{23} = |\mathbf{z}_2^n|$$

$$r_{14} = \left|\frac{2}{3}\mathbf{z}_1^n - \frac{1}{3}\mathbf{z}_2^n - \mathbf{z}_3^n\right|, \quad r_{24} = \left|-\frac{1}{3}\mathbf{z}_1^n + \frac{2}{3}\mathbf{z}_2^n - \mathbf{z}_3^n\right|,$$

$$r_{34} = \left|-\frac{1}{3}\mathbf{z}_1^n - \frac{1}{3}\mathbf{z}_2^n - \mathbf{z}_3^n\right|$$

while the electron–nucleus distances are

$$r'_{1j} = \left| \mathbf{z}_j - \frac{2}{3}\mathbf{z}_1^n + \frac{1}{3}\mathbf{z}_2^n + \mathbf{z}_3^n \right|, \quad r'_{2j} = \left| \mathbf{z}_j + \frac{1}{3}\mathbf{z}_1^n - \frac{2}{3}\mathbf{z}_2^n + \mathbf{z}_3^n \right|$$

$$r'_{3j} = \left| \mathbf{z}_j + \frac{1}{3}\mathbf{z}_1^n + \frac{1}{3}\mathbf{z}_2^n + \mathbf{z}_3^n \right|, \quad r'_{4j} = |\mathbf{z}_j|$$

The electron–nucleus distances here are expressed in terms of $\mathbf{z}_i$ which have their origin at the nitrogen nucleus. But were the electronic origin to be placed at the center of nuclear mass, they would have the forms

$$r'_{1j} = \left| \mathbf{z}_j - \frac{2}{3}\mathbf{z}_1^n + \frac{1}{3}\mathbf{z}_2^n + \frac{m_N}{M}\mathbf{z}_3^n \right|, \quad r'_{2j} = \left| \mathbf{z}_j + \frac{1}{3}\mathbf{z}_1^n - \frac{2}{3}\mathbf{z}_2^n + \frac{m_N}{M}\mathbf{z}_3^n \right|$$

$$r'_{3j} = \left| \mathbf{z}_j + \frac{1}{3}\mathbf{z}_1^n + \frac{1}{3}\mathbf{z}_2^n + \frac{m_N}{M}\mathbf{z}_3^n \right|, \quad r'_{4j} = \left| \mathbf{z}_j - \frac{M_3}{M}\mathbf{z}_3^n \right|$$

In all these expressions $|\mathbf{z}_3^n|$ is the distance between the center-of-mass of the three protons and the nitrogen nucleus, while $|\mathbf{z}_1^n|$ and $|\mathbf{z}_2^n|$ are a pair of interproton separations r_1 and r_2.

These expressions are quite clumsy and may be developed in terms of the internal coordinates, r_1, r_2, θ and the components of $\mathbf{z}_3^n$ explicitly, only at the cost of making them even more clumsy.

3. The Dipole Moment Expression in a Frame Fixed in the Body for NH_3

In the coordinates used in this section the effective electronic charge, $\tilde{Z}$ is 1, and effective nuclear charges are given by (2.85) so that

$$\mathbf{d}(\mathbf{q}, \mathbf{z}) = -e \left[\frac{10M_3}{M} \begin{pmatrix} r_1\cos a\theta + r_2\cos(1-a)\theta \\ -r_1\sin a\theta + r_2\sin(1-a)\theta \\ 0 \end{pmatrix} \right.$$

$$\left. +3 \begin{pmatrix} z_{x3}^n \\ z_{y3}^n \\ z_{z3}^n \end{pmatrix} + \sum_{i=1}^{L} \begin{pmatrix} z_{xi} \\ z_{yi} \\ z_{zi} \end{pmatrix} \right] \tag{3.70}$$

where the coefficient of the leftmost term is approximately $\frac{30}{17}$.

F. Alternative Forms of the Kinetic-Energy Operator Expressed in the Frame Fixed in the Body

The internal coordinate part of the inverse metric matrix arising from (3.20) and (3.21) is

$$\begin{pmatrix} \mathbf{I} & \mathbf{y} \\ \mathbf{y}^T & \mathbf{f} \end{pmatrix} \tag{3.71}$$

The component matrices are

$$\mathbf{I} = \sum_{i,j=1}^{H-1} \mu_{ij}^n \hat{\mathbf{z}}_i^{nT} \hat{\mathbf{z}}_j^n \tag{3.72}$$

and this matrix is clearly the form of the instantaneous inertia tensor in a frame fixed in the body, while

$$\mathbf{y} = \sum_{i,j=1}^{H-1} \mu_{ij}^n \hat{\mathbf{z}}_i^{nT} \tilde{\mathbf{Q}}^j, \qquad \mathbf{f} = \sum_{i,j=1}^{H-1} \mu_{ij}^n \tilde{\mathbf{Q}}^{iT} \tilde{\mathbf{Q}}^j \tag{3.73}$$

in which the matrix μ^n, inverse to the matrix defined as explained immediately following (2.40) has elements

$$\mu_{ij}^n = \sum_{k=1}^{H} m_k ((\mathbf{V}^n)^{-1})_{ik} ((\mathbf{V}^n)^{-1})_{jk}, \qquad i,j = 1,2,\ldots,H-1 \tag{3.74}$$

By standard matrix manipulations

$$\begin{aligned} \mathbf{f}^{-1} &= (\mathbf{g} - \tau \kappa^{-1} \tau^T) & \mathbf{I}^{-1} &= (\kappa - \tau^T \mathbf{g}^{-1} \tau) \\ \mathbf{g}^{-1} &= (\mathbf{f} - \mathbf{y}^T \mathbf{I}^{-1} \mathbf{y}) & \kappa^{-1} &= (\mathbf{I} - \mathbf{y} \mathbf{f}^{-1} \mathbf{y}^T) \end{aligned} \tag{3.75}$$

and

$$\tau \kappa^{-1} + \mathbf{f}^{-1} \mathbf{y}^T = \mathbf{0}_{3H-6,3} \tag{3.76}$$

With these results the kinetic-energy operator is put into a form analogous to that found by Eckart [2] by introducing the Coriolis coupling operator

$$\hat{\pi} = \frac{\hbar}{i} \kappa^{-1} \tau^T \frac{\partial}{\partial \mathbf{q}} \tag{3.77}$$

The sum of the kinetic-energy operators (3.37) and (3.38) may be rewritten as

$$\hat{K}_a(\boldsymbol{\phi}, \mathbf{q}, \mathbf{z}) + \hat{K}_a(\mathbf{q}, \mathbf{z})$$

with

$$
\hat{K}_a(\boldsymbol{\phi}, \mathbf{q}, \mathbf{z}) = \frac{1}{2} \left[\sum_{\alpha\beta} \kappa_{\alpha\beta} (\hat{L}_\alpha + \hat{\pi}_\alpha + \hbar \hat{l}_\alpha)(\hat{L}_\beta + \hat{\pi}_\beta + \hbar \hat{l}_\beta) \right.
$$
$$
\left. + \hbar \sum_\alpha \bar{\nu}_\alpha (\hat{L}_\alpha + \hbar \hat{l}_\alpha) \right] \tag{3.78}
$$

where

$$\bar{\nu}_\alpha = \frac{\nu_\alpha}{i} + \lambda'_\alpha$$

and

$$
\hat{K}_a(\mathbf{q}, \mathbf{z}) = -\frac{\hbar^2}{2} \left(\sum_{k,l=1}^{3H-6} f_{kl}^{-1} \frac{\partial^2}{\partial q_k \partial q_l} + \sum_{k=1}^{3H-6} \bar{h}_k \frac{\partial}{\partial q_k} \right) +
$$
$$
- \hbar^2 \sum_{j=1}^{L} \sum_\alpha \left(\sum_{k=1}^{3H-6} g'_{\alpha k} \frac{\partial^2}{\partial q_k \partial z_{\alpha j}} + h'_\alpha \frac{\partial}{\partial z_{\alpha j}} \right) \tag{3.79}
$$

in which

$$\bar{h}_k = h_k - \sum_\beta \left(\tau^{\mathrm{T}} \frac{\partial}{\partial \mathbf{q}} \right)_\beta (\kappa^{-1} \tau^{\mathrm{T}})_{\beta k} \tag{3.80}$$

G. Internal and Orientational Coordinates Expressed Directly in Terms of the Original Coordinates

Clearly any expressions for the angular and internal coordinates in terms of translationally invariant coordinates can be reexpressed using (2.5) in terms of the laboratory-fixed coordinates. So it is always possible to pass directly from laboratory-fixed coordinates to these coordinates without an explicit choice of translationally invariant coordinates. However, if this is done and it is desired to preserve the forms of the operators already derived here, care must be taken to ensure that appropriate permutational restrictions are enforced. Assuming that this has been done, the derivatives of the angular and internal coordinates with respect to $x^n_{\alpha i}$ and $x^e_{\beta j}$ may be developed much as above. The internal coordinate part of the Jacobian now has an extra $3H$ by 3 partition that consists of the H 3×3 matrices, $M^{-1}m_i\mathbf{E}_3$ for $i = 1$, $2, \ldots, H$, and there is a similar extension to the inverse Jacobian consisting of H repetitions of $\mathbf{E}_3$. The sums (3.23) now extend up to H, and the equivalent extension of (3.24) is

$$\Omega^i \hat{\mathbf{z}}^{n\mathrm{T}}_j + \mathbf{Q}^i \tilde{\mathbf{Q}}^{j\mathrm{T}} = \delta_{ij}\mathbf{E}_3 - M^{-1}m_i \tag{3.81}$$

The extra requirements arising from the product of the enlarged Jacobian and its inverse are that the following relationships are satisfied as identities:

$$\sum_{i=1}^{H}\Omega^i = \mathbf{0}_3 \qquad \sum_{i=1}^{H}\mathbf{Q}^i = \mathbf{0}_{3,3H-6}$$

$$\sum_{i=1}^{H}m_i\hat{\mathbf{z}}^n_i = \mathbf{0}_3 \qquad \sum_{i=1}^{H}m_i\tilde{\mathbf{Q}}^i = \mathbf{0}_{3,3H-6} \tag{3.82}$$

which are easily seen to be satisfied as identities for any translationally invariant choice of angular and internal coordinates.

The metric and its inverse are constructed using $m_i^{-1}\delta_{ij}$ in place of $1/\mu^n_{ij}$ and $m_i\delta_{ij}$ in place of μ^n_{ij}, respectively. The metric matrix then becomes $3H \times 3H$ with an extra 3×3 block on the diagonal with elements M^{-1}; similarly, the inverse metric has an extra 3×3 block with elements M. Only null blocks connect these to the rest of the matrix and the components of the rest of the matrix generalize in an obvious way with sums over particle indices running to H rather than $H - 1$.

1. Example: The Eckart Choice of a Frame Fixed in the Body

In Eckart's original account of fixing a frame in a molecule, the electrons were considered simply as providing a potential for nuclear motion and in the Hamiltonian as usually written, there is no mention of the electrons. For clarity and for comparison purposes, in our discussion of this embedding therefore, we, too, shall neglect the electronic contributions to the equations, although we shall assume that electronic coordinates are treated relative to the center of nuclear mass so that the permutational restrictions (2.52)–(2.54) are effectively satisfied. This neglect is not, however, an intrinsic matter, and as in the earlier account of NH_3, the electronic terms could be included if wished.

First, a redundant set of Cartesian coordinates for the nuclei is defined in the frame fixed in the body by

$$\mathbf{x}_i^{\text{n}} - \mathbf{X} = \mathbf{C}\mathbf{z}_i^{\text{n}}$$

where $\mathbf{X}$ is the center-of-nuclear-mass coordinate. Thus

$$\sum_{i=1}^{H} m_i \mathbf{z}_i^{\text{n}} = 0 \qquad (3.83)$$

and a matrix $\mathbf{C}$ is chosen as in (3.2) to define a set of Cartesians in the frame fixed in the body such that the reference structure of the nuclear framework is specified by $\mathbf{z}_i^{\text{n}} = \mathbf{a}_i$, where the $\mathbf{a}_i$ are constant matrices and, by definition, $\sum_{i=1}^{H} m_i \mathbf{a}_i = 0$. The reference structure is, of course, usually chosen to be the classical molecular geometry. It is assumed chosen to reflect the equilibrium geometry of the molecule as it would be at the minimum of the potential. However, for purposes of carrying through the derivations, there is no need to specify it more closely than as a reference configuration.

Defining the displacement coordinates as $\boldsymbol{\rho}_i = \mathbf{z}_i^{\text{n}} - \mathbf{a}_i$, the specification of the matrix $\mathbf{C}$ is completed by requiring that

$$\sum_{i=1}^{H} m_i \hat{\mathbf{a}}_i \boldsymbol{\rho}_i = \mathbf{0} \equiv \sum_{i=1}^{H} m_i \vec{a}_i \times \vec{\rho}_i = \vec{0} \qquad (3.84)$$

provided the $\mathbf{a}_i$ do not define a line. Traditionally the displacements are regarded as usually small, but this choice is not intrinsic to the discussion of this section.

These constraint conditions are on the components of the mass weighted sum over all the vectors, of the vector products of the reference vectors with the displacement vectors. In classical mechanics the vanishing of these components would be interpreted as the system having no internal angular momentum at the reference geometry. The two conditions, (3.83) and (3.84) defining the embedded frame are often called the *Eckart conditions*. Here $\mathbf{z}_i^{\text{n}}$ is used for Watson's $\mathbf{r}_i$ and $\mathbf{a}_i$ for Watson's $\mathbf{r}_i^0$.

The $\mathbf{z}_i^{\text{n}}$ are completely expressible in terms of a set of $3H - 6$ internal coordinates. The internal coordinates are expressed in terms of displacements from the reference geometry by

$$q_k = \sum_{i=1}^{H} \sum_{\alpha} b_{\alpha i k} \rho_{\alpha i} \equiv \sum_{i=1}^{H} \mathbf{b}_{ik}^{\text{T}} \boldsymbol{\rho}_i, \qquad k = 1, 2, \ldots, 3H - 6 \qquad (3.85)$$

where the elements $b_{\alpha i k}$ are simply constants that may be regarded as components of a column matrix $\mathbf{b}_{ik}$. The range of these coordinates is $(-\infty, \infty)$.

Internal coordinates must be linearly independent, and so the $\mathbf{b}_{ik}$ must be linearly independent, and for an inverse transformation to exist between the internal coordinates and the coordinates in the frame fixed in the laboratory, it is [45] required that

$$\sum_{i=1}^{H} \mathbf{b}_{ik} = \mathbf{0}, \qquad \sum_{i=1}^{H} \hat{\mathbf{a}}_i \mathbf{b}_{ik} = \mathbf{0} \equiv \sum_{i=1}^{H} \vec{a}_i \times \vec{b}_{ik} = \vec{0}$$

Using the two Eckart conditions given above, it follows that

$$\mathbf{C} = \mathbf{B}(\mathbf{B}^{\mathrm{T}}\mathbf{B})^{-(1/2)} \tag{3.86}$$

where

$$\mathbf{B} = \sum_{i=1}^{H} m_i(\mathbf{x}_i^{\mathrm{n}} - \mathbf{X})\mathbf{a}_i^{\mathrm{T}} \tag{3.87}$$

The 3×3 matrix $\mathbf{B}^{\mathrm{T}}\mathbf{B}$ is symmetric and therefore diagonalizable, so functions of it may be properly defined in terms of its eigenvalues. There are, in principle, eight (2^3) possible distinct square-root matrices, all such that their square yields the original matrix product. Consistency (see Ezra [41]) requires, however, that the positive square roots of each eigenvalue be chosen. The eigenvalues cannot be negative, but one or more might be zero; if this is so, the Eckart frame cannot be properly defined. This is a matter to which we shall return later.

If we denote the $3 \times H$ matrix of the $\mathbf{x}_i^{\mathrm{n}} - \mathbf{X}$ as $\mathbf{w}$ and the similar collection of all the $\mathbf{z}_i^{\mathrm{n}}$ and thus the $\mathbf{a}_i$ as $\mathbf{z}^{\mathrm{n}}$ and $\mathbf{a}$ then, from (3.87), $\mathbf{B}$ may be written as

$$\mathbf{B} = \mathbf{wma}^{\mathrm{T}} = \mathbf{Cz}^{\mathrm{n}}\mathbf{ma}^{\mathrm{T}} \tag{3.88}$$

where $\mathbf{m}$ is an $H \times H$ diagonal matrix with the nuclear masses along the diagonal. The second Eckart condition (3.84) can be manipulated to show that it is equivalent to the requirement that the matrix

$$\mathbf{A} = \sum_{i=1}^{H} m_i \mathbf{z}_i^{\mathrm{n}} \mathbf{a}_i^{\mathrm{T}} \equiv \mathbf{z}^{\mathrm{n}}\mathbf{ma}^{\mathrm{T}} \tag{3.89}$$

is symmetric. Thus $\mathbf{B}^{\mathrm{T}}\mathbf{B} \equiv \mathbf{A}^{\mathrm{T}}\mathbf{A} = \mathbf{A}^2$, and so from (3.86), we obtain $\mathbf{C} = \mathbf{B}(\mathbf{A}^2)^{-(1/2)}$.

Whatever the precise specification made of the inverse square root, if $\mathbf{A}$ is singular, then $\mathbf{C}$ is undefined and the Eckart specification fails. Thus, as mentioned above, if the $\mathbf{a}_i$ together specify a line—for example, if all the a_{yi} and a_{zi} vanish—then $\mathbf{A}$ is clearly singular. But there is a special case. If the $\mathbf{a}_i$ together specify a planar figure—for example, if all the a_{zi} vanish—then $\mathbf{A}$ is again singular, but in this case the Eckart conditions may be satisfied by requiring that the z_{zi}^{n} vanish, too. The last row and column of $\mathbf{A}$ are null, and there is a 2×2 nonvanishing block, and, provided this block is nonsingular, the problem remains well defined. Infact, there is no need in practice to treat this special case explicitly. The second Eckart condition implicitly orients the third axis to be perpendicular to the defined plane and, as long as the 2×2 block of $\mathbf{A}$ is nonsingular, the planarity constraints are subsumed.

It is possible to write the Cartesians expressed in the Eckart frame directly in terms of the coordinates of the frame fixed in the laboratory as

$$\mathbf{z}^{\mathrm{n}} = \mathbf{C}^{\mathrm{T}}\mathbf{w} = (\mathbf{amw}^{\mathrm{T}}\mathbf{wma}^{\mathrm{T}})^{-(1/2)}\mathbf{amw}^{\mathrm{T}}\mathbf{w}$$

and because the ijth element of $\mathbf{w}^{\mathrm{T}}\mathbf{w}$ is a scalar product, $(\mathbf{x}_i^{\mathrm{n}} - \mathbf{X})^{\mathrm{T}}(\mathbf{x}_j^{\mathrm{n}} - \mathbf{X})$, the internal cartesians are clearly invariant under any orthogonal transformation, including inversion, of the original nuclear coordinates.

Using (3.17) and (3.84), it follows that

$$\Omega^i = m_i\hat{\mathbf{a}}_i\mathbf{I}''^{-1} \tag{3.90}$$

with

$$\mathbf{I}'' = \sum_{j=1}^{H} m_j\hat{\mathbf{z}}_j^{\mathrm{nT}}\hat{\mathbf{a}}_j \tag{3.91}$$

and using (3.14), (3.16), and (3.85), we obtain

$$Q_{\gamma k}^i = b_{\gamma ik} + \left(\Omega^i \sum_{j=1}^{H} \hat{\mathbf{z}}_j^{\mathrm{n}}\mathbf{b}_{jk}\right)_\gamma = b_{\gamma ik} + (\Omega^i\mathbf{d}_k)_\gamma \tag{3.92}$$

where it is clear that the expressions are suitably invariant under translations and orthogonal transformations of the coordinates and satisfy the appropriate Jacobian relations.

Using the extension of (3.39), (3.90), and (3.91)

$$\boldsymbol{\kappa} = \mathbf{I}''^{-1}\mathbf{I}^0\mathbf{I}''^{-1}$$

where $\mathbf{I}^0$ is the inertia tensor for the molecule at the reference geometry

$$\mathbf{I}^0 = \sum_{i=1}^{H} m_i \hat{\mathbf{a}}_i^{\mathrm{T}} \hat{\mathbf{a}}_i$$

and so is a constant matrix. From the extension of (3.46) with (3.92)

$$g_{kl} = \sum_{i=1}^{H} m_i^{-1} \mathbf{b}_{ik}^{\mathrm{T}} \mathbf{b}_{il} + \mathbf{d}_k^{\mathrm{T}} \boldsymbol{\kappa} \mathbf{d}_l$$

and from the extension of (3.43) using (3.90) and (3.92), it follows that $\tau_{k\alpha} = (\boldsymbol{\kappa}\mathbf{d}_k)_\alpha$, so that from (3.75)

$$(\mathbf{f}^{-1})_{kl} = \sum_{i=1}^{H} m_i^{-1} \mathbf{b}_{ik}^{\mathrm{T}} \mathbf{b}_{il}$$

and from (3.77), the Coriolis coupling operator is

$$\hat{\boldsymbol{\pi}} = \frac{\hbar}{i} \sum_{k=1}^{3H-6} \mathbf{d}_k \frac{\partial}{\partial q_k}$$

To realize these forms fully, the $\mathbf{z}_i^n$ must be expressed in terms of the q_k. Provided the $\boldsymbol{\rho}_i$ satisfy the two Eckart conditions, (3.85) can be inverted to yield

$$\rho_{\alpha i} = \sum_{k=1}^{3H-6} m_{\alpha ik} q_k, \qquad m_{\alpha ik} = m_i^{-1} \sum_{l=1}^{3H-6} b_{\alpha il} f_{lk}$$

where

$$\sum_{\alpha} \sum_{i=1}^{H} b_{\alpha il} m_{\alpha ik} = \delta_{kl}$$

This is not, of course, a complete inverse, and

$$\sum_{k=1}^{3H-6} m_{\alpha ik} b_{\beta jk} = \delta_{ij}\delta_{\alpha\beta} - m_j M^{-1} - m_j (\hat{\mathbf{a}}_i^{\mathrm{T}} \mathbf{I}^{0^{-1}} \hat{\mathbf{a}}_j)_{\alpha\beta}$$

Expressing $\mathbf{d}_k$ in terms of the q_k gives

$$\mathbf{d_k} = \sum_{l=1}^{3H-6} \sum_{i=1}^{H} \hat{\mathbf{m}}_{il} \mathbf{b}_{ik} q_l = \sum_{l=1}^{3H-6} \zeta_{kl} q_l$$

The Hamiltonian derived using the general internal coordinates is really very cumbersome and extremely difficult to use (see, e.g., Ref. 45), but Watson [35] showed (see also Louck [43] for an approach rather more like ours) that it could be simplified by the particular choice of orthogonal internal coordinates and by incorporating the internal coordinate part of the Jacobian into the operator as discussed in Section III.D. The resulting operator is often called the *Eckart–Watson Hamiltonian*. In this realisation and notation, the $\mathbf{b}_{ik}$ are written $\sqrt{m_i}\mathbf{l}_{ik}$, and it is required that

$$\sum_{i=1}^{H} \mathbf{l}_{ik}^{\mathrm{T}} \mathbf{l}_{im} = \delta_{km}, \qquad \sum_{i=1}^{H} \sqrt{m_i} \mathbf{l}_{ik} = 0, \qquad \sum_{i=1}^{H} \sqrt{m_i} \hat{\mathbf{a}}_i \mathbf{l}_{ik} = 0$$

The internal coordinates so defined will still be written as q_k where now, explicitly

$$q_k = \sum_{i=1}^{H} \sqrt{m_i} \mathbf{l}_{ik}^{\mathrm{T}} \boldsymbol{\rho}_i$$

The Watson form of the Eckart kinetic energy operators for the molecule can be written as

$$\hat{K}_a(\boldsymbol{\phi}, \mathbf{q}) = \frac{1}{2} \sum_{\alpha\beta} \mu_{\alpha\beta} (\hat{L}_\alpha + \hat{\pi}_\alpha)(\hat{L}_\beta + \hat{\pi}_\beta)$$

$$\hat{K}_a(\mathbf{q}) = -\frac{\hbar^2}{2} \sum_{k=1}^{3H-6} \frac{\partial^2}{\partial q_k^2} - \frac{\hbar^2}{8} \sum_{\alpha} \mu_{\alpha\alpha} \qquad (3.93)$$

in which $\boldsymbol{\mu}$ is the inverse generalized inertia tensor $\boldsymbol{\kappa}$ given above. This is the standard notation, but $\boldsymbol{\mu}$ here is not to be confused with any of the reduced mass matrices used elsewhere. The operator $\hat{\pi}_\alpha$ is the Coriolis coupling operator as given above becomes

$$\hat{\pi}_\alpha = \frac{\hbar}{i} \sum_{k,l=1}^{3H-6} \zeta_{kl}^\alpha q_k \frac{\partial}{\partial q_l}$$

in which

$$\zeta_{kl}^{\alpha} = \sum_{i=1}^{H} (\hat{\mathbf{l}}_{ik}\mathbf{l}_{il})_{\alpha}$$

This operator is often called a *component of internal angular momentum*, but this is rather a misnomer, as it has none of the properties of angular momentum.

In the original presentation by Watson, the components of the total angular momentum operators are denoted $\hat{\Pi}_{\alpha}$ and obey the anomalous commutation conditions. Here we have followed Louck and used angular momentum operators that obey the standard commutation conditions. The Watson form is achieved from the present one by making the substitution

$$(\hat{L}_{\alpha} + \hat{\pi}_{\alpha}) \rightarrow -(\hat{\Pi}_{\alpha} - \hat{\pi}_{\alpha})$$

IV. SYMMETRY PROPERTIES REALISED IN A FRAME FIXED IN THE BODY

As noted at the end of our initial discussion of fixing a frame in the body, the symmetry properties of the angular and internal coordinates necessary to the fixing are not at once apparent. Here we shall attempt to discuss these matters.

A. Rotation-reflection symmetry

The full rotation-reflection symmetry of the Hamiltonian is that of the full orthogonal group in three dimensions, $O(3)$, but the symmetric-top functions (3.29) are a basis only for representations of the rotation group $SO(3)$; that is, parity is ignored. To introduce parity we consider the appropriate extension by considering the construction of the angular functions using the orientation matrix $\mathbf{C}$. Using the elements of $\mathbf{D}^1$ from (3.34), the elements of the matrix $\mathbf{D}^2$ can be obtained by vector coupling

$$D_{mm'}^{2} = \sum_{ss'} \langle 11ss'|2m'\rangle D_{ps}^{1}D_{p's'}^{1}, \qquad m = p + p'$$

in which $\langle 11ss'|2m'\rangle$ is a Clebsch–Gordan coefficient as defined in Brink and Satchler [36]. $\mathbf{D}^3$ can be constructed by coupling the elements of $\mathbf{D}^1$ and $\mathbf{D}^2$ and so on, at every stage coupling to the maximum allowed J value, the so-called *fully stretched* coupling scheme. Coupling the indices p and p' above is also possible because (see Appendix 2 of Ezra [41]), the matrices $\mathbf{D}^J$ are strictly spherical double tensors. We do not bother to do this because we are uninterested in the z component of angular momentum in the laboratory frame. The general matrix $\mathbf{D}^J$ can then be obtained and hence expressions for the angular momentum eigenfunctions directly in terms of the elements of $\mathbf{C}$ as can expressions for $\mathscr{D}^J$. (For details, see Section 6.19 of Biedenharn and Louck [37].) The relationship is

$$D^J_{mm'}(\mathbf{C}) = (-1)^{m'} \mathscr{D}^{J*}_{m-m'}(\mathbf{C}), \qquad |\mathbf{C}| = +1$$

If the matrix $\mathbf{C}$ is replaced by the matrix product $\mathbf{CU}$, where $\mathbf{U}$ is also an orthogonal matrix then from (3.34)

$$
\begin{aligned}
|1Mk\rangle &\rightarrow \left(\frac{3}{8\pi^2}\right)^{1/2} (\mathbf{X}^{\mathrm{T}}\mathbf{C}\mathbf{X})_{Mk} \\
&= \left(\frac{3}{8\pi^2}\right)^{1/2} (\mathbf{X}^{\mathrm{T}}\mathbf{C}\mathbf{X}\mathbf{X}^{\dagger}\mathbf{U}\mathbf{X})_{Mk} \\
&= \sum_{n=-1}^{+1} |1Mn\rangle \mathscr{D}^1_{nk}(\mathbf{U})
\end{aligned}
\tag{4.1}
$$

so that the general result is

$$|JMk\rangle \rightarrow \sum_{n=-J}^{+J} |JMn\rangle \mathscr{D}^J_{nk}(\mathbf{U}) \tag{4.2}$$

in which $\mathscr{D}^J(\mathbf{U})$ is the matrix made up from the elements of $\mathbf{U}$ in exactly the same way that $\mathscr{D}^J$ is made up from the elements of $\mathbf{C}$. A precise account of how this is to be done is given in Section 6.19 of Biedenharn and Louck [37]. If $\mathbf{U}$ is a constant matrix, then $\mathscr{D}^J(\mathbf{U})$ is a constant matrix and (4.2) simply represents a linear combination. If $\mathbf{U}$ is a unit matrix, then $|JMk\rangle$ is invariant.

The group $O(3)$ is the direct product of the inversion group C_i with the special orthogonal group in three dimensions, $SO(3)$. The inversion group consists of the identity operator $\hat{E}$ and the inversion operator $\hat{I}$. The operations of $SO(3)$ may be realized in three-dimensional coordinate space by proper 3×3 orthogonal matrices $\mathbf{R}$, and the inversion may be realized by $-\mathbf{E}_3$. Thus for every matrix $\mathbf{R}$ in $SO(3)$ there is a companion matrix $-\mathbf{R}$ in $O(3)$. The matrix $-\mathbf{R}$ in general represents a reflection such that if a proper rotation by π is performed about the normal to the reflection plane and this operation is followed by the inversion, the matrix $-\mathbf{R}$ results. From the analysis above, because of the fully stretched form of the coupling, given a Wigner matrix composed from an orthogonal matrix $\mathbf{R}$

$$\mathscr{D}^J(-\mathbf{R}) = (-1)^J \mathscr{D}^J(\mathbf{R}) \tag{4.3}$$

so the matrices constructed in this way provide irreducible representations of even parity for J even but of odd parity for J odd. Thus, in a colloquial sense at least, they provide only half the representations that there should be for $O(3)$ because there should also be representations of odd parity for J even and of even parity for J odd since the group manifold consists of two disconnected but isomorphic sheets. This defect can be remedied however, by following the work of Biedenharn and Louck [37] and of Ezra [41] and defining

$$\mathscr{D}^{0J}(\mathbf{R}) = \mathscr{D}^J(\mathbf{R}), \qquad \mathscr{D}^{1J}(\mathbf{R}) = |\mathbf{R}|\mathscr{D}^J(\mathbf{R}) \tag{4.4}$$

where $|\mathbf{R}|$ denotes the determinant of $\mathbf{R}$. Thus it follows that

$$\mathscr{D}^{rJ}(-\mathbf{R}) = (-1)^{r+J}\mathscr{D}^{rJ}(\mathbf{R}), \qquad r = 0, 1 \tag{4.5}$$

The angular momentum eigenfunctions are generalized to include parity as

$$|JMkr\rangle = |\mathbf{C}|^r |JMk\rangle \tag{4.6}$$

then it is easily shown that under the change $\mathbf{C} \to \mathbf{CU}$ (4.2) generalizes to

$$|JMkr\rangle \to \sum_{n=-J}^{+J} |JMnr\rangle \mathscr{D}^{rJ}_{nk}(\mathbf{U}) \tag{4.7}$$

This establishes the position of the matrices $\mathscr{D}^{rJ}$ as representation matrices for the general orthogonal transformation in three dimensions and in particular if $\mathbf{U} = -\mathbf{E}_3$, then $\mathscr{D}^{rJ}(-\mathbf{E}_3) = (-1)^{r+J}\mathbf{E}_3$ as is required.

The underlying integral for the normalization of $|JMkr\rangle$ must now extend over both sheets of $O(3)$ and is [37]

$$\int \mathscr{D}^{rJ'}_{M'k'}(\mathbf{C})\mathscr{D}^{rJ}_{Mk}(\mathbf{C})|\mathbf{D}|^{-1}d\phi + \int \mathscr{D}^{rJ'}_{M'k'}(-\mathbf{C})\mathscr{D}^{rJ}_{Mk}(-\mathbf{C})|\mathbf{D}|^{-1}d\phi \tag{4.8}$$

where $|\mathbf{D}|^{-1}$ is the angular part of the Jacobian and the integrals are each over the whole range of the Eulerian angles. If $r' = r$, then both integrals are the same and each has the value $\delta_{J'J}\delta_{M'M}\delta_{k'k}8\pi^2/(2J+1)$; if $r' \neq r$, then both integrals have the same absolute value but opposite signs and hence they cancel. Thus the generalized rotation-reflection eigenfunctions are orthonormal according to

$$\langle J'M'k'r' \mid JMkr \rangle = \delta_{J'J}\delta_{M'M}\delta_{k'k}\delta_{r'r} \qquad (4.9)$$

In a discussion of inversion and parity toward the end of Chapter 19 of his book on group theory, Wigner [47] notes that in two- and three-body systems (and in these systems only) it is possible to realize the effect of an inversion by means of a sequence of proper rotations because each of these cases have rather special features. Thus the matrix $\mathbf{C}$ that is used to transform to the coordinate system fixed in the body can be chosen in the three-body case such that all three particles lie in a plane. If that plane is, say, the x–y one, the z coordinates of the particles will all be zero and the effect of an inversion can be achieved by performing a proper rotation by π about the z axis. Two-body systems have not been considered in this chapter for reasons explained earlier, but three-body systems have not, so far, been excluded. However the only genuinely three-body systems of interest are such systems as the helium atom and the hydrogen molecule ion, and these are best dealt with by quite special methods that are specific to each problem. Thus three-body systems can also be ignored without much loss of generality. To do so is not to ignore triatomic molecules because in these systems there will be four or more particles, as at least one electron must be present for binding. Thus the foregoing discussion can be regarded as the general treatment of parity for our purposes.

In the discussion of internal coordinates made below (i.e., in text immediately following) (3.2), it was noted that internal coordinates composed of scalar triple products of translationally invariant coordinates would not be used in the present work since it was desired that the internal coordinates be invariant not only under rotations but also under rotation-reflections too. This seems to be the sensible thing to do because it makes the most complete separation possible between the internal and orientation parts of the wavefunction, but it is not a choice that is universally made.

It is now strictly necessary to extend (3.52) as

$$\Psi^{J,M}(\mathbf{t}^n, \mathbf{t}^e) \rightarrow \Psi^{J,M,r}(\boldsymbol{\phi}, \mathbf{q}, \mathbf{z}) = \sum_{k=-J}^{+J} \Phi_k^{J,r}(\mathbf{q}, \mathbf{z})|JMkr\rangle \qquad (4.10)$$

This form allows for the possibility that states of different parity can, in principle, have different energies. There are, however, no parity-dependent terms in the Hamiltonian, so that there is no reason to expect that the functions $\Phi_k^{J,r}(\mathbf{q}, \mathbf{z})$ that differ just in r value, are in any way different. Thus the energies of such a parity pair would be degenerate. This degeneracy must however be "accidental," in the sense that it is not due to $O(3)$ symmetry. This is because $O(3)$ is the direct product $C_i \times SO(3)$ so that there will be two distinct irreducible representations of $O(3)$ for every representation of $SO(3)$. This is precisely analogous to the way in which distinct g and u representations arise in point groups such as C_{6h} or D_{6h}, which are direct products $C_i \times C_6$ or $C_i \times D_6$. The fact that the two representations are distinct means that there is no group theoretical reason to suppose that the two states $|JMkr\rangle$ for $r = 0, 1$ *should* be degenerate.

If states of identical form with each r value were allowed, however, it would be possible to understand the presence of a permanent dipole moment by attributing it to the mixing of the degenerate states by the small perturbing electric field used for the measurement. The observed dipole moment would then be analogous to that observed in the excited states of the hydrogen atom, which is customarily attributed to the mixing by the electric field of the accidentally degenerate s and p orbitals.

B.   Rotation Reflections Defined in the Frame Fixed in the Body

Any chosen set of translationally invariant coordinates can be reexpressed in terms of another choice of the matrix that fixes a frame in the body according to

$$\mathbf{t}^n = \mathbf{C}\mathbf{z}^n \rightarrow \overline{\mathbf{C}}\,\overline{\mathbf{z}}^n$$

with

$$\overline{\mathbf{C}} = \mathbf{C}\mathbf{R}^T, \qquad \overline{\mathbf{z}}^n = \mathbf{R}\mathbf{z}^n \tag{4.11}$$

and where $\mathbf{R}$ is an orthogonal matrix whose elements are at most functions of the internal coordinates q_k. Clearly the matrix $\mathbf{R}$ can be chosen to produce, from any given frame fixed in the body, any other frame that is desired. Such a change, while changing the $\mathbf{z}^n$ as given above, will not change the internal coordinates as these depend only upon scalar products of the $\mathbf{t}_i^n$. Of course, since the electronic variables expressed in the frame fixed in the body are defined by

$$\mathbf{z}_i = \mathbf{C}^T \mathbf{t}_i^e, \qquad i = 1, 2, \ldots, L \tag{4.12}$$

it follows that these variables will change on the transformation made above according to $\mathbf{z}_i \rightarrow \overline{\mathbf{z}}_i$. The changes induced in the dipole operator by these transformations are perfectly apparent and will not be explicitly considered here, but the derivative operators do need to attention. Retaining the original choice of Eulerian angles and internal coordinates it can readily be shown that.

$$\frac{\partial}{\partial \mathbf{t}_i^n} = \overline{\mathbf{C}} \left(\overline{\Omega}^i \overline{\mathbf{D}} \frac{\partial}{\partial \phi} + \overline{\mathbf{Q}}^i \frac{\partial}{\partial \mathbf{q}} + \overline{\Omega}^i \sum_{j=1}^{L} \hat{\mathbf{z}}_j \frac{\partial}{\partial \overline{\mathbf{z}}_j} \right) \tag{4.13}$$

and

$$\frac{\partial}{\partial \mathbf{t}_i^e} = \overline{\mathbf{C}} \frac{\partial}{\partial \overline{\mathbf{z}}_i} \tag{4.14}$$

Here

$$\overline{\Omega}^i = |\mathbf{R}| \mathbf{R} \Omega^i \mathbf{R}^T \tag{4.15}$$
$$\overline{\mathbf{D}} = |\mathbf{R}| \mathbf{R} \mathbf{D} \tag{4.16}$$

and

$$\overline{\mathbf{Q}}^i = \mathbf{R} \mathbf{Q}^i \tag{4.17}$$

so that

$$\hat{\overline{\mathbf{L}}}(\phi) = \frac{\hbar}{i} |\mathbf{R}| \mathbf{R} \mathbf{D} \frac{\partial}{\partial \phi} = |\mathbf{R}| \mathbf{R} \hat{\mathbf{L}}(\phi) \tag{4.18}$$

Note that in this form the components of $\hat{\overline{\mathbf{L}}}$ do not commute with $\partial/\partial q_k$ unless it happens that $\mathbf{R}$ is a constant matrix. It would be possible in principle to define a new set of Eulerian angles in which to express the angular components of $\hat{\overline{\mathbf{L}}}$ so that they did commute with $\partial/\partial q_k$. But the present form is the one most useful for our purposes. Carrying these changes through to the construction of the kinetic-energy operator, it turns out that $\hat{K}(\mathbf{q}, \mathbf{z})$ is invariant while $\hat{K}(\phi, \mathbf{q}, \mathbf{z})$ becomes

$$\frac{1}{2} \left(\sum_{\alpha\beta} \overline{\kappa}_{\alpha\beta} \hat{\overline{L}}_\alpha \hat{\overline{L}}_\beta + \hbar \sum_{\alpha} \overline{\overline{\lambda}}_\alpha \hat{\overline{L}}_\alpha \right) \tag{4.19}$$

where

$$\overline{\kappa} = \mathbf{R} \kappa \mathbf{R}^T, \qquad \overline{\tau} = |\mathbf{R}| \tau \mathbf{R}^T, \qquad \overline{v} = |\mathbf{R}| v \mathbf{R}^T$$

and so

$$\bar{\bar{\lambda}}_\alpha = |\mathbf{R}| \sum_\beta R_{\alpha\beta} \bar{\lambda}_\beta$$

When $\mathbf{R} = -\mathbf{E}_3$ and thus represents a simple inversion, $\hat{K}(\phi, \mathbf{q}, \mathbf{z})$ remains invariant, as would have been expected from our earlier discussion.

Any changes induced in $\hat{K}(\phi, \mathbf{q}, \mathbf{z})$ by $\mathbf{R}$ are in form alone. All that has happened is that the axes about which the angular momentum is defined have been changed. But since the original choice was one of convention, the changes are of no physical significance. Of course, one particular choice might be much better than another for the purposes of approximate calculation.

The invariance of the terms h_k and h'_α in (3.38) is, at first sight, surprising because both apparently depend, via Ω^i, on the choice of Eulerian angles. In fact, since they arise from terms that are scalar products of derivatives, they cannot actually so depend, as has been explicitly demonstrated. The fact that the h_k in (3.48) do not depend in any way on the embedding choice, whereas the v_α do, is consistent with these results. This invariance has interesting consequences when considering only states with $J = 0$. One can then choose any convenient embedding to derive the Ω^i and compute both expressions using it. Perhaps the easiest one to choose is the three particle one as exemplified above, although in general one would have to reexpress the internal coordinates chosen with this embedding in terms of the ones chosen for the actual problem. This observation is entirely consistent with the observation made by Lukka [30], that a purely conventional choice of angular variables is sufficient to derive the rotationally invariant part of the kinetic-energy operator.

While considering invariances of this kind it is perhaps appropriate to note here as well that, in principle, once the part of the Hamiltonian independent of the Eulerian angles has been obtained in any set of internal coordinates, it may be transformed to any other chosen set of internal coordinates within their common domains. Of course, since the relationship between the two sets of internal coordinates will usually be nonlinear and not invertible in closed form, this observation is worth much less in practice than in theory.

1. Example: Rotation Reflections in a Frame Fixed in the Body

It would be rather tiresome and long-winded to give very general examples, but it is hoped that the following rather simple examples may prove convincing and suitably illustrative. In all cases the example of ammonia provided in Section III.E will be used.

Consider first the rotation by $\pi/2$ about the x axis of the frame fixed in the body represented by the matrix

$$\mathbf{R} = \begin{pmatrix} 1 & 0 & 0 \\ 0 & 0 & 1 \\ 0 & -1 & 0 \end{pmatrix}$$

This maps the array (3.63) into

$$\mathbf{z}^n = \begin{pmatrix} r_1 \cos a\theta & r_2 \cos(1-a)\theta \\ 0 & 0 \\ r_1 \sin a\theta & -r_2 \sin(1-a)\theta \end{pmatrix} \tag{4.20}$$

according to (4.11). Thus the three protons are placed in the x–z plane with the x axis dividing the bond vectors. Under this transformation

$$\bar{\bar{\lambda}}_x = \bar{\lambda}_x = 0, \qquad \bar{\bar{\lambda}}_y = \bar{\lambda}_z, \qquad \bar{\bar{\lambda}}_z = -\bar{\lambda}_y = 0$$

and hence the roles of the z and y components are exchanged. Similarly

$$\bar{\kappa}_{xx} = \kappa_{xx}, \qquad \bar{\kappa}_{xz} = -\kappa_{xy}, \qquad \bar{\kappa}_{yy} = \kappa_{zz}, \qquad \bar{\kappa}_{zz} = \kappa_{yy}$$

The results of these transformations may be compared with the directly derived results given in Sutcliffe and Tennyson [48].

Next consider the invariance of h_k. If we choose

$$\mathbf{R} = \begin{pmatrix} \cos a\theta & -\sin a\theta & 0 \\ \sin a\theta & \cos a\theta & 0 \\ 0 & 0 & 1 \end{pmatrix}$$

This maps the array (3.63) into

$$\mathbf{z}^n = \begin{pmatrix} r_1 & r_2 \cos \theta \\ 0 & r_2 \sin \theta \\ 0 & 0 \end{pmatrix}$$

a form that may also be achieved by the special choice in (3.63) of $a = 0$. With this form the 1–3 bond vector $\mathbf{t}_1$ lies along the x axis and the Ω^i have components

$$\Omega^1_{zx} = \frac{\cos \theta}{r_1 \sin \theta}, \qquad \Omega^2_{zx} = \frac{-1}{r_2 \sin \theta}$$

$$\Omega^1_{zy} = \frac{1}{r_1}, \qquad \Omega^2_{zy} = 0$$

$$\Omega^1_{xz} = 0, \qquad \Omega^2_{xz} = 0$$

$$\Omega^1_{yz} = \frac{-1}{r_1}, \qquad \Omega^2_{yz} = 0$$

The $\mathbf{Q}^i$ are also changed to have components

$$Q^1_{x1} = 1, \qquad Q^1_{y1} = 0$$

$$Q^2_{x2} = \cos\theta, \qquad Q^2_{y2} = \sin\theta$$

$$Q^1_{x3} = 0, \qquad Q^1_{y3} = \frac{-1}{r_1}$$

$$Q^2_{x3} = -\frac{\sin\theta}{r_2}, \qquad Q^2_{y3} = \frac{\cos\theta}{r_2\sin\theta}$$

and it is easily seen that the h_k constructed from these components according to (3.48) are unchanged from those obtained originally, as required.

To consider changing the internal coordinates alone, two examples will be used. First, we shall transform from the set r_1, r_2, θ to the set r_1, r_2, r_{12} while leaving $\mathbf{z}^n_3$ unchanged. (The transformed set is the same set used by Hylleraas in his helium calculation.) Perfectly standard algebra and geometry may be used to show that

$$\frac{\partial}{\partial r_1} = \frac{\partial}{\partial r_1} + \frac{(r_1^2 - r_2^2 + r_{12}^2)}{2r_1 r_{12}} \frac{\partial}{\partial r_{12}}$$

$$\frac{\partial}{\partial r_2} = \frac{\partial}{\partial r_2} + \frac{(r_2^2 - r_1^2 + r_{12}^2)}{2r_2 r_{12}} \frac{\partial}{\partial r_{12}}$$

$$\frac{\partial}{\partial\theta} = \frac{\sqrt{(r_1 + r_2 + r_{12})(r_1 + r_2 - r_{12})(r_1 - r_2 + r_{12})(r_2 - r_1 + r_{12})}}{2r_{12}} \frac{\partial}{\partial r_{12}}$$

In this case it is possible to obtain all the original coordinate expressions explicitly in terms of the new ones. In fact, the rather nasty square-root term in the expression for the derivative with respect to (original coordinate) θ is eliminated when the second derivatives are constructed.

Now consider transforming from the set r_1, r_2, θ, again leaving $\mathbf{z}^n_3$ unchanged, to the three internal coordinates $q_k, k = 1, 2, 3$ that would be chosen in the standard Watson–Eckart approach to this part of the problem discussed earlier. The components of the translationally invariant coordinates in terms of the q_k coordinates are, when expressed in the frame fixed in the body by the Eckart conditions

$$z_{\alpha i}^{n} = (a_{\alpha i} - a_{\alpha 3}) + \sum_{k=1}^{3}(\sqrt{m_i}l_{\alpha i k} - \sqrt{m_3}l_{\alpha 3 k})q_k, \qquad i = 1, 2$$

so that, for example

$$r_i^2 = \sum_{\alpha}[(a_{\alpha i} - a_{\alpha 3}) + \sum_{k=1}^{3}(\sqrt{m_i}l_{\alpha i k} - \sqrt{m_3}l_{\alpha 3 k})q_k]^2, \qquad i = 1, 2$$

It is thus easy to obtain expressions for the original internal coordinates in terms of the q_k, but it is not possible to invert the resulting expressions to obtain in closed form the q_k in terms of the original internal coordinates. Indeed, this is almost obvious from the expression given above for r_i^2. It is thus not possible to obtain expressions for the original partial derivatives in terms of the new coordinates (q_k) as was possible in the first example given. Because of the many desirable properties of the Eckart embedding for describing molecular spectra, much effort has been expended on reexpressing results calculated in some embedding into their equivalents in the Eckart embedding; some examples of these are described in the literature [49–52].

C. Permutational Symmetry

We now consider the behavior of both the internal coordinates and the Eulerian angles under the permutation of identical nuclei. Because of the choices made in deriving equation (2.63), the permutation of electrons is standard and need not be explicitly considered.

Let the (redundant) set of $(H-1)^2$ scalar products of the $\mathbf{t}_i$ be denoted by the square matrix $\mathbf{S}$, of dimension $H-1$. Then, using (2.64), it is seen that a permutation

$$\mathscr{P}\mathbf{t}^{n} = \mathbf{t}^{n}\mathbf{H} = \mathbf{t}'^{n} \tag{4.21}$$

so that

$$\mathbf{S}' = \mathbf{H}^{T}\mathbf{S}\mathbf{H} \tag{4.22}$$

Making explicit the functional dependencies, (3.2) may be written as

$$\mathbf{t}^{n} = \mathbf{C}(\boldsymbol{\phi})\mathbf{z}^{n}(\mathbf{q}) \tag{4.23}$$

and using (4.21) and (4.22) two different expressions for the permuted translationally invariant coordinates may be obtained. The first follows at once from (4.23) and (4.24)

$$\mathbf{t'}^n = \mathbf{t}^n \mathbf{H} = \mathbf{C}(\phi)\mathbf{z}^n(\mathbf{q})\mathbf{H} \tag{4.24}$$

and this gives the $\mathbf{t'}_i^n$ as functions of ϕ and $\mathbf{q}$.

Alternatively, the Eulerian angles and the internal coordinates can be expressed directly as functions of the $\mathbf{t}^n$ and hence of the $\mathbf{t'}^n$ according to

$$\phi_m(\mathbf{t}^n) \rightarrow \phi_m(\mathbf{t}^n \mathbf{H}^{-1}) = \overline{\phi}_m(\phi, \mathbf{q}) \tag{4.25}$$

and

$$q_k(\mathbf{S}) \rightarrow q_k(\mathbf{H}^{-T}\mathbf{S}\mathbf{H}^{-1}) = \overline{q}_k(\mathbf{q}) \tag{4.26}$$

To avoid overloading the notation, the convention has been adopted in which the transformed function is indicated as induced from the original function by the inverse transformation of the original variables.

Note that the effect of the permutation on q_k can at most produce a function of the q_k. This is because any set of translationally invariant coordinates may be linearly transformed into any other such set, and hence the collection of scalar products in one set is related by a similarity transformation to the collection of scalar products of the other set. Thus internal coordinates of one kind must be functions of internal coordinates of the other. However, the effect of the permutation on ϕ_m can produce a function of both the ϕ_m and the q_k. If the permuted internal coordinates and Eulerian angles are now used in (4.23), the resulting expression will be for the permuted translation-free variables thus

$$\mathbf{t'}^n = \mathbf{C}(\overline{\phi}(\phi, \mathbf{q}))\mathbf{z}^n(\overline{\mathbf{q}}(\mathbf{q})) \tag{4.27}$$

so that

$$\mathbf{t'}^n = \overline{\mathbf{C}}(\phi, \mathbf{q})\overline{\mathbf{z}}^n(\mathbf{q}) \tag{4.28}$$

Equating (4.24) and (4.28) it follows that

$$\overline{\mathbf{z}}^n = \overline{\mathbf{C}}^T \mathbf{C}\mathbf{z}^n \mathbf{H} \tag{4.29}$$

Naturally some care must be taken concerning the domain in which the orthogonal matrix $\overline{\mathbf{C}}^T \mathbf{C}$ exists, but since it can be at most a function of the internal coordinates, it follows that, where it exists, its elements are, at most, functions of the internal coordinates. Denoting this matrix by $\mathbf{U}$ (and from now on, since they will always be the original ones fixed in the body, the variables will not be explicitly given), it follows that

$$\bar{\mathbf{z}}^n = \mathbf{U}\mathbf{z}^n\mathbf{H} \tag{4.30}$$

and

$$\overline{\mathbf{C}} = \mathbf{C}\mathbf{U}^T \tag{4.31}$$

giving a relationship (albeit implicit) between the permuted and unpermuted variables fixed in the body. It is as well to state explicitly that there will be such a relationship for every distinct permutation, and so strictly the matrices should carry a designation to indicate which permutation is being considered. But that would be to overload the notation in a way that is not necessary here, and so it will not be done.

Now that these relationships have been established, the effects of a permutation on the various parts of the wavefunction must be worked out. Using the convention mentioned above, the variable change (4.30) will be

$$\mathbf{z}^n \rightarrow \mathbf{U}^T\mathbf{z}^n\mathbf{H}^{-1} \tag{4.32}$$

while (4.31) will be

$$\mathbf{C} \rightarrow \mathbf{C}\mathbf{U} \tag{4.33}$$

when considering the change in a function the change of variables. Thus the angular eigenfunctions change under nuclear permutations according to (4.7).

It is rather difficult to say anything precise about the change induced in the q_k under the permutation. The internal coordinates are expressible entirely in terms of scalar products, and so they are invariant under inversion, which simply causes the $\mathbf{t}_i^n$ to change sign. So it is only the nuclear permutation group and not the permutation-inversion group that is relevant here. The scalar products of the $\mathbf{t}_i^n$ are identical to the scalar products of the $\mathbf{z}_i^n$ and the change is that given in (4.26), namely

$$\mathbf{q}(\mathbf{S}) \rightarrow \mathbf{q}(\mathbf{H}^{-T}\mathbf{S}\mathbf{H}^{-1}) \tag{4.34}$$

where the notation of (4.32) has been used and where $\mathbf{S}$ is regarded as a function of the q_k. However, the result has no general form, and so the best that can be said is that a permutation of nuclei induces a general function change

$$\Phi_k^{J,r}(\mathbf{q}, \mathbf{z}) \rightarrow \Phi_k'^{J,r}(\mathbf{q}, \mathbf{z}) \tag{4.35}$$

where the precise nature of the function change depends on the permutation, the chosen form of the internal coordinates, and the chosen functional form. Thus the general change induced in (4.10) by $\mathscr{P}$ is

$$\Psi^{J,M,r}(\boldsymbol{\phi}, \mathbf{q}, \mathbf{z}) \rightarrow \sum_{k=-J}^{+J} \sum_{n=-J}^{+J} \mathscr{D}_{nk}^{rJ}(\mathbf{U}) \Phi_k'^{J,r}(\mathbf{q}, \mathbf{z}) |JMnr\rangle$$

$$= \sum_{n=-J}^{+J} \overline{\Phi}_n^{J,r}(\mathbf{q}, \mathbf{z}) |JMnr\rangle \tag{4.36}$$

From the discussion in Section II.J the function (4.10) should carry an additional label $\mathbf{s}$ to specify to which irreps of the various symmetric groups in the problem it belongs. If the function is such that it belongs to a one-dimensional irrep of the symmetric group containing the permutation $\mathscr{P}$, the resulting function (4.36) can differ from the original one by at most a sign change. In the case of a multidimensional representation, the resulting function (4.36) will be at most a linear combination of the set of degenerate functions providing a basis for the irrep. So, in spite of possible coordinate mixing, there are no difficulties in principle. However, in practice one must construct approximate wavefunctions that are not immediately adapted to the permutational symmetry of the problem and that must be explicitly adapted by, for example, the use of projections. In these circumstances coordinate mixing can cause tremendous complications. The expression (4.36) will clearly be very difficult to handle, for not only will a $\mathbf{U}$ be difficult to determine but also one must be found for each distinct permutation of the identical nuclei, and in a problem of any size there will be a very large number of such permutations.

It should be noted here that this coupling of rotations by the permutations can mean that certain rotational states are not allowed by the Pauli principle. Whether this is the case has to be determined in any particular occurrence by the changes induced according to (4.36), and this would be exceptionally tricky, in general. However, Ezra [41] has discussed the problem in detail in some special cases. This possibility is relevant in assigning statistical weights to rotational states.

An account of the changes induced in the various operators by the permutations is much the same as that given in Section IV.B in describing rotation reflections in a frame fixed in the body. The only additional complications arise from the presence of $\mathbf{H}$ in (4.30) as compared with the second term in (4.11). In fact, this matrix takes care of the permutational invariance of the quantities such as $\boldsymbol{\mu}^{-1}$, which arise from the translationally invariant forms and otherwise the manipulations go through just as before.

1. Example: The Effects of Nuclear Permutations on the Internal and Orientational Coordinates in NH_3

In order to simplify and as clarify as much possible, we shall consider the special case where $a = 0$ in (3.63) so that the 1–3 bond vector $\mathbf{t}_1$ lies along the x axis and

$$\mathbf{z}^n = \begin{pmatrix} r_1 & r_2 \cos \theta \\ 0 & r_2 \sin \theta \\ 0 & 0 \end{pmatrix}$$

and just the two permutations of proton coordinates (12) and (132). So we shall treat the problem as if it involved only the two translationally invariant coordinates $\mathbf{t}_1^n$ and $\mathbf{t}_2^n$ because the other coordinate $\mathbf{t}_3^n$ is invariant under permutations of the proton coordinates. It is the inverse of these permutations that produces the required mappings of the $\mathbf{z}_i^n$,

The inverse permutation for the transposition is just the transposition, as transpositions are involutory. Then $\mathbf{t}_1 \rightarrow \mathbf{t}_2$ and $\mathbf{t}_2 \rightarrow \mathbf{t}_1$. The internal coordinate changes are then

$$r_1 \rightarrow r_2, \qquad r_2 \rightarrow r_1, \qquad \theta \rightarrow -\theta$$

where the negative sign for θ is chosen to reflect the fact that the transposition changes the direction of the z axis. To determine the changes induced in $\mathbf{C}$, it is noted that

$$\frac{t_{\alpha 1}}{r_1} = C_{\alpha x}, \qquad \frac{t_{\alpha 2}}{r_2} = C_{\alpha x} \cos \theta + C_{\alpha y} \sin \theta$$

and after a little algebra

$$\mathbf{C} \rightarrow \mathbf{C} \begin{pmatrix} \cos \theta & -\sin \theta & 0 \\ \sin \theta & \cos \theta & 0 \\ 0 & 0 & 1 \end{pmatrix} = \mathbf{CU}$$

where the last column of the transformation matrix is determined by the requirements of orthonormality and righthandedness. The changes to the internal coordinates here are quite straightforward, and in the case when

$J = 0$, the changes induced in the internal motion function can be easily formed. Even in the higher J cases the terms arising from the matrix $\mathbf{U}$ can be determined and incorporated.

The inverse of (132) is (123) and so $\mathbf{t}_1 \rightarrow -\mathbf{t}_1 + \mathbf{t}_2$ and $\mathbf{t}_2 = \rightarrow -\mathbf{t}_1$. The internal coordinate changes are

$$r_1 \rightarrow \sqrt{(r_1^2 + r_2^2 - 2r_1 r_2 \cos \theta)} = g_1, \qquad r_2 \rightarrow r_1$$

and

$$\cos \theta \rightarrow \frac{r_1 - r_2 \cos \theta}{g_1} = \frac{p_1}{g_1}, \qquad \sin \theta \rightarrow \frac{r_2 \sin \theta}{g_1}$$

The change induced in $\mathbf{C}$ is

$$\mathbf{C} \rightarrow \mathbf{C} \begin{pmatrix} -\dfrac{p_1}{g_1} & -\dfrac{r_2 \sin \theta}{g_1} & 0 \\[2mm] \dfrac{r_2 \sin \theta}{g_1} & -\dfrac{p_1}{g_1} & 0 \\[2mm] 0 & 0 & 1 \end{pmatrix} = \mathbf{CU}$$

It is easily checked that both the transformation matrices $\mathbf{U}$ are orthogonal.

The remaining permutations, (13), (23) and (132), not explicitly considered here, will lead to transformations like that for (123) and hence to complicated internal coordinate changes, making the permutations difficult to incorporate properly even in the case when $J = 0$. However, in all cases the matrices $\mathbf{U}$ are defined in the same domain as that of the original coordinate transformation, ensuring that no difficulties of principle arise.

Had it been possible to use the bond lengths and bond angle internal coordinates arising from the bond vector choice of translationally invariant coordinates, then the $\mathbf{U}$ matrices would have all been constant matrices arising from simple coordinate renaming. But as explained earlier at the beginning of Section III.E, this choice is not open to us because of the dependence of the angular coordinates.

2. Example: The Effects of Nuclear Permutations on the Eckart Coordinates

If we represent a particular permutation of identical nuclei by $\mathbf{P}$, we can express any permutation of nuclei by

$$\mathbf{x}^n \rightarrow \mathbf{x}^n \mathbf{P}$$

and we see that this permutation induces in $\mathbf{B}$ (3.88) the change

$$\mathbf{B} \rightarrow \mathbf{wP}^T\mathbf{ma}^T = \mathbf{wm}(\mathbf{aP})^T = \mathbf{Cz}^n\mathbf{m}(\mathbf{aP})^T = \mathbf{CA}^P$$

because the permutation is of identical nuclei, and so its representative matrix must commute with the mass matrix. Thus the permutation *seems* to act as if it changed the reference vectors, but these are, of course, quantities fixed at the time of problem specification and hence unchanging parameters once a choice is made. From (3.86) it follows that under this permutation

$$\mathbf{C} \rightarrow \mathbf{CA}^P(\mathbf{A}^{PT}\mathbf{A}^P)^{-1/2} \equiv \mathbf{CU}$$

If and where the matrix $\mathbf{A}^P$ is singular, $\mathbf{U}$ will not exist, the change in $\mathbf{C}$ will be undefined, and hence the Eckart frame embedding will fail. So no such permutation can be realized in the Eckart formulation, and a fundamental symmetry of the problem is broken. In this formulation, therefore, difficulties of principle and not merely of practice could arise in incorporating permutational symmetry into the problem.

If the permutation is such is that $\mathbf{A}^P$ is non-singular and symmetric, it can be diagonalized by an orthogonal matrix $\mathbf{V}$ with eigenvector columns $\mathbf{v}_i$ and eigenvalues a_i^P. In dealing with the square root, consistency again requires the choice of the positive value and $\mathbf{U}$ can therefore be written as

$$\mathbf{U} = \sum_{i=1}^{3} \text{sgn}(a_i^P)\mathbf{v}_i\mathbf{v}_i^T$$

If all the a_i^P are positive, then $\mathbf{U}$ is just the unit matrix; if all negative, then $\mathbf{U}$ is the negative unit matrix and represents an inversion. The form that it actually takes depends on the precise nature of the permutation.

If $\mathbf{U}$ is a constant matrix, separation between angular and internal coordinates is essentially preserved. The most that can happen is that the internal coordinates go into linear combinations of themselves and the angular momentum components go into linear combinations of themselves. A special case of such a constant matrix is if the permutation is such that

$$\mathbf{aP} \equiv \mathbf{a}^P = \mathbf{Sa}, \qquad \mathbf{S}^T\mathbf{S} = \mathbf{E}_3 \qquad (4.37)$$

then

$$\mathbf{C} \rightarrow \mathbf{CS}^T, \qquad \mathbf{z}^n \rightarrow \mathbf{Sz}^n\mathbf{P}^T \qquad (4.38)$$

The matrix $\mathbf{S}$ could arise as the orthogonal representation matrix of a point group operation on the molecular framework, for the point group of the molecular framework is isomorphic with a subgroup of the full permutation group of the molecule. The subgroup might be the full group, but that is only rarely the case. However, a relation such as (4.37) might be possible for a permutation or set of permutations that are unconnected with any point group operations. When the permutations can be realised by proper or improper rotation matrices, they are often called *perrotations*, a name introduced by Gilles and Philippot [53]. These and related matters are discussed in Louck and Galbraith [54] and in Chapters 2 and 3 of Ezra's monograph [41].

In general, the permutation will yield an unsymmetric $\mathbf{A}^P$, but provided it is nonsingular, then $\mathbf{U}$ will be well defined and orthogonal. To illustrate these features, we shall consider ethene, which also illustrates the planar special case. We do not consider ammonia because its point group (C_{3v}) is isomorphic with the full permutation group (S_3) of the problem and so all the permutations in this case are perrotations and in such a case no principled difficulties with $\mathbf{U}$ can arise.

$$
\begin{array}{c}
\text{H}_1 \qquad\qquad \text{H}_4 \\
\diagdown \qquad\nearrow \\
\text{C}_1\!=\!\text{C}_2 \\
\diagup \qquad\diagdown \\
\text{H}_2 \qquad\qquad \text{H}_3
\end{array}
$$

with the following choice of $\mathbf{a}_i$

$$
\mathbf{a}_1 = \begin{pmatrix} -a \\ b \\ 0 \end{pmatrix} \quad
\mathbf{a}_2 = \begin{pmatrix} -a \\ -b \\ 0 \end{pmatrix} \quad
\mathbf{a}_3 = \begin{pmatrix} a \\ -b \\ 0 \end{pmatrix} \quad
\mathbf{a}_4 = \begin{pmatrix} a \\ b \\ 0 \end{pmatrix}
$$

$$
\mathbf{a}_5 = \begin{pmatrix} -d \\ 0 \\ 0 \end{pmatrix} \quad
\mathbf{a}_6 = \begin{pmatrix} d \\ 0 \\ 0 \end{pmatrix}
$$

where $\mathbf{a}_1 \cdots \mathbf{a}_4$ represent the reference positions of the protons, each with mass m_p and $\mathbf{a}_5$ and $\mathbf{a}_6$ the reference positions of the carbons, each with mass m_c. The system is specified to lie in the x–y plane, and we shall be concerned only with the 2×2 nonvanishing submatrix of $\mathbf{A}$. The point group of the reference figure is D_{2h}, using the Schönfliess notation, which is standard in molecular spectroscopy. The full permutation group is $S_4 \times S_2$, assuming the nuclei to be identical isotopes of hydrogen and of carbon, respectively. The point group contains 8 operations while the permutation group contains 48 operations.

It is convenient to rewrite $\mathbf{A}$ from (3.89) as

$$m_p \mathbf{z}_p^n \mathbf{a}_p^T + m_c \mathbf{z}_c^n \mathbf{a}_c^T \equiv \mathbf{A}_p + \mathbf{A}_c$$

where the split has been made into the proton and the carbon parts of the problem and permutations will occur only within each part. The carbon part of the problem is always of the form

$$\begin{pmatrix} \pm m_c (z_{x6}^n - z_{x5}^n)d & 0 \\ 0 & 0 \end{pmatrix} = \pm \mathbf{A}_c$$

where the minus sign represents the identity permutation and the plus sign the transposition of the carbons. If $\mathbf{A}_p$ is nonsingular, then

$$\mathbf{A}_p \pm \mathbf{A}_c = \mathbf{A}_p (\mathbf{E}_2 \pm \mathbf{A}_p^{-1} \mathbf{A}_c)$$

and even though $\mathbf{A}_c$ is singular, there is no reason to expect the second term on the right to be singular, except perhaps for particular values of the coordinates, and we can assume that such regions of coordinate space may be treated as unvisited for our present purposes. It is possible that even if $\mathbf{A}_p$ is singular, $(\mathbf{A}_p \pm \mathbf{A}_c)$ might not be, but this again will happen only in rather special regions. What has been said in relation to $\mathbf{A}_p$ here, will hold equally for any $\mathbf{A}_p^P$. We would anticipate therefore that that any permutation of the protons that yields a nonsingular $\mathbf{A}_p^P$ would yield a nonsingular $\mathbf{A}^P$ when taken with either of the possibilities for $\mathbf{A}_c$. If the permuted form is symmetric, then $\mathbf{A}^P$ will be symmetric, although there is no reason to expect this to be generally the case. With these matters in mind we shall concentrate on the proton permutations in what follows.

Looking at the effect of the point group operations on the framework vectors it is easy to see that

$$C_{2x} \equiv \sigma_{zx} \equiv (12)(34)(5)(6), \quad C_{2y} \equiv \sigma_{yz} \equiv (14)(23)(56),$$
$$C_{2z} \equiv i \equiv (13)(24)(56)$$

while σ_{xy} is equivalent to the identity operation in the point group and the identity permutation. Thus the effect of these permutations on $\mathbf{A}_p$ is

$$\mathbf{A}_p \overset{C_{2x}}{\to} \mathbf{A}_p \begin{pmatrix} 1 & 0 \\ 0 & -1 \end{pmatrix} \mathbf{A}_p \overset{C_{2y}}{\to} \mathbf{A}_p \begin{pmatrix} -1 & 0 \\ 0 & 1 \end{pmatrix}$$

$$\mathbf{A}_p \overset{C_{2z}}{\to} \mathbf{A}_p \begin{pmatrix} -1 & 0 \\ 0 & -1 \end{pmatrix}$$

Adjoining the relevant transformed A_c to each of these matrices allows us to separate the constant matrix as S, and we obtain precisely the results for U anticipated from (4.37).

The preceding proton permutations are the complete set of double transpositions for the protons, and we would expect each of them, even when joined with the alternative carbon permutations to those of the point group, to yield nonsingular A^P. Hence the definition of the Eckart frame would be maintained even for such non–point group permutations. However, one would not in general be able to factor out a common constant matrix for such permutations as one can for perrotations, and one would expect the relevant U to be functions of the internal coordinates.

It is difficult to analyze the problem further at this level of generality. However, if we choose the reference values for the z_i^n, we can go a little further, and if, with this special choice, the Eckart frame definition fails, then the frame certainly cannot be defined for small-displacements, which form the most important region for the traditional Eckart approach. For present purposes, therefore, we shall regard A as if composed ama^T and the transformation as $ama^T \rightarrow am(aP)^T$. With these restrictions

$$A_p = \begin{pmatrix} 4m_p a^2 & 0 \\ 0 & 4m_p b^2 \end{pmatrix} \quad \text{and} \quad A_c = \begin{pmatrix} 2m_c d^2 & 0 \\ 0 & 0 \end{pmatrix}$$

Looking at the six proton transpositions, four yield singular A_p^P but two, (13) and (24), do not. The permutation (12) gives

$$a_p m_p a_p^T \rightarrow a_p m_p a_p^{P_{12}\,T} = \begin{pmatrix} 4m_p a^2 & 0 \\ 0 & 0 \end{pmatrix}$$

and the matrix for (34) is the same. In neither case can the singularity be removed with the aid of A_c. The permutation (14) gives

$$a_p m_p a_p^T \rightarrow a_p m_p a_p^{P_{14}\,T} = \begin{pmatrix} 0 & 0 \\ 0 & 4m_p b^2 \end{pmatrix}$$

and the matrix for (23) is the same. In both cases the singularity can be removed with the aid of A_c. The permutation (24) yields

$$a_p m_p a_p^T \rightarrow a_p m_p a_p^{P_{24}\,T} = \begin{pmatrix} 0 & -4m_p ab \\ -4m_p ab & 0 \end{pmatrix}$$

and (13) is the same but with positive off-diagonal elements. Both matrices are nonsingular, and the Eckart frame can certainly be well defined in the case of either of these permutations associated with either of the carbon permutations.

The eight permutations involving just three protons all lead to singular A_p^P of the form

$$\begin{pmatrix} 0 & \pm 4m_p ab \\ 0 & 0 \end{pmatrix} \quad \text{or} \quad \begin{pmatrix} 0 & 0 \\ \pm 4m_p ab & 0 \end{pmatrix}$$

so for none of these permutations can a satisfactory Eckart embedding be defined even when A_c is considered.

Of the six permutations involving all four protons, two—(1324) and (1423)—yield singular A_p^P of the form

$$\begin{pmatrix} -4m_p a^2 & 0 \\ 0 & 0 \end{pmatrix}$$

and two, (1243) and (1342), yield singular A_p^P such as

$$\begin{pmatrix} 0 & 0 \\ 0 & -4m_p b^2 \end{pmatrix}$$

This matrix can be rendered nonsingular with the aid of A_c. The permutation (1234) yields the nonsingular but unsymmetric A_p^P

$$\begin{pmatrix} 0 & 4m_p ab \\ -4m_p ab & 0 \end{pmatrix}$$

and (4321) yields the negative of this matrix.

Thus in the region of the reference geometry, of the 24 possible proton permutations, 16 lead to singular A_p^P, and of these only 4 can lead to nonsingular A^P. The remaining 8 proton permutations form a group and all lead to nonsingular A^P combined with either of the possible carbon permutations. This group has itself a subgroup of order 4, which gives rise to the perrotations discussed above.

We see, therefore, that in the Eckart approach, in some cases it may not be possible, even in principle, to treat the permutational symmetry of the problem fully. This point seems first to have been noted in print by Berry [55], who regarded it as a consequence of making the Born–Oppenheimer approximation. But why typical broken symmetry solutions to the clamped nuclei problem turn out to be so effective in practice has been a vexing puzzle from the very beginning of the molecular quantum mechanics discipline. It has occasioned an enormous amount of work, particularly since the publication in 1963 of a paper by Longuet-Higgins [56] in which permutations were divided into *feasible* and *unfeasible* types and in which it was argued that it was necessary to consider only the (often rather small) set of feasible permutations in a given problem. A feasible permutation is one for which the equivalent physical exchange of nuclei is regarded as possible. This is generally expressed by saying that, in the appropriate energy range, for feasible permutations the potential barrier for exchange is low. A summary of much of the relevant work in the area of molecular spectroscopy is reviewed in the monographs by Ezra [41] and Bunker [57] and in a more general context in Kaplan [58] and Maruani and Serre [59].

In this context it is perhaps appropriate to note that permutation of identical particles in the theory is an operation that occurs in the mathematics. It is distinct from any supposed physical exchange of particles. So it is not obvious precisely how the idea of feasibility, with its physical setting, can actually be relevant to the mathematics. Of course, the idea of an reference nuclear geometry is in fact completely compatible with the idea of an equilibrium structure arising from clamped nuclei electronic structure calculations, and such calculations are generally used to construct a potential surface for calculations using the Eckart Hamiltonian. However, no separation of electronic from nuclear motion is *necessary* in choosing the Eckart conditions in order to define a frame fixed in the molecule. The discussion in Section III.G 1 does not depend on any such separation and could easily be extended to include the electronic motions explicitly. In the absence of a separation of electronic from nuclear motions, it is difficult to see how feasibility could even be defined.

V. REMOVING ROTATIONAL MOTION IN THE FRAME FIXED IN THE BODY

From (4.10) it is seen that one can eliminate angular motion from the problem by allowing the operator to work on the function and multiplying from the left by its complex conjugate and integrating out over the angular variables. This yields an effective operator within any (J, M, k, r) rotation-reflection manifold that depends only on the internal coordinates. Of course, any such effective operator will be defined only where the Jacobian (3.60) does not vanish, and this restriction will be implicitly assumed from here on. For ease of exposition in what follows, the parity index will be suppressed initially and then considered separately.

A. The Internal Motion Effective Hamiltonian Operator in Permutationally Restricted Coordinates

To remove the rotational motion, we write (3.35) as

$$\hat{K}_I(\mathbf{q}, \mathbf{z}) + \hat{K}_R(\boldsymbol{\phi}, \mathbf{q}, \mathbf{z}) \tag{5.1}$$

in which the first term, $\hat{K}_I$, consists of the first two terms in (3.35) and the subscripts "I" and "R" denote internal and rotational, respectively. The matrix elements with respect to the angular variables of the operators that depend only on the q_k and the $\mathbf{z}_i$ are trivial. Thus

$$\langle J'M'k' \mid \hat{K}_I + V \mid JMk \rangle = \delta_{J'J}\delta_{M'M}\delta_{k'k}(\hat{K}_I + V) \tag{5.2}$$

In what follows explicit allowance for the diagonal requirement on J and M will be assumed and the indices suppressed to save writing. Similarly, the fact that the integration implied is over $\boldsymbol{\phi}$ only will be left implicit.

To treat the second term in (5.1) is much more complicated and best done by reexpressing the components of $\hat{\mathbf{L}}$ in terms of $\hat{L}_\pm(\boldsymbol{\phi})$ and $\hat{L}_z(\boldsymbol{\phi})$ and using (3.28) and (3.30). When this is done, we obtain

$$\begin{aligned}
\langle JMk' \mid \hat{K}_R \mid JMk \rangle = {} & \frac{\hbar^2}{4}(b_{+2}C_{Jk+1}^+ C_{Jk}^+ \delta_{k'k+2} + b_{-2}C_{Jk-1}^- C_{Jk}^- \delta_{k'k-2}) \\
& + \frac{\hbar^2}{4}(C_{Jk}^+(b_{+1}(2k+1) + \bar{\lambda}_+)\delta_{k'k+1} \\
& + C_{Jk}^-(b_{-1}(2k-1) + \bar{\lambda}_-)\delta_{k'k-1}) \\
& + \frac{\hbar^2}{2}((J(J+1) - k^2)b + b_0 k^2 + \bar{\lambda}_0 k)\delta_{k'k} \tag{5.3}
\end{aligned}$$

In this expression

$$\begin{aligned}
b_{\pm 2} &= \frac{\kappa_{xx} - \kappa_{yy}}{2} \pm \frac{\kappa_{xy}}{i} \\
b_{\pm 1} &= \kappa_{xz} \pm \frac{\kappa_{yz}}{i} \\
b &= \frac{\kappa_{xx} + \kappa_{yy}}{2}, \qquad b_0 = \kappa_{zz} \tag{5.4}
\end{aligned}$$

and in terms of the $\bar{\lambda}_\alpha$ in (3.40) $\bar{\lambda}_0$ is $\bar{\lambda}_z$ and the $\bar{\lambda}_\pm$ are

$$\bar{\lambda}_\pm = \bar{\lambda}_x \pm \frac{\bar{\lambda}_y}{i} \tag{5.5}$$

The apparently odd positioning of the complex unit as $1/i$, when i might have been expected, is because the standard commutation conditions have been chosen for the internal angular momentum components.

Thus within any rotational manifold it is the eigensolutions of the effective Hamiltonian given by (5.2) and (5.3), which are invariant to orthogonal transformations, and it is these functions that will be used to consider the separation of electronic and nuclear variables. Inversion leaves the kinetic-energy, the potential-energy, and the angular momentum operators invariant so that (5.2) generalizes to

$$\langle J'M'k'r' \mid \hat{K}_\mathrm{I} + V \mid JMkr \rangle = \delta_{r'r}\delta_{J'J}\delta_{M'M}\delta_{k'k}(\hat{K}_\mathrm{I} + V) \qquad (5.6)$$

and (5.3) generalizes in a similar manner given that the $|JMkr\rangle$ remain eigenfunctions of the angular momentum operators.

B.   The Internal Motion Effective Dipole Moment Operator in Permutationally Restricted Coordinates

The form (3.57) of the dipole operator enables matrix elements to be calculated over the angular functions by use of straightforward angular momentum algebra. Thus, using the standard expression for a Gaunt coefficient, we obtain

$$\langle J'M'k' \mid \mathscr{D}_{mp}^{1^*} \mid JMk \rangle$$

$$= (-1)^{M'+k}[(2J'+1)(2J+1)]^{1/2}\begin{pmatrix} J' & 1 & J \\ -M' & m & M \end{pmatrix}\begin{pmatrix} J' & 1 & J \\ k' & p & -k \end{pmatrix} (5.7)$$

where the $3-j$ symbols are as defined in Brink and Satchler [36]. The matrix elements of the dipole moment connect the *two* rotational manifolds because of the Gaunt coefficient in (5.7) so that

$$\langle \Psi' \mid d_m^s(\mathbf{t}^n, \mathbf{t}^e) \mid \Psi \rangle = [(2J'+1)(2J+1)]^{1/2}(-1)^{M'}\begin{pmatrix} J' & 1 & J \\ -M' & m & M \end{pmatrix}$$

$$\sum_{p=-1}^{+1}\sum_{k'=-J'}^{+J'}\sum_{k=-J}^{+J}(-1)^k\begin{pmatrix} J' & 1 & J \\ k' & p & -k \end{pmatrix}\langle \Phi_{k'}^{J'} \mid d_p^s(\mathbf{q}, \mathbf{z}) \mid \Phi_k^J \rangle$$

$$(5.8)$$

Since it is usual to express selection rules in terms of the Cartesian components of the dipole, it is often useful to rewrite the last term in (5.8) using (3.55) as

$$\langle \Phi_{k'}^{J'} \mid d_t^s(\mathbf{q}, \mathbf{z}) \mid \Phi_k^J \rangle = \sum_\alpha (\mathbf{X})_{\alpha t}\langle \Phi_{k'}^{J'} \mid d_\alpha(\mathbf{q}, \mathbf{z}) \mid \Phi_k^J \rangle$$

It is clear from (5.8) that a $J = 0$ state cannot have a permanent dipole moment, or can there be transitions between pairs of states with $J = 0$. Thus, on the face of it, it seems that if a "purely vibrational" transition is impossible. For the equivalent expression using the more conventional angular momentum eigenfunctions, see Section 7.10 of Biedenharn and Louck [37]. The expressions differ only by a phase factor.

Under inversion, the dipole moment operator changes sign, so (3.54) should be rewritten with $\mathscr{D}^1$ replaced by $\mathscr{D}^{01}$, and (5.7) generalizes as

$$
\langle J'M'k'r' \mid \mathscr{D}^{01*}_{mp} \mid JMkr \rangle
$$

$$
= (1 - \delta_{ss'})(-1)^{M'+k}[(2J' + 1)(2J + 1)]^{1/2}
$$

$$
\times \begin{pmatrix} J' & 1 & J \\ -M' & m & M \end{pmatrix} \begin{pmatrix} J' & 1 & J \\ k' & p & -k \end{pmatrix} \tag{5.9}
$$

where the parity s is $(-1)^{J+r}$ and similarly for r' and thus (5.8) generalizes accordingly. When the initial and final states are the same, the formula yields the expectation value of the dipole for that state. However because the parity factors in the wavefunctions are the same and the parity of the dipole operator is odd, the expectation value must be zero, so it would seem that no molecule can have a permanent dipole moment, whatever its rotational state, a somewhat counterintuitive result.

C. Example: Removing the Rotational Motion from the NH₃ Operators

Of the operators developed for the NH_3 molecule in Section III.E, only the kinetic-energy operator requires explicit consideration. In this operator the first three terms of (3.67) comprise K_I in (5.2) and so need not be considered further, either. It is seen that $b_{\pm 1}$ vanish and that we may write

$$
\hat{K}_{VR} = \delta_{k'k\pm2} \frac{\hbar^2}{4} C^{\pm}_{Jk\pm1} C^{\pm}_{Jk} b_{\pm2} + \delta_{k'k\pm1} \frac{\hbar^2}{4} C^{\pm}_{Jk} \lambda^{\pm}
$$

$$
+ \delta_{k'k} \frac{\hbar^2}{2} (b(J(J + 1) - k^2) + b_0 k^2 + \lambda_0 k)
$$

In which

$$
\lambda^{\pm} = 2 \left[(\kappa \hat{\mathbf{j}})_x \pm \frac{(\kappa \hat{\mathbf{j}})_y}{i} \right]
$$

and

$$
\lambda_0 = \frac{2}{i} \left[\left(\frac{(1-a)}{\mu_1 r_1^2} - \frac{a}{\mu_2 r_2^2} \right) \left(\frac{\partial}{\partial \theta} + \frac{\cot \theta}{2} \right) \right.
$$

$$
+ \frac{(2a-1)}{\mu_{12} r_1 r_2} \left(\cos \theta \frac{\partial}{\partial \theta} + \frac{\cos \theta}{2} \right)
$$

$$
\left. + \frac{\sin \theta}{\mu_{12}} \left(\frac{a}{r_2} \frac{\partial}{\partial r_1} - \frac{(1-a)}{r_1} \frac{\partial}{\partial r_2} + \frac{(1-2a)}{r_1 r_2} \right) \right] + 2(\boldsymbol{\kappa} \hat{\mathbf{j}})_z
$$

while

$$
b_{+2} = \frac{e^{2ia\theta}}{2 \sin^2 \theta} \left(\frac{e^{-2i\theta}}{\mu_1 r_1^2} + \frac{1}{\mu_2 r_2^2} - 2 \frac{e^{-i\theta}}{\mu_{12} r_1 r_2} \right) = b_{-2}^*
$$

and

$$
b = \frac{1}{2 \sin^2 \theta} \left(\frac{1}{\mu_1 r_1^2} + \frac{1}{\mu_2 r_2^2} - 2 \frac{\cos \theta}{\mu_{12} r_1 r_2} \right)
$$

with

$$
b_0 = \frac{(1-a)^2}{\mu_1 r_1^2} + \frac{a^2}{\mu_2 r_2^2} + 2 \frac{a(1-a) \cos \theta}{\mu_{12} r_1 r_2}
$$

In this form it is clear that the operators should be treated with great care where $\sin \theta$, and hence the Jacobian for the transformation, vanishes.

VI. CONSTRUCTING EFFECTIVE OPERATORS FOR NUCLEAR MOTION IN A PRODUCT FUNCTION BASIS

Returning to (3.52), it is seen that the expansion such as (3.1) that applies to the internal motion part of the problem and that is, hence, an approximate solution to the effective Hamiltonian in (5.2), is expressed in terms of a sum of products of the form

$$
\Phi_{kp}^J(\mathbf{q}) \psi_p(\mathbf{q}, \mathbf{z}) \tag{6.1}
$$

where p labels the electronic state and the sum is over p. The explicit variables in which ψ_p is imagined expressed are electron nucleus separation variables. Technically this means that ψ_p is an implicit function of the $\mathbf{q}$ and the $\mathbf{z}$. But it is probably clearer to use the terminology that is appropriate when a fixed value is chosen for $\mathbf{q}$ and say that the explicit variables are the $\mathbf{z}$ and that the $\mathbf{q}$ are parameters in the function.

In this approach the approximating functions including angular momentum are taken to be of the form.

$$\psi_p(\mathbf{q}, \mathbf{z}) \sum_{k=-J}^{+J} \Phi_{kp}^J(\mathbf{q}) |JMk\rangle \qquad (6.2)$$

To simplify writing, the parity index of the angular momentum eigenfunctions is suppressed and will be noted only when needed.

The function $\psi_p(\mathbf{q}, \mathbf{z})$ is assumed known, just as $|JMk\rangle$ is assumed to be known and the effective nuclear motion Hamiltonian is obtained in terms of matrix elements of the effective internal motion Hamiltonian between the $\psi_p(\mathbf{q}, \mathbf{z})$ with respect to the variables $\mathbf{z}$, just as the effective internal motion Hamiltonian itself is expressed in terms of matrix elements of the full Hamiltonian between the $|JMk\rangle$ with respect to the ϕ. The effective nuclear motion Hamiltonian then contains the electronic state labels p as parameters, in much the same way that the full effective Hamiltonian for internal motion contains the angular momentum labels k. Of course, the analogy between the two derivations is simply a formal one. There is no underlying symmetry structure in the effective nuclear problem; nor is the sum over p of definite extent as is the sum over k.

A. The Effective Angular Momentum Operator for Nuclear Motion

Because the electronic part of the wavefunction has no dependence on the Eulerian angles, the angular momentum operator has no effect on it. So the effective angular momentum operator for nuclear motion is simply the complete one.

B. The Effective Hamiltonian for Nuclear Motion

If we consider the effective internal motion operator (5.2) working on a product function, we get

$$\Phi_{kp}^J(\mathbf{q})(\hat{K}(\mathbf{z}) + V^e(\mathbf{z}) - V^{ne}(\mathbf{q}, \mathbf{z}) + V^n(\mathbf{q}))\psi_p(\mathbf{q}, \mathbf{z}) + \hat{K}(\mathbf{q}, \mathbf{z})\Phi_{kp}^J(\mathbf{q})\psi_p(\mathbf{q}, \mathbf{z}) \qquad (6.3)$$

As explained below (5.2), this equation is, within any rotational manifold, diagonal in J, M, k, and r, but this will be left implicit from here on. The nuclear function has been moved through the expression as far as possible, but the electronic function may not be moved at all, since it depends on all the variables in the problem at this stage.

By analogy with the standard approach it could be stipulated that the set of known functions, $\psi_p(\mathbf{q}, \mathbf{z})$, were to be chosen as exact solutions of the electronic problem

$$(\hat{K}(\mathbf{z}) + V^{\mathrm{e}}(\mathbf{z}) - V^{\mathrm{ne}}(\mathbf{q}, \mathbf{z}))\psi_p(\mathbf{q}, \mathbf{z}) \equiv \hat{H}^{\mathrm{elec}}(\mathbf{q}, \mathbf{z})\psi_p(\mathbf{q}, \mathbf{z}) = E_p(\mathbf{q})\psi_p(\mathbf{q}, \mathbf{z})$$

(6.4)

Because there are no terms in this equation that involve derivatives with respect to the q_k, there is no development with respect to $\mathbf{q}$ in $E_p(\mathbf{q})$ or $\psi_p(\mathbf{q}, \mathbf{z})$. Thus the $\mathbf{q}$ act here simply as parameters that can be chosen at will. Such solutions are therefore just like solutions to the clamped nuclei electronic problem, in which the nuclear positions $\mathbf{a}_i$ are parameters.

In fact it is not absolutely essential for what follows to require the ψ_p to be eigenfunctions of $\hat{H}^{\mathrm{elec}}$. A reasonably concise and useful form can be obtained simply by requiring that

$$\int \psi_{p'}^*(\mathbf{q}, \mathbf{z})\psi_p(\mathbf{q}, \mathbf{z})d\mathbf{z} \equiv \langle \psi_{p'} | \psi_p \rangle_{\mathbf{z}} = \delta_{p'p}$$

(6.5)

and, using the preceeding abbreviation to denote integration over all $\mathbf{z}$ only

$$\langle \psi_{p'} | \hat{H}^{\mathrm{elec}} | \psi_p \rangle_{\mathbf{z}} = \delta_{p'p}E_p(\mathbf{q})$$

(6.6)

The requirements (6.5) and (6.6) can be met in a simple and practical way by requiring the ψ_p to be solutions of a linear variation problem with matrix elements determined by integration over the $\mathbf{z}$ alone, for each and every value assigned to $\mathbf{q}$. Indeed, it could be argued that the linear variation account is more convincing than the direct one, because the assumed basis can always be extended to include functions that are capable of providing an L^2 approximation to the continuum. Hence in such an account, the continuum states can be approximately included, if it is so desired.

Whichever of these approaches is favoured to specify ψ_p, it is defined only up to a phase factor of the form

$$\exp\, iw^p(\mathbf{q})$$

where w^p is any twice differentiable real function of the nuclear coordinates. So any chosen electronic wave function may be multiplied by such a phase factor and all such products yield the same the electronic energy. For the moment, we shall simply imagine a particular choice of phase to have been made.

Assuming that a set of electronic functions that satisfy (6.5) and (6.6) have been chosen, we may work through (6.3), remembering that the product rule must be used when considering the effect of derivative operators with respect to the q_k because *both* terms in the product (6.1) depend on the $\mathbf{q}$ variables. The effective nuclear motion Hamiltonian, depending only on the $\mathbf{q}$, can be obtained by multiplying the resulting expression from the left by by $\psi_{p'}(\mathbf{q},\mathbf{z})$ and integrating over the $\mathbf{z}$. Doing this yields an equation rather like (5.2) but with coupling between different electronic states, labeled p. The resulting effective operator is

$$\delta_{p'p}\delta_{k'k}(\hat{K}_H + E_p(\mathbf{q}) + V^n(\mathbf{q})) + \delta_{k'k}\gamma_{p'p}(\mathbf{q}) \tag{6.7}$$

where the designation of the angular integration variables has been left implicit as before, as have the diagonal requirement on J, M, and r. The term $\hat{K}_H$ consists of the first group of terms from (3.38):

$$\hat{K}_H = -\frac{\hbar^2}{2}\left(\sum_{k,l=1}^{3H-6} g_{kl}\frac{\partial^2}{\partial q_k \partial q_l} + \sum_{k=1}^{3H-6} h_k \frac{\partial}{\partial q_k}\right)$$

If the last term in (6.7) is ignored, then we obtain a nuclear motion problem in a potential specified by $E_p(\mathbf{q}) + V^n(\mathbf{q})$.

The last term in (6.7) is

$$
\begin{aligned}
\gamma_{p'p}(\mathbf{q}) = \frac{\hbar^2}{2}&\left[\sum_{\alpha\beta}\langle\psi_{p'}|\hat{l}_\alpha\hat{l}_\beta|\psi_p\rangle_\mathbf{z}\kappa_{\alpha\beta} + \sum_\alpha\langle\psi_{p'}|\hat{l}_\alpha|\psi_p\rangle_\mathbf{z}\lambda_\alpha\right.\\
&+ \sum_\alpha\langle\psi_{p'}|\lambda'_\alpha\hat{l}_\alpha|\psi_p\rangle_\mathbf{z}\\
&- \sum_{k,l=1}^{3H-6} g_{kl}\left(\left\langle\psi_{p'}\left|\frac{\partial^2}{\partial q_k \partial q_l}\right|\psi_p\right\rangle_\mathbf{z} + \left\langle\psi_{p'}\left|\frac{\partial}{\partial q_k}\right|\psi_p\right\rangle_\mathbf{z}\frac{\partial}{\partial q_l}\right.\\
&\left.+ \left\langle\psi_{p'}\left|\frac{\partial}{\partial q_l}\right|\psi_p\right\rangle_\mathbf{z}\frac{\partial}{\partial q_k}\right)\\
&- 2\sum_{j=1}^{L}\sum_\alpha\sum_{k=1}^{3H-6} g'_{\alpha k}\left(\left\langle\psi_{p'}\left|\frac{\partial^2}{\partial q_k \partial z_{\alpha j}}\right|\psi_p\right\rangle_\mathbf{z} + \left\langle\psi_{p'}\left|\frac{\partial}{\partial z_{\alpha j}}\right|\psi_p\right\rangle_\mathbf{z}\frac{\partial}{\partial q_k}\right)\\
&- 2\sum_{j=1}^{L}\sum_\alpha\sum_{k=1}^{3H-6} h'_k\left\langle\psi_{p'}\left|\frac{\partial}{\partial z_{\alpha j}}\right|\psi_p\right\rangle_\mathbf{z}\\
&\left.+ \sum_{k=1}^{3H-6}\left(\frac{2}{i}\left\langle\psi_{p'}\left|(\tau\hat{\mathbf{l}})_k\frac{\partial}{\partial q_k}\right|\psi_p\right\rangle_\mathbf{z} - h_k\left\langle\psi_{p'}\left|\frac{\partial}{\partial q_k}\right|\psi_p\right\rangle_\mathbf{z}\right)\right] \tag{6.8}
\end{aligned}
$$

This can be written in a more compact form as

$$\gamma_{p'p}(\mathbf{q}) = g_{p'p}^0 + \sum_{k=1}^{3H-6} g_{p'p}^k \frac{\partial}{\partial q_k} \tag{6.9}$$

with

$$
\begin{aligned}
g_{p'p}^0 = \frac{\hbar^2}{2} \Bigg(& \sum_{\alpha\beta} \langle \psi_{p'} | \hat{l}_\alpha \hat{l}_\beta | \psi_p \rangle_{\mathbf{z}} \kappa_{\alpha\beta} + \frac{1}{i} \sum_\alpha \langle \psi_{p'} | \hat{l}_\alpha | \psi_p \rangle_{\mathbf{z}} \nu_\alpha \\
& + \sum_\alpha \langle \psi_{p'} | \lambda'_\alpha \hat{l}_\alpha | \psi_p \rangle_{\mathbf{z}} - \sum_{k,l=1}^{3H-6} g_{kl} \left\langle \psi_{p'} \left| \frac{\partial^2}{\partial q_k \partial q_l} \right| \psi_p \right\rangle_{\mathbf{z}} \\
& - 2 \sum_{j=1}^{L} \sum_\alpha \sum_{k=1}^{3H-6} \left(g'_{\alpha k} \left\langle \psi_{p'} \left| \frac{\partial^2}{\partial q_k \partial z_{\alpha j}} \right| \psi_p \right\rangle_{\mathbf{z}} + h'_k \left\langle \psi_{p'} \left| \frac{\partial}{\partial z_{\alpha j}} \right| \psi_p \right\rangle_{\mathbf{z}} \right) \\
& + \sum_{k=1}^{3H-6} \left(\frac{2}{i} \left\langle \psi_{p'} \left| (\tau \hat{\mathbf{l}})_k \frac{\partial}{\partial q_k} \right| \psi_p \right\rangle_{\mathbf{z}} - h_k \left\langle \psi_{p'} \left| \frac{\partial}{\partial q_k} \right| \psi_p \right\rangle_{\mathbf{z}} \right) \Bigg)
\end{aligned}
\tag{6.10}
$$

$$
g_{p'p}^k = \hbar^2 \left(\frac{1}{i} \sum_\alpha \langle \psi_{p'} | \hat{l}_\alpha | \psi_p \rangle_{\mathbf{z}} \tau_{k\alpha} - \sum_{l=1}^{3H-6} g_{lk} \left\langle \psi_{p'} \left| \frac{\partial}{\partial q_l} \right| \psi_p \right\rangle_{\mathbf{z}} - \sum_{j=1}^{L} \sum_\alpha g'_{\alpha k} \left\langle \psi_{p'} \left| \frac{\partial}{\partial z_{\alpha j}} \right| \psi_p \right\rangle_{\mathbf{z}} \right)
\tag{6.11}
$$

The expression (5.3) can be developed in a similar fashion now using $\hat{K}_R(\boldsymbol{\phi}, \mathbf{q}, \mathbf{z})$ and integrating out over the $\mathbf{z}$ as well as the $\boldsymbol{\phi}$. This gives

$$
\begin{aligned}
\langle JMk'p' | \hat{K}_R | JMkp \rangle_{\mathbf{z}} = {} & \frac{\hbar^2}{4} (b_{+2} C_{Jk+1}^+ C_{Jk}^+ \delta_{k'k+2} + b_{-2} C_{Jk-1}^- C_{Jk}^- \delta_{k'k-2}) \delta_{p'p} \\
& + \frac{\hbar^2}{4} (C_{Jk}^+ (b_{+1}(2k+1) + \lambda_+) \delta_{k'k+1} \\
& + C_{Jk}^- (b_{-1}(2k-1) + \lambda_-) \delta_{k'k-1}) \delta_{p'p} \\
& + \frac{\hbar^2}{4} (C_{Jk}^+ \gamma_{p'p}^+(\mathbf{q}) \delta_{k'k+1} + C_{Jk}^- \gamma_{p'p}^-(\mathbf{q}) \delta_{k'k-1}) \\
& + \frac{\hbar^2}{2} ((J(J+1) - k^2)b + b_0 k^2 + \lambda_0 k) \delta_{k'k} \delta_{p'p} \\
& + \frac{\hbar^2}{2} \delta_{k'k} k \gamma_{p'p}^0(\mathbf{q})
\end{aligned}
\tag{6.12}
$$

The γ terms are specified by

$$\gamma^\alpha_{p'p}(\mathbf{q}) = 2(\langle\psi_{p'}|(\kappa\hat{\mathbf{l}})_\alpha|\psi_p\rangle_\mathbf{z} + \frac{1}{i}\sum_{k=1}^{3H-6}\tau_{k\alpha}\left\langle\psi_{p'}\left|\frac{\partial}{\partial q_k}\right|\psi_p\right\rangle_\mathbf{z} + \langle\psi_{p'}|\lambda'_\alpha|\psi_p\rangle_\mathbf{z})$$

(6.13)

where γ^0 and $\gamma^\pm$ are defined in terms of the γ^α in a manner analogous to the definition of the equivalent λ quantities in (5.5).

Examination of (6.8) and (6.13) shows that a pair of electronic wavefunctions that differ only in internal coordinate dependent phase factors can be distinguished between at this stage as giving rise to different forms for the derivative expressions in q_k. This, in turn, means that any such phase factor will be accommodated to in the form of the resulting nuclear motion wavefunction on solution of the full problem. This is to be expected, for the wavefunction for the full problem, although defined only up to a phase factor, the factor must be a constant one and cannot depend on any variables of the problem. However, if the last term in (6.7) is ignored and the nuclear motion problem is solved using only the first part, then, because the potential is independent of any choice of phase factor, the resulting nuclear motion functions will not depend on the phase choice made. Thus any nuclear motion wavefunction determined from this part alone will be somewhat underspecified.

It is perhaps worthwhile to reemphase that the forms given here for the various γ terms arise from the kinetic-energy operators specified in (3.37) and (3.38). If, however, the Jacobian were to be incorporated, in whole or in part, the form of these operators, might well change and thus, correspondingly, so would the forms of the γ.

C. The Effective Dipole Moment for Nuclear Motion

The internal coordinate part of the dipole matrix element (5.8) changes to

$$\langle\Phi^{J'}_{k'}(\mathbf{q},\mathbf{z}) \mid d^s_t(\mathbf{q},\mathbf{z}) \mid \Phi^J_k(\mathbf{q},\mathbf{z})\rangle \to \langle\Phi^{J'}_{k'p'}(\mathbf{q}) \mid d^{f_{p'p}}_t(\mathbf{q}) \mid \Phi^J_{kp}(\mathbf{q})\rangle \quad (6.14)$$

where the effective dipole operator results from the integral over the electronic coordinates

$$\mathbf{d}^{f_{p'p}}(\mathbf{q}) = \langle\psi_{p'}(\mathbf{q},\mathbf{z}) \mid \mathbf{d}^s(\mathbf{q}) - \mathbf{d}^s(\mathbf{z})) \mid \psi_p(\mathbf{q},\mathbf{z})\rangle_\mathbf{z} \quad (6.15)$$

D. The Domain of the Effective Nuclear Motion Operators

To remain in the the domain of the internal coordinate operators, where the Jacobian for the coordinate transformation (3.60) does not vanish, special care must be taken with product functions of the kind envisaged as trial functions in this section. The electronic functions contain the internal coordinates q_k without any reference to a particular choice of coordinate frame fixed in the body. This is possible because the internal coordinates can be defined directly in terms of scalar products of the translationally invariant coordinates, so they can be defined without reference to a full coordinate transformation. To ensure that the product is well defined for a particular embedding, the nuclear motion functions $\Phi_{kp}^J(\mathbf{q})$ must be chosen so as to make a vanishing contribution where the Jacobian vanishes because the electronic functions alone will not generally vanish there. So the electronic Hamiltonian is perfectly well defined and is in principle solvable for all values (within their ranges) of the internal coordinates and hence for arbitrary nuclear geometries.

However, a particular choice of internal coordinates may lead to difficulty when trying to describe asymptotic behavior of the electronic Hamiltonian, assuming the trial function to be described in the same coordinate set. This is because the electron–nucleus distance, given generally by

$$|\mathbf{x}_i^n - \mathbf{x}_j^e| \equiv = r'_{ij}(\mathbf{q}, \mathbf{z}) = \left| \sum_{k=1}^{H-1} \mathbf{z}_k^n(\mathbf{q})((\mathbf{V}^n)_{ki}^{-1} - b_k) + \frac{a}{1 + La} \sum_{k=1}^{L} \mathbf{z}_k - \mathbf{z}_j \right|$$

may admit only certain forms of asymptotes as a consequence of the permutational invariance restrictions imposed by relations (2.53) and (2.54). This may perhaps be appreciated without detailed analysis if it is remembered that these two conditions require that the electrons have a common origin in the nuclei and that the origin is such that all identical nuclei enter its specification in the same way. Thus if there are two or more identical nuclei, it is impossible that an electron–nucleus distance refers to only one of them, so an asymptote involving an electron and only one of the nuclei is not accessible. Or if there is a unique nucleus used as the electronic origin, the only accessible asymptote is the one in which the electrons are referred to this nucleus.

It should be made absolutely clear that this observation in no way restricts the domain of definition of the electronic Hamiltonian. Any possible imagined nuclear geometry can be expressed in terms of a set of independent internal coordinates. However, not every interaction can be effectively described in a particular set. The situation has perhaps some analogies to that of attempting the central field problem in rectangular Cartesian coordinates. The problem is perfectly well specified in these coordinates, but it is impossible to get the desired asymptotic behavior for the operator when expressed in them. To solve the problem for the desired behavior, spherical polars have to be used.

Thus it is difficult to suppose that the present formulation would be any help at all in justifying the separation of electronic and nuclear motions in the treatment of scattering problems. To allow for such aysmptotes it is necessary to abandon the settled distinction and to permit the mixing of all coordinate types under permutations. However if the settled distinction is abandoned then it is very difficult to see how to cope with permutational symmetry other than by treating all the particles, electrons and nuclei, on the same footing.

Although any given nuclear geometry can be expressed in terms of an arbitrary set of $3H - 6$ independent internal coordinates, certain choices of internal coordinates are such that two distinct geometries can generate the same internal coordinate values. The set of internuclear distances forms a proper set of internal coordinates, and in the general case, there are $H(H - 1)/2$ of these, of which only $3H - 6$ can be independent. When there are either just three or just four nuclei, then there are as many internuclear distances as there are independent coordinates. In this case the internuclear distances can be used as internal coordinates directly. However, when there are five or more nuclei, the number of internuclear distances exceeds the number of independent internal coordinates; that is, they form a redundant set. If they are to be used, an independent subset of them must be chosen. But it always possible to construct two (or more) distinct figures for the nuclear geometry in which all the chosen independent internuclear distances are the same. Of course, not all the internuclear distances will be the same, and there will be distinct electronic energies for the same specification of internal coordinates. Thus with such a choice of internal coordinates, there will be regions in which the electronic energy function is not an analytic function of the internal coordinates.

It is very doubtful if it is in general possible to find a set of internal coordinates in terms of which the electronic energy is analytic globally. It certainly would be safest to assume that it is not possible and to recognize that a particular choice will have only a local validity. A global covering would then involve local patches, a situation analogous to that of constructing an atlas of local charts to keep the coordinate transformations well defined. There is, however, no reason to suppose that the charts for the coordinate transformation are the same as the patches in which the electronic wavefunction is analytic.

The electronic wavefunctions must also be such that, within the region of interest of the internal coordinates, the coupling term $\gamma_{p'p}(\mathbf{q})$ as given in (6.8), and its analog given in (6.13), remain properly defined. Furthermore, they must be small if it is wished to confine attention to a single product function. However, it seems unlikely that either can generally be the case globally. To see this, suppose that the $\psi_p(\mathbf{q}, \mathbf{z})$ are eigenfunctions of the electronic Hamiltonian (6.4). It is then easy to show, using the turnover rule, that the first derivative terms in (6.8) and (6.13) can be written as

$$\langle \psi_{p'} | \frac{\partial}{\partial q_k} | \psi_p \rangle_{\mathbf{z}} = [E_{p'}(\mathbf{q}) - E_p(\mathbf{q})]^{-1} \langle \psi_{p'} | \frac{\partial V^{ne}}{\partial q_k} | \psi_p \rangle_{\mathbf{z}} \qquad (6.16)$$

Insofar as this is a valid approach, in a region where the electronic wavefunction is an analytic function of the q_k, it is seen that the coupling term must be divergent if and where there exists a geometry for which the two electronic energies are the same, unless the integral on the right in (6.16) happens to vanish. It is perfectly possible that there is a nuclear geometry at which the energy of two electronic states are equal, and there is no reason to believe that the electronic wavefunctions generally vanish there or cause the integral to vanish, either. But wherever divergences occur, the expansion in terms of sums of products becomes undefined. Thus it is strictly necessary to regard particular trial product functions as effective only in the limited domain in which the electronic energies of adjacent states are well separated. However, this restriction is a very unwelcome one. Many "effects" in ordinary chemical parlance, such as the Jahn–Teller effect or the Renner–Teller effect, are described in terms of actual or possible surface crossings, and much contemporary discussion of reaction mechanisms invokes such crossings, too. Of course, the experimental phenomena accounted for in these methods are what they are, but it is difficult to see that they can be properly described in such terms if the present discussion is at all accurate. We shall attempt a more extended discussion later, but for the time being we shall assume that a sum of products solution is effective only in the intersection of the domains specified by the nonvanishing of the Jacobian and the nondivergence of the coupling terms.

To get a clearer idea of what is involved, an example is perhaps helpful.

1. Example: The Electronic Hamiltonian for NH_3

Looking at the electron–nucleus distances as given by the $r'_{ij}, i = 1, 2, 3, 4$ in Section III.E.2 it is seen that they seldom vanish, or behave in any special way, when the angle θ between $\mathbf{z}_1^n$ and $\mathbf{z}_2^n$ is either 0 or π and hence where the Jacobian vanishes. Of course, the separations vanish when the electron is coincident with the particular nucleus involved, but this is a consequence of the basic structure of the equations and not an artifact of transformation.

To see the asymptotic problems that can arise when a particular choice of internal coordinates is considered, suppose that the asymptote is the one in which the protons are infinitely separated from the nitrogen nucleus. In this case $|\mathbf{z}_3^n|$ will tend to infinity. In the coordinate choice in which the nitrogen nucleus is the electronic origin, the electron–nucleus distances r'_{ij}, where $i = 1, 2, 3$ will become infinitely large while r'_{4j} will remain unchanged for all j. Thus the electron–proton interaction terms will become vanishingly small and the asymptotic form of the electronic Hamiltonian will involve just a nitrogen nucleus with all 10 of the electrons, infinitely separated from a collection of three protons, devoid of any electrons.

The situation is, if anything, more problematic in the case where the centre of nuclear mass is chosen as the electronic origin and the same asymptote is considered. In this case *all* the electron–nucleus distances become very large and hence all the attraction terms become very small. The asymptotic form in this case is presumably a completely ionized system.

If the internal coordinates were transformed to a set appropriate to the translationally invariant forms chosen in (2.36), then other asymptotic behavior would be possible, including behavior that yields a Hamiltonian that describes a hydrogen atom infinitely separated from an NH_2 radical. However, this would be bought at the cost of having the interelectron distance $1 - j, j = 2, 3, \ldots, 10$ taking the form $|\mathbf{z}_1 - \mathbf{z}_j + \mathbf{z}_1^n|$, where $|\mathbf{z}_1^n|$ is the separation between the proton relative to which electron 1 is positioned and the nitrogen nucleus. This form of the interelectron separation breaks the coordinate division on which our discussion so far has been based and would vitiate most of our conclusions up to now.

It is not possible to illustrate the ambiguity in the internal coordinate choice simply using NH_3 because only four nuclei are involved, but following Collins and Parsons [60], an example for five nuclei may be given. So let us imagine ammonia in a definite geometry specified by the three protons $\mathbf{a}_i, i = 1, 2, 3$ and the nitrogen as $\mathbf{a}_4$ approached by an arbitrary nucleus at position $\mathbf{a}_5$.

Let NH_3 be specified by

$$
\mathbf{a}_1 = \begin{pmatrix} -a \\ -\dfrac{b}{\sqrt{3}} \\ 0 \end{pmatrix}, \quad
\mathbf{a}_2 = \begin{pmatrix} a \\ -\dfrac{b}{\sqrt{3}} \\ 0 \end{pmatrix}, \quad
\mathbf{a}_3 = \begin{pmatrix} 0 \\ \dfrac{2b}{\sqrt{3}} \\ 0 \end{pmatrix}, \quad
\mathbf{a}_4 = \begin{pmatrix} 0 \\ 0 \\ c \end{pmatrix}
$$

and imagine the fifth atom placed equivalently either above or below the plane of the protons as

$$
\mathbf{a}_5 = \begin{pmatrix} d_{x5} \\ d_{y5} \\ \pm d_{z5} \end{pmatrix}
$$

If we choose the nine interparticle coordinates $r_{12}, r_{13}, r_{14}, r_{15}, r_{23}, r_{24}, r_{25}$, r_{34}, and r_{35} as the independent internal coordinates, we that the two possible positions of $\mathbf{a}_5$ lead to exactly the same values of the chosen internal coordinates. Of course, if r_{45} were considered, the two geometries would be distinguishable and naturally the electronic energy of the system at the two geometries would be different. So the electronic energy is not an analytic function of these coordinates where this happens.

VII. THE CLAMPED NUCLEI OPERATORS

As has already been pointed out, overwhelmingly the most usual way of accounting for molecular structure in quantum mechanical terms is by using the clamped nuclei Hamiltonian (1.4) in which the nuclei are treated as classical fixed points in the operator forms appropriate to a coordinate system expressed in a frame fixed in the laboratory. Because its role is so central to understanding molecular structure, we shall begin by considering this operator and only later, in the light of our considerations of the Hamiltonian, look at the other operators.

A. The Clamped Nuclei Hamiltonian

The operator itself is

$$\hat{H}^{cn}(\mathbf{a}, \mathbf{x}^e) = -\frac{\hbar^2}{2m} \sum_{i=1}^{L} \nabla^2(\mathbf{x}_i^e) + \frac{e^2}{8\pi\epsilon_0} \sum_{i,j=1}^{N}{}' \frac{1}{|\mathbf{x}_i^e - \mathbf{x}_j^e|}$$
$$-\frac{e^2}{4\pi\epsilon_0} \sum_{i=1}^{H} \sum_{j=1}^{L} \frac{Z_i}{|\mathbf{x}_j^e - \mathbf{a}_i|}$$

and it has solutions of the form

$$\hat{H}^{cn}(\mathbf{a}, \mathbf{x}^e) \psi_p^{cn}(\mathbf{a}, \mathbf{x}^e) = E_p^{cn}(\mathbf{a}) \psi_p^{cn}(\mathbf{a}, \mathbf{x}^e) \qquad (7.1)$$

The origin of coordinates for this form of the Hamiltonian is not explicitly specified, but in practice it is chosen as a point within the nuclear framework and quite commonly one of the $\mathbf{a}_i$ is chosen. However the choice is made, it is assumed to be the same for the $\mathbf{x}^e$ and for the $\mathbf{a}$. Thus the equation is such that

$$E_p^{cn}(\mathbf{a}) \rightarrow E_p^{cn}(\mathbf{a}), \quad \text{as } \mathbf{a}_i \rightarrow \mathbf{a}_i + \mathbf{d}, \quad \text{and/or} \quad \mathbf{a}_i \rightarrow U\mathbf{a}_i, \quad i = 1, 2 \ldots, H$$

if **d** is a constant column matrix and **U** is a constant orthogonal matrix. Thus if $E_p^{cn}(\mathbf{a})$ is to be presented as a function of the $\mathbf{a}_i$, for single valuedness, the domain of the $\mathbf{a}_i$ must be confined to those geometries that are distinct under uniform translations and/or rigid rotation reflections.

In a sequence of calculations for different choices of the nuclear positions, it is usual to keep the origin fixed so that no one calculation differs from another merely by a uniform translation. Strategies to keep distinctness under rigid rotation reflections are less easily specified generally. In a sequence of calculations on triatomic systems, the nuclear positions are always chosen so as to remain in a plane and in the limit to define a line, and thus to generate only distinct configurations. For tetratomic and more extensive systems, choices specific to the system can be made to maintain the required distinctness, but it is always possible to treat any three particles as a triatomic system is treated, and to maintain rotational invariance that way.

The bottom of the continuous spectrum of this operator is clearly the first ionisation energy just as it is in an atom, and, as shown in Thirring [13], the solutions to the clamped nuclei problem have many formal similarities with the solutions of the atomic problem. In particular, if the system is neutral or positively charged, it has an infinite number of bound states below the first ionization energy. The relative positions of the $\mathbf{a}_i$ will affect only the nature of these states and the numeric values of the energies associated with them. For a neutral system in the limit in which all the nuclei coalesce, the spectrum is exactly that of the united atom. When the nuclei are separated in such a way as to yield a set of distinct neutral atom Hamiltonians, the spectrum of the operator is the direct sum of the spectra of the atomic operators.

The clamped nuclei Hamiltonian is a well-defined entity in its own right, and its solutions may be studied independently of any view that might be taken about its physical significance. However, it is usually treated as if its solutions fitted into solutions of the full problem rather as we have indicated that solutions to the electronic problem specified by $\hat{H}^{elec}(\mathbf{q}, \mathbf{z})$ are to be fitted. The explicit form of $\hat{H}^{elec}$ that arises from (6.4) using (3.36) and (3.50) is

$$\hat{H}^{elec}(\mathbf{q}, \mathbf{z}) = -\frac{\hbar^2}{2\mu} \sum_{i=1}^{L} \nabla^2(\mathbf{z}_i) + \frac{e^2}{8\pi\epsilon_0} \sum_{i,j=1}^{L}{}' \frac{1}{|\mathbf{z}_j - \mathbf{z}_i|}$$

$$- \frac{e^2}{4\pi\epsilon_0} \sum_{i=1}^{H} \sum_{j=1}^{L} \frac{Z_i}{r'_{ij}(\mathbf{q}, \mathbf{z})} - \frac{\hbar^2}{2\mu} \sum_{i,j=1}^{L}{}' \vec{\nabla}(\mathbf{z}_i) \cdot \vec{\nabla}(\mathbf{z}_j) \qquad (7.2)$$

Here, from (3.51)

$$r'_{ij}(\mathbf{q}, \mathbf{z}) = |\mathbf{x}_i^n - \mathbf{x}_j^e| = \left| \sum_{k=1}^{H-1} \mathbf{z}_k^n(\mathbf{q})((\mathbf{V}^n)_{ki}^{-1} - b_k) - \mathbf{z}_j + \frac{a}{1 + La} \sum_{k=1}^{L} \mathbf{z}_k \right|$$

(7.3)

whereas from (2.71)

$$\frac{1}{\mu} = m^{-1}a(2 + La) + \sum_{k=1}^{H} m_k^{-1} v_k^2$$

and from (2.57)

$$\mathbf{b} = -\frac{(\mathbf{V}^n)^{-1}\mathbf{v}}{(1 + La)}$$

The difficulty is to decide on the proper way in which the clamped nuclei problem should be matched with the electronic problem. In what follows we shall consider the clamped nuclei Hamiltonian as a given and determine what coordinate choices may be made in the electronic Hamiltonian to achieve an effective and convincing matching of the two operators. In doing this we shall regard the $\mathbf{a}_i$ as particular values of the $\mathbf{x}_i^n$. With such an interpretation it is possible to generate a set of translationally invariant nuclear coordinates according to (2.21)

$$(\mathbf{t}^n(\mathbf{a})\mathbf{X}(\mathbf{a})) = \mathbf{a}\mathbf{V}^n$$

(7.4)

with any of the choices of $\mathbf{V}^n$ made earlier. Under this interpretation it follows from (2.62) and (3.2) that

$$\mathbf{t}_k^n(\mathbf{a}) = \mathbf{C}(\mathbf{a})\mathbf{z}_k^n(\mathbf{q}(\mathbf{a}))$$

and

$$\sum_{k=1}^{H-1} \mathbf{z}_k^n(\mathbf{q}(\mathbf{a}))(\mathbf{V}^n)_{ki}^{-1} = \mathbf{C}^T(\mathbf{a})(\mathbf{a}_i - \mathbf{X}(\mathbf{a})) \equiv \mathbf{c}_i$$

and from the definition of $\mathbf{b}$ that

$$\sum_{k=1}^{H-1} \mathbf{z}_k^n(\mathbf{q}(\mathbf{a}))b_k = \frac{1}{(1 + La)} \sum_{j=1}^{H} \mathbf{c}_j v_j$$

which is just a constant matrix. It should be noted that the a_i used to specify the reference structure in the Eckart approach are not exactly the clamped nuclei a_i used here. In fact they are the c_i but since at the reference structure C is a unit matrix, the required c_i are actually the a_i specified with respect to the fixed center-of-nuclear mass $X(a)$. This is implied in (3.83) and assumed in the discussion of the Eckart embedding.

The electron–nucleus separation term appropriate to the electronic Hamiltonian at a fixed nuclear geometry may now be written as

$$r'_{ij}(\mathbf{q}(\mathbf{a})), \mathbf{z}) = \left| \mathbf{c}_i - \mathbf{z}_j - \frac{1}{(1 + La)} \sum_{j=1}^{H} \mathbf{c}_j v_j + \frac{a}{(1 + La)} \sum_{k=1}^{L} \mathbf{z}_k \right| \quad (7.5)$$

If the direct correspondences $m \to \mu$ and $\mathbf{x}_i^e \to \mathbf{z}_i$ are made in the clamped nuclei Hamiltonian, then the kinetic-energy and electron-repulsion terms can be matched with the first and second terms in (7.2). The $\mathbf{x}_i^e$ and the $\mathbf{z}_i$ are free variables with same range, and so integrals over their whole range will be precisely the same. It is also usual in clamped nuclei calculations to treat m as the unit of mass, and the mass correspondence can be achieved by a redefinition of the unit. But there is no term in the clamped nuclei Hamiltonian that corresponds to the kinetic-energy term here. (This term is often called the *mass polarization* or sometimes the *Hughes–Eckart* term.) Because there is a definite relation between $\mathbf{c}_i$ and $\mathbf{a}_i$, we shall for the moment regard any expression in the clamped nuclei problem involving $\mathbf{a}_i$ as matched if there is an identical expression in the electronic problem involving $\mathbf{c}_i$. Thus one can match the first term in (7.5) with nuclear position term in the clamped nucleus expression for electron–nucleus separation, but no obvious correspondences arise for the third and fourth terms.

By placing the electronic origin at the center of nuclear mass scaled by α as in (2.76), the choice $\alpha = (M/M_T)^{1/2}$ causes $1/\bar{\mu}$ to vanish. Thus μ becomes m, and there is no mass polarization term in (7.2). This choice also causes the b_k to vanish, and hence the third term in (7.5) vanishes, to yield

$$r'_{ij}(\mathbf{q}(\mathbf{a}), \mathbf{z}) = \left| \mathbf{c}_i - \mathbf{z}_j + \frac{a_R}{1 + La_R} \sum_{k=1}^{L} \mathbf{z}_k \right| \quad (7.6)$$

where

$$a_R = \frac{M^{1/2} - M_T^{1/2}}{M_T^{1/2}}$$

This choice of electronic origin is analogous to the Radau choice of heliocentric coordinates [61], but here the center of nuclear mass plays the part of the distinguished coordinate. However, there seems to be no way to map (7.6) onto the clamped nuclei electron–nucleus attraction term, through the correspondence $x_i^e \rightarrow z_i$, because of the presence of the second term involving a_R. Although this term may be expected to be small, because $M_T = M + Lm$ and Lm will be small compared with M, and although such a correspondence may be an extremely good approximation, it is not a strictly exact one.

It might, on the other hand, seem reasonable to ignore the mass polarization term as a first approximation. It can always be incorporated later by extending the definition of $\gamma_{p'p}(\mathbf{q})$ so that the operator there includes the term neglected here. In that case it is possible to choose $a = 0$, and the electron–nucleus separation would become

$$r'_{ij}(\mathbf{q}(\mathbf{a}), \mathbf{z}) = \left| \mathbf{c}_i - \mathbf{z}_j - \sum_{j=1}^{H} \mathbf{c}_j v_j \right| \qquad (7.7)$$

and the effective coupling mass would become

$$\frac{1}{\mu} = \sum_{k=1}^{H} m_k^{-1} v_k^2$$

If the electronic origin is now chosen simply as the center of nuclear mass, the third term in (7.7) vanishes and the electron–nucleus separation becomes effectively

$$r'_{ij}(\mathbf{q}(\mathbf{a}), \mathbf{z}) = |\mathbf{c}_i - \mathbf{z}_j| \qquad (7.8)$$

So, given a suitable interpretation of $\mathbf{c}_i$, the correspondence $x_i^e \rightarrow z_i$ leads to an exact match between the electron–nucleus attraction terms in the clamped nuclei and in the electronic Hamiltonian. Here, as shown below (2.79) $\bar{\mu}$ becomes M, the total nuclear mass. With this choice of electronic origin the terms in (6.8) marked with a prime, such as h'_α, vanish. These terms multiply electronic derivative operators, and so many such operators are absent from the coupling terms with this choice. However, coupling terms involving the $\hat{\mathbf{l}}$ electronic operator do survive.

The appropriate extension of $\gamma_{p'p}(\mathbf{q})$ is now

$$\gamma_{p'p}(\mathbf{q}) \Rightarrow \gamma_{p'p}(\mathbf{q}) - \frac{\hbar^2}{2M} \sum_{i,j=1}^{L} {}' \langle \psi_{p'} | \vec{\nabla}(\mathbf{z}_i) . \vec{\nabla}(\mathbf{z}_j)) | \psi_p \rangle_{\mathbf{z}} \qquad (7.9)$$

This seems a reasonable enough extension for the integral in (7.9) is multiplied by the reciprocal of the total nuclear mass and the added term might be hoped to be the smallest of the terms there. In any case, its inclusion in the diagonal terms will produce at most a nuclear-mass-dependent constant energy shift for any electronic state and will not lead to any new analytic complications in the term coupling electronic states.

Now let us consider in rather more detail the status of the c_i. The set of c_i is a set that differs from the set of a_i by at most a uniform translation and/or a rigid rotation reflection, so the set of all c_i define a geometric object that differs from that defined by the set of all a_i at most by a uniform constant translation and a constant rigid rotation reflection. Thus a clamped nuclei calculation done at the c_i will yield exactly the same clamped nuclei electronic energy as one done at the a_i. It is also the case that $q(a)$ is identical with $q(c)$. So our treatment in this context of the c_i and a_i as effectively the same quantities is perfectly valid where the electronic origin is identified with the center of nuclear mass.

Thus to map the results of a standard clamped nuclei calculation onto an electronic Hamiltonian solution, one must treat the unit of electronic mass as $mM/(M + m)$ rather than m and then $E^{cn}(a) \rightarrow E^{elec}(q(c))$ and $\psi_p^{cn}(a, x^e) \rightarrow \psi_p(q(c), z))$. Provided one is concerned only with a single calculation or with a sequence of calculations on a single electronic state, this correspondence is a distinction without a difference.

However, if electronic derivatives are required, the distinction is important. Remembering that it is supposed that the trial function explicitly involves electron nucleus separation variables, consider the following two derivatives:

$$\frac{\partial}{\partial x_{\alpha i}^e} \psi_p^{cn}(a, x^e) = g^{cn}(a, x^e)$$

$$\frac{\partial}{\partial z_{\alpha i}} \psi_p(q(c), z) = g^{elec}(c, z)$$

Although the functions to be derived match exactly, $g^{cn}(a, x^e)$ does not match $g^{elec}(c, z)$ on making the same correspondences. This is because the explicit variables are electron–nucleus separation variables, so the differentiation process will throw up constant terms involving individual a_i components in the first form but individual c_i components in the second form. These will be different numbers in general and the geometric equivalence of the a and the c will not be sufficient to make a match. So different values will result from integrating over the relevant electronic variable in each case.

In clamped nuclei calculations, derivatives with respect to nuclear positions, the so-called geometric derivatives, are usually determined directly by differentiating the trial functions with respect to the components of the a_i. For particular trial function forms, it is possible to obtain analytic expressions for the derivatives. These derivative terms are most commonly used to construct energy or property gradients on for a single electronic state. It is sometimes also possible for analytical forms of second derivatives to be constructed, [62]. However this may be, such derivatives can, in principle, be utilized in the realization of a clamped nuclei function as an electronic function by writing

$$\frac{\partial}{\partial q_k} = \sum_\alpha \sum_{i=i}^H \frac{\partial x_{\alpha i}^n}{\partial q_k} \frac{\partial}{\partial x_{\alpha i}^n} \tag{7.10}$$

In a standard clamped nuclei calculation, the components of the a_i would be treated as components of the x_i^n, but to fit into an electronic function calculation, the a in the clamped nuclei function should be replaced with the c. The derivative of $x_{\alpha i}^n$ with respect to q_k may in principle be evaluated as

$$\frac{\partial(X_{\alpha i} + \sum_\gamma C_{\alpha\gamma} z_{\gamma i}^n)}{\partial q_k} = \sum_\gamma C_{\alpha\gamma} \frac{\partial z_{\gamma i}^n}{\partial q_k} \tag{7.11}$$

and this may readily be evaluated if the z_i^n are known as functions of the q_k.

To summarise, it appears to be possible to map a properly constructed sequence of solutions from standard clamped nuclei calculations onto solutions of an electronic Hamiltonian that can be specified in terms of a solution to the full problem. To do so involves no change the way in which clamped nuclei calculations are usually carried out and has no effect on the way in which a potential is usually calculated or expressed. The matching is important only when it is desired to compute correction terms for the effect of electronic motion on nuclear motion.

What has been accomplished here is to rationalize the position of the clamped nuclei Hamiltonian in a particular context. The context is one in which the assumed nuclear motion functions for the full problem vanish suitably so as to keep the full problem well defined in terms of both the nonvanishing of the Jacobian and the convergence of the electronic coupling terms. The coordinates used to describe the trial wavefunctions for the full problem must also be of a permutationally restricted class, and the electronic wavefunction must be an analytic function of internal coordinates used in the range in which the nuclear motion wavefunction is suitably defined. The context is thus a very limited one and liable to be very local in spatial extent.

The potential computed from clamped nuclei calculations is often used as a potential in scattering problems. In such cases the potential is usually spatially extensive with the asymptotes appropriate to separated atoms. The scattered entities themselves are sometimes regarded as atoms or sometimes as nuclei. However this may be, there is nothing in what has gone before that would enable such a procedure to be put in the context of the full problem. This is not to say that such a procedure cannot be put in context, but simply to note that it is not done here. The clamped nuclei problem itself is sui generis. Its solutions are well defined in their own right and do not depend for their validity on any context provided by the solution of the full problem. However, if they are to be used in solutions of the full problem, a suitable context must be supplied. We have simply supplied one such context and exhibited its boundaries. Other contexts are undoubtedly possible.

1. Example: The Clamped Nuclei Operator for NH_3

The only quantities that need explicit consideration here are the electron–nucleus and nucleus–nucleus separations, and let us consider these in the case where the clamped nuclei geometry of the NH_3 molecule is specified as in Section VI.D. 1, namely, as

$$
\mathbf{a}_1 = \begin{pmatrix} -a \\ b \\ -\dfrac{}{\sqrt{3}} \\ 0 \end{pmatrix}, \qquad
\mathbf{a}_2 = \begin{pmatrix} a \\ b \\ -\dfrac{}{\sqrt{3}} \\ 0 \end{pmatrix}, \qquad
\mathbf{a}_3 = \begin{pmatrix} 0 \\ \dfrac{2b}{\sqrt{3}} \\ 0 \end{pmatrix}, \qquad
\mathbf{a}_4 = \begin{pmatrix} 0 \\ 0 \\ c \end{pmatrix}
$$

The translationally invariant nuclear coordinates are defined according to (2.67) with inverse as in (2.68) Thus

$$
\mathbf{t}_1^n(\mathbf{a}) = \begin{pmatrix} -a \\ -\sqrt{3}b \\ 0 \end{pmatrix}, \qquad
\mathbf{t}_2^n(\mathbf{a}) = \begin{pmatrix} a \\ -\sqrt{3}b \\ 0 \end{pmatrix}
$$

and

$$
\mathbf{t}_3^n(\mathbf{a}) = \begin{pmatrix} 0 \\ 0 \\ c \end{pmatrix}
$$

This last equation arises because the choice for $\mathbf{a}_i$, $i = 1, 2, 3$ places the center of proton mass at the origin. The full center of nuclear mass is then

$$\mathbf{X}(\mathbf{a}) = M^{-1} m_N \begin{pmatrix} 0 \\ 0 \\ c \end{pmatrix}$$

The frame is fixed in the body according to (3.63), and so in this case the matrix $\mathbf{C}(\mathbf{a})$ is the unit matrix and the $\mathbf{t}^n$ may be treated directly as the $\mathbf{z}^n$. Thus the $\mathbf{c}$ become, suppressing the explicit functional dependence on $\mathbf{a}$ to save writing

$$\mathbf{c}_1 = +\frac{2}{3}\mathbf{z}_1^n - \frac{1}{3}\mathbf{z}_2^n - \frac{m_N}{M}\mathbf{z}_3^n,$$

$$\mathbf{c}_2 = -\frac{1}{3}\mathbf{z}_1^n + \frac{2}{3}\mathbf{z}_2^n - \frac{m_N}{M}\mathbf{z}_3^n$$

$$\mathbf{c}_3 = -\frac{1}{3}\mathbf{z}_1^n - \frac{1}{3}\mathbf{z}_2^n - \frac{m_N}{M}\mathbf{z}_3^n, \mathbf{c}_4 = \frac{M_3}{M}\mathbf{z}_3^n$$

It is immediately checked that each of the $\mathbf{c}_i$ is $\mathbf{a}_i - \mathbf{X}(\mathbf{a})$ as required, and so

$$\mathbf{c}_1 = \begin{pmatrix} -a \\ -\dfrac{b}{\sqrt{3}} \\ -\dfrac{m_N}{M}c \end{pmatrix}, \quad \mathbf{c}_2 = \begin{pmatrix} a \\ -\dfrac{b}{\sqrt{3}} \\ -\dfrac{m_N}{M}c \end{pmatrix}, \quad \mathbf{c}_3 = \begin{pmatrix} 0 \\ \dfrac{2b}{\sqrt{3}} \\ -\dfrac{m_N}{M}c \end{pmatrix}, \quad \mathbf{c}_4 = \begin{pmatrix} 0 \\ 0 \\ \dfrac{M_3}{M}c \end{pmatrix}$$

and the geometric equivalence of the figures formed from the $\mathbf{a}$ and from the $\mathbf{c}$ is obvious. It is easily seen that using the $\mathbf{c}$ together with v_k chosen as m_k/M, the third term in (7.7) vanishes and the form (7.8) is achieved. The results may be compared with the relevant expressions in Section III.E.2. Had the electronic origin been chosen at the nitrogen nucleus, that would have meant choosing $v_k = 0$, $k = 1, 2, 3$ and $v_4 = 1$, and (7.7) would have become

$$r'_{ij} = |\mathbf{c}_i - \mathbf{z}_j - \mathbf{c}_4|$$

This is obviously not a satisfactory form for matching purposes, but its correspondence with the relevant expressions in Section III.E. 2 can again be checked.

If it is supposed that a clamped nuclei calculation has been performed in the usual linear combination of atomic orbitals (LCAO) approximation using Gaussian orbitals, then in the resulting trial function it is perfectly possible that we have a term involving an s Gaussian $N \exp(-b|\mathbf{x}_i^e - \mathbf{a}_j|^2)$, and its electronic equivalent would be $N \exp(-b|\mathbf{z}_i - \mathbf{c}_j|^2)$. Differentiating with respect to $\mathbf{x}^e_{\alpha i}$ ($z_{\alpha i}$) yields

$$-2b(x^e_{\alpha i} - a_{\alpha j})N \exp(-b|\mathbf{x}^e_i - \mathbf{a}_j|^2) \quad \text{or} \quad -2b(z_{\alpha i} - c_{\alpha j})N \exp(-b|\mathbf{z}_i - \mathbf{c}_j|^2)$$

and these two functions are different even when the electronic variables are matched. Thus integration over the appropriate electronic variable will result in two different values. Derivatives with respect to the components of $\mathbf{a}_i$ ($\mathbf{c}_i$) lead to analogous forms, and the need to work in terms of the $\mathbf{c}_i$ for matching purposes is similarly apparent.

In the present case, if analytic derivatives are available for the trial function, it is easy to form the partial derivatives in (7.11) to make up (7.10) since the $\mathbf{z}^n_i$ are known as explicit functions of the q_k from (7.7) for $q_1 \equiv r_1, q_2 \equiv r_2$, and $q_3 \equiv \theta$ together with the three components of $\mathbf{z}^n_3$ as q_4, q_5, and q_6. The values of the internal coordinates will be known from the molecular geometry at at which the derivatives are taken, as will the elements of $\mathbf{C}$.

B.   The Clamped Nuclei Electric Dipole

For ease of exposition in this section, the dipole will be expressed in terms of its Cartesian components. The expected value of the dipole operator computed for a single electronic state in the clamped nuclei approximation is by convention

$$\mathbf{d}^{cn}(\mathbf{a}) = e \sum_{i=1}^{H} Z_i \mathbf{a}_i - e \sum_{i=1}^{L} \langle \psi^{cn}(\mathbf{a}, \mathbf{x}^e) \mid \mathbf{x}^e_i \mid \psi^{cn}(\mathbf{a}, \mathbf{x}^e) \rangle_{\mathbf{x}^e} \qquad (7.12)$$

Clearly this equation seldom vanishes, and is usually taken to specify the molecular dipole at the geometry $\mathbf{a}$. This form has to be matched with (6.15) given the choice of the center of nuclear mass for origin. From (3.53), in cartesian form and making the same suppositions about coordinates as for the clamped nuclei Hamiltonian

$$\mathbf{d}(\mathbf{q}) \rightarrow e \sum_{i=1}^{H-1} \tilde{Z}_i \mathbf{z}^n_i(\mathbf{a}) = e \sum_{i=1}^{H-1} \mathbf{z}^n_i(\mathbf{a}) \sum_{j=1}^{H} ((\mathbf{V}^n)^{-1})_{ij} Z_j = e \sum_{j=1}^{H} Z_j \mathbf{c}_j \qquad (7.13)$$

so that for a single electronic state of a neutral molecule the Cartesian component form of (6.15) can be written as

$$\mathbf{d}^{f_p}(\mathbf{c}) = \left\langle \psi_p(\mathbf{c}, \mathbf{z}) \left| e \sum_{j=1}^{H} Z_j \mathbf{c}_j - e \sum_{i=1}^{L} \mathbf{z}_i \right| \psi_p(\mathbf{c}, \mathbf{z}) \right\rangle_{\mathbf{z}}$$

and, assuming the electronic wavefunction normalized to unity

$$\mathbf{d}^{f_p}(\mathbf{c}) = \sum_{j=1}^{H} Z_j \mathbf{c}_j - e \left\langle \psi_p(\mathbf{c}, \mathbf{z}) \left| \sum_{i=1}^{L} \mathbf{z}_i \right| \psi_p(\mathbf{c}, \mathbf{z}) \right\rangle_{\mathbf{z}} \qquad (7.14)$$

Provided the molecule is neutral, the constant origin shift from the $\mathbf{a}_i$ to the $\mathbf{c}_i$ is of no consequence. The constant arising from first term in (7.14) is exactly canceled by the constant arising from the second term after integration over the free variable. So the vector $\mathbf{c}_i$ can always be regarded as being of the same length as $\mathbf{a}_i$. Then $\mathbf{c}$ differs from $\mathbf{a}$ by only a rigid rotation or rotation reflection. But this difference is not at all important as it merely changes the relative sizes of the components of the Cartesian coordinates, and this has no effect on the selection rules or the intensity formulas.

If, for a particular electronic state p, the nuclear configuration specified by $\mathbf{a}$, such that the electronic wavefunction vanishes strongly as $\mathbf{q}$ departs from $\mathbf{q} = \mathbf{0} \equiv \mathbf{q}(\mathbf{a})$ then, to a first approximation, the components of (6.15) can be written as

$$d_t^{f_p}(\mathbf{q}) = d_t^{f_p}(\mathbf{0}) + \sum_{l=1}^{3H-6} \left(\frac{\partial d_t^{f_p}}{\partial q_l} \right)_{\mathbf{q}=0} q_l \dots \qquad (7.15)$$

Such a form is most likely to be appropriate about a minimum in the potential and the linear form for the dipole is consistent with the potential being quadratic in the displacements q_k. This approximation is widely invoked in discussion of molecular spectra.

C.   The Clamped Nuclei Form of the Angular Momentum Operators

As would be anticipated, there is no clamped nuclei angular momentum operator. However, there is a Hamiltonian operator for a rigid body, which might be considered the equivalent of a clamped nuclei problem for angular momentum. We shall consider this in the discussion of the molecule described in a spectroscopic context.

D.   The Molecule as Described by Molecular Spectroscopy

If we take the view that a single product function of the form (6.1) is sufficient for our explanatory purposes and we believe the potential arising from that product to have a deep minimum in it (in terms of the energy range to be considered) around an equilibrium molecular geometry specified by $\mathbf{a}$, then the clamped nuclei Hamiltonian would plausibly be the appropriate electronic Hamiltonian with which to describe the electronic structure of the molecule close to that minimum. The potential for the nuclear motion problem locally can then be written in good approximation as the sum of the clamped nuclei electronic and the classical nuclear repulsion energy. If difficulties that might arise from failures of permutational invariance can be ignored, the bound states of the associated nuclear motion problem are then probably adequately described by means of the Eckart Hamiltonian discussed in Section III.G.1 with the equilibrium molecular geometry taken as specifying the reference structure.

If the displacements are sufficiently small, then the potential for nuclear motion can be expanded about the equilibrium geometry as a power series in the q_k about $\mathbf{q}(\mathbf{a}) = \mathbf{0}$

$$V(\mathbf{q}) = V(\mathbf{0}) + \frac{1}{2} \sum_{kl=1}^{3H-6} F_{kl} q_k q_l + \cdots$$

The first term in this expression is the sum of the electronic and the nuclear repulsion energy at the equilibrium geometry. The linear term is absent because the first derivatives of the potential vanish at the equilibrium geometry, by definition. The matrix $\mathbf{F}$ is composed of the second derivatives of the potential evaluated at the equilibrium geometry. The matrix of second derivatives is, in the context of power series, often called the *Hessian matrix*, and $\mathbf{F}$ is often called the *Hessian at the minimum*. It is a symmetric positive definite matrix and its elements are usually called the *force constants* for the problem. This usage arises because they have units of [force][distance]$^{-1}$ and in the classical texts on spectroscopy are always quoted in millidynes per Angström (equivalent to newtons per centimeter). If it is sufficient simply to consider the quadratic approximation to the potential, then it is possible to construct a special set of internal coordinates that diagonalize the force constant matrix and maintain the kinetic-energy operator forms given in (3.93). These coordinates, called *normal* coordinates, are often denoted as Q_k to distinguish them from the Eckart–Watson q_k.

If the displacements are small, it is plausible as well to treat the coordinates q_k as vanishingly small, and so the Coriolis coupling operator vanishes and the inverse generalized inertia tensor becomes $\boldsymbol{\mu}^0 = \mathbf{I}^{0^{-1}}$, the inverse of the equilibrium inertia tensor. In these circumstances one can choose the frame fixed in the body so that the equilibrium inertia tensor is diagonal (the principal axis choice) with $\mu_{\alpha\beta}^0 = \delta_{\alpha\beta}/I_{\alpha\alpha}^0$.

If this is the case, the equations for the kinetic-energy (3.93) simplify to yield, together with the quadratic potential, a Hamiltonian

$$\hat{H}^0(\boldsymbol{\phi}, \mathbf{Q}) = \hat{K}^0(\boldsymbol{\phi}) + \hat{H}^0(\mathbf{Q})$$

in which

$$\hat{K}^0(\boldsymbol{\phi}) = \frac{1}{2}\sum_\alpha \mu^0_{\alpha\alpha}\hat{L}^2_\alpha \equiv \frac{1}{2}\sum_\alpha \frac{\hat{L}^2_\alpha}{I^0_{\alpha\alpha}} \tag{7.16}$$

$$\hat{H}^0(\mathbf{Q}) = -\frac{\hbar^2}{2}\sum_{k=1}^{3H-6}\frac{\partial^2}{\partial Q_k^2} + \frac{1}{2}\sum_{k=1}^{3H-6}\lambda_k Q_k^2 \tag{7.17}$$

where the zero of the potential energy in (7.17) has been chosen to incorporate the constant terms. It has also been assumed that the coordinates are normal ones, chosen to diagonalise the quadratic approximation to the potential. Thus the λ_i are the eigenvalues of the quadratic form $\mathbf{F}$. All the eigenvalues must be positive for a stable molecule because the matrix of second derivatives of the internal coordinates evaluated at a minimum (the Hessian at the minimum) must be positive definite. If for any reason the expansion is made not about the equilibrium geometry, but rather at some other geometry, then it is possible to get negative values of λ_i and hence imaginary vibration frequencies.

Because the elements of $\boldsymbol{\mu}^0$ are constants, (7.16) is simply the Hamiltonian for an asymmetric top. This may be regarded as the rotational problem which is equivalent to the clamped nuclei Hamiltonian. The rotational matrix element in (5.3) simplifies to

$$\langle JMk' \mid \hat{K}^0(\boldsymbol{\phi}) \mid JMk\rangle$$

$$= \frac{\hbar^2}{4}\left(\frac{(\mu^0_{xx} - \mu^0_{yy})}{2}\left(C^+_{Jk+1}C^+_{Jk}\delta_{k'k+2} + C^-_{Jk-1}C^-_{Jk}\delta_{k'k-2}\right)\right)$$

$$+ \frac{\hbar^2}{2}\left(\frac{(\mu^0_{xx} + \mu^0_{yy})}{2}\left(J(J+1) - k^2\right) + \mu^0_{zz}k^2\right)\delta_{k'k} \tag{7.18}$$

The $(2J+1)$-dimensional secular problem composed of these matrix elements cannot generally be solved to give an energy expression in closed form, but the rotational wavefunctions solutions are of the form

$$^M\chi^J_\tau(\boldsymbol{\phi}) = \sum_{k=-J}^{k=J} c^J_{\tau k}|JMk\rangle, \qquad \tau = -J, -J+1, \ldots, J$$

The $c^J_{\tau k}$ are constant coefficients and each rotational wavefunction is associated with an energy $E_{J\tau}$.

If two of the equilibrium moments of inertia are the same (the symmetric-top case), these may be designated as x and y and the first term in (7.18) vanishes. The energy is then given by the last term in (7.18) and the $|JMk\rangle$ are individually angular eigenfunctions. Thus, for the symmetric top, k is a good quantum number. These matters are treated in standard texts on molecular spectroscopy.

The Hamiltonian (7.17) simply represents a sum of noninteracting Harmonic oscillators, each with a wavefunction of the standard form

$$\psi_{n_i}(Q_i) = N_i e^{-(\alpha_i Q_i^2/2)} H_{n_i}(\sqrt{\alpha_i}Q_i)$$

where

$$\alpha_i = \frac{\sqrt{\lambda_i}}{\hbar} \equiv \frac{\omega}{\hbar}$$

and the energy of the oscillator is

$$\epsilon_{n_i} = \left(n_i + \frac{1}{2}\right)\hbar\omega \equiv \left(n_i + \frac{1}{2}\right)h\nu$$

The full vibrational wavefunction is then usually written

$$\Psi(\mathbf{Q}) = \prod_{i=1}^{3H-6} \psi_{n_i}(Q_i)$$

and the total vibrational energy of the system is just

$$E_v = \sum_{i=1}^{3H-6} \epsilon_{n_i}$$

The assumption here is that it is not necessary to consider explicitly changes induced in the normal coordinates by permutations of identical particles. Further, it is not necessary to consider nuclear spin statistics. Thus the normal coordinates are regarded as identifiable entities and a product form for the wavefunction is acceptable.

The wavefunction for the nuclear motion part of the problem arises from the simplification of (6.2) and is a product

$$\phi(\mathbf{a}, \mathbf{z})\,^M\chi_\tau^J(\phi)\Psi(\mathbf{Q})$$

in which $\phi(\mathbf{a}, \mathbf{z})$ is the electronic wavefunction taken at the equilibrium nuclear geometry The wavefunction for the full problem in the single product approximation is

$$\psi(\mathbf{x}^n, \mathbf{x}^e) \Rightarrow T(\mathbf{X}_T) \, {}^M\chi^J_\tau(\boldsymbol{\phi})\Psi(\mathbf{Q})\phi(\mathbf{a}, \mathbf{z})$$

The total energy of the molecule in this approximation is

$$E = E_T + E_{J\tau} + E_v + V(\mathbf{a})$$

The translational energy E_T is usually ignored as is the translational wavefunction and the fact that in this approximation the energy is the sum of an electronic and a rotational and a vibrational part is often said to specify the Born–Oppenheimer approximation.

VIII. THE CURRENT POSITION

This chapter has provided only a rather tentative and limited relationship between the full Schrödinger problem for a molecule and the clamped nuclei problem. The outstanding difficulties that have not been resolved in our approach can be described under two headings: coordinate problems and domain problems.

There is perhaps a general problem in relation to internal coordinates as to whether a set can actually be found in which the electronic energy can be expressed globally as an analytic function in terms of them. Although it seems unlikely, we have, however, not been able to determine any general results, and so we cannot really discuss it further, except to provide a cautionary note as a context for the discussion below. This is of the problems that arise in attempting to realize the changes induced in the coordinates, by permutations of identical nuclei. If it were possible to find a prescription by which certain of the identical nuclei could be "identified" in some way, and for permutations involving them, to be ignorable, then these problems could be avoided. The identification of certain translationally invariant coordinates with electronic and with nuclear motions, does not in itself complicate matters.

The domain problems are largely of our own making in seeking to attempt solutions of the full problem in terms of an expansion of product functions of perceived electronic and nuclear motion types.

A. Coordinate Problems

The exact solutions to the full Schrödinger problem are classifiable according to the symmetric groups of identical particles with the accessible irreps determined by the spin symmetry requirements. It is thus strictly necessary to identify and to restrict the symmetric group symmetries of any trial functions. But even if we had not made the distinction that we did between electronic and nuclear coordinates, we should have had coordinate problems in attempting to construct functions of the correct symmetry type after the separation of translational motion. Any set of translationally invariant coordinates will generally go into linear combinations of themselves under a permutation of the original coordinate set. In principle, it is possible to label all translationally invariant trial functions according to the irreps of the symmetric groups of the problem. This is fairly easily done when products of one-variable (three dimensional) functions such as orbitals are used as trial functions and the permutations induce only variable label changes. This is the situation encountered in standard electronic structure calculations where the electronic wavefunction is built from orbital products. However, one-variable functions of a useful type, seldom have nice properties on replacing the explicit variable with a linear combination of variables. In the present exposition we have shown how this problem can be confined to a subset of the variables that we have identified with the nuclei, but we have been unable to suggest any general solution.

Similarly, no matter how we had made the translationally invariant coordinate division, we should still generally have had problems keeping the internal coordinates and the angular coordinates separate under permutations. When an allowed permutation can mix angular and internal coordinates, the idea of internal motion being separate from rotational motion becomes a difficult one to maintain. In particular, it becomes very difficult to maintain a consistent description of a system rotating more or less rigidly, while executing small amplitude vibrations. When it is possible to define an orthogonal transformation to a frame fixed in the body in such a way that all the variables in the problem enter into the definition in a symmetric way, then the resulting Eulerian angles will go, at most, into functions of each other under a permutation, and a separation of internal motion and rotation coordinates will be possible. One such definition can be made by choosing the orthogonal matrix so as to diagonalize the instantaneous inertia tensor for the system. This choice, made in the context of separate identification of electronic and nuclear coordinates, leads, however, to a Hamiltonian in which it proves impossible to describe a system rigidly rotating while performing small-amplitude vibrations, even though the coordinate choice describes an invariant division between angular and internal coordinates. Indeed, two of the first attempts to separate rotational from vibrational motion [63,64] adopted precisely this approach (and it has been a choice made subsequently on many occasions; see, e.g., Buck et al. [65]). It turns out that in this approach the definition of the frame embedded in the system depends on the reciprocals of differences between instantaneous moments of inertia and therefore fails whenever two moments are equal. Thus the Hamiltonian could not be used to describe any symmetric top molecule, such as ammonia. Indeed, it turns out more generally, inspite of some heroic efforts by van Vleck [66], that the Hamiltonian so derived is largely ineffective in describing molecules in terms of their traditional geometric structures and thus has found no use in the elucidation of molecular spectra.

It is thus a choice that, whatever its mathematical merits, make it an inappropriate one with which to describe molecules. Apart from this principal axis choice, it is not easy to imagine any other into which all the coordinates enter in a symmetric fashion, except in very special cases.

As we have seen, a perfectly reasonable nuclear motion Hamiltonian for ammonia is such that proton permutations mix the angular and internal coordinates. The Eckart Hamiltonian, which seems to be the appropriate one with which to describe at lowish energies, molecular vibration and rotation motion, is one that often allows the mixing of internal and rotational coordinates under permutations. This Hamiltonian can even exhibit broken symmetry under permutations. An additional aspect of this puzzle arises because in investigating the spectrum of a molecule and the spectra of those molecules that differ from it only by isotopic substitutions (isotopomers), it appears possible to interpret the spectra of the isotopomers as if they had the same nuclear motion wavefunction as the original molecule, but with changed masses for the replaced nuclei. However, the nuclear motion wavefunctions for the molecule and its isotopomers should generally have completely different permutational symmetry structures, if only because of nuclear spin statistics. So rather more radical changes might be needed to interpret the different spectra than seem actually to be necessary. The success of such approaches to the interpretation of molecular spectra, however, encourages the belief that somehow such coordinate mixing or even symmetry breaking permutations are of no effect.

The division of the translationally invariant coordinates into a set associated with electrons and a set associated with nuclei does not complicate the problems just considered. Indeed, it simplifies them to the extent that the electronic coordinates behave under permutations just as do electronic coordinates defined in the frame fixed in the laboratory. However, it is unfortunate that in order to make the division settled under permutations, only a rather limited description of asymptotic behavior is possible. There seems to be no way, however, in which settled division can be maintained while allowing an arbitrary origin for the translationally invariant electronic coordinates and permitting all proper permutations. It has not proved possible, therefore, to present one here.

1. How Coordinate Problems Might be Avoided

If it can be argued that at particular internuclear separations in particular energy ranges, permutations that involve a particular coordinate can be disregarded, then many of these problems need not arise. We have given reasons for doubting that the idea of feasibility can provide a theoretically satisfactory resolution of the puzzles arising from attempts to maintain permutational symmetry, but it might perhaps be possible to reinterpret the idea without invoking the occurrence of actual nuclear motions, in such a way as to provide a suitable account.

Let us consider the situation when we describe electrons. The Pauli principle asserts that all electronic wavefunctions shall be antisymmetric with respect to the interchange of space and spin coordinates of a pair of particles. The principle does not contain any restrictions about the proximity of the electrons to be considered. Thus one might, perhaps rather perversely, consider a hydrogen atom in the laboratory and another on the moon as candidates to form a hydrogen molecule. One might then attempt a trial molecular electronic wavefunction in the Heitler-London form

$$[1s_a(\mathbf{x}_1)1s_b(\mathbf{x}_2) + 1s_a(\mathbf{x}_2)1s_b(\mathbf{x}_1)](\alpha(s_1)\beta(s_2) - \alpha(s_2)\beta(s_1))$$

where a denotes the orbital centered in the lab and b, that centered on the moon and indicating the electron spin variables by s_i. However, on calculating the electronic energy with this trial wavefunction at the internuclear distance appropriate to the separation, the result is so close to the sum of the electronic energies of the separated atoms as to make the antisymmetry requirement, although not ignored, actually of no effect. As a less extreme choice than this, when we consider intermolecular interactions at rather large separations, we usually argue that in practice, it is not necessary to use antisymmetric electronic functions to describe the intermolecular behavior, because a similar result to that found for the hydrogen atoms above would be found for the separated molecule energies. It is believed that the results would be essentially the product function results, although there has been no extensive computational testing in this context. But this sort of argument is convincing, provided the properly symmetrized trial function provides a decent description of the system. It is not that permutations are neglected, it is just that, in relation to energy measurements, the proper wavefunction can be approximated with a high degree of accuracy by a simple product form. To make such an approximation is not to deny quantum entanglement as it arises in consideration of the arguments of Einstein, Rosen, and Podolsky and of Bell, nor is it to minimize the importance of Aspect's experimental results on large-distance quantum correlations. (For a discussion of quantum entanglement and related matters see, e.g., Mermin [67] and Rae [68].) It suffices to say that in the context of this category of measurement, the wavefunction with the correct permutational symmetry yields results that differ hardly at all from a well-chosen product form in which the correct symmetry is imposed only within the separated parts. To establish the validity of any such argument, it is strictly necessary to perform comparable calculations using the properly formed *and* the product wavefunction. But it might be hoped often possible, to guess the circumstances in which such an argument might have force, without actually performing detailed calculations.

Any equivalent argument for nuclear variables would certainly have to be very strongly structured, for there are circumstances in which one wishes to account for results in terms of entities that are individually distinguishable although they are, in a quantum-mechanical context, indistinguishable and proximate. To appreciate this, one only has to think of interpreting an IR spectrum to identify the fingerprint of an OH functional group in a substituted hydrocarbon, or of interpreting a proton NMR spectrum to identify the location of a proton in a system. If it were possible to construct such a strongly structured account, it would encompass the resolution of any problems arising from the Eckart structure choice and from the interpretation of isotopomer spectra. Let us call the permutations that we are able or would wish to handle in any given formulation the "plausible" permutations.

If one could attempt a solution $\Psi(\mathbf{t})$ to the problem specified by $\hat{H}(\mathbf{t})$ in the form specified in (3.1), then

$$\Psi(\mathbf{t}) = \sum_p \Phi_p(\mathbf{t}^n)\psi_p(\mathbf{t}^n, \mathbf{t}^e)$$

where the electronic part of the problem belongs to the totally symmetric representation of each of the relevant nuclear permutation groups and is an angular momentum eigenfunction with $J = 0$. If, further, the nuclear motion function can be written in terms of "orbital products" of the "orbitals" $\xi_j(\mathbf{t}_k^n)$, then, provided the orbitals could be chosen to behave in a decent manner under permutations, one might in principle consider a properly symmetrized function and various categories of simple product functions, in calculations to determine the relative merits of each choice. It is easy to imagine how, in principle, such calculations might be carried out for ammonia using the translationally invariant bond vector coordinates. However, it seems that the only calculations of this kind that have ever been done, (for examples, see Thomas [69], Petit and Dancura [70], and Monkhorst [71]) have taken the electronic origin as fixed so that they are essentially one centre calculations. The results are insufficient to enable any firm conclusions to be drawn. It would seem to be essential to get a much improved description of the electronic part of the problem, by allowing a functional form depending on explicit electron–nucleus separation coordinates before any sensible conclusions might be drawn. But this would involve formidable computational difficulties and would be a project not to be undertaken lightly. Although if one were to undertake it for ammonia, the outcome might enable one to account for the similarities in the spectra of NH_3, NH_2D, ND_2H, and ND_3, it is difficult to see how one might attempt a comparable endeavor for ethene, where there are many plausible choices of translationally invariant coordinates. In any case, such an approach does not map easily onto an account given in a frame fixed in the body of the molecule, and from the point of view of interpreting spectra, it is such an account that matters. One might therefore guess that attempts to view the problem from just a translationally invariant perspective might be a largely wasted effort, particularly since it would prove so difficult to get a decent description of the system with a properly symmetrized function.

Suppose that one had chosen a particular frame fixed in the body of the molecule exactly as has been described earlier and that a particular product function form had been chosen as a trial function to describe the nuclear motion within any given electronic state. It is not assumed that this trial function has proper permutational symmetry, and it will be properly defined only in a limited domain. But it is imagined, in some sense, to provide a decent description of the molecule, at least in a limited region of the internal coordinates $\mathbf{q}$ in the vicinity of a particular choice of nuclear geometry. Thus in this approximation, we shall use (6.2) as the function in (4.36) to yield the form change induced by a particular permutation as

$$\psi_p(\mathbf{q}, \mathbf{z}) \sum_{k=-J}^{+J} \Phi_{kp}^J(\mathbf{q}) |JMk\rangle \rightarrow \psi_p'(\mathbf{q}, \mathbf{z}) \sum_{k=-J}^{+J} \sum_{n=-J}^{+J} \mathscr{D}_{nk}^J(\mathbf{U}) \Phi_{kp}'^J(\mathbf{q}) |JMn\rangle$$

$$= \psi_p'(\mathbf{q}, \mathbf{z}) \sum_{n=-J}^{+J} \overline{\Phi}_{pn}^J(\mathbf{q}) |JMn\rangle \tag{8.1}$$

where the parity index has been suppressed to save writing. It is assumed that the permutation does not break the symmetry of the problem so that the rotational functions remain well defined. The effect of any permutation on the internal coordinates alone will be perfectly well defined, so changes in the electronic wavefunction will be defined even in the case of permutations that break the symmetry. Note that the changed function has been expressed in terms of the original variables. This is always possible in principle, but, as explained earlier, it might prove very difficult in practice because the variables are usually defined in a nonlinear manner in terms of the translationally invariant Cartesians.

From any given trial function we can in principle generate a function that belongs to a particular irrep of the relevant symmetric groups of the nuclei by using the standard projection operator formalism with a suitable realisation of (8.1) for each permutation involved. In the case of ethene this would result in a function composed as a sum of 48 terms. Let us write such a function as

$$\sum_{i=0}^{s_p} c^{(i)} \psi_p^{(i)}(\mathbf{q}, \mathbf{z}) \, ^{(i)}\Phi_p^J(\mathbf{q}, \phi)$$

where the sum extends over all the required permutations including the identity one, which is labeled (0). The terms $^{(i)}\Phi_p^J(\mathbf{q}, \phi)$ are the full nuclear motion functions, composed as in the sum portion of (8.1). The weights $c^{(i)}$ are those arising from the projection to produce the required irrep. Some terms in the sum might well be the same and enter with identical weights. This might occur when at a particular choice of $\mathbf{q}$ there is point group symmetry and the permutation is a perrotation. In this case the electronic wavefunctions, at least, would be identical.

We can then, again in principle, calculate an expectation value of the Hamiltonian with the full function of the chosen symmetry. In so doing it should be borne in mind that the electronic functions will not be orthogonal with respect to integration over the $\mathbf{z}$ and that in general

$$\langle \psi_p^{(i)} | \hat{H}^{\text{elec}} | \psi_p^{(i)} \rangle_{\mathbf{z}} \neq \langle \psi_p^{(j)} | \hat{H}^{\text{elec}} | \psi_p^{(j)} \rangle_{\mathbf{z}} \qquad (8.2)$$

where $\hat{H}^{\text{elec}}$ is as defined in (6.4) and that

$$\langle \psi_p^{(i)} | \hat{H}^{\text{elec}} | \psi_p^{(j)} \rangle_{\mathbf{z}} = E_p^{(ij)}(\mathbf{q}) \qquad (8.3)$$

is not necessarily zero.

Using methods much as described in Section VI.B, the rotationally independent part of the nuclear motion contribution to the matrix element can be calculated. Ignoring the effect of the nuclear operators on the electronic wavefunction, each contribution to the full matrix element is rather like the first term in (6.7)

$$\langle \, ^{(i)}\Phi_p^J(\mathbf{q}, \phi) | \langle \psi_p^{(i)} | \psi_p^{(j)} \rangle_{\mathbf{z}} (\hat{K}_H + +V^{\text{n}}(\mathbf{q})) + E_p^{(ij)}(\mathbf{q}) | \, ^{(j)}\Phi_p^J(\mathbf{q}, \phi) \rangle \qquad (8.4)$$

where the exterior integration is over the internal and rotational coordinates.

Let the diagonal contributions be ranked according to $E_p^{(ij)}(\mathbf{q})$ for a particular choice of $\mathbf{q}$ corresponding to a molecular geometry for which the lowest energy form is associated with the original function, the identity permutation (0). If it turns out that only those functions arising from "plausible" permutations have comparably low or, preferably identical electronic energies to the original function and further if only they have large overlaps or, best, unit overlaps with the original function and large values of $E_p^{(ij)}(\mathbf{q})$, then one could argue that only these "plausible" permutations will need further consideration. If at this level any symmetry breaking permutations turn out to be ignorable, then the whole argument can be rerun including these permutations, for they would drop out at this stage.

In this approach it is assumed that in some region of the internal coordinates $\mathbf{q}$ around a particular molecular geometry, a particular function that does not necessarily have the correct symmetry properties can be chosen to provide a decent description of the molecule. If a function with the full symmetry is projected from this initial function, it is hoped that the terms associated with "implausible" permutations can be neglected. It can be seen that what really matters in the permutational weighting in this formulation is the electronic energy of the permuted electronic problem. And this at once enables a connection to be effected with the idea of "feasible" permutations in a fairly obvious way.

The account given above of the effect of permutations on functions defined in a frame fixed in the body is by no means an account of how the exact wavefunction for the problem at a particular energy seems to behave as if certain permutations could be ignored. It is not even an account of how a decent approximation with the full permutational symmetry can be suitably restricted. It is, at best, an account of why, in a particular approximation, it might not be necessary to consider certain permutations. The domain limitations arising from both fixing the frame in the body and avoiding energy crossings mean that even the putative validity of the approximation is likely to hold only in a very restricted range. It does, however, wrest the idea of "feasibility" away from the idea of actual exchange of nuclei and place it in a proper context.

B. Domain Problems

The domain restrictions arising from the Jacobian for the coordinate transformation are irreducible and irremovable. They are actually not important restrictions on the transformation to a translationally invariant set; rather, they are for the transformation to a coordinate system defined in a frame fixed in the body. In this last case, however, they can always in principle be handled and the full space can be covered by standard topologic techniques. Since no problems of principle are involved, such restrictions will not be further discussed here. We shall simply assume that any chosen function is such as to make a vanishing contribution to an expected value wherever the Jacobian vanishes.

The real problems arise when using product form trial functions in which the electronic function also has nuclear coordinate dependence. Problems would arise in any formulation in these terms even using only translationally invariant coordinates, but we shall just consider the formulation in a frame fixed in the body. From what was said earlier about phases for the electronic wavefunction even if we confine attention to just one electronic surface, we ought not to ignore the diagonal term γ_{pp} in the "potential" part of (6.7). Including that term in these circumstances leads to what Ballhausen and Hansen [11] call the "Born–Huang adiabatic approximation." For diatomic one- and two-electron systems, the effects of including this term in the nuclear motion problem were reviewed by Kołos [72]. In this review he also considered its effects on the nature of the variational bounds. For diatomic one-electron systems this term has also been studied by Bishop and his co-workers [73] and from a slightly different point of view by Kohl and Shipsey [74] and also by Moss [75].

From these investigations it appears that the term is a small one (of the order of tens or hundreds of reciprocal centimeters) close to the equilibrium internuclear distance, but is quite a sensitive function of the distance. For certain values of the electronic angular momentum, the term becomes infinite as the distance decreases, but sometimes cancellation of terms removes the divergences. It seems possible that what is being seen here is an example of the problems that can arise in circumstances where the Jacobian vanishes. (The internuclear distance is a spherical polar coordinate, and so the Jacobian involves its square.) Therefore, there might well be analogous behavior in the more general case given by (6.7).

If we now consider the case of a number of surfaces, in the case of diatomic systems, rather special considerations apply because of the noncrossing rule, so the potentially divergent terms need cause no problems [76], but, in general, if such an expansion were attempted, it would not be possible to discount these terms. If one adopts a conventional Eckart-like approach to the nuclear motion part of the problem, then the body-fixed frame appropriate to the ground electronic state is defined by means of an equilibrium nuclear geometry. If the electronic state is actually degenerate at that geometry, then, in the present approach, one would expect divergent coupling terms between the degenerate partner states. It was this problem to which Jahn and Teller addressed themselves in 1937. (For a review, see Bersuker [77].) Their conclusion was that such a state of affairs would not arise because a minimum in the potential surface could not occur at a geometry that gave rise to degenerate electronic states. This is usually glossed by saying that a degenerate state would not arise because the coupling would cause distortion of the assumed equilibrium geometry in such a way as to lift the degeneracy. This picture is used to explain many observations on transition-metal complexes and, when invoked to account for a particular result, is called the *Jahn–Teller effect*. At a linguistic level one can see why it is a bit odd to call it an "effect" but at the theoretical level the approach, assuming this account appropriate, actually avoids the problem of divergence. But the divergence does not go away. It persists at the assumed equilibrium geometry even if this is not really the observed one.

A similar discussion of the Renner–Teller effect in the spectra of triatomic molecules could be advanced. This is supposed to originate in the possibility of a sufficiently energetic vibration, making the molecule linear in a degenerate electronic state.

1. How Domain Problems Might be Avoided

If we consider just the $J = 0$ case, the problems could be avoided if one could transform the electronic wavefunctions so that in the transformed basis, the matrix γ with elements $\gamma_{p'p}$, were diagonal. Previous discussions of this sort of possibility have not been cast explicitly in terms of a frame fixed in the body, as here, but the main results of these discussions apply equally in the present formulation and we shall continue to call the coupling matrix γ, although as will be seen, there are small differences.

Smith [78] derived a set of differential equations that the elements of the coupling matrix must satisfy to achieve this diagonal property. However, Mead and Truhlar [79] showed that, if the electronic functions were functions of the nuclear coordinates at all then, except in certain very special cases, these equations would have no useful solutions. They demonstrated this by introducing a kind of "vector potential" and showing that its curl could not, in general vanish. (The "vector potential" idea has proved to be very fruitful in analyzing Berry's phase for electronic wavefunctions with surface crossings, and reviews have been given by Frey and Davidson [80] and by Mead [81].) However, it is often possible to diagonalize γ if one can assume that the curl term is negligibly small. The approximate diabatic bases obtained in this way are often used to avoid the divergences encountered at surface crossings in the adiabatic approach, and there has been considerable interest in numerical methods to calculate the $\gamma_{p'p}$ in this context. An example of their calculation for MCSCF Multi-configuration (self-consistent field) electronic wavefunctions can be found in the work of Bak et al. [82]. Reviews have been provided by Lengsfield and Yarkony [83] and by Sidis [84].

It could, of course, be argued that it is actually unnecessary to get oneself into difficulties here because the $\psi_p(\mathbf{q}, \mathbf{z})$ form a complete set in the space of electronic variables for any fixed choice of $\mathbf{q} = \mathbf{q}^0$, say. It is therefore possible to construct a product expansion but using in place of $\psi_p(\mathbf{q}, \mathbf{z})$ the functions $\psi_p(\mathbf{q}^0, \mathbf{z}) \equiv \psi_p{}^0(\mathbf{z})$. Ballhausen and Hansen [11] call this the *Longuet–Higgins* representation. Here the electronic solutions have no nuclear coordinate dependence but are simply functions centered so as to achieve a permutationally settled division between electronic and nuclear coordinates. A single such function cannot be an eigenfunction of an electronic Hamiltonian nor the clamped nuclei Hamiltonian, but a set of such one-center functions can be chosen as a complete discrete set in the electronic variables. Thus a linear combination of these functions can, in the limit, represent an electronic or a clamped nuclei solution. Although the trial functions do not depend on the electron–nucleus separation, the electronic and the clamped nuclei Hamiltonian do so depend. Calculations performed with such functions at different nuclear geometries will therefore lead to different electronic energies, and so a sort of potential surface can still be constructed as a sum of electronic and nuclear repulsion energies as a function of geometry.

The essential completeness of the electronic function set leads to an expansion in product function form of the full wavefunction, which can, again in the limit, represent the full function exactly. If it is used, many of the problems encountered earlier, do not arise. The coupling terms between electronic states shown in (6.8) and (6.13), which come from the derivatives with respect to the q_k of the electronic wavefunction, vanish identically. In (6.10) this leaves only the first, second, third, and fifth terms terms in $g_{p'p}^0$ contributing and in (6.11), the first and third terms in $g_{p'p}^k$. If the electronic origin is chosen as the center of nuclear mass, scaled or not, only the first two terms remain in $g_{p'p}^0$ and only the first term in $g_{p'p}^k$. In (6.13) only the first and third terms survive and only the first remains if a center-of-nuclear-mass-related origin is chosen for the electronic coordinates.

With this choice, although the coupling of the electronic states does not vanish, the absence of derivatives with respect to the q_k of the electronic wavefunctions ensures that no divergent terms can arise from the couplings. So in this form there is no need to further limit the domain to avoid divergences, and the domain of the product function is the fundamental domain of the transformation. A formulation in which the coupling of electronic states is altogether absent is often called a *diabatic* formulation, as opposed to the one that we have exhibited earlier in (6.7) and (6.12), which is called an *adiabatic* formulation. The present formulation is an almost diabatic one. The coupling terms that remain when a center-of-nuclear-mass origin is chosen for the electrons, arise from the interaction of the electronic angular momentum with the rotation of the system as a whole.

Although such a choice of product function is attractive, all experience goes to show that an expansion in such products is so slowly convergent as to be useless to describe systems with nuclear geometries corresponding to typical molecules and more extended structures. And although it can yield a sort of potential-energy surface, this surface will bear little resemblance to those usually calculated. It is thus probably not an approach to offer any way forward in practice.

One could also avoid domain problems by avoiding the separation of electronic and nuclear motions and treating the problem as a whole. This sort of thing has been attempted in work referred to earlier [69,70,71], although not in coordinates fixed in the body but simply in translation-free coordinates. Although it has been quite successful in few-electron diatomics [85], it has not been successful in polyatomics. The only examples of full calculations made in coordinates fixed in the body are on the hydrogen molecule, and it is difficult to generalize the results obtained to polyatomic systems.

IX. CONCLUSIONS

On the basis of experience in interpreting experimental results in molecular physics, one cannot help but feel that if one were vouchsafed a vision of the full exact wavefunction for a neutral molecule, one would see that in a significant region of nuclear configuration space it could be written in excellent approximation as a product of electronic and nuclear factors. To go the other way, however, and attempt to use product functions to build up the full exact function to account for experimental results, seems to result rather more in a nightmare than a vision. From the standpoint of the present discussion, it seems that few general conclusions are possible. Rather, specific instances must be investigated with care to see what inferences may be drawn.

It seems likely from the present discussion, that the results of clamped nuclei electronic structure calculations for neutral systems not too far from their equilibrium geometries can be safely interpreted in the usual way, provided no electronic degeneracies are close by. The use of such solutions in interpreting molecular spectra in the usual way is, however, not secured in the present approach, because the way in which permutational invariance is to be satisfied remains unsettled.

Unfortunately, nothing presented here is sufficient to allow a secure discussion of scattering on a potential-energy surface. There is a fundamental problem of coordinate choice, which must be prior to any discussion of the separation of electronic from nuclear motion. Not only is there a problem of permutational invariance but of global analyticity as well. Whether coordinate choices more satisfactory for a discussion of scattering than those made here, can be made, remains an open question. However neither the work of Klein [86] on the scattering of two atoms nor the work reviewed in [87] on global internal coordinates seem to encourage optimism in these matters.

References

1. M. Born and J. R. Oppenheimer, *Ann. Phys.* (Leipzig) **84**, 457 (1927).
2. C. Eckart, *Phys. Rev.* **47**, 552 (1935).
3. G. Hunter, *Int. J. Quantum Chem.* **9**, 237 (1975).
4. J. C. Slater, *Proc. Natl. Acad. Sci.* **13**, 423 (1927).
5. M. Born and K. Huang, *Dynamical Theory of Crystal Lattices*, Oxford Univ. Press, Oxford, 1955.
6. J. M. Combes, *Acta Physica Austriaca*, Suppl. XVII, 139 (1977).
7. M. Klein, A. Martinez, R. Seiler, and X. P. Wang, *Commun. Math. Phys.* **143**, 607 (1992).
8. G. Hagedorn, *Commun. Math. Phys.* **77**, 1 (1980).
9. G. Hagedorn, *Commun. Math. Phys.* **136**, 433 (1991).
10. H. Essén, *Int. J. Quantum Chem.* **XII**, 721 (1977).
11. C. J. Ballhausen and A. E. Hansen, *Annu. Rev. Phys. Chem.* **23**, 15 (1972).
12. M. Reed and B. Simon, *Methods of Modern Mathematical Physics, Vol. IV, Analysis of Operators*, Academic Press, New York (1978).
13. W. Thirring, *A Course in Mathematical Physics*, Vol. 3, *Quantum Mechanics of Atoms and Molecules* (E. M. Harrell trans.), Springer, New York, 1981.
14. G. M. Zhislin, *Trudy. Mosk. Mat. Obsc.* **9**, 82 (1960).
15. J. Uchiyama, *Pub. Res. Inst. Math. Sci. Kyoto* **A2**, 117 (1967).
16. B. Simon, *Quantum Mechanics for Hamiltonians Defined as Quadratic Forms*, Princeton Univ. Press, Princeton, NJ, 1971.
17. G. M. Zhislin, *Theor. Math. Phys.* **7**, 571 (1971).
18. R. Nyden Hill, *J. Math. Phys.* **18**, 2316 (1977).
19. M. P. Ruskai and J. P. Solovej, in *Schrödinger Operators*, Lecture Notes in Physics, Vol. 403, E. Balslev, ed., Springer, Berlin, 1992, p. 153.
20. M. B. Ruskai, *Ann. Inst. Henri Poincaré* **52**, 397 (1990).
21. M. B. Ruskai, *Commun. Math. Phys.* **137**, 553 (1991).
22. B. Simon, *Helv. Phys. Acta* **43**, 607 (1970).
23. S. A. Vugal'ter and G. M. Zhislin, *Theor. Math. Phys.* **32**, 602 (1977).
24. W. D. Evans, R. T Lewis, and Y. Saito, *Phil. Trans. Roy. Soc. Lond.* A **338**, 113 (1992).
25. J.- M. Richard, J. Fröhlich, G.-M. Graf, and M. Seifert, *Phys. Rev. Lett.* **71**, 1332 (1993).
26. A. Nauts and X. Chapuisat, *Mol. Phys.* **55**, 1287 (1985).
27. M. Menou and X. Chapuisat, *J. Mol. Spectrosc.* **159**, 300 (1993).
28. N. C. Handy, *Mol. Phys.* **61**, 207 (1987).
29. A. T. Császár and N. C. Handy, *Mol. Phys.* **86**, 959 (1995).
30. T. J. Lukka, *J. Chem. Phys.* **102**, 3945 (1995).
31. B. T. Sutcliffe, in *Methods of Computational Chemistry*, Vol. 4, S. Wilson, ed., Plenum Press, New York, 1991, p. 33.
32. B. T. Sutcliffe, *J. Chem. Soc., Faraday Trans.* **89**, 2321 (1993).

33. B. T. Sutcliffe, in *Conceptual Trends in Quantum Chemistry*, edited by E. S. Kryachko and J. L. Calais, eds., Kluwer Academic, Dordrecht, 1994, p. 53.

34. G. O. Sørensen, in *Topics in Current Chemistry*, Vol. 82, Springer, Berlin, 1979, p. 99.

35. J. K. G. Watson, *Mol. Phys.* **15**, 479 (1968).

36. D. M. Brink and G. R. Satchler, *Angular Momentum* , 2nd ed., Clarendon Press, Oxford, 1968.

37. L. C. Biedenharn and J. D. Louck, *Angular Momentum in Quantum Physics*, Addison-Wesley, Reading, MA, 1982.

38. E. U. Condon and G. H. Shortley, *The Theory of Atomic Spectra*, Cambridge Univ. Press, Cambridge, UK, 1935.

39. J. M. Brown and B. J. Howard, *Mol. Phys.* **31**, 1517 (1976).

40. R. N. Zare, *Angular Momentum*, Wiley, New York, 1988, Chapter 3.4.

41. G. Ezra, *Symmetry Properties of Molecules* , Lecture Notes in Chemistry, Vol. 28, Springer, Berlin, 1982.

42. B. Schutz, *Geometrical Methods of Mathematical Physics*, Cambridge Univ. Press, Cambridge, UK, 1980.

43. J. D. Louck, *J. Mol. Spec.* **61**, 107 (1976).

44. E. C. Kemble, *The Fundamental Principles of Quantum Mechanics*, McGraw-Hill, New York, 1937.

45. R. A. J. Malhiot and S. M. Ferigle, *J.Chem. Phys.* **22**, 717 (1954).

46. B. T. Sutcliffe, in *Quantum Dynamics of Molecules*, R. G. Woolley, ed., Plenum Press, New York, 1980, p. 1.

47. E. P. Wigner, *Group Theory*, Academic Press, New York, 1959.

48. B. T. Sutcliffe and J. Tennyson, *Int. J. Quantum Chem.* **39**, 183 (1991).

49. W. C. Ermler and B. J. Krohn, *J. Chem. Phys.* **67**, 1360 (1977).

50. S. M. Adler-Golden and G. R. Carney, *Chem. Phys. Lett.* **113**, 582 (1985).

51. C. R. le Sueur, S. Miller, J. Tennyson, and B. T. Sutcliffe, *Mol. Phys.* **76**, 1147 (1992).

52. H. Wei and T. Carrington Jr., *J. Chem. Phys.* **107**, 2813,9493 (1997).

53. J. M. F. Gilles and J. Philippot, *Int. J. Quantum Chem.* **6**, 225 (1972).

54. J. D. Louck and H. W. Galbraith, *Rev. Mod. Phys.* **48**, 69 (1976).

55. R. S. Berry, *Rev. Mod. Phys.* **32**, 447 (1960).

56. H. C. Longuet-Higgins, *Mol. Phys.* **6**, 445 (1963).

57. P. R. Bunker, *Molecular Symmetry and Spectroscopy*, Academic Press, London, 1979.

58. I. G. Kaplan, *Symmetry of Many-Electron Systems*, Academic Press, London, 1975.

59. J. Maruani and J. Serre, eds., *Symmetries and Properties of Non-Rigid Molecules*, Elsevier, Amsterdam, 1983.

60. M. A. Collins and D. F. Parsons, *J. Chem. Phys.* **99**, 6756 (1993).

61. F. T. Smith, *Phys. Rev. Lett.* **45**, 1157 (1980).

62. J. Gauss and D. Cremer, *Adv. Quantum Chem.* **23**, 206 (1992).

63. C. Eckart, *Phys. Rev.* **46**, 384 (1934).

64. J. O. Hirschfelder and E. Wigner, *Proc. Natl. Acad. Sci.* (USA) **21**, 113 (1935).

65. B. Buck, L. C. Biedenharn, and R. Y. Cusson, *Nucl. Phys.* **A317**, 215 (1979).

66. J. H. van Vleck, *Phys. Rev.* **47**, 487 (1935).

67. N. D. Mermin, *Boojums All the Way Through*, Cambridge Univ. Press, Cambridge, UK, 1990.

68. A. Rae, *Quantum Physics: Illusion or Reality?*, Cambridge Univ. Press, Cambridge, UK, 1986.

69. I. L. Thomas, *Phys. Rev.* **185**, 90 (1969); *Phys. Rev. A.* **2**, 1200 (1970); *Phys. Rev. A* **3**, 1565 (1971).

70. B. A. Petit and W. Dancura, *J. Phys. B* **20**, 1899. (1987).

71. H. J. Monkhorst, *Phys. Rev. A* **36**, 1544 (1987).

72. W. Kołos, *Adv. Quantum Chem.* **5**, 99 (1970).

73. J. N. Silverman, and D. N. Bishop, *Chem. Phys. Lett.* **130**, 132 (1986).

74. D. A. Kohl and E. J. Shipsey, *J. Chem. Phys.* **84**, 2707 (1986).

75. R. E. Moss, *Mol. Phys.* **78**, 371 (1993).

76. P. Quadrelli, K. Dressler, and L. Wolniewicz, *J. Chem. Phys.* **92**, 7461 (1990).

77. I. B. Bersuker, *Coord. Chem. Rev.* **14**, 357 (1975).

78. F. T. Smith, *Phys. Rev.* **179**, 111 (1969).

79. C. A. Mead and D. G. Truhlar, *J. Chem. Phys.* **77**, 6090 (1982).

80. R. G. Frey and E. R. Davidson, in *Advances in Molecular Electronic Structure Theory*, Vol. 1 T. H. Dunning, ed., JAI Press, London, 1990, p. 213.

81. C. A. Mead, *Rev. Mod. Phys.* **64**, 51 (1992).

82. K. L. Bak, P. Jørgensen, H. J. A. Jensen, J. Olsen, and T. Helgacker, *J. Chem. Phys.* **97**, 7573 (1992).

83. B. H. Lengsfield, and D. R. Yarkony, *Adv. Chem. Phys.* **82**, Part 2, 1 (1992).

84. V. Sidis, *Adv. Chem. Phys.* **82**, Part 2, 73 (1992).

85. C. A. Traynor, J. B. Anderson, and B. H. Boghosian, *J. Chem. Phys.* **94**, 3657 (1991).

86. M. Klein in A. F. Sax, ed "Potential Energy Surface", Lecture Notes in Chemistry, **71**, Springer, Berlin, 1999, 215.

87. M. A. Collins and K. C. Thompson in D. Bonchev and D. Rouvray eds "Chemical Group Theory: Techniques and Applications", Gordon and Breach, Reading, MA, 1995, 191.

ASSOCIATION, DISSOCIATION, AND THE ACCELERATION AND SUPPRESSION OF REACTIONS BY LASER PULSES

MOSHE SHAPIRO

Department of Chemical Physics, The Weizmann Institute of Science, Rehovot, Israel

CONTENTS

I. INTRODUCTION

The dissociation of molecules by light is complicated because of the existence of a multiplicity of strongly coupled continua ("channels"), arising from the different internal states of the molecular fragments. When the dissociation also involves irradiation by strong (or even moderately

Advances in Chemical Physics, Volume 114, Edited by I. Prigogine and Stuart A. Rice.
ISBN 0-471-39267-7 © 2000 John Wiley & Sons, Inc.

strong) pulses, the problem becomes doubly complicated because of the existence of an additional type of (strong) interaction, that between the light pulse and the molecule. Since we are interested in both the time evolution of the photon-induced processes and the long-time populations of each the dissociative continuum, we cannot neglect any of the abovementioned aspects of the problem.

In the weak-field regime one can solve the photodissociation problem by concentrating on the "material" part. In essence, one uses exact propagation methods, to deal with the presence of the many material channels [1–10], and perturbation theories, to describe the way the (weak) field is absorbed by the molecular system.

The multichannel aspect of the problem becomes more involved when dealing with the dissociation of molecules [11–21], or the ionization of atoms [22–29] in the strong-pulse regime. Few approaches used in solving such problems try to separate out the multichannel material part from the time-dependent part. Rather, "brute force" numeric solutions of the multi-dimensional time-dependent or time-independent field matter Schrödinger equation are performed.

It is, however, possible, even in the strong-field regime, to treat first the material multichannel aspect of the problem and then the field-dependent part. The advantage in so doing is that, having invested effort in obtaining the material dipole matrix, we can perform the calculations of the effects of the field more easily and repeat them for many pulse intensities and configurations at relatively little cost.

In this chapter we examine a number of situations in which knowledge of the purely material dipole matrix elements allows one to formulate analytically solvable molecule–pulse interaction models. The ability to formulate analytic theories stems from the fact that in many situations the radiative coupling between the continua is mediated by only a few bound states. Under these circumstances (which preclude the ultra-strong-field regime where transitions within the continuum are abundant) use of the material bound-continuum matrix elements, obtained by solving the nonradiative multichannel scattering problem, enables the solution of the entire radiation matter problem.

We present four studies: the one-photon and two-photon dissociation of a single state (Fig. 1a,b); controlled photoassociation (Fig. 1c); and laser catalysis (Fig. 1d).

II. ONE-PHOTON DISSOCIATION OF A SINGLE-PRECURSOR STATE BY A STRONG LASER PULSE

Our first example deals with molecular dissociation by absorption of one photon from a strong laser pulse. We develop a number of analytic and

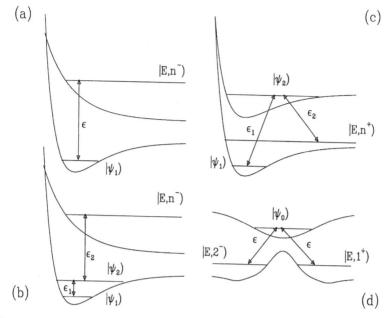

Figure 1. Energy levels and pulses pertaining to (*a*) one-photon dissociation; (*b*) resonantly enhanced two-photon dissociation; (*c*) resonantly enhanced two-photon association; (*d*) laser catalysis.

semianalytic approaches, including the "flat" or "slowly varying" continuum approximation (SVCA) [30–33] and the pole expansion [32] for dealing with structured continua. The main issues that will interest us in this section is the extent to which dissociation or ionization processes are irreversible and the extent to which transitions to multiple continua can be controlled.

A. Theory

We consider the action of a single pulse, described by a classical time-evolving electric field of polarization $\hat{\epsilon}$

$$\vec{\varepsilon} = \hat{\epsilon}\varepsilon(t)\cos(\omega_1 t) \tag{2.1}$$

on a bound molecular system. Given the total radiation matter Hamiltonian in the "electric dipole" approximation, we obtain

$$H_{\text{tot}} = H - \vec{\mu} \cdot \vec{\varepsilon}(t) \tag{2.2}$$

where $\vec{\mu}$ is the dipole operator, our aim is to solve the time-dependent Schrödinger equation:

$$\frac{i\hbar\partial|\Psi\rangle}{\partial t} = H_{\text{tot}}|\Psi\rangle \qquad (2.3)$$

The situation depicted in Fig. 1a is that of a field in near resonance with a band of transition frequencies from an initial bound state $|\psi_1\rangle$ to a multiple continuum of $|E, \mathbf{n}^-\rangle$ states. Both $|\psi_1\rangle$ and $|E, \mathbf{n}^-\rangle$ are eigenstates of H:

$$[E_1 - H]|\psi_1\rangle = [E - H]|E, \mathbf{n}^-\rangle = 0 \qquad (2.4)$$

The index $\mathbf{n}$, differentiating one continuum from another, encompasses a set of quantum numbers specifying the final $(t \to \infty)$ internal (vibrational, rotational, etc.) states of the dissociated (polyatomic) fragments.

Given the radiation-free states described above, we expand the full time-dependent wavefunction as [3,34,35]

$$|\Psi(t)\rangle = b_1(t)|\psi_1\rangle\exp\frac{-iE_1t}{\hbar} + \sum_{\mathbf{n}}\int dE b_{E,\mathbf{n}}(t)|E, \mathbf{n}^-\rangle\exp\frac{-iEt}{\hbar}. \qquad (2.5)$$

Substitution of this expansion into the time-dependent Schrödinger equation and use of the orthogonality of the basis functions results in a set of first-order differential equations for the expansion coefficients

$$\frac{d}{dt}b_1 = i\int dE \sum_{\mathbf{n}} \Omega_{1,E,\mathbf{n}}(t)b_{E,\mathbf{n}}(t)\exp(-i\Delta_E t) \qquad (2.6a)$$

$$\frac{d}{dt}b_{E,\mathbf{n}} = i\Omega_{1,E,\mathbf{n}}^*(t)\exp(i\Delta_E t)b_1(t), \quad \text{for each } E \text{ and } \mathbf{n} \qquad (2.6b)$$

where we have retained only the "rotating waves" terms. The detuning, Δ_E, is defined as

$$\Delta_E \equiv \omega_{E,1} - \omega_1, \quad \text{with} \quad \omega_{E,1} \equiv \frac{E - E_1}{\hbar} \qquad (2.7)$$

and $\Omega_{1,E,\mathbf{n}}(t)$, the (time-varying) "Rabi frequency," is defined as

$$\Omega_{1,E,\mathbf{n}}(t) \equiv \frac{\langle\psi_1|\mu|E, \mathbf{n}^-\rangle\varepsilon(t)}{\hbar} \qquad (2.8)$$

We proceed [30] by integrating the $b_{E,\mathbf{n}}$ continuum coefficients of Eq. (2.6) over time, while imposing the boundary condition that none of the continua are populated as we start process $[b_{E,\mathbf{n}}(t \to -\infty) = 0]$

$$b_{E,\mathbf{n}}(t) = i \int_{-\infty}^{t} dt' \Omega_{1,E,\mathbf{n}}^{*}(t') b_1(t') \exp(i\Delta_E t') \qquad (2.9)$$

The state-specific photodissociation probability, $P_{\mathbf{n}}(E)$, is the long-time probability, at fixed energy E, of observing a particular internal state $|\mathbf{n}\rangle$ of the dissociated fragments. It is given using Eq. (2.9) as

$$P_{\mathbf{n}}(E) = |b_{E,\mathbf{n}}(t \to \infty)|^2 = \left| \frac{1}{\hbar} \langle \psi_1 | \mu | E, \mathbf{n}^- \rangle \int_{-\infty}^{\infty} dt' \varepsilon^*(t') b_1(t') \exp(i\Delta_E t') \right|^2 \qquad (2.10)$$

It follows that the final branching ratio between two internal fragment states is given as

$$\frac{P_{\mathbf{n}}(E)}{P_{\mathbf{m}}(E)} = \left| \frac{b_{E,\mathbf{n}}(\infty)}{b_{E,\mathbf{m}}(\infty)} \right|^2 = \left| \frac{\langle \psi_1 | \mu | E, \mathbf{n}^- \rangle}{\langle \psi_1 | \mu | E, \mathbf{m}^- \rangle} \right|^2 \qquad (2.11)$$

We see that the relative probabilities of populating different asymptotic states at a fixed energy E are *independent of the laser pulse attributes*. This result, which coincides with that of perturbation theory, holds true irrespective of the laser power, provided only *one* initial ("precursor") state $|E_1\rangle$ is coupled to all the continua. It holds true even when the rotating-waves approximation assumed above breaks down, because even if that happens, we can write the photodissociation probability as

$$P_{\mathbf{n}}(E) = |b_{E,\mathbf{n}}(t \to \infty)|^2 = \left| \frac{1}{\hbar} \langle \psi_1 | \mu | E, \mathbf{n}^- \rangle \int_{-\infty}^{\infty} dt' \{\varepsilon^*(t') \exp[i\Delta_E t']. \right.$$
$$\left. + \varepsilon(t') \exp[i(\omega_{E,1} + \omega_1)t']\} b_1(t') \right|^2 \qquad (2.12)$$

and the pulse attributes still cancel out when the $P_{\mathbf{n}}(E)/P_{\mathbf{m}}(E)$ branching ratio is evaluated. We conclude that under the one-precursor state condition no radiative *control* over the energy partitioning between the various fragment states is possible.

Control can, however, be attained if we move away from the single initial (precursor) state situation. For example, we can start with a linear superposition of two initial states:

$$|\Phi(t)\rangle = b_1|\psi_1\rangle \exp\frac{-iE_1 t}{\hbar} + b_2|\psi_2\rangle \exp\frac{-iE_2 t}{\hbar} \qquad (2.13)$$

Under these circumstances we have, in analogy to Eq. (2.9)

$$b_{E,\mathbf{n}}(t \to \infty) = \frac{i}{\hbar} \{ \langle E, \mathbf{n}^-|\mu|\psi_1\rangle \int_{-\infty}^{\infty} dt' \varepsilon(t') \exp(-i\omega_{E,1}t')b_1(t')$$

$$+ \langle E, \mathbf{n}^-|\mu|\psi_2\rangle \int_{-\infty}^{\infty} dt' \varepsilon(t') \exp(-i\omega_{E,2}t')b_2(t') \} \qquad (2.14)$$

In first-order perturbation theory, $b_1(t)$ and $b_2(t)$ are constant; hence in the weak-field regime

$$b_{E,\mathbf{n}}(t \to \infty) \approx \frac{2\pi i}{\hbar} \{ \langle E, \mathbf{n}^-|\mu|\psi_1\rangle \bar{\varepsilon}(\omega_{E,1})b_1 + \langle E, \mathbf{n}^-|\mu|\psi_2\rangle \bar{\varepsilon}(\omega_{E,2})b_2 \}$$

$$(2.15)$$

where

$$\bar{\varepsilon}(\omega) \equiv (1/2\pi) \int_{-\infty}^{\infty} dt \varepsilon(t) \exp(-i\omega t) \qquad (2.16)$$

Recognizing that $\bar{\varepsilon}(\omega)$ has a phase, we can write

$$\bar{\varepsilon}(\omega_{E,1}) = |\bar{\varepsilon}(\omega_{E,1})|e^{-i\theta(\omega_{E,1})}, \qquad \bar{\varepsilon}(\omega_{E,2}) = |\bar{\varepsilon}(\omega_{E,2})|e^{-i\theta(\omega_{E,2})}, \qquad (2.17)$$

and transform Eq. (2.15) into

$$b_{E,\mathbf{n}}(\infty) = \frac{2\pi i}{\hbar} \{ \langle E, \mathbf{n}^-|\mu|\psi_1\rangle |\bar{\varepsilon}(\omega_{E,1})|e^{-i\theta(\omega_{E,1})}b_1$$

$$+ \langle E, \mathbf{n}^-|\mu|\psi_2\rangle |\bar{\varepsilon}(\omega_{E,2})|e^{-\theta(\omega_{E,2})}b_2 \} \qquad (2.18)$$

The probability of seeing product state $\mathbf{n}$ at infinite time is therefore now given as

$$P_{\mathbf{n}}(E) = \frac{4\pi^2}{\hbar^2} |\langle E, \mathbf{n}^-|\mu|\psi_1\rangle |\bar{\varepsilon}(\omega_{E,1})|e^{-i\theta(\omega_{E,1})}b_1$$

$$+ \langle E, \mathbf{n}^-|\mu|\psi_2\rangle |\bar{\varepsilon}(\omega_{E,2})|e^{-i\theta(\omega_{E,2})}b_2|^2$$

We see that in this configuration the pulse attributes have been "entangled" with the material matrix elements. As a result, by changing the pulse attributes [such as the relative phase $\theta(\omega_{E,2}) - \theta(\omega_{E,1})$ or the relative amplitude $|\bar{\epsilon}(\omega_{E,2})/\bar{\epsilon}(\omega_{E,1})|$], we *can* change the branching ratios to different channels. The mechanism described above serves as the basis for the "bichromatic" coherent control scenario [36].

In order to calculate the effect of the pulse on the bound state part, we substitute Eq. (2.9) in Eq. (2.6a) to obtain a first-order integrodifferential equation for b_1;

$$\frac{db_1}{dt} = \frac{-1}{\hbar^2} \int dE \sum_n |\langle E, \mathbf{n}^- |\mu|\psi_1\rangle|^2 \varepsilon(t) \int_{-\infty}^t dt' \varepsilon(t') \exp[-i\Delta_E(t - t')] b_1(t')$$

(2.20)

This equation can be solved numerically in a straightforward fashion. Nevertheless, it is instructive to analyze it in terms of $F_1(t - t')$, the "spectral autocorrelation function" [30,31,35,37], defined as the Fourier transform of the absorption spectrum

$$F_1(t - t') = \int dE A_1(E) \exp[-i\omega_{E,1}(t - t')]$$

(2.21)

where $A_i(E)$, the absorption spectrum from the ith state, is given as

$$A_i(E) \equiv \sum_n |\langle E, \mathbf{n}^- |\mu|\psi_i\rangle|^2$$

(2.22)

With this definition for $F_1(t - t')$, we can rewrite Eq. (2.20) as

$$\frac{db_1}{dt} = \frac{-\varepsilon(t)}{\hbar^2} \int_{-\infty}^t dt' \varepsilon(t') F_1(t - t') \exp[i\omega_1(t - t')] b_1(t')$$

(2.23)

We see that the value of the ground-state coefficient at time t is determined by its past history at $t' < t$ via the "memory Kernel" $\varepsilon(t)\varepsilon(t')F_1(t - t')$.

The simplest (albeit approximate) solution of Eq. (2.23) is obtained by assuming that all the continua are "flat," that is, that the bound-continuum matrix elements vary slowly with energy and can be replaced by their value at some average energy, say, $E_L = E_1 + \hbar\omega_1$:

$$\sum_n |\langle E, \mathbf{n}^- |\mu|\psi_1\rangle|^2 \approx \sum_n |\langle E_L, \mathbf{n}^- |\mu|\psi_1\rangle|^2$$

(2.24)

This approximation, called the *slowly varying continuum approximation* (SVCA) [30–33], localizes the autocorrelation function in time, since by Eqs. (2.24) and (2.21)

$$F_1(t - t') = 2\pi\hbar A_1(E_L)\delta(t - t')$$

(2.25)

Substituting Eq. (2.25) in Eq. (2.23) and performing the integration over E and t', we obtain

$$\frac{db_1}{dt} = -\Omega(t)b_1(t) \qquad (2.26)$$

hence

$$b_1(t) = b_1(-\infty)\exp\left[-\int_{-\infty}^{t} \Omega(t')dt'\right] \qquad (2.27)$$

where $\Omega(t)$, the "imaginary Rabi frequency," is defined as

$$\Omega(t) \equiv \frac{\pi A_1(E_L)\varepsilon(t)^2}{\hbar} = \pi \sum_{\mathbf{n}} \frac{|\langle E_L, \mathbf{n}^-|\mu|\psi_1\rangle\varepsilon(t)|^2}{\hbar} \qquad (2.28)$$

The factor of $\frac{1}{2}$ relative to Eq. (2.25) comes about because the integration over $t' - t$ in Eq. (2.23) is performed over the $[-\infty, 0]$ range and not over the usual $[-\infty, +\infty]$ range.

It follows from Eq. (2.27) that a "slowly varying" continuum acts as a irreversible "perfect absorber," since in this approximation $b_1(t)$ decreases monotonically (but not necessarily purely exponentially) with time. In many cases the continuum may have structures that are narrower than the effective bandwidth of the pulse (which depends on its frequency profile *and* its intensity). Such structures may be due to either the natural spectrum of the nonradiative Hamiltonian [38,39] or the interaction with the strong external field [40,41]. Under such circumstances we expect the SVCA approximation to break down, yielding nonmonotonic decay dynamics.

Using the SVCA we can now write an analytic formula for "bichromatic control" that goes beyond perturbation theory. Allowing the initial coefficients to decay according to Eq. (2.27), we obtain, from Eq. (2.14)

$$b_{E,\mathbf{n}}(t \to \infty) = \frac{i}{\hbar}\left\{ \langle E, \mathbf{n}^-|\mu|\psi_1\rangle b_1 \right.$$

$$\times \int_{-\infty}^{\infty} dt'\varepsilon_1(t')\exp\left[-i\omega_{E,1}t' - \frac{\pi}{\hbar}A_1(E_L)\int_{-\infty}^{t'} |\varepsilon_1(t'')|^2 dt''\right]$$

$$+ \langle E, \mathbf{n}^-|\mu|\psi_2\rangle b_2 \int_{-\infty}^{\infty} dt'\varepsilon_2(t')\exp\left[-i\omega_{E,2}t' - \frac{\pi}{\hbar}A_2(E_L)\int_{-\infty}^{t'} |\varepsilon_2(t'')|^2 dt''\right]\right\}$$

$$\qquad (2.29)$$

where $b_i \equiv b_i(-\infty), i = 1, 2$. Therefore, the probability of observing a particular channel $\mathbf{n}$ is given as

$$P_{\mathbf{n}}(E) = \frac{4\pi^2}{\hbar^2} |\langle E, \mathbf{n}^- | \mu | \psi_1 \rangle | \bar{\eta}(\omega_{E,1}) | e^{-i\theta(\omega_{E,1})} b_1$$

$$+ \langle E, \mathbf{n}^- | \mu | \psi_2 \rangle | \bar{\eta}(\omega_{E,2}) | e^{-i\theta(\omega_{E,2})} b_2 |^2 \qquad (2.30)$$

where

$$\bar{\eta}(\omega) \equiv \frac{1}{2\pi} \int_{-\infty}^{\infty} dt \varepsilon(t) \exp\left[-\frac{\pi}{\hbar} A_1(E_L) \int_{-\infty}^{t} |\varepsilon(t')|^2 dt' \right] \exp(-i\omega t). \qquad (2.31)$$

We obtain a form that, although correct (within the range of validity of the SVCA) to all field strengths, resembles the weak-field bichromatic control result of Eq. (2.19). The only difference is that instead of the Fourier transform of the pulse electric field, Eq. (2.30) depends on the Fourier transform of the product of the pulse electric field and the $\exp\left[-(\pi)/\hbar(A_1(E_L)) \times \int_{-\infty}^{t} |\varepsilon(t')|^2 dt'\right]$ decaying factor, describing the depletion of the initial state(s) due to the action of the pulse.

We now examine in some detail the range of validity of the SVCA and how to go beyond it. We first study the outcome of allowing the continuous spectrum to have a single Lorentzian form:

$$A_1(E) = \frac{\Gamma_s}{(E - \mathcal{E}_s)^2 + \Gamma_s^2/4} \qquad (2.32)$$

In this case, using the fact that $t > t'$, we have

$$\exp[i\omega_1(t - t')]F_1(t - t') = f_s^+(t)f_s^-(t') \qquad (2.33)$$

where

$$f_s^{\pm}(t) = \sqrt{2\pi} \exp[\mp i\chi_s t] \qquad (2.34)$$

with

$$\chi_s \equiv \Delta_s - i\frac{\Gamma_s}{2\hbar}, \qquad \Delta_s \equiv \frac{\mathcal{E}_s - E_1}{\hbar} - \omega_1 \qquad (2.35)$$

Using Eq. (2.33) we can transform Eq. (2.23) into two coupled first-order differential equations:

$$\frac{d}{dt}b_1 = \frac{i}{\hbar}\varepsilon(t)f_s^+(t)B_s(t) \tag{2.36a}$$

$$\frac{d}{dt}B_s = \frac{i}{\hbar}\varepsilon(t)f_s^-(t)b_1(t) \tag{2.36b}$$

which can be solved in a routine way (although care must be taken to renormalize the exponentially growing $f_s^-(t)$ (function).

We next examine the more general case in which the bound-continuum dipole matrix elements $\langle\psi_1|\mu|E,\mathbf{n}^-\rangle$ can be fitted by a sum of complex poles [32]

$$\langle\psi_1|\mu|E,\mathbf{n}^-\rangle = \sum_{s=1}^{N}\frac{i\mu_{sn}\Gamma_s/2}{E-\mathcal{E}_s+i\Gamma_s/2} \tag{2.37}$$

The spectrum is now written as

$$A_1(E) = \sum_{\mathbf{n}}|\langle\psi_1|\mu|E,\mathbf{n}^-\rangle|^2 = \sum_{s's}\frac{\mu_{s's}\Gamma_s\Gamma_{s'}/4}{(E-\mathcal{E}_s+i\Gamma_s/2)(E-\mathcal{E}_{s'}-i\Gamma_{s'}/2)} \tag{2.38}$$

where $\mu_{s's} \equiv \sum_{\mathbf{n}}\mu_{sn}\mu_{s'n}^*$. This expansion is quite general, as depending on the values of $\mathcal{E}_s$ and Γ_s, it enables us to fit both a highly structured set of narrow resonances or a smooth, slightly bumpy, continuum. If we only keep the diagonal $(s = s')$ terms, $A_1(E)$ becomes a sum of Lorentzians. The off-diagonal terms allow for interferences between overlapping resonances. An illustration of a typical spectrum obtained from Eq. (2.38) is given in Fig. 2.

The Fourier transform of $A_1(E)$ (using the fact that in Eq. (2.23) $t > t'$), now becomes

$$\exp[i\omega_1(t-t')]F_1(t-t') = 2\pi\sum_{s}\overline{\mu}_s^2\exp[-i\chi_s(t-t')]$$

$$= \sum_{s}\overline{\mu}_s^2 f_s^+(t)f_s^-(t') \tag{2.39}$$

where

$$\overline{\mu}_s^2 \equiv \sum_{s'}\frac{-i\mu_{s's}\Gamma_s\Gamma_{s'}/4}{\mathcal{E}_s-\mathcal{E}_{s'}-i(\Gamma_s+\Gamma_{s'})/2)} \tag{2.40}$$

and $f_s^\pm$ are defined in Eq. (2.34).

Total Cross-Section

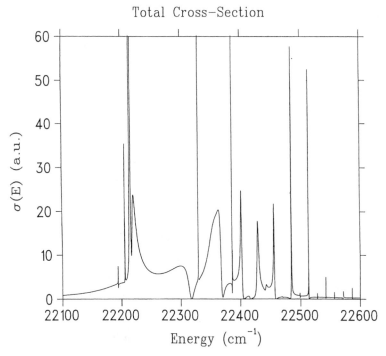

Figure 2. Absorption spectrum resulting from a sum of overlapping resonances. The model spectrum was generated by considering excitation from a ground harmonic oscillator (of frequency of $200\,\mathrm{cm}^{-1}$ and equilibrium distance 2.35 au), to an excited harmonic oscillator (of frequency of $15\,\mathrm{cm}^{-1}$, equilibrium distance 2.5 au, with vertical excitation energy of 0.1011 au), coupled by a constant coupling of $100\,\mathrm{cm}^{-1}$ to a linearly decaying potential (of slope of -1.35 au) intersecting the excited harmonic potential at $E = 87.5\,\mathrm{cm}^{-1}$ above the excited potential minimum.

With Eq. (2.39) we can transform Eq. (2.20) into a discrete set of coupled differential equations:

$$\frac{d}{dt}b_1 = \frac{i}{\hbar}\varepsilon(t)\sum_s \overline{\mu_s^2}f_s^+(t)B_s(t) \tag{2.41a}$$

$$\frac{dB_s}{dt} = \frac{i}{\hbar}\varepsilon(t)f_s^-(t)b_1(t), \qquad s = 1,\ldots,N \tag{2.41b}$$

These last two equations can be solved in a routine way using a variety of propagation methods. Once $b_1(t)$ is known, the continuum coefficient $b_{E,\mathbf{n}}$ can be computed by a straightforward quadrature, using Eq. (2.9).

Alternatively, as discussed in Appendix A, we can solve Eqs. (2.41) by the uniform or primitive semiclassical approximations.

We finally verify that the SVCA is obtained in the large Γ_s limit of the pole expansion. Parameterizing all the Γ_s widths as

$$\Gamma_s = \Gamma \gamma_s, \tag{2.42}$$

and letting $\Gamma \to \infty$, we obtain

$$F_1(t - t') = 2\pi \sum_s \overline{\mu_s^2} \exp \left(\frac{-i\mathcal{E}_s - \Gamma \gamma_s}{2} \frac{t - t'}{\hbar} \right) \xrightarrow{\Gamma \to \infty} 2\pi \hbar \bar{\mu} \delta(t - t') \tag{2.43}$$

where, using Eqs. (2.42) and (2.40), we obtain

$$\bar{\mu} = \sum_{s's} \frac{2\mu_{s's} \gamma_{s'}}{\gamma_s + \gamma_{s'}} \tag{2.44}$$

Equation (2.27) is obtained with the imaginary Rabi frequency $\Omega(t)$ given as $\Omega(t) = (\pi/\hbar)\bar{\mu}\varepsilon^2(t)$.

B. Computational Examples

We now demonstrate the outcome of the preceding model for a number of molecular continua and pulse configurations. We first study the effect of the pulse intensity on transition probabilities to a slowly varying continuum by considering a continuum composed of single broad Lorentzian of width $\Gamma_s = 2000 \text{cm}^{-1}$, excited by a 120-cm^{-1} pulse (i.e., a pulse of $\sim 80\,\text{fs}$ duration). The center frequency of the pulse is tuned to the center of the continuum ($\Delta_s = 0$). In Figs. 3a–3c we present the $|b_{E,\mathbf{n}}(t)|$ continuum coefficients as a function of time, at different intensities. The onset of off-resonance processed is typified by a nonmonotonic behavior. At off-pulse center energies, the continuum coefficients rise and fall with the pulse, and with the effect becomes more pronounced the farther away from the line center the continuum energy levels are. In the far wings of the pulse the continuum coefficients are zero at the end of the pulse, giving rise to a pure transient, otherwise known as a "virtual" state.

As we increase the field strength, the photodissociation probabilities lineshapes [given as $|b_{E,\mathbf{n}}(t = \infty)|$] broaden. This broadening is due to saturation of the continuum population, which is greater for continuum states near the pulse center than at the pulse wings. For example, for the relatively weak pulse of peak height of 0.01 au (atomic unit), shown in Fig. 3a, the photodissociation linewidth is $\sim 100 \text{ cm}^{-1}$, i.e. which is roughly that

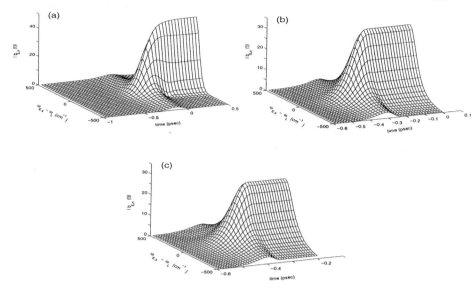

Figure 3. Temporal evolution of the continuum coefficient $b_{E,n}$ for different pulse intensities at the center of the absorption spectrum [the spectral width (Γ_s) is fixed at $2000\,\mathrm{cm}^{-1}$, the laser bandwidth is fixed at $120\,\mathrm{cm}^{-1}$, the transition dipole-moment is fixed at 2.8×10^{-3} au]: (a) peak intensity = 0.01 au; (b) peak intensity = 0.1 au; (c) peak intensity = 0.5 au.

of the pulse itself. As we increase the pulse field to the peak value of 0.1 au, the photodissociation lineshapes broaden beyond the pulse linewidth, assuming the width of $\sim 200\,\mathrm{cm}^{-1}$. At peak heights of 0.5 au the photodissociation linewidth is already $\sim 300\,\mathrm{cm}^{-1}$.

A slowly varying continuum is almost a perfect absorber. Therefore, as we increase the pulse intensity, we empty the initial state $(|\psi_1\rangle)$ faster and the dissociation is over before any recurrence can occur. For example, in the 0.01-au peak-height case, the continuum levels reach their final population by the time the pulse peaks (at $t = 0$). This time gets successively shorter as we increase the field strength. This shortening of the lifetime of the initial state, which causes the pulse to be effectively shorter, also contributes to the power broadening of the continuum lineshapes.

The situation is quite different for a *structured* continuum. Figure 4, where the strong-pulse induced transition to a narrow continuum $(\Gamma_s = 50\,\mathrm{cm}^{-1})$ is displayed, exhibits an intermediate behavior between a "flat" continuum and a discrete set of levels. We see that *centerline*, $\omega_{E,1} - \omega_1 \approx 0$, continuum levels display recurrences, or Rabi oscillations, similar, although not identical, to those of discrete two-level systems. In contrast, continuum states at the pulse wings rise and fall smoothly with the pulse, as in the slowly varying continuum case.

MOSHE SHAPIRO

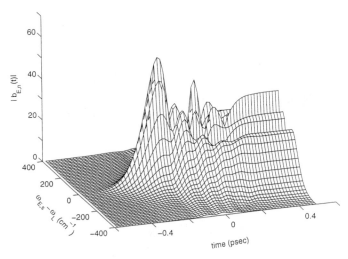

Figure 4. Temporal evolution of the continuum coefficients at the center of the absorption spectrum for a narrow absorption band, $\Gamma_s = 50\,\mathrm{cm}^{-1}$. Other parameters are pulse bandwidth $= 120\,\mathrm{cm}^{-1}$, peak intensity $= 0.05$ au., transition dipole strength $= 5.7 \times 10^{-5}$.

Extending the study of narrowband continua by looking at a continuum composed of *two* distinct diffuse features (see Fig. 5a), we now study the way a strong pulse alters the lineshapes of neighboring resonances. In Figs. 5a–c we study pulse absorption by a sum of two poles. As shown in Figs. 5b, c, as we switch on the pulse, the two initially separated lines begin to merge. For moderate laser powers (Fig. 5b), this merging is a signature of saturation; the continuum states at the center of the absorption lines cease to rise while the population of continuum states between the line centers continues to increase. At higher laser intensities (Fig. 5c), we see the effect of the Rabi cycling: The populations of the continuum states at the line centers oscillate at a higher frequency than the populations at the wings of the lines. It may happen, as shown in Fig. 5c, that the line-center continuum states execute a 2π cycle and are empty at the end of the pulse whereas continuum states away from the line centers execute only a π cycling and are highly populated. Thus, under the action of the strong laser pulse, the lines are reversed; the absorption is effectively zero at the resonance $(\mathcal{E}_s)$ positions. In addition, the optically induced interference between the lines causes the formation of "dark states" [2,42], which is the cancellation of the absorption at the end of the pulse of continuum states lying midway between the two line centers. As a result, we see three holes in the continuum populations: two transparent lines at the center of the resonances due to 2π cycling of these continuum states, accompanied by a third transparent line

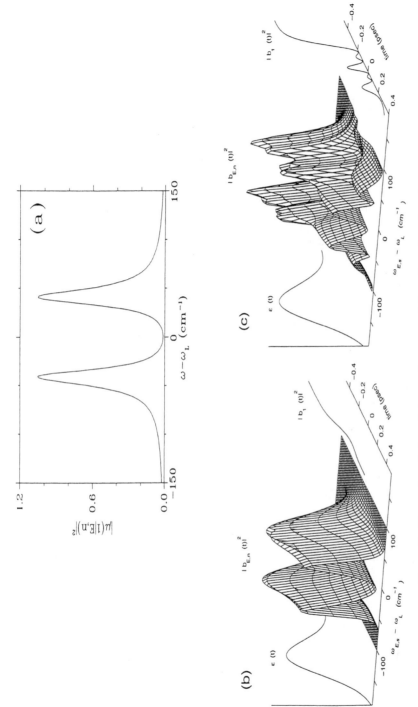

Figure 5. Temporal evolution of the continuum populations for a bound-continuum spectrum composed of two overlapping resonances: (a) weak-field absorption spectrum; (b) $|b_{E,n}|^2$ as a function of t and E, for $\varepsilon_L = 5 \times 10^{-3}$ au; (c) the same as in (b), for $\varepsilon_L = 5 \times 10^{-2}$ au.

137

residing midway between the resonances, due to destructive interference between the resonances.

III. ADIABATICITY IN TWO-PHOTON DISSOCIATION

In this section we treat the problem of resonantly enhanced two-photon dissociation by two laser pulses. In particular, we shall examine the validity of the adiabatic approximation for treating this problem.

The adiabatic approximation has proved to be a very useful and attractive way of understanding many coherent multiphoton processes. Its usefulness derives from its simplicity that often lends itself to the abolity to solve a given problem in a closed analytic form. In particular, attention has been devoted to the phenomenon of optical adiabatic passage (AP), which builds on a similar phenomenon in magnetic resonance [43]. The essence of AP is the locking of a material system onto an eigenstate of the combined field matter Hamiltonian. By forcing the system to follow such eigenstates, an external field, which is made to change slowly enough or is intense enough, guides the system from an initial state to the target state.

The development of the theory [44] of optical AP originates in the generalization of the Feynman Vernon Hellwarth (FVH) [45,46] vector model to three- and more-level systems [47–50]. The use of AP in a three-level system in which the intermediate level is higher in energy than the initial and final levels (the Λ configuration), was extensively investigated theoretically [50–56]. It was demonstrated experimentally [52–59] first by Bergmann et al. [52] that AP enables, under certain conditions, the *complete* transfer of population from one level to another.

A natural extension of three-level AP is to consider AP to a final continuum. If the behavior of such a system is similar to a Λ system, then complete population transfe to the continuum, hence complete ionization or dissociation would become possible [30].

A. Theory of Resonantly Enhanced Two-Photon Dissociation

The formulation developed in Section II can be extended to treating the two-photon dissociation by two pulses. We consider a molecule, initially (the initial time is now assigned as $t = 0$) in state $|\psi_1\rangle$, being excited to a continuum of states $|E, \mathbf{n}^-\rangle$, due to the combined action of *two* laser pulses of central frequencies ω_1 and ω_2. We assume that ω_1 is in near resonance with an intermediate bount state—$|\psi_2\rangle$ and that ω_2 is in near resonance with the transition from $|\psi_2\rangle$ to the continuum. The situation is depicted in Fig. 1b.

The total Hamiltonian in the dipole approximation is now written as

$$H_{\text{tot}} = H - 2\vec{\mu}_1 \cdot \hat{\epsilon}_1 \varepsilon_1(t)\cos(\omega_1 t) - 2\vec{\mu}_2 \cdot \hat{\epsilon}_2 \varepsilon_2(t)\cos(\omega_2 t) \qquad (3.1)$$

where $\hat{\epsilon}_1$ and $\hat{\epsilon}_2$ are the polarization directions, $\varepsilon_1(t)$ and $\varepsilon_2(t)$ are "slowly varying" electric field amplitudes, and $\vec{\mu}_1$ and $\vec{\mu}_2$ are the (electronic) transition dipoles.

Given that there is now an additional eigenstate of the nonradiative Hamiltonian, $|\psi_2\rangle$

$$[E_2 - H]|\psi_2\rangle = 0 \tag{3.2}$$

we expand the total wavefunction as

$$|\Psi(t)\rangle = b_1|\psi_1\rangle\exp\frac{-iE_1t}{\hbar} + b_2|\psi_2\rangle\exp\frac{-iE_2t}{\hbar}$$
$$+ \sum_{\mathbf{n}} \int dE b_{E,\mathbf{n}}(t)|E,\mathbf{n}^-\rangle\exp\frac{-iEt}{\hbar} \tag{3.3}$$

In analogy with Section II, we obtain a set of first-order differential equations, which is now of the form

$$\frac{d}{dt}b_1 = i\Omega_1^*(t)\exp(-i\Delta_1 t)b_2(t)$$
$$\frac{d}{dt}b_2 = i\Omega_1(t)\exp(i\Delta_1 t)b_1(t) + i\int dE \sum_{\mathbf{n}} \Omega_{2,E,\mathbf{n}}(t)\exp(-i\Delta_E t)b_{E,\mathbf{n}}(t)$$
$$\frac{d}{dt}b_{E,\mathbf{m}} = i\Omega_{2,E,\mathbf{m}}^*(t)\exp(i\Delta_E t)b_2(t) \qquad \text{for all } E \text{ and } \mathbf{m} \tag{3.4}$$

where

$$\Omega_1(t) \equiv \frac{\langle\psi_2|\mu_1|\psi_1\rangle\varepsilon_1(t)}{\hbar}, \qquad \Omega_{2,E,\mathbf{m}}(t) \equiv \frac{\langle\psi_2|\mu_2|E,\mathbf{m}^-\rangle\varepsilon_2(t)}{\hbar},$$
$$\Delta_1 \equiv \frac{E_2 - E_1}{\hbar} - \omega_1, \qquad \Delta_E \equiv \frac{E - E_2}{\hbar} - \omega_2. \tag{3.5}$$

As in Section II, we eliminate the continuum equations by substituting the formal solution of of Eq. (3.4c)

$$b_{E,\mathbf{n}}(t) = i\int_0^t dt' \Omega_{2,E,\mathbf{n}}^*(t')\exp(i\Delta_E t')b_2(t') \tag{3.6}$$

into Eq. (3.4b), to obtain

$$\frac{d}{dt}b_2 = i\Omega_1(t)\exp(i\Delta_1 t)b_1(t) - \sum_{\mathbf{n}} \int dE \Omega_{2,E,\mathbf{n}}(t)\exp(-i\Delta_E t)$$
$$\times \int_0^t dt' \Omega_{2,E,\mathbf{n}}^*(t')\exp(i\Delta_E t')b_2(t') \tag{3.7}$$

By invoking the SVCA [Eq. (2.24)] we obtain the two-photon analog of Eq. (2.26):

$$\frac{db_2}{dt} = i\Omega_1(t)\exp(i\Delta_1 t)b_1(t) - \Omega_2(t)b_2(t) \tag{3.8}$$

where

$$\Omega_2(t) = \pi \sum_{\mathbf{n}} |\frac{\langle E_L, \mathbf{n}^- |\mu_2|\psi_2\rangle \varepsilon_2(t)|^2}{\hbar} \tag{3.9}$$

B.   The Adiabatic Approximation for a Final Continuum Manifold

Equations (3.8) and (3.7), can be expressed in matrix notation

$$\frac{d}{dt}\mathbf{b} = i\mathbf{H} \cdot \mathbf{b}(t) \tag{3.10}$$

where

$$\mathbf{b} \equiv (\exp(i\Delta_1 t)b_1, b_2) \tag{3.11}$$

and

$$\mathbf{H} = \begin{pmatrix} \Delta_1 & \Omega_1^* \\ \Omega_1 & i\Omega_2 \end{pmatrix} \tag{3.12}$$

Assuming that Ω_1 is real, we obtain the adiabatic solutions to Eq. (3.10) by diagonalizing the $\mathbf{H}$ matrix

$$\mathbf{U} \cdot \mathbf{H} = \hat{\mathcal{E}} \cdot \mathbf{U} \tag{3.13}$$

where $\hat{\mathcal{E}}$, the eigenvalue matrix, is given as

$$\mathcal{E}_{1,2} = \frac{1}{2}\{\Delta_1 + i\Omega_2 \pm [(\Delta_1 - i\Omega_2)^2 + 4\Omega_1^2]^{1/2}\} \tag{3.14}$$

Since $\mathbf{H}$ is a complex symmetric matrix, it is diagonalizable by a complex orthogonal matrix $\mathbf{U}$, satisfying the equation

$$\mathbf{U}(t) \cdot \mathbf{U}^\mathsf{T}(t) = \mathbf{I} \tag{3.15}$$

Where $\mathbf{U}$ must be nonunitary on physical grounds to allow flux loss to the continuum.

In the 2×2 case the U matrix can be parameterized in terms of a *complex* "mixing angle" θ

$$\mathsf{U} = \begin{pmatrix} \cos\theta & \sin\theta \\ -\sin\theta & \cos\theta \end{pmatrix} \qquad (3.16)$$

where

$$\theta(t) = \frac{1}{2}\arctan\left(\frac{2\Omega_1}{i\Omega_2 - \Delta_1}\right) \qquad (3.17)$$

Operating with $\mathsf{U}(t)$ on Eq. (3.10), and defining

$$\mathbf{a}(t) = \mathsf{U}(t) \cdot \mathbf{b}(t) \qquad (3.18)$$

we obtain

$$\frac{d}{dt}\mathbf{a} = \{i\hat{\mathcal{E}}(t) + \mathsf{A}\} \cdot \mathbf{a} \qquad (3.19)$$

where

$$\mathsf{A} \equiv \frac{d\mathsf{U}(t)}{dt} \cdot \mathsf{U}^{\mathsf{T}} = \begin{pmatrix} 0 & \dot{\theta} \\ -\dot{\theta} & 0 \end{pmatrix} \qquad (3.20)$$

is the "nonadiabatic" coupling matrix.

The adiabatic approximation amounts to ignoring A. This can be done whenever the rate of change of U with time is slow. Equation (3.19) then becomes

$$\frac{d}{dt}\mathbf{a} = i\hat{\mathcal{E}}(t) \cdot \mathbf{a}(t) \qquad (3.21)$$

yielding the adiabatic solutions

$$\mathbf{a}(t) = \exp\left\{i\int_0^t \hat{\mathcal{E}}(t')dt'\right\}\mathbf{a}(0) \qquad (3.22)$$

Using Eqs. (3.11) and (3.18), and imposing the initial condition, $\mathbf{b}(0) = (1,0)$ we obtain for the $b_1(t)$ and $b_2(t)$ coefficients

$$b_1(t) = \{U_{1,1}(t)\exp[i\int_0^t \mathcal{E}_1(t')dt']U_{1,1}(0) + U_{2,1}(t)$$

$$\times \exp[i\int_0^t \mathcal{E}_2(t')dt']U_{1,2}(0)\}\exp(-i\Delta_1 t),$$

$$b_2(t) = U_{1,2}(t)\exp[i\int_0^t \mathcal{E}_1(t')dt']U_{1,1}(0) + U_{2,2}(t)\exp[i\int_0^t \mathcal{E}_2(t')dt']U_{1,2}(0)$$

$$\qquad (3.23)$$

If both lasers are assumed to be off initially, that is, if $\varepsilon_1(0) = \varepsilon_2(0) = 0$, we have that $\theta(0) = 0$. Hence $U_{1,1}(0) = 1, U_{1,2}(0) = 0$, and

$$\begin{pmatrix} b_1(t) \\ b_2(t) \end{pmatrix} = \begin{pmatrix} \exp(-i\Delta_1 t)\cos\theta(t) \\ \sin\theta(t) \end{pmatrix} \exp\left\{ i \int_0^t \mathcal{E}_1(t')dt' \right\} \qquad (3.24)$$

Once $b_2(t)$, is known, the (channel-specific) continuum coefficients $b_{E,n}(t)$ are obtained directly via Eq. (3.6).

C. Going Beyond the Adiabatic Approximation

As discussed in Appendix B, because of the presence of a continuum, there is at least one point in time at which the adiabatic approximation breaks down. Mathematically speaking, this arises because of the existence of an imaginary Rabi frequency that makes the denominator of Eq. (B.7) vanish whenever $2\Omega_1(t) = \Omega_2(t)$. Thus, unless we also have that $(d\Omega_1/dt) = (d\Omega_2/dt)$, the adiabatic condition cannot be fulfilled at that point. If U is nearly diagonal, the failure of the adiabatic approximation at one timepoint is of very little consequence. If, however, U has substantial off-diagonal elements, the timepoints satisfying the $2\Omega_1(t) = \Omega_2(t)$ condition may cause the true solutions to deviate substantially from the adiabatic ones. It is possibel to salvage the situation and make use of the adiabatic solutions by adopting the following procedure. Using the adiabatic solutions [Eq. (3.22)] as zero-order approximants

$$\mathbf{a}^{(0)}(t) = \exp\left[\frac{i}{\hbar} \int_0^t \hat{\mathcal{E}}(t')dt' \right] \cdot \mathbf{a}^{(0)}(0) \qquad (3.25)$$

we improve the $\mathbf{a}(t)$ vector as

$$\frac{d}{dt}\mathbf{a}_v^{(1)} = \left\{ i\mathcal{E}_v(t) + \sum_{v'} A_{v,v'}\mathbf{a}_{v'}^{(0)}/\mathbf{a}_v^{(0)} \right\} \mathbf{a}_v^{(1)}. \qquad (3.26)$$

The second iterant is obtained by replacing $\mathbf{a}^{(0)}$ of Eq. (3.26) with the resulting $\mathbf{a}^{(1)}$ solution, Iteration, whose general step is written as

$$\mathbf{a}_v^{(n+1)}(t) = \mathbf{a}_v^{(0)}(t)\exp\left[\int_0^t dt' \frac{1}{\mathbf{a}_v^{(n)}(t')} \sum_{v'} A_{v,v'}(t')\mathbf{a}_{v'}^{(n)}(t') \right] \qquad (3.27)$$

is continued until $|\mathbf{a}_v^{(n+1)}(t) - \mathbf{a}_v^{(n)}(t)| < \epsilon$, where ϵ is a preset tolerance, at which point it is easily verified that Eqs. (3.19) are satisfied (to an accuracy determined by the value of ϵ).

In the 2×2 case, we can write the iteration step in terms of the mixing angle as

$$a_{1,2}^{(n+1)}(t) = a_{1,2}^{(0)}(t)\exp\left[\pm i \int^t \dot{\theta}(t') \frac{a_{2,1}^{(n)}(t')}{a_{1,2}^{(n)}(t')} dt'\right] \tag{3.28}$$

In numeric applications precaution should be taken so that none of the solutions $a_i^{(n)}(t)$ is identically zero, because in that case the scheme will diverge. For example, with the initial condition $|\Psi(0)\rangle = |\psi_1\rangle$, one has $a_1(0) = 1, a_2(0) = 0$ and therefore $a_2^{(0)}(t)$ is identically zero. To avoid divergence, the iteration scheme should be executed with two linearly independent sets of initial conditions, both giving nonzero solutions, such as, $a_2^+(0) = 1, a_2^+(0) = 1$ and $a_1^-(0) = 1, a_2^-(0) = -1$. The solution satisfying the correct initial conditions is then given as

$$a_i(t) = \frac{a_i^+(t) + a_i^-(t)}{2}$$

D. Computational Examples

We illustrate the procedure described in Section III.C by studying the case of resonantly enhanced two-photon dissociation by two Gaussian laser pulses. The energy gap between the two bound states, $E_2 - E_1$, was chosen to be $20,000\,\text{cm}^{-1}$, and the gap between E_2 and the continuum threshold was chosen as $19,000\,\text{cm}^{-1}$. The goodness and the rate of convergence of the iterative procedure was judged by calculating the distance:

$$d \equiv \lim_{t\to\infty} \sum_{i=1}^{2} \frac{1}{t} \int_0^t |a_i^{(n+1)}(t') - a_i^{(n)}(t')| dt' \tag{3.29}$$

The iteration was stopped when $d < 10^{-4}$. Reference solutions were obtained by direct numerical integration of Eq. (3.10) with a Runge–Kutta–Merson (RKM) algorithm.

Having obtained the $a(t)$ vector, the probabilities of observing $|\psi_1\rangle$ and $|\psi_2\rangle$, given respectively as

$$P_1 \equiv |b_1|^2 = |a_1(t)\cos\theta(t) + a_2(t)\sin\theta(t)|^2$$
$$P_2 \equiv |b_2|^2 = |-a_1(t)\sin\theta(t) + a_2(t)\cos\theta(t)|^2 \tag{3.30}$$

and the overall dissociation probability

$$P_d \equiv 1 - (P_1 + P_2) \tag{3.31}$$

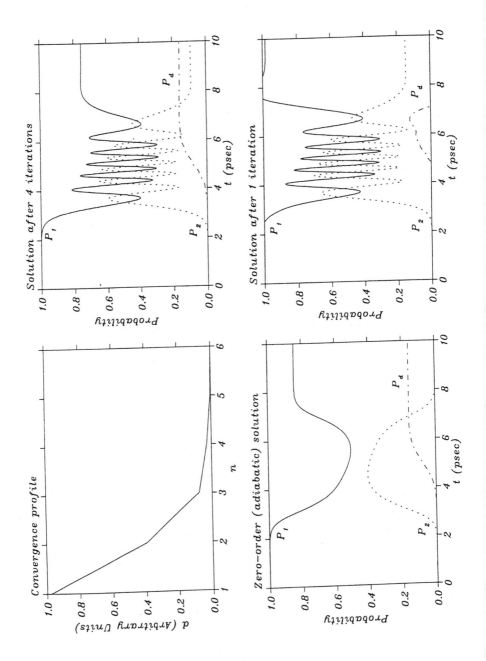

were computed as a function of time. A typical convergence profile of the iteration procedure is shown in the upper left panel of Fig. 6. Also shown are the adiabatic (zero-order) solutions (lower left), the solution after one iteration (lower right), and the solution after four iteration steps (upper right). We see that the convergence is extremely rapid. In fact, after only one iteration an oscillatory pattern, very similar to the final solution, is seen to emerge. This pattern can be no means be depicted by the adiabatic single-channel solution given in Eq. (3.24), as it is a result of the interference between the two adiabatic channels.

Adiabatic, iterative, and numeric solutions (by the RKM method) for coincident laser pulses are shown in Figs. 7–9. For wide ($>30\,\mathrm{ps}$), slowly varying pulses (Fig. 7), the adiabatic approximation is satisfactory and no iterative improvement is necessary. As the pulses get shorter than $\sim 30\,\mathrm{ps}$, the numeric solutions show an oscillatory pattern (resulting from nonadiabatic channel interference) that is completely absent in the adiabatic solution. As shown in Fig. 8 for a 10-ps pulse, the iterative approach is very successful in generating the correct oscillatory pattern, even after very few iterations.

The effect of decreasing the detuning Δ_1 is demonstrated in Fig. 9. All pulse parameters but Δ_1 which $1\,\mathrm{cm}^{-1}$, are identical to those of Fig. 7. Comparing the adiabatic and the RKM solutions, we see that a decrease of Δ_1 by an order of magnitude results in the complete breakdown of adiabaticity, in accordance with Eq. (B.25), even though $\Omega^0 \Delta\tau \gg 1$ (as suggested by the large number of oscillations during pulse time). As in the two previous examples, the iteration scheme converges well to the RKM solution.

IV. RESONANTLY ENHANCED TWO-PHOTON ASSOCIATION

The set of equations used above for two-photon dissociation problems can be used to solve another problem – that of resonantly enhanced two-photon *association*, depicted schematically in Fig. 1c.

The significance of resonantly enhanced two-photon association stems from the possibility of using it to form ultracold molecules, the production of which still presents an imposing challenge [60,61]. Laser cooling

---

Figure 6. Convergence measure of the iterative scheme versus iteration number (upper left), along with populations of the initial (P_1), intermediate (P_2), and continuum (P_d) states versus time, in the zeroth-order (adiabatic) approximation (lower left), after one iterative step (lower right) and after four iterations (upper right). Pulses were coincident and detuned ($\Delta_1 = 10\,\mathrm{cm}^{-1}$).

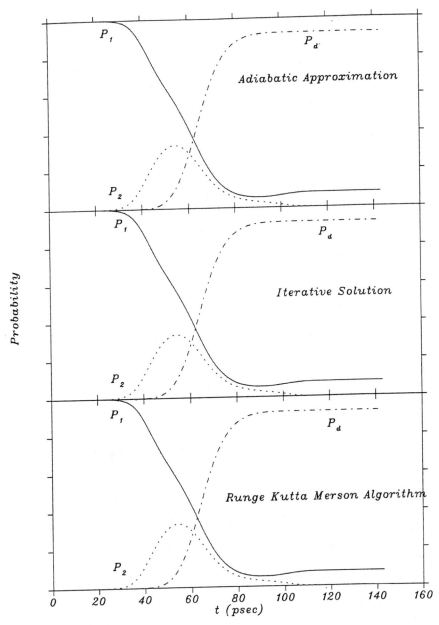

Figure 7. Comparison of the adiabatic approximation solution (above), iterative solution (middle), and direct integration solution (below) for 100-ps, coincident pulses. Detuning is the same as in Fig. 6.

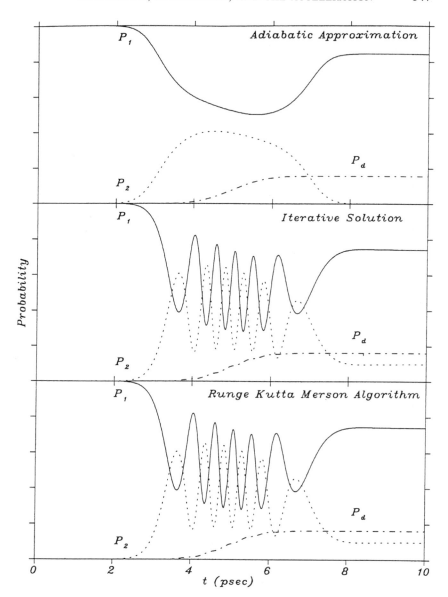

Figure 8.   The same as in Fig. 7, for 10-ps pulses.

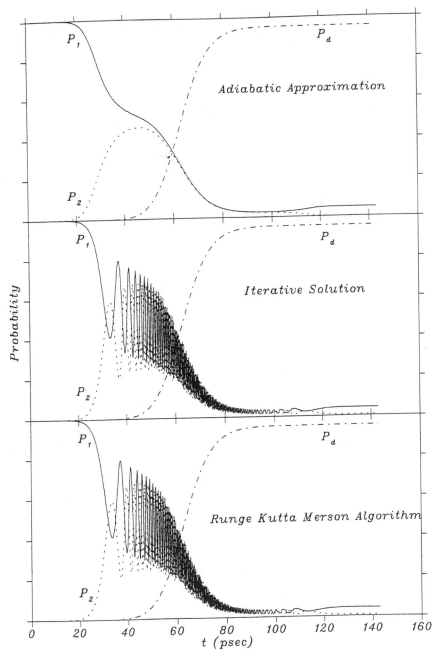

Figure 9. The same as in Fig. 7, with $\Delta_1 = 1\text{cm}^{-1}$.

schemes that work for atoms [62,63] fail for molecules, due mainly to the existence of many near-resonance lines and the (not uncorrelated)-fact that in addition to translation, other degrees of freedom (rotations, vibrations, etc.) must be cooled. Using off-resonance inhomogeneous effects allows aligning [64–70] and even trapping [71] of molecules, but such off-resonant processed do not lead to cooling.

Rather than cool existing molecules, one can try to *synthesize* cold molecules via the association of cold atoms. The molecules formed as a result of photoassociation are expected to maintain the translational temperature of the recombining atoms, because the center-of-mass motion remains unchanged in the process (except for the little momentum imparted by the photon). This idea was first proposed by Julienne et al. [72,73], who envisioned a multistep association, first involving bound-continuum excitation of translational continuum states of cold trapped atoms to an excited vibrational level in an excited electronic molecular state, followed by a bound–bound spontaneous emission to the ground electronic state.

An undesired feature of this scheme is that due to the spontaneous nature of the second step, the molecules thus produced might end up populating a large range of vibrational levels. In some cases, because of favorable Franck–Condon factors this need not be the case, as shown experimentally by Pillet et al. [74]. In general, however, the use of *stimulated* emission is preferable because in this way one can tune in the final molecular state of preference [75].

A.   Theory of Photoassociation of a Coherent Wavepacket

We consider a pair of colliding atoms described by multiple continua of scattering states $|E, \mathbf{n}^+\rangle$, where $\mathbf{n}$ incorporates the quantum indices specifying the electronic states of the separated atoms and E is the total collision energy. The $+$ notation signifies, in accordance with scattering theory conventions, that, contrary to the dissociation cases studied above, the *initial* internal state of the fragments is known.

As depicted in Fig. 1c for a Λ-type configuration (i.e., a configuration in which the intermediate-resonance $|\psi_2\rangle$ is higher in energy than the initial $|E, \mathbf{n}^+\rangle$ or final $|\psi_1\rangle$ states), under the combined action of two laser pulses, of central frequencies ω_1 and ω_2, the atoms may recombine to form a bound molecular state $|\psi_1\rangle$. We assume that ω_2 is in near resonance with the transition from the continuum to an intermediate bound state $|\psi_2\rangle$ and that ω_1 is in near resonance with the transition from $|\psi_2\rangle$ to $|\psi_1\rangle$.

With the total Hamiltonian of the system given by Eq. (3.1) and the material wavefunction of the system expanded as in Eq. (3.3), a set of first-order differential equations for the expansion coefficients, essentially identical to Eq. (3.4), is obtained. The only difference is that the input bound-continuum dipole matrix elements involve the $+$ rather than the $-$ continuum states.

In the photoassociation case, contrary to the dissociation cases discussed above, the continuum is initially populated. Hence the formal solution of Eq. (3.4c) is now of the form

$$b_{E,\mathbf{n}}(t) = b_{E,\mathbf{n}}(t = 0) + i \int_0^t dt' \Omega^*_{2,E,\mathbf{n}}(t') \exp(i\Delta_E t') b_2(t') \qquad (4.1)$$

Substituting this solution into Eq. (3.4b), we obtain

$$\frac{d}{dt} b_2 = i\Omega_1(t) \exp(i\Delta_1 t) b_1(t) + i \sum_{\mathbf{n}} \int dE\, \Omega_{2,E,\mathbf{n}}(t) \exp(-i\Delta_E t) b_{E,\mathbf{n}}(t = 0)$$

$$- \sum_{\mathbf{n}} \int dE \int_0^t dt' \Omega_{2,E,\mathbf{n}}(t) \Omega^*_{2,E,\mathbf{n}}(t') \exp[-i\Delta_E(t - t')] b_2(t') \qquad (4.2)$$

If the molecular continuum is unstructured (as in the Na–Na continuum as threshold energies, where the bound-continuum dipole matrix elements vary with energy by less than 1% over a typical nanosecond-pulse bandwidth), we can safely invoke the SVCA and replace the energy-dependent bound-continuum dipole matrix elements by their value at the pulse center, given (in the Λ configuration of Fig. 1c) as $E_L = E_2 - \hbar\omega_2$. We obtain

$$\frac{db_2}{dt} = i\Omega_1(t) \exp(i\Delta_1 t) b_1(t) - \Omega_2(t) b_2(t) + iF(t) \qquad (4.3)$$

where $\Omega_2(t)$ is defined in Eq. (3.9). The (known) source term $F(t)$ is given as

$$F(t) = \frac{\varepsilon_2(t)\bar{\mu}_2(t)}{\hbar} \qquad (4.4)$$

where

$$\bar{\mu}_2(t) = \sum_{\mathbf{n}} \int dE \langle \psi_2 | \mu_2 | E, \mathbf{n}^+ \rangle \exp(-i\Delta_E t) b_{E,\mathbf{n}}(t = 0)$$

Equations (4.3) and Eq. (3.7), can be expressed in matrix notation as

$$\frac{d}{dt} \mathbf{b} = i\{\mathbf{H} \cdot \mathbf{b}(t) + \mathbf{f}\} \qquad (4.5)$$

where

$$f(t) \equiv \begin{pmatrix} 0 \\ F(t) \end{pmatrix} \tag{4.6}$$

b is defined in Eq. (3.11) and H by Eq. (3.12).

The term $P(t)$, the *net association rate*, defined as $(d/dt)(|b_1|^2 + |b_2|^2)$, that is the rate of population change in the bound manifold, can be written, using Eq. (4.5) and its complex conjugate, as

$$
\begin{aligned}
P(t) &= \frac{d}{dt}(|b_1|^2 + |b_2|^2) = \frac{d}{dt}|b|^2 = b^\dagger \cdot (\frac{d}{dt}b) + (\frac{d}{dt}b^\dagger) \cdot b \\
&= i\{b^\dagger \cdot (H - H^\dagger) \cdot b + b^\dagger \cdot f - f^\dagger \cdot b\} \\
&= 2\,\mathrm{Im}\,[F^*(t)b_2(t)] - 2\Omega_2(t)|b_2(t)|^2
\end{aligned}
\tag{4.7}
$$

The first term in Eq. (4.7) represents the association rate

$$P_{\mathrm{rec}}(t) \equiv 2\mathrm{Im}[F^*(t)b_2(t)] \tag{4.8}$$

and the second term the backdissociation rate

$$P_{\mathrm{diss}}(t) \equiv 2\Omega_2(t)|b_2(t)|^2 \tag{4.9}$$

We see that the net association rate is, indeed, the difference between the association and backdissociation rates.

As in Eq. (3.13), we can solve Eq. (4.5) adiabatically by diagonalizing the H matrix. Operating with $U(t)$ on Eq. (4.5), with $a(t)$ defined as in Eq. (3.18), we obtain

$$\frac{d}{dt}a = \{i\hat{\mathcal{E}}(t) + A\} \cdot a + ig \tag{4.10}$$

where the source vector g is given as

$$g(t) = \begin{pmatrix} F(t)U_{1,2}(t) \\ F(t)U_{2,2}(t) \end{pmatrix} = \begin{pmatrix} F(t)\sin\theta(t) \\ F(t)\cos\theta(t) \end{pmatrix} \tag{4.11}$$

When we invoke the adiabatic approximation, which amounts to ignoring the A matrix of Eq. (3.20), we obtain, from Eq. (4.10)

$$\frac{d}{dt}a = i\hat{\mathcal{E}}(t) \cdot a(t) + ig(t) \tag{4.12}$$

In the association process the initial conditions are such that

$$a(t = 0) = 0 \tag{4.13}$$

hence the adiabatic solutions are of the form

$$a(t) = v(t) \cdot \phi(t) \tag{4.14}$$

where

$$v(t) = \exp\left\{ i \int_0^t \hat{\mathcal{E}}(t') dt' \right\} \tag{4.15}$$

and

$$\phi(t) = i \int_0^t v^{-1}(t') \cdot g(t') dt' \tag{4.16}$$

Using Eqs. (3.18) and (3.11), we obtain for the $b_1(t)$ and $b_2(t)$ coefficients:

$$
\begin{aligned}
b_1(t) &= i \left\{ \cos\theta(t) \int_0^t \exp\left[i \int_{t'}^t \mathcal{E}_1(t'') dt'' \right] F(t') \sin\theta(t') dt' \right. \\
&\quad \left. - \sin\theta(t) \int_0^t \exp\left[i \int_{t'}^t \mathcal{E}_2(t'') dt'' \right] F(t') \cos\theta(t') dt' \right\} \exp(-i\Delta_1 t) \\
b_2(t) &= i \left\{ \sin\theta(t) \int_0^t \exp\left[i \int_{t'}^t \mathcal{E}_1(t'') dt'' \right] F(t') \sin\theta(t') dt' \right. \\
&\quad \left. + \cos\theta(t) \int_0^t \exp\left[i \int_{t'}^t \mathcal{E}_2(t'') dt'' \right] F(t') \cos\theta(t') dt' \right\}
\end{aligned} \tag{4.17}
$$

Given $b_2(t)$, the (channel-specific) continuum coefficient $b_{E,n}(t)$ are obtained directly via Eq. (4.1).

It is instructive to study the adiabatic solution when there is insignificant temporal overlap between the two laser pulses. Assuming in that case that the ω_2 pulse comes before the ω_1 pulse, we have during the ω_2 pulse that $\varepsilon_2 \ll \varepsilon_1$, hence that, $\mathcal{E}_1 = \Delta_1, \mathcal{E}_2 = i\Omega_2$ and $\theta(t) = 0$. Substituting these values into Eq. (4.17), we obtain the following during the ω_2 pulse:

$$b_1(t) = 0; \qquad b_2(t) = i \int_0^t \exp\left\{ -\int_{t'}^t \Omega_2(t'') dt'' \right\} F(t') dt' \tag{4.18}$$

From Eq. (4.4) it is clear that the source term $F(t)$ is linearly proportional to the pulse amplitude. On the other hand, since $\Omega_2 > 0$ and $t' < t$, the $\exp\{-\int_{t'}^{t}\Omega_2(t'')dt''\}$ factor (depicting dissociation back to the continuum) decays exponentially with increasing intensity. Thus, the mere increase in laser power does not necessarily increase the association yield. There exists some optimal intensity, beyond which the association probability decreases. In the next section we probe some pulse configuration for a realistic case of photoassociation.

B. Photoassociation of a Coherent Na + Na Wavepacket

We now apply the preceding formulation to the study of the pulsed photoassociation of a coherent wavepacket of cold Na atoms [75]. We describe the colliding atoms by an (energetically narrow) normalized Gaussian packet of $J = 0$ radial waves:

$$|\Psi(t = 0)\rangle = \int dE b_E(t = 0)|E, 3s + 3s\rangle \qquad (4.19)$$

where $|E, 3s + 3s\rangle$ are the free Na–Na s waves, and

$$b_E(t = 0) = (\delta_E^2\pi)^{-1/4}\exp\left\{\frac{-(E - E_{col})^2}{2\delta_E^2} + i\Delta_E t_0\right\} \qquad (4.20)$$

where t_0 marks the instant of maximum overlap of the Na+Na wavepacket with the $|\psi_2\rangle$ state. In our simulations we have chosen E_{col}, the mean collision energy, to vary as $E_{col} = 0.00695$–$0.0695\,cm^{-1} \approx 0.01$–$0.1K$ and the wavepacket widths, δ_E, to vary as $\delta_E = 10^{-4}$–$10^{-3}\,cm^{-1}$.

Although the initial ensemble may contain many radial waves, the fact that the bandwidth of the wavepacket of Eq. 4.20 and that of the lasers used is much smaller than the rotational spacings allows one to tune the ω_2 laser frequency such that only the $E_2(v, J = 1) \leftarrow E$ transitions are in resonance with the laser central frequency. As a result, for parallel transitions [as in the $Na_2(A) \leftarrow Na_2(X)$ case], only the $J = 0$ (and the $J = 2$) partial waves are absorbed by the ω_2 pulse, resulting in a natural impact-parameter selectivity.

As depicted in Figs. 1c and 10, the combined effect of the two laser pulses of central frequencies ω_2 and ω_2 [taken to be in resonance with the $(X^1\sum_g^+, v = 0, J = 0)$ to $(A^1\sum_u^+, v' = 34, J = 1)$ transition], is the transfer of population from the continuum to be the ground vibrotational (vibrational–rotational) state $(X^1\sum_g^+, v = 0, J = 0)$, with the bound $(A^1\sum_u^+, v' = 34, J = 1)$ state acting as an intermediate resonance.

In order to perform the calculation we need to compute the dipole matrix of Eq. (3.5). Given the ab initio electronic dipole moments and potential

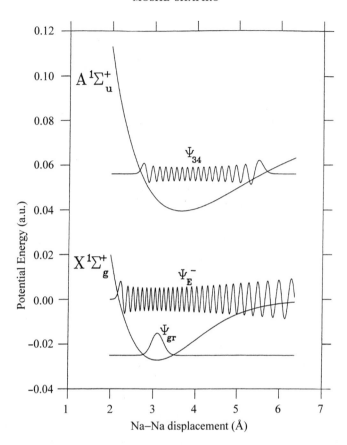

Figure 10. Potentials and vibrational wavefunctions used in simulation of the $Na + Na$ two-photon association.

curves of Schmidt et al. [75], depicted in Fig. 10, the bound eigenfunctions and eigenenergies are obtained using the renormalized Numerov method [76], and the continuum wavefunctions obtained, to excellent accuracy, in terms of the uniform Airy functions [77–82]. The bound–bound dipole matrix elements are calculated using Simpson quadrature and the bound-continuum dipole matrix elements are evaluated using a high-order Gauss–Legendre quadrature.

As we verified numerically, the $\langle \psi_1 | \mu_2 | E, \mathbf{n}^+ \rangle$ bound-continuum matrix elements do not vary appreciably with kinetic energy over the pulse spectral band-width, thus justifying the use of the SVCA of Eq. (2.24) in which $\langle \psi_2 | \mu_2 | E, \mathbf{n}^+ \rangle$ is replaced by $\langle \psi_2 | \mu_2 | E_L, \mathbf{n}^+ \rangle$ at the average energy E_L. We can therefore rewrite Eq. (4.4) as

$$F(t) = \Omega_{E_L}(t) \int dE \exp[-i\Delta_E t] b_E(t = 0) \tag{4.21}$$

where

$$\Omega_{E_L}(t) \equiv \varepsilon_2(t) \sum_{\mathbf{n}} \frac{\langle \psi_2 | \mu_2 | E_L, \mathbf{n}^+ \rangle}{\hbar} \tag{4.22}$$

and obtain from Eqs. (4.20) and (4.21)

$$F(t) = (4\delta_E^2 \pi)^{1/4} \Omega_{E_L}(t) \exp\left\{ -\frac{\delta_E^2(t - t_0)^2}{2\hbar^2} - i\Delta_{E_L}(t - t_0) \right\} \tag{4.23}$$

Choosing Gaussian pulses of the form

$$\varepsilon_{1,2}(t) = \varepsilon_{1,2}^0 \exp\left\{ -\frac{(t - t_2)^2}{\Delta t_2^2} \right\} \tag{4.24}$$

we can write Eq. (4.23) as

$$F(t) = \frac{(4\delta_E^2 \pi)^{1/4}}{\hbar} \sum_{\mathbf{n}} \langle 2 | \mu_2 | E_L, \mathbf{n}^+ \rangle \varepsilon_2^0$$

$$\times \exp\left\{ -\frac{(t - t_2)^2}{\Delta t_2^2} - i\Delta_{E_{\text{col}}}(t - t_0) - \frac{\delta_E^2(t - t_0)^2}{2\hbar^2} \right\} \tag{4.25}$$

Having computed all the input matrix elements, we can solve the dynamical equations using the adiabatic solutions [Eqs. (4.17)]. This method of solution is checked against direct integration of the full nonadiabatic equation [Eq. (3.10)] by the RKM algorithm. For $\varepsilon_{1,2}^0$ and $\Delta t_{1,2}$ pulse parameters of relevance to this work, the adiabatic solutions were found to be practically *indistinguishable* from the numerically exact RKM solutions.

To minimize spontaneous emission losses, we hve concentrated on the "counterintuitive" [52] pulse sequence scheme. Contrary to the "intuitive" scheme (see Fig. 1c), where the "dump" $\varepsilon_1(t)$ pulse is applied *after* the "pump" $\varepsilon_2(t)$ pulse, in the "counterintuitive" scheme, the "dump" pulse is applied *before* the "pump" pulse. As shown by Bergmann et al. [52] for bound–bound Λ-type configurations, the "counterintuitive" scheme enables 100% population transfer from the initial state to the final state without ever populating the intermediate resonance, by an adiabatic passage (AP) process.

As discussed in Section III and Appendix B, the situation in photo-dissociation, for which the final state is in the continuum, is not so straight-forward [30,33]. Although adiabaticity may still work in this case [33], the ability to execute AP, in which population is transferred *monotonically* from one level to another without populating the intermediate level, although possible [30], is more limited. In the photodissociation case we show, however, that with the proper choice of pulse parameters, it is possible to transfer the entire population contained in the continuum wavepacket to the ground state, while keeping at all times the intermediate state population low. The process depicted here, although not a perfect adiabatic *passage*, is nevertheless adiabatic, since the adiabatic solutions [Eqs. (4.17)] to Eq. (3.10) are in perfect agreement with the exact-numeric RKM solutions.

A typical population evolution is shown in Fig. 11, which is obtained with pulse intensities of order $10^8\,\text{W/cm}^2$ and pulse durations of several nano-seconds. Such pulse intensities are sufficiently small to avoid unwanted pho-toionization, photodissociation, and other strong-field parasitic processes.

The disadvantage of working with pulses is that only atoms sufficiently close to one another during the laser pulse can be recombined. In other words, the initial wavepacket of continuum states considered here must be synchronized in time and in duration with the recombining pulses. It is therefore of interest to see whether it is possible to employ longer pulses (of lower intensity) in order to increase the absolute number of recombining atoms and the overall duty cycle of the process.

Use of pulses of different intensity and different durations is illustrated for the "counterintuitive" scheme in Figs. 12a,b, where the rates of association [P_{rec} of Eq. (4.8)], backdissociation [P_{diss} of Eq. (4.9)], and the net association rate [$P(t)$ of Eq. (4.7)], are plotted as a function of time. A short-pulse case is shown in Fig. 12a and a long-pulse case, with a more spread-out wavepacket, is shown in Fig. 12b. Both figures appear identical, although in Fig. 12b the x-axis is scaled up by a factor of 10 and the y-axis is scaled down by a factor of 10.

The behavior demonstrated in Figs. 12 is due to the existence of an exact scaling relations in Eq. (4.5). This scaling is obtained when the initial wavepacket width and the pulse intensities are scaled down as

$$\delta_E \to \frac{\delta_E}{s}, \qquad \varepsilon_1^0 \to \frac{\varepsilon_1^0}{s}, \qquad \varepsilon_2^0 \to \frac{\varepsilon_2^0}{\sqrt{s}} \qquad (4.26)$$

and the durations of both pulses are scaled up as

$$\Delta t_{1,2} \to \Delta t_{1,2} s \qquad (4.27)$$

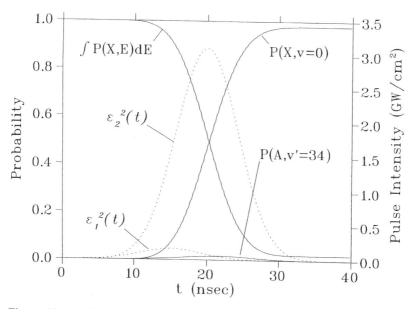

Figure 11. Results of the "counterintuitive" pulse sequence. Shown are integrated population of the wavepacket of initial continuum states, the population of the $v = 34, J = 1$ intermediate state and the population of the $v = 0, J = 0$ final ground states versus time. Dashed lines are the intensity profiles of the two Gaussian pulses whose central frequencies are $\omega_1 = 18{,}143.775 \text{cm}^{-1}$ and $\omega_2 = 12{,}277.042 \text{cm}^{-1}$ (i.e., $\Delta_1 = \Delta_{E_{\text{col}}} = 0$). The maximum intensity of the dump pulse is $1.6 \times 10^8 \text{W/cm}^2$, and that of the pump pulse $3.1 \times 10^9 \text{W/cm}^2$. Both pulses last 8.5 ns. The pump pulse peaks at the peak of the Na + Na wavepacket ($t_0 = 20$ nsec) and the dump pulse peaks 5 ns before that time. The initial kinetic energy of the Na atoms is 0.0695cm^{-1} (or 0.1 K).

It follows from Eqs. (4.25) and (3.9) that under these transformations

$$F(t) \rightarrow \bar{F}_2(t) = \frac{F(t/s)}{s}; \qquad \Omega_{1,2}(t) \rightarrow \bar{\Omega}_{1,2}(t) = \frac{\Omega_{1,2}(t/s)}{s}$$

and Eq. (4.5) becomes

$$\frac{d}{dt/s}\bar{\mathbf{b}} = i\left\{\mathsf{H}\left(\frac{t}{s}\right) \cdot \bar{\mathbf{b}} + f\left(\frac{t}{s}\right)\right\} \tag{4.28}$$

where $\bar{\mathbf{b}}$ denotes the vector of solutions of the scaled equations. We see that the scaled coefficients at time t are identical to the unscaled coefficients at time t/s.

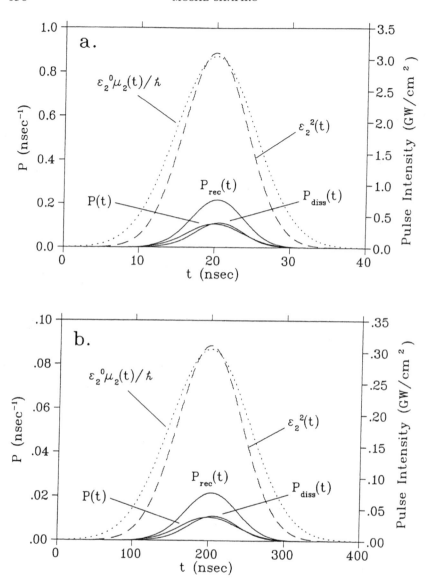

Figure 12.   Rates of association, backdissociation, and total molecule formation versus t in the "counterintuitive" scheme. Dashed lines indicate pulse intensity profile, dotted lines denote the effective Rabi frequency $\varepsilon_2^0 \bar{\mu}_2(t)/\hbar$, where ε_2^0 is the peak pulse intensity. (a) Initial wavepacket width of $\delta_E = 10^{-3}\mathrm{cm}^{-1}$; pulse parameters are as in Fig. 11. (b) Initial wavepacket width of $\delta_E = 10^{-4}\mathrm{cm}^{-1}$. Both pulses last 85 ns. The pump pulse peaks at the peak of the Na + Na wavepacket ($t_0 = 200\,\mathrm{ns}$), and the dump pulse peaks at 50 ns before that time. Peak intensity of the dump pulse is $1.6 \times 10^6\,\mathrm{W/cm}^2$ and of the pump pulse is $3.1 \times 10^8\,\mathrm{W/cm}^2$.

One of the results of these scaling relations is that the pulses' durations can be made longer and their intensities concomitantly scaled down, without changing the final population-transfer yields. As mentioned above, lengthening of the pulses is beneficial because it causes more atoms to recombine within a given pulse.

We now examine the range of pulse parameters (such as the pulse area, $\Omega_{2,E_L} \Delta t_2$) that maximizes the association yield for a *fixed* initial wavepacket. The net association yields versus the pulse area is plotted in Figs. 13a for the "intuitive" scheme and in Fig. 13b for the "counterintuitive" scheme. In both cases a clear maximum occurs at a specific pulse area; the mere increase in pulse intensity does not lead to an improved association yield. We can attribute this behavior to the fact that the association rate [P_{rec} of Eq. (4.8)] increases linearly with increasing pulse intensity, whereas the dissociation rate [P_{diss} of Eq. (4.9)] increases exponentially with the intensity. Hence, as long as the energetic width of the initial wavepacket stays fixed, the association yield turns over with increasing pulse area. The turnover point is different for the two pulse schemes—in the counterintuitive case it occurs at a much higher intensity (area).

The existence of a window of intensities for efficient association explains why it is not possible to increase the pulse durations ad infinitum, that is, to work with continuous-wave (CW) light. As $\Delta t_{1,2}$ increases, it follows by Eq. (4.26) that $|\varepsilon_2/\varepsilon_1|^2$ must also increase. Since $|\varepsilon_1|^2$ cannot vanish, $|\varepsilon_2|^2$ must diverge if one is to stay within the windows of intensities for efficient association in the CW limit. We conclude that radiative association as described in this section cannot take place in the CW regime.

V. LASER CATALYSIS

During the 1980s and 1990s a number of scenarios for laser acceleration and suppression of dissociation processes and chemical reactions were proposed [83–102]. A theme common to many of these schemes is that lasers affect chemical reactivity by forming in conjunction with the molecule new ("dressed") potentials. The dressed potentials may be more amenable to promoting a given reaction. The main difference between laser enhancement of chemical reactions and ordinary photochemistry is that the former is envisaged to involve no net absorption of laser photons. The concept of "laser catalysis" [95–97]—a process in which a laser field, after altering a reaction, returns to its *exact* initial state—is a refinement of such scenarios.

Most of the laser enhancement schemes require very high laser powers because the dressing of the potential surfaces [87,88,90] is ineffective because of the rather weak (collisionally, or optically) induced nuclear

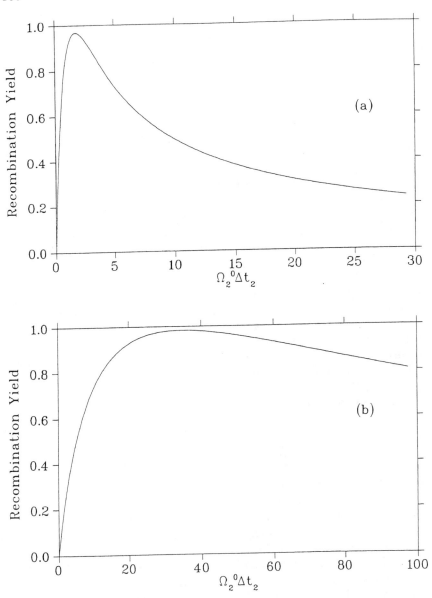

Figure 13. The association yield versus pump pulse intensity for a fixed wavepacket spectral width; (a) "intuitive" case, where pulse widths, central frequencies, and delays are as in Fig. 11; (b) "counterintuitive" case, where pulse widths, central frequencies, and delays are as in Fig. 11.

dipole moments involved [91,93]. Thus, the use of IR radiation to overcome reaction barriers on the ground electronic surface [91,93], necessitates powers in the order of TW/cm^2 (terawatts per square centimeter). At these powers nonresonant multiphoton absorption, which invariably leads to ionization and/or dissociation, becomes dominant, drastically reducing the yield of the reaction of interest.

Continuum–continuum transitions involving excited electronic states [92] ought to require less power than those occurring on the ground state, because the laser in this case couples to strong electronic transition dipoles. However, even in this case the continuum–continuum nuclear factors reduce the transition dipole matrix elements, and moreover, once the system is deposited on an unbounded excited electronic surface, it is impossible to prevent the reaction on that surface and the resultant retention of the absorbed photon from taking place. Such a chain of events resembles that of conventional (weak-field) photochemistry where the laser is used to impart energy to the reaction and not to catalyze it.

Scenarios [95–97] employing transitions between scattering states on the ground electronic surface and bound excited electronic states may ease the above power requirements, primarily because of the involvement of the much stronger bound-continuum nuclear factors. For an excited surface possessing reaction well(s), such schemes give rise to "laser catalysis" [95–97] because the reagents, once excited, remain in the transition-state region and shuttle freely between the reactants' side and the products' side of the ground-state barrier. If the energy available to the nuclei on the excited state is insufficient to break any bond, the system, not being able to escape the transition-state region, eventually relaxes (radiatively or nonradiatively) back to the ground state. In doing so it has a priori similar probabilities of landing on the products' side as on the reactants' side of the barrier. If the laser is strong enough, the stimulated radiative relaxation route, yielding back the same photon absorbed, overcomes the nonradiative channels, resulting in true laser catalysis.

We show below that the simple methodology developed in the previous sections can account for many of the phenomena listed above. The only modification of the models discussed above needed is the consideration of the coupling by a laser pulse of *two* continua (that of the reactants and that of the products) to a bound manifold.

A. Coupling of Two Continua to a Bound State by a Laser Pulse

We consider an $A + BC \rightarrow AB + C$ exchange reaction described by a smooth one-dimensional potential barrier along the reaction coordinate. The eigenstates of the system form a continuum of "outgoing" scattering states $|E, 1^+\rangle$ and $|E, 2^+\rangle$, where E is the total collision energy. In accordance with our general notation, the 1^+ and 2^+ indices are reminders that the reaction has originated in either arrangement channel 1, the A + BC channel, or in channel 2, the AB + C channel. The situation is depicted in Fig. 1d.

The asymptotic behavior of the $|E, 1^+\rangle$ and $|E, 2^+\rangle$ states is given by

$$\lim_{x \to -\infty} \langle x|E, 1^+\rangle = \sqrt{\frac{m}{k_1 h}} \exp(ik_1 x) + R_1(E)\exp(-ik_1 x) \qquad (5.1a)$$

$$\lim_{x \to +\infty} \langle x|E, 1^+\rangle = T_1(E)\exp(ik_2 x) \qquad (5.1b)$$

and

$$\lim_{x \to \infty} \langle x|E, 2^+\rangle = \sqrt{\frac{m}{k_2 h}} \exp(-ik_2 x) + R_2(E)\exp(ik_2 x) \qquad (5.2a)$$

$$\lim_{x \to -\infty} \langle x|E, 2^+\rangle = T_2(E)\exp(-ik_1 x), \qquad (5.2b)$$

where $k_{1,2} = \left[\sqrt{2m(E - V(\mp\infty))}\right]/\hbar$.

The laser catalysis scenario is shown in Fig. 1d. Under the action of a laser pulse of central frequency ω, assumed to be in near resonance with the transition from the continuum to an intermediate bound state $|\psi_0\rangle$, population is transferred from states $|E, 1^+\rangle$ to a set of "incoming" scattering states $|E, 2^-\rangle$, with the asymptotic behavior

$$\lim_{x \to \infty} \langle x|E, 2^-\rangle = \sqrt{\frac{m}{k_2 h}} \exp(ik_2 x) + R_2^*(E)\exp(-ik_2 x) \qquad (5.3a)$$

$$\lim_{x \to -\infty} \langle x|E, 2^-\rangle = T_2^*(E)\exp(ik_1 x) \qquad (5.3b)$$

With the total Hamiltonian of the system given by Eq. (3.1) we can expand the material wavefunction of the system as in Eq. (2.5)

$$|\Psi(t)\rangle = b_0|\psi_0\rangle\exp\frac{-iE_0 t}{\hbar} + \int dE(b_{E,1}(t)|E, 1^+\rangle$$
$$+ b_{E,2}(t)|E, 2^+\rangle)\exp\frac{-iEt}{\hbar} \qquad (5.4)$$

where $|\psi_0\rangle$ and $|E, \mathbf{n}^+\rangle$ satisfy the material Schrödinger equation

$$[E_0 - H]|\psi_0\rangle = [E - H]|E, \mathbf{n}^+\rangle = 0, \mathbf{n} = 1, 2 \qquad (5.5)$$

The main difference from Eq. (2.5) is that, $|\psi_0\rangle$, the bound state, is not the initial state.

Substitution of the expansion of Eq. (5.4) into the time-dependent Schrödinger equation, and use of the orthogonality of the $|\psi_0\rangle, |E, 1^+\rangle$, and $|E, 2^+\rangle$ basis states, results in a set of first-order differential equations similar to Eq. (3.4)

$$\frac{d}{dt}b_0 = i \int dE \sum_{n=1,2} \Omega_{0,E,n}(t)\exp(i\Delta_E t)b_{E,n}(t) \qquad (5.6a)$$

$$\frac{d}{dt}b_{E,m} = i\Omega_{0,E,m}^*(t)\exp(-i\Delta_E t)b_0(t), \qquad m = 1, 2 \qquad (5.6b)$$

where

$$\Omega_{0,E,n}(t) \equiv \frac{\langle 0|\mu|E, n^+\rangle \epsilon(t)}{\hbar}, \qquad n = 1, 2 \qquad (5.7)$$

and

$$\Delta_E \equiv \frac{E_0 - E}{\hbar} - \omega. \qquad (5.8)$$

Substituting the formal solution of Eq. (5.6b)

$$b_{E,n}(t) = b_{E,n}(t_0) + i \int_{t_0}^{t} dt' \Omega_{0,E,n}^*(t')\exp(-i\Delta_E t')b_0(t') \qquad (5.9)$$

into Eq. (5.6a), we obtain

$$\frac{d}{dt}b_0 = i \sum_{n=1,2} \int dE \Omega_{0,E,n}(t)\exp(i\Delta_E t)b_{E,n}(t = 0)$$

$$- \sum_{n=1,2} \int dE \int_{t_0}^{t} dt' \Omega_{0,E,n}(t)\Omega_{0,E,n}^*(t')\exp[i\Delta_E(t - t')]b_0(t') \qquad (5.10)$$

As above, we now invoke the SVCA, which in the context of the two-photon Λ configuration of Fig. 1d, reads as

$$\sum_n |\langle E, n^+|\mu|\psi_0\rangle|^2 \approx \sum_n |\langle E_L, n^+|\mu|\psi_0\rangle|^2 \qquad (5.11)$$

where $E_L = E_0 - \hbar\omega$. On substitution of Eqs. (3.5) and (5.11) into Eq. (5.10), we obtain

$$\frac{db_0}{dt} = iF(t) - \sum_{n=1,2} \Omega_n(t)b_0(t) \qquad (5.12)$$

where

$$\Omega_\mathbf{n}(t) \equiv \frac{\pi|\langle E_L, \mathbf{n}^+|\mu|\psi_0\rangle\epsilon(t)|^2}{\hbar}, \qquad \mathbf{n} = 1, 2 \qquad (5.13)$$

The source term $F(t)$ is given as

$$F(t) = \sum_{\mathbf{n}=1,2} F_\mathbf{n}(t) = \epsilon(t) \sum_{\mathbf{n}=1,2} \frac{\bar{\mu}_\mathbf{n}(t)}{\hbar} \qquad (5.14)$$

where

$$\bar{\mu}_\mathbf{n}(t) = \int dE \langle 0|\mu|E, \mathbf{n}^+\rangle \exp(i\Delta_E t) b_{E,\mathbf{n}}(t_0), \qquad \mathbf{n} = 1, 2 \qquad (5.15)$$

As in the photoassociation case, we can obtain analytic solutions of Eq. (5.12)

$$b_0(t) = v(t) \cdot \phi(t) + b_0(t_0)v(t) \qquad (5.16)$$

where

$$v(t) = \exp\left\{ -\int_{t_0}^t [\Omega_1(t') + \Omega_2(t')]dt' \right\} \qquad (5.17)$$

and

$$\phi(t) = i \int_{t_0}^t \frac{F(t')}{v(t')} dt' \qquad (5.18)$$

In the laser catalysis process, the initial conditions are such that $b_0(t_0) = 0$ and $b_{E,2}(t_0) = 0$ for all E. Therfore, we obtain the following for the $b_0(t)$ coefficient:

$$b_0(t) = i \int_{t_0}^t F_1(t')\exp\left\{ -\int_{t'}^t [\Omega_1(t'') + \Omega_2(t'')]dt'' \right\}dt' \qquad (5.19)$$

Given $b_0(t)$, the continuum population distributions $b_{E,1}(t)$ and $b_{E,2}(t)$ are obtained directly by Eq. (5.9). Typical potentials and eigenfunctions used to simulate one-photon laser catalysis are plotted in Fig. 14.

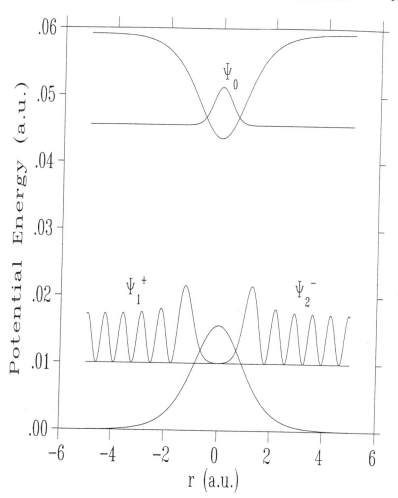

Figure 14. Eckart potentials and wavefunctions used in the simulation of the laser catalysis process. Potential parameters were $A = 0$ au, $B = 6.247$ au, $l = 4.0$ au, and $m = 1060.83$ au.

B. Pulsed Laser Catalysis with a Pair of Eckart Potentials

As an illustration of the above formulation, we consider laser catalysis with an Eckart potential [107,108],

$$V_{\text{ground}}(x) = V[\xi(x)] = -\frac{A\xi}{1-\xi} - \frac{B\xi}{(1-\xi)^2}; \quad \xi = -\exp(2\pi x/l). \quad (5.20)$$

for the ground state and an inverted Eckart potential

$$V_{\text{excited}}(x) = \mathcal{E} - V[\xi(x)] \qquad (5.21)$$

having a well, for the excited state. The asymptotic values for the Eckart potential are

$$V(x = -\infty) = 0, \qquad V(x = +\infty) = A \qquad (5.22)$$

and the barrier height is given as

$$V_{\text{max}} = \frac{(A + B)^2}{4B} \qquad (5.23)$$

The Schrödinger equation with the Eckart potential has well-known analytic solutions for both bound and continuum $|E, \mathbf{n}^+\rangle$ states [107–109]. The parameters of Eq. (5.20) chosen for the ground Eckart potential are $A = 0, B = 6.247 \times 10^{-2}$ au, $l = 4.0$ au. The particle's mass m was chosen as that of the $H + H_2 \rightarrow H_2 + H$ reaction, specifically, $m = 1060.83$ au. The intermediate state was the $v = 0$ level of the inverse Eckart potential given in Eq. (5.21), with the same parameters. Potential parameters were chosen to resemble the energy profile along the reaction path for the linear $H + H_2 \rightarrow H_2 + H$ reaction [110,111]. The resulting potential curves are plotted in Fig. 14. Given these parameters, eigenfunctions and eigenenergies were obtained using the formulas in Ref. 109, and the bound-continuum dipole matrix elements $\langle \psi_0 | \mu | E, \mathbf{n}^+ \rangle$ that enter Eq. (3.5) were calculated using high-order Gauss–Legendre quadrature.

The initial state of the system is described by a normalized Gaussian wavepacket of + scattering states

$$|\Psi(t = 0)\rangle = \int dE b_{E,1}(t = 0)|E, 1^+\rangle \qquad (5.24)$$

where $b_{E,1}(t = 0)$ is given by Eq. (4.20). Simulations were made for initial collision energies of $E_{\text{col}} = 0.005 - 0.03$ au and wavepacket widths $\delta_E = 10^{-4} - 10^{-3} \text{cm}^{-1}$. Our calculations show that the bound-continuum matrix elements do not vary appreciably with kinetic energy over these widths; hence we can simplify Eq. (5.14) as we have done in Eq. (4.21), and be choosing a Gaussian pulse as in Eq. (4.24), we obtain a form similar to Eq. (4.23) for $F_1(t)$.

Having computed all input matrix elements, the dynamics, embodied in Eq. (5.12), is solved using either a RKM algorithm for direct numeric integration or the exact expression of Eq. (5.19). Both methods give identical results, thus confirming the validity of the analytic solution. The resulting $b_0(t)$ coefficient is then used to calculate the continuum population distributions $b_{E,1}(t)$ and $b_{E,2}(t)$ according to Eq. (3.6).

The energy dependence of the nonradiative reaction probability is presented in Fig. 15. The time dependences of the expansion coefficients are shown in Fig. 16. Initial collision energy for this calculation was 0.01 au. From Fig. 15 it is evident that the nonradiative reaction probability at this energy is negligible. We see that the effect of the laser pulse is to induce a near-complete (>99%) population transfer from the wavepacket of $|E, 1^+\rangle$ states (localized to the left of the potential barrier) to a wavepacket of $|E, 2^-\rangle$ states (localized to the right of the barrier), while keeping the population of the $|\psi_0\rangle$ states to a bare minimum. In this way spontaneous emission losses are essentially eliminated.

Because of the invariance of Eq. (5.12) on the rescaling of Eq. (4.26), it is possible to freely vary the pulse durations and intensity, as long as the integrated pulse power $|\varepsilon^0|^2 \times \Delta t$ is kept fixed. This behavior is demonstrated in Figs. 17a,b. A short-pulse case is illustrated in Fig. 17a and a long-pulse case, with a narrower bandwidth (longer duration) initial wavepacket, is shown in Fig. 17b. It is evident that the time evolution of the system is scaled up by a factor of 10, whereas pulse intensity is scaled down

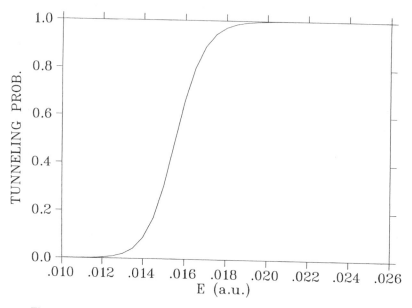

Figure 15.   Nonradiative reactive probability as a function of collision energy.

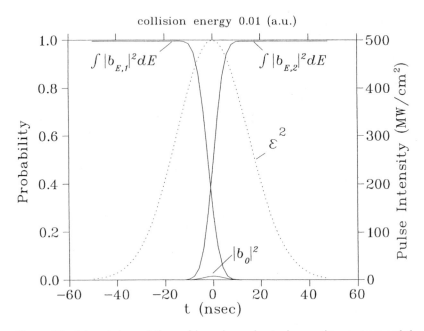

Figure 16. Integrated populations of incoming and outgoing continuum states and the population of the $v = 0$ intermediate state versus time. Dashed line indicates the intensity profile of the Gaussian pulse whose maximum intensity is 5×10^8 W/cm^2. The full width at half-maximum (FWHM) of the pulse is 30 ns and the central frequency was chosen so that $\Delta_{E_{\text{col}}} = 0$. The initial reactant collision energy is 0.01 au and initial wavepacket width is $\delta_E = 10^{-3}$ cm^{-1}.

by the same factor. The advantage of long pulses is that only reactants that will collide during the laser pulse will react. Thus, longer pulses would increase the number of product molecules formed within a single pulse duration. The disadvantage is that the power requirements become increasingly difficult to fulfill the longer the pulse, because the peak power must go down exactly as $1/\Delta t$ whereas in most practical devices the power goes down much faster with increasing pulse durations.

As mentioned above, be keeping the population of the intermediate resonance low (as is the case in Fig. 16), we effectively eliminate the spontaneous emission losses. In Fig. 18 we plot the intermediate-level population as a function of t at four different pulse intensities. Radiative reaction probability for all plotted intensities is near unity. However, it is evident that the intermediate-state population throughout the process decreases with increasing pulse intensity. Thus, to avoid spontaneous emission losses, high pulse intensities should be used.

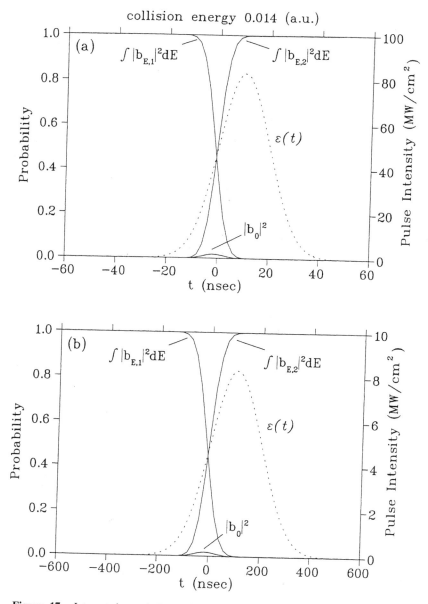

Figure 17. Integrated populations of incoming and outgoing continuum states and population of the $v = 0$ intermediate state versus time. Dashed lines denote pulse intensity profile. The initial reactant collision energy is 0.014 au. (*a*) Initial wavepacket width of $\delta_E = 10^{-3}$cm^{-1}. Pulse intensity is 83 MW/cm^2, and pulse FWHM is 20 ns. (*b*) Initial wavepacket width of $\delta_E = 10^{-4}$cm^{-1}. Pulse intensity is 8.3 MW/cm^2 and pulse FWHM is 200 ns.

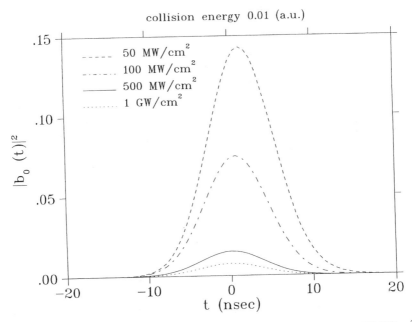

Figure 18. Intermediate-state population versus time at pulse intensities of $50\,MW/cm^2$, $100\,MW/cm^2$, $500\,MW/cm^2$, and $1\,GW/cm^2$. Pulse duration and central frequency and initial kinetic energy of the reactants are as in Fig. 17.

Calculated reactive lineshapes (i.e., the reaction probability as a function of the pulse center frequency) at three pulse intensities are shown in Fig. 19. The initial collision energy is 0.014 au, which is slightly closer to the barrier maximum than before. According to Fig. 15, the nonradiative reaction probability is now about 9%. This causes the lineshapes to assume an asymmetric form due to the interference between the nonradiative tunneling pathway and the laser-catalyzed pathway. We see that the reaction probability is enhanced for a positive (blue) detuning and suppressed for a negative (red) detuning. This result is similar to the findings in the CW case [12,96], except that the power requirements can be easily met and spontaneous emission is essentially nonexistent.

The effects noted above are absent in the weak-field limit; specifically, when the field is too weak, the maximal reaction probability is less than unity. As shown in Fig. 20, before reaching saturation, marked by unit reaction probability at the right frequency, the reaction probability increases monotonically with increasing laser intensity. This is in contrast with the CW results [12,96] where the sole effect of the reduction in laser power is to narrow down the asymmetric lineshapes of Fig. 19 while leaving one point of perfect transmission at the center of the line.

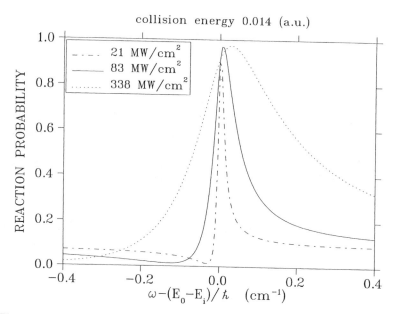

Figure 19. Calculated reactive lineshapes at 21 MW/cm², 83 MW/cm², and 338 MW/cm². The FWHM of the pulse is 20 ns. Reactants collision energy is 0.014 au and initial wavepacket width is $\delta_E = 10^{-3}$ cm⁻¹.

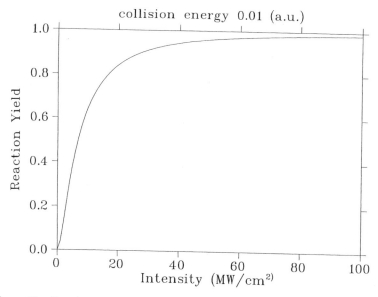

Figure 20. Reactive probability versus pulse intensity for a fixed wavepacket spectral width. Pulse width and central frequency are as in Fig. 17.

The existence of a point where the reaction probability assumes the value of 1 is best understood by adopting the "(photon) dressed states" picture. When the 2 × 2 dressed-potential matrix composed of the (diagonal) dressed potentials and the (off-diagonal) field dipole coupling terms is diagonalized, the two field matter eigenvalues shown in Fig. 21 result. As demonstrated in Fig. 21, the ground field matter eigenvalue assumes the shape of a double-barrier potential and the excited eigenvalue assumes the shape of a double-well potential. The separation between these eigenvalues increases as the coupling field strength is increased.

In the adiabatic approximation, particles starting out in the remote past in the ground state remain on the lowest eigenvalue at all times. These particles experience resonance scattering by a double-barrier potential, admitting tunneling probability of unity, irrespective of the details of the potential [112–117], when the incident energy is near a bound state of the well within the barriers. Similar phenomena have been noted for semiconductor devices

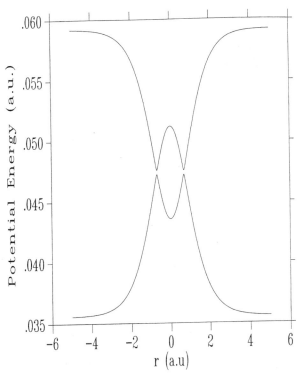

Figure 21. Dressed-state potentials for the laser catalysis process at maximum pulse intensity. Initial kinetic energy is 0.01 au.

[112–117] in the context of the Ramsauer-Townsend effect [118] and for Fabry–Perot interferometers [119]. In addition, field-induced transparency, observed when a high-frequency field acts on a single barrier, was explained as a result of the emergence of an average field-dressed double-barrier potential [120].

The point of tunneling suppression (i.e., when the tunneling probability of Fig. 19 assumes the value of 0) appears, in the dressed-states picture, as a result of the breakdown of the adiabatic approximation: At the energy of tunneling suppression, the flux leaking to the excited double-well eigenvalue interferes destructively with the flux remaining on the low double-barrier potential.

VI. CONCLUSIONS

In the first model considered we have recast the problem of one net photon absorption to a continuum from a laser pulse as a single integral equation. Using this methodology we have shown that a transition to continuum is not irreversible if the continuum possesses features narrower than the effective width (reflecting the pulse strength and width) of the pulse. A smooth transition of a given continuum from a perfectly absorbing entity to a set of levels, each executing Rabi oscillations with a precursor bound state, was shown to occur as the pulse strength is increased. As a result of such oscillations, spectral migrations and formations of transparent lines, are shown to occur. The field-induced interference between neighboring lines has also been investigated.

We see that the reversibility or irreversibility of a process is a function of the mode of preparation of the initial state. A continuum that appears perfectly flat to a nanosecond pulse, giving rise to a monotonic decay of the ground state, may appear "bumpy" to a femtosecond pulse, causing as a result transitions from the continuum levels back to the ground precursor state.

The fact that transitions to the continuum can be reversed was also demonstrated by our study of pulsed two-photon *association*. In this study we have shown that one can effectively recombine ultracold atoms to form ultracold molecules. We have performed detailed computations on the $Na + Na \rightarrow Na_2$ process and have demonstrated that $\leqslant 97\%$ of $J = 0$ colliding atom pairs can be recombined per (commercially available nanosecond laser) pulse.

It is worth noting that our results can be extended to computing radiative association for a thermal ensemble of cold atoms. We have shown [106] that under such circumstances the yield per pulse is related to the pulse length Δt_2, the velocity $v(T)$ and the density n in the trap as,

$\eta(T) = [\pi n \hbar^2 / 4m^2 v(t)] \Delta t_2$. For 20-ns pulses and typical trap densities, and microkelvin-range ensemble temperatures, the association yield per pulse is as high as 6×10^{-6}; thus, with a 100-Hz laser source, half of the entire ensemble of cold atoms can be transferred to an ensemble of cold molecules within ~ 25 min.

Finally, we have shown that laser catalysis with pulses is a very sensitive way of controling tunneling processes. The emphasis on pulses in this work stems from the fact that the power requirements for the process, although reasonable for nanosecond pulses, are beyond most commercial CW laser sources. For cold atoms or molecules, the presence of a barrier guarantees that no reactive events in the absence of the laser pulse would take place. We therefore expect that laser catalysis would first be observed for cold-molecule collisions, where one could most readily take advantage of the low (or even zero) nonradiative background signal.

ACKNOWLEDGMENTS

The author acknowledges fruitful collaborations with Einat Frishman and Ami Vardi and thanks the US-Israel BSF (grant no. 96-00432), the GIF (grant no. I-410-017.05), the James Franck Programme, and the German-Israeli Strategic Cooperation Project (D.I.P) for supporting this work.

APPENDIX A.   THE UNIFORM AND WKB APPROXIMATIONS FOR DISSOCIATION BY A STRONG PULSE

When the features in the continuum are not much sharper than the laser bandwidth, or the separation between the resonances is larger than the laser bandwidth, the set of coupled equations [Eqs. (2.41)] can be reduced to a single second-order equation. By differentiating the second of Eqs. (2.41) we obtain

$$\frac{d^2 b_1}{dt^2} = \frac{d \ln \varepsilon(t)}{dt} \frac{db_1}{dt} - \frac{2\pi \varepsilon^2(t)}{\hbar^2} \sum_s \overline{\mu_s^2} b_1 - \frac{\varepsilon(t)}{\hbar} \sum_s \overline{\mu_s^2} \chi_s f_s^+(t) B_s(t) \quad \text{(A.1)}$$

where we have used the explicit form of f_s^+ [Eq. (2.34)]. We now define $\chi(t)$ as

$$\sum_s \overline{\mu_s^2} \chi_s f_s^+ B_s = \chi(t) \sum_s \overline{\mu_s^2} f_s^+ B_s \quad \text{(A.2)}$$

using which we obtain the following (where $\alpha = \sum_s \overline{\mu_s^2}$) from Eqs. (2.41) and (A.1);

$$\frac{d^2 b_1}{dt^2} = \left(\frac{d\ln\varepsilon(t)}{dt} - i\chi(t)\right)\frac{db_1}{dt} - \frac{2\pi}{\hbar^2}\varepsilon^2(t)\alpha b_1 \tag{A.3}$$

Denoting

$$g_1(t) = -\frac{d\ln\varepsilon(t)}{dt} + i\chi(t) \tag{A.4}$$

$$g_0(t) = \frac{2\pi}{\hbar^2}\varepsilon^2(t)\alpha \tag{A.5}$$

and writing Eq. (A.3) as

$$\frac{d^2 b_1}{dt^2} + g_1(t)\frac{db_1}{dt} + g_0(t)b_1 = 0 \tag{A.6}$$

we obtain a Schrödinger-like equation

$$\left[\frac{d^2}{dt^2} - W(t)\right]c(t) = 0 \tag{A.7}$$

for the transformed coefficient

$$c(t) = \exp\left[\frac{1}{2}\int^t g_1(t')dt'\right]b_1(t) = \varepsilon(t)^{-1/2}\exp\left[\frac{i}{2}\int^t \chi(t')dt'\right]b_1(t) \tag{A.8}$$

The value $W(t)$ of Eq. (A.7) is given as

$$W(t) = \frac{1}{2}g_1'(t) + \frac{1}{4}g_1^2(t) - g_0(t)$$

$$= -\frac{d\varepsilon(t)}{2dt}\frac{d^2\ln\varepsilon(t)}{dt^2} + \frac{1}{4}\left(\frac{d\ln\varepsilon(t)}{dt} - i\chi(t)\right)^2 - \frac{2\pi}{\hbar^2}\varepsilon^2(t)\alpha \tag{A.9}$$

The "time-dependent potential," $W(t)$, is analogous to minus the local momentum squared, $-p^2(x)$, of the time-independent Schrödinger equation. The term $W(t)$ can be complex, with the imaginary part depending on $\Delta(= \sum_s \overline{\mu_s^2}\Delta_s f_s^+ B_s)$, the average detuning of the laser's center frequency ω_1 with respect to the various resonances that contribute to the continuum.

For a single resonance Eq. (A.2) is an identity, and we of course know what the (time-independent) χ function is. When there are more resonances, $\chi(t)$ cannot be determined without knowing $b_1(t)$, since by Eq. (2.41)

$$\chi(t) = \frac{\sum_s \overline{\mu_s^2} \chi_s f_s^+ \int^t dt' \varepsilon(t') f_s^-(t') b_1(t')}{\sum_s \overline{\mu_s^2} f_s^+ \int^t dt' \varepsilon(t') f_s^-(t') b_1(t')} \tag{A.10}$$

In practical applications we find that we can approximate $\chi(t)$ as

$$\chi(t) \approx \chi^0(t) = \frac{\sum_s \overline{\mu_s^2} \chi_s f_s^+ \int^t dt' \varepsilon(t') f_s^-(t')}{\sum_s \overline{\mu_s^2} f_s^+ \int^t dt' \varepsilon(t') f_s^-(t')} \tag{A.11}$$

Thus, for a Gaussian pulse envelope centered about t_0

$$\varepsilon(t) = \varepsilon_L \exp\left[-\left(\frac{t-t_0}{2\delta}\right)^2\right] \tag{A.12}$$

we can calculate χ^0 via the identity [35]:

$$\int^t dt' \varepsilon(t') f_s^-(t') = \pi\sqrt{2}\varepsilon_L\, \delta \exp(-\delta^2\chi_s^2)\left\{1 + \mathrm{erf}\left(\frac{t}{2\delta - i\delta\chi_s}\right)\right\} \tag{A.13}$$

The time-dependent potential assumes the form

$$W(t) = \frac{-(t-t_0)}{8\delta^4}\varepsilon_L\exp\left[-\left(\frac{t-t_0}{2\delta}\right)^2\right]$$
$$- \frac{2\pi}{\hbar^2}\alpha\varepsilon_L\exp\left[-2\left(\frac{t-t_0}{2\delta}\right)^2\right] - \frac{1}{4}\left(\frac{t-t_0}{\delta^2} - i\chi^0(t)\right)^2$$

In complete analogy to the case of the time-independent Schrödinger equation, we call the timepoints satisfying the $W(t^*) = 0$ equation, the "turning points." If there is only one turning point, the solutions of Schrödinger-like equation [Eq. (A.7)] can be written to an excellent approximation in terms of the Uniform regular and irregular Airy functions $A_i(T)$ abd $B_i(T)$ [121]

$$C_{\mathrm{uni}}(t) = \left[\frac{T(t)}{-W(t)}\right]^{1/4}\{C_a Ai(-T(t)) + C_b B_i(-T(t))\} \tag{A.15}$$

where the complex argument T is defined as

$$T(t) = \left[\frac{3}{2}\int_{t^*}^{t} \sqrt{-W(t')}dt'\right]^{2/3} \tag{A.16}$$

and C_a and C_b are constants determined by the initial conditions, $b_s(-\infty) = 1$ and $b_s'(-\infty) = 0$. If there are no turning points on the real-time axis, we choose the complex turning point that is closest to the relevant time range.

The Uniform approximation is the exact solution of the equation

$$\left[\frac{d^2}{dt^2} - W(t) - \eta(t)\right]c_{\text{uni}}(t) = 0 \tag{A.17}$$

where

$$\eta(t) = [T'(t)]^{1/2}\frac{d^2}{dt^2}[T'(t)]^{-1/2} \tag{A.18}$$

It is an excellent approximation to Eq. (A.7), because usually $|\eta(t)| \ll |W(t)|$. If there is more than one turning point, the Airy functions can still be used (provided the turning points do not coalesce), by writing the solutions of Eq. (A.15) for each time interval containing a turning point and matching these solutions and their derivatives across the time intervals. Usually no more than two turning points exist.

The uniform semiclassical approximation can be simplified by introducing the (zeroth-order) WKB approximation, according to which

$$b_t(t) = \exp\left[-\frac{1}{2}\int^{t} g_1(t')dt'\right]\left\{C_a\exp\left[\int_{t^*}^{t}[W(t')]^{1/2}dt'\right]\right.$$
$$\left. + C_b\exp\left[-\int_{t^*}^{t}[W(t')]^{1/2}dt'\right]\right\} \tag{A.19}$$

As shown in Fig. 22, if the field is low to moderately high, so that b_1 does not decay too much, the Schrödinger-like equation works extremely well. At very high fields, when the ground state begins to undergo Rabi oscillations, the approximation begins to break down.

For a single resonance the Schrödinger-like equation is exact, and we can use it to study further approximations based on it. In Fig. 23a we demonstrate the goodness of the Uniform approximation at the center of the absorption line (Δ). This is a case in which there is only one effective

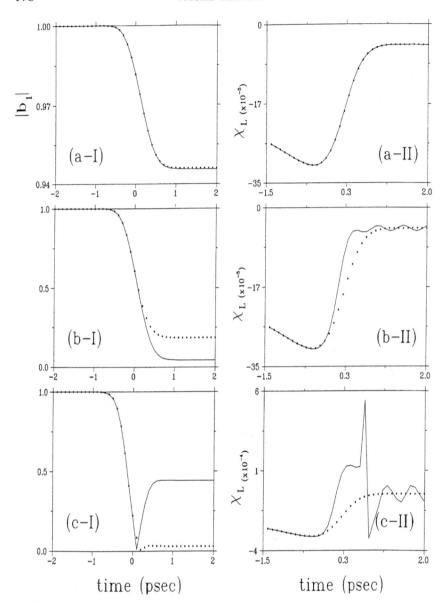

Figure 22. Comparison of $b_1(t)$, the bound state coefficient, obtained numerically and by the single Schrödinger-like equation for different pulse amplitudes. Exact result (full line), single Schrödinger-like equation results (dotted line). The pulse bandwidth is $50\,\mathrm{cm}^{-1}$. The spectrum is composed of three resonances with, $\mu_{1n} = \mu_{2n} = \mu_{3n} = 1$, $\Gamma_1 = \Gamma_2 = \Gamma_3 = 20\,\mathrm{cm}^{-1}, \Delta_1 = 0\,\mathrm{cm}^{-1}, \Delta_2 = -60\,\mathrm{cm}^{-1}$, $\Delta_3 = 60\,\mathrm{cm}^{-1}$. (a) $\varepsilon_L = 1 \times 10^{-3}$ au, (b) $\varepsilon_L = 5 \times 10^{-3}$ au, (c) $\varepsilon_L = 8 \times 10^{-3}$ au.

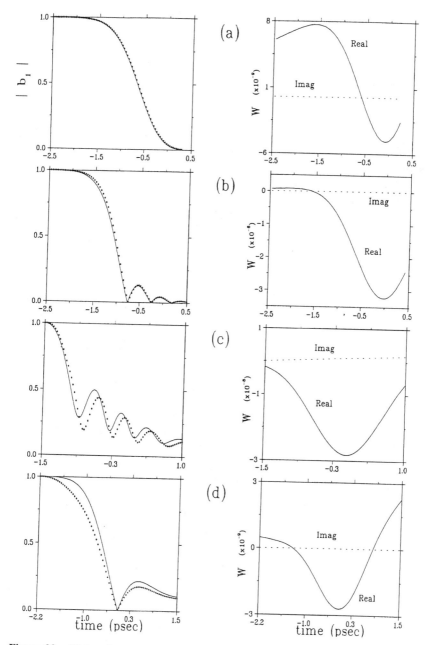

Figure 23. Right column: the time-dependent potential, $W(t)$. Left column: $b_1(t)$, the bound-state coefficient. Exact results (full line) and the uniform approximation (dotted lines). The pulse bandwidth is $20\,\text{cm}^{-1}$ and $\mu_{s's} = 1$. (a) Pulse amplitude $\varepsilon_L = 0.005$ au, spectral width $\Gamma_s = 100\,\text{cm}^{-1}, \Delta_2 = 0$; (b) $\varepsilon_L = 0.01$ au, $\Gamma_s = 50\,\text{cm}^{-1}, \Delta_s = 0$; (c) $\varepsilon_L = 0.01$ au, $\Gamma_s = 20\,\text{cm}^{-1}, \Delta_s = 10\,\text{cm}^{-1}$; (d) $\varepsilon_L = 0.005$ au, $\Gamma_s = 20\,\text{cm}^{-1}, \Delta_s = 0$.

real-axis turning point. Although the time-dependent potential, $W(t)$, shown in Fig. 23a, possesses two turning points, one at $t^* \approx -0.5$ ps and one at $t^* \approx 0.5$ ps, the $t^* \approx 0.5$-ps turning point is not expressed because at that time the b_1 coefficient is essentially 0. Therefore, the problem maps to a single-turning-point Uniform approximation, which, as shown in Fig. 23a, faithfully reproduces the numerical solution of Eq. (2.36). As we increase the field intensity, a case shown in Fig. 23b, the ψ_1 population begins to show an oscillatory behavior. As shown in Fig. 23b, this is still a single real-axis turning point case and the Uniform approximation reproduces perfectly the oscillatory pattern.

In Fig. 23c we display a case with relatively small Γ_s and a Δ detuning, for which $W(t)$ possess only complex turning points. In the case shown in Fig. 23c, the turning point occurs at $t^* = -1.0967 - 0.1083i$. Here too we obtain excellent agreement between the uniform approximation and the exact numerical solutions.

When a second turning point exists while the laser is still on, we need to construct the uniform Airy solutions around each turning point and match the two solutions at some intermediate point. Such a case is shown in Fig. 23d. Comparison with the exact solution demonstrates that the uniform approximation, although not perfect, works quite well even in this case.

The Uniform approximation is more complicated than the primitive WKB method. In most situations it is the only approximation that can be used because it correctly generates solutions across the problematic turning point region(s). In contrast, the zeroth-order WKB solutions of Eq. (A.19) do not conserve flux and the first-order WKB solutions diverge at the turning points [due to the $W(t)^{-1/4}$ term]. To show the need for the uniform approximation, a series of computations based on the zeroth-order WKB approximation for different absorption bandwidths is displayed in Fig. 24 and compared to the exact result. We see that the zeroth-order WKB method yields inferior results as compared with the uniform approximation, although at times the zeroth-order WKB is a reasonable approximation. It has an additional flaw in that it critically depends on the choice of the initial integration time.

The residual term in the zeroth-order WKB equation, given as $2|dW/dt|/|W^{3/2}|$, can be shown to be roughly proportional to Γ_s^{-2}. Hence, the error in the zeroth-order WKB approximation is expected to increase with decreasing Γ_s. Thus, choosing a set of pulse intensities such that the zeroth-order WKB method works well for $\Gamma_s > 500 \text{cm}^{-1}$, we see that the quality of the approximation deteriorates as Γ_s dips below that value, culminating in the complete breakdown of the approximation when $\Gamma_s \leq 200 \text{cm}^{-1}$. In contrast, the Uniform approximation remains valid for values of Γ_s well below 200cm^{-1}.

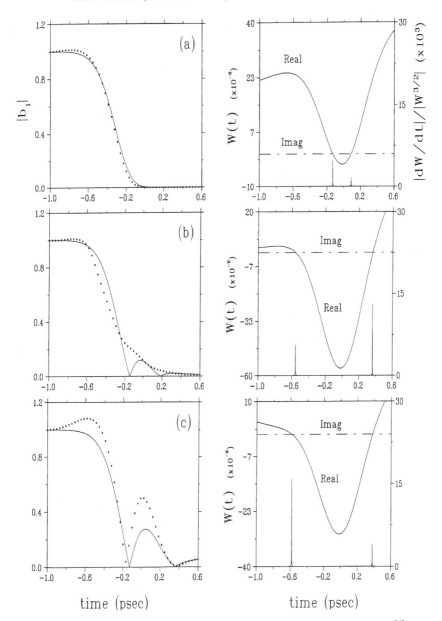

Figure 24. Right column: the time-dependent potential, $W(t)$ and $|W'(t)/\,[W(t)]^{3/2}|$. Left column: $b_1(t)$, the bound-state coefficient. Exact results (full line), the zero-order WKB approximation (dotted lines). The laser is tuned to the center of the absorption spectrum. Pulse amplitude = 0.01 au, pulse bandwidth = $50\,\mathrm{cm}^{-1}$: (a) $\Gamma_s = 500\,\mathrm{cm}^{-1}$; (b) $\Gamma_s = 100\,\mathrm{cm}^{-1}$; (c) $\Gamma_s = 50\,\mathrm{cm}^{-1}$.

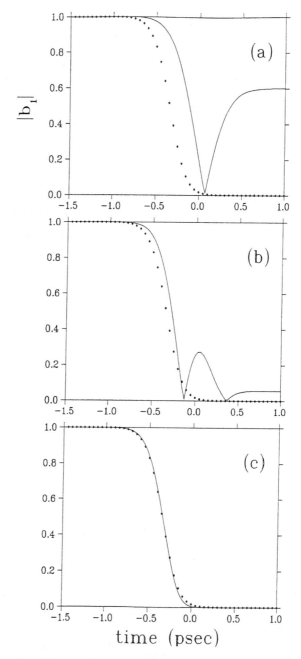

Figure 25. The SVCA at different spectral widths. $\mu_{s's} = 1, \varepsilon_L = 0.01$ au, pulse bandwidth $= 50\,\mathrm{cm}^{-1}$. The laser is tuned to the center of the absorption spectrum. Exact results are shown as full lines and the SVCA values are denoted by as dotted lines. (a) $\Gamma_s = 10\,\mathrm{cm}^{-1}$; (b) $\Gamma_s = 50\,\mathrm{cm}^{-1}$; (c) $\Gamma_s = 500\,\mathrm{cm}^{-1}$.

A different limit of the uniform approximation leads to the SVCA. In Figs. 25a–c we examine the effect of the spectral width, at the center of the absorption spectrum ($\Delta = 0$). The SVCA, which follows the exact solution faithfully for large values of Γ_s, begins to fail as Γ_s is decreased.

APPENDIX B. THE ADIABATIC APPROXIMATION IN THE PRESENCE OF A CONTINUUM

The existence of an imaginary term in the effective Hamiltonian matrix [Eq. (3.12)], arising from the (perfectly absorbing) continuum, has far-reaching consequences for the validity of the adiabatic approximation.

In order for adiabaticity to hold the time derivative of the mixing angle must be small with respect to the gap between the eigenvalues:

$$|\dot{\theta}(t)| \ll |\mathcal{E}_2(t) - \mathcal{E}_1 t)| \qquad \text{(B.1)}$$

The condition expressed in Eq. (B.1) can be rewritten more explicitly, using Eqs. (3.17) and (3.14) as

$$\frac{1}{2}\left|\frac{1}{\left(\dfrac{2\Omega_1}{i\Omega_2 - \Delta_1}\right)^2 + 1}\frac{d}{dt}\left(\frac{2\Omega_1}{i\Omega_2 - \Delta_1}\right)\right| \ll |(4\Omega_1^2 + (\Delta_1 - i\Omega_2)^2)^{1/2}| \qquad \text{(B.2)}$$

We now examine two limiting cases:

1. $\Delta_1 = 0$ *(Resonant Excitation).* In this case Eq. (B.2) becomes

$$\left|\frac{1}{1 - \left(\dfrac{2\Omega_1}{\Omega_2}\right)^2}\frac{d}{dt}\left(\frac{\Omega_1}{\Omega_2}\right)\right| \ll \left|(4\Omega_1^2 - \Omega_2^2)^{1/2}\right| \qquad \text{(B.3)}$$

In a two-pulse configuration three different temporal domains must be considered:

1a. $\Omega_2 \gg 2\Omega_1$. In this region the adiabaticity condition (B.3) takes the form

$$\left|\frac{d}{dt}\left(\frac{\Omega_1}{\Omega_2}\right)\right| \ll |\Omega_2| \qquad \text{(B.4)}$$

1b. $2\Omega_l \gg \Omega_2$. In this domain, Eq. (B.3) becomes

$$\left| \frac{d}{dt}\left(\frac{\Omega_1}{\Omega_2}\right) \right| \ll \left| \left(\frac{2\Omega_1}{\Omega_2}\right)^2 2\Omega_1 \right| \tag{B.5}$$

and as $|2\Omega_1/\Omega_2| \gg 1$, if suffices to demand that

$$\left| \frac{d}{dt}\left(\frac{\Omega_1}{\Omega_2}\right) \right| \leq |2\Omega_1| \tag{B.6}$$

1c. $2\Omega_l \sim \Omega_2$. Here essential breakdown of adiabaticity occurs, as

$$\frac{1}{1 - \left(\dfrac{2\Omega_1}{\Omega_2}\right)^2} \to \infty \tag{B.7}$$

while the righthand side of Eq. (B.3) vanishes:

$$(4\Omega_1^2 - \Omega_2^2)^{1/2} \to 0 \tag{B.8}$$

Thus, adiabatic behavior can be achieved only if $d\Omega_1/dt = d\Omega_2/dt$ at that timepoint. This breakdown of adiabaticity is due to the existence of an imaginary Rabi frequency in H of Eq. (3.12), resulting in the possibility of the vanishing of the denominator of Eq. (B.7).

2. $|\Delta_l| \gg \Omega_2$ (Off-Resonance Excitation). In this case Eq. (B.2) assumes the form

$$\left| \frac{\Delta_1^2}{4\Omega_1^2 + \Delta_1^2} \frac{d}{dt}\left(\frac{\Omega_1}{\Delta_1}\right) \right| \ll |\Delta_1| \tag{B.9}$$

This condition is simplified by the inequality

$$\left| \frac{\Delta_1^2}{4\Omega_1^2 + \Delta_1^2} \right| \leq 1 \tag{B.10}$$

hence it suffices to require that

$$\left| \frac{d}{dt}\left(\frac{\Omega_1}{\Delta_1}\right) \right| \ll |\Delta_1| \tag{B.11}$$

we choose to demonstrate the adiabaticity criteria formulated above for the case of Gaussian pulses:

$$\Omega_{1,2}(t) = \Omega_{1,2}^0 \exp\left[-\frac{1}{2}\left(\frac{t - t_{1,2}}{\delta\tau_{1,2}}\right)^2\right] \tag{B.12}$$

The time derivative of Ω_1/Ω_2, needed to estimate the resonance case conditions 1a–1c is given explicitly as

$$\frac{d}{dt}\left(\frac{\Omega_1}{\Omega_2}\right) = \frac{\Omega_1^0}{\Omega_2^0}\left[-\frac{t - t_1}{\delta\tau_1^2} + \frac{t - t_2}{\delta\tau_2^2}\right]\exp\frac{1}{2}\left[-\left(\frac{t - t_1}{\delta\tau_1}\right)^2 + \left(\frac{t - t_2}{\delta\tau_2}\right)^2\right] \tag{B.13}$$

Assuming $\Omega_1^0 = \Omega_2^0 = \Omega^0$ and $\delta\tau_1 = \delta\tau_2 = \delta\tau$ and defining the reduced delay

$$\eta \equiv \frac{t_1 - t_2}{\delta\tau} \tag{B.14}$$

Eq. (B.13) becomes

$$\frac{d}{dt}\left(\frac{\Omega_1}{\Omega_2}\right) = \frac{\eta}{\delta\tau}\exp\left[\frac{\eta}{\delta\tau}\left(t - \frac{t_1 + t_2}{2}\right)\right] \tag{B.15}$$

Thus, after some rearrangement, adiabaticity condition (B.3) assumes the form

$$\frac{\Omega^0\delta\tau}{|\eta|} \gg \frac{\exp\left\{\dfrac{\eta}{\delta\tau}\left(t - \dfrac{t_1 + t_2}{2}\right)\right\}}{\left|4\exp\left\{-\left(\dfrac{t - t_1}{\delta\tau}\right)^2\right\} - \exp\left\{-\left(\dfrac{t - t_2}{\delta\tau}\right)^2\right\}\right|^{3/2}} \tag{B.16}$$

Unless $\eta = 0$, this condition is violated at $t \to \pm\infty$. This may be of little physical importance if adiabaticity is maintained during effective radiation time. However, essential breakdown of adiabaticity occurs also at $t = t_0$, defined as the point where

$$4\exp\left\{-\left(\frac{t_0 - t_1}{\delta\tau}\right)^2\right\} = \exp\left\{-\left(\frac{t_0 - t_2}{\delta\tau}\right)^2\right\}. \tag{B.17}$$

Therefore, adiabaticity in the resonant-excitation case cannot be held throughout the pulse sequence, unless $\eta \ll \delta\tau$ [i.e., $\Omega_1(t)/\Omega_2(t) \sim$ const for all t].

It is instructive to evaluate Eq. (B.16) at three typical times:

1. $t = t_2$. At this time, condition (B.16) assumes the form

$$\frac{\Omega_0 \delta\tau}{|\eta|} \gg \frac{\exp\left(-\dfrac{\eta^2}{2}\right)}{|4\exp(-\eta^2) - 1|^{3/2}} \tag{B.18}$$

corresponding, if η is sufficiently large to case 1a discussed earlier. For $\eta = 1$, this means adiabaticity at this time is maintained if

$$\Omega^0 \delta\tau \gg 1.87 \tag{B.19}$$

2. $t = t_1$. Condition (B.16) at this point is evaluated as

$$\frac{\Omega_0 \delta\tau}{|\eta|} \gg \frac{\exp\left\{\dfrac{\eta^2}{2}\right\}}{|4 - \exp(-\eta^2)|^{3/2}} \tag{B.20}$$

which, for large enough η corresponds to case 1b. For $\eta = 1$, this means adiabaticity at this time is maintained if

$$\Omega^0 \delta\tau \gg 0.24 \tag{B.21}$$

3. $t = 0.5(t_1 + t_2)$. Adiabatic behavior at this time is maintained if

$$\frac{\Omega_0 \delta\tau}{|\eta|} \gg \frac{1}{\left[3\exp\left\{-\left(\dfrac{\eta}{2}\right)^2\right\}\right]^{3/2}} \tag{B.22}$$

and for $\eta = 1$ this condition is evaluated as

$$\Omega^0 \delta\tau \gg 0.28 \tag{B.23}$$

Off-resonance adiabaticity condition for Gaussian pulses is readily obtained as

$$\frac{d}{dt}\left(\frac{\Omega_1}{\Delta_1}\right) = \frac{\Omega_1^0}{\Delta_1 \delta\tau_1}\left(-\frac{t - t_1}{\delta\tau_1}\right)\exp\left[-\frac{1}{2}\left(\frac{t - t_1}{\delta\tau_1}\right)^2\right] \leq \frac{\Omega_1^0}{\Delta_1 \delta\tau_1} \tag{B.24}$$

hence condition 2 assumes the form

$$\frac{\Omega_1^0}{\delta\tau_1\Delta_1^2} \ll 1 \qquad (B.25)$$

It follows from Eq. (B.25) that unless the detuning Δ_1 is large, the adiabatic approximation is expected to fail for pulses of high intensity and short duration. In other words, with the increase in the amplitude of the Rabi cycling (given by Ω_1/Δ_1), the range of adiabaticity decreases to longer and longer pulses. For Δ_1 of $10\,\mathrm{cm}^{-1}$, if we choose $\Omega_{1,2}^0 = 10^{-4}$ au, the adiabatic approximation is valid for pulses as short as $\delta\tau_{1,2} \gg 2$ ps, for which $\Omega_{1,2}^0/(\delta\tau_{1,2}\Delta_1^2) = 0.58$.

Condition (B.25) has a somewhat different variation of the off-resonance adiabaticity condition derived by Bergmann et al. [52]:

$$\Omega^0\Delta\tau \gg F \qquad (B.26)$$

where $F = 1$ for $\Delta_1 = 0$ and $F = \Delta_1/\Omega_1^0$ for $|\Delta_1| \gg \Omega^0$, which requires increasing power for shorter interaction times or increasing detuning [as opposed to "decreasing power for shorter interaction time or decreasing detuning" as stated in Eq. (B.25) and demonstrated numerically in Section III]. The reason for this discrepancy, as well as for the inequivalence between the on-resonance adiabaticity conditions, is the inequivalence of the mixing angles and eigenvalue gaps involved. Bergmann et al. treated a *real* initial–final mixing angle $\Theta \equiv \arctan(\Omega_1/\Omega_2)$, completely independent of Δ_1 while the angle defined in Eq. (III.17) is a *complex* initial–intermediate mixing angle explicitly dependent on Δ_1. When $\Delta_1 = 0$ (resonance contitions), $\theta \equiv \arctan(-i[2\Omega_1/\Omega_2])$ has a pole in $2\Omega_1/\Omega_2 = 1$ resulting in the breakdown of adiabaticity, while Θ is analytic everywhere.

References

1. M. Shapiro and R. D. Levine, *Chem. Phys. Lett.* **5**, 499 (1970).
2. M. Shapiro, *J. Chem. Phys.* **56**, 2582 (1972).
3. M. Shapiro, *Isr. J. Chem.* **11**, 691 (1973).
4. O. Atabek, J. A. Beswick, R. Lefebvre, S. Mukamel, and J. Jortner, *J. Chem. Phys.* **65**, 4035 (1976); O. Atabek and R. Lefebvre, *Chem. Phys.* **23**, 51 (1997); *idem., Chem. Phys.* **55**, 395 (1981).
5. M. D. Morse, K. F. Freed, and Y. B. Band, *J. Chem. Phys.* **70**, 3620 (1979); Y. B. Band, K. F. Freed, and D. J. Kouri, *J. Chem. Phys.* **74**, 4380 (1981).
6. K. C. Kulander and J. C. Light, *J. Chem. Phys.* **73**, 4337 (1980).
7. G. G. Balint-Kurti and M. Shapiro, *Chem. Phys.* **61**, 137 (1981); *idem., Adv. Chem. Phys.* **60**, 403 (1985).

8. M. Shapiro and R. Bersohn, *Annu. Rev. Phys. Chem.* **33**, 409 (1982).

9. E. Segev and M. Shapiro, *J. Chem. Phys.* **73**, 2001 (1980); *idem.*, **77**, 5601 (1982).

10. V. Engel, R. Schinke, and V. Stämmler, *Chem. Phys. Lett.* **130**, 413 (1986); P. Andresen, and R. Schinke, in *Molecular Photodissociation Dynamics*, M. N. R. Ashfold and J. E. Baggott, eds., The Royal Society of Chemistry, London, 1987, p. 61; R. Schinke, *Photodissociation Dynamics*, Cambridge Univ. Press, Cambridge, UK, 1992.

11. M. Shapiro and H. Bony, *J. Chem. Phys.* **83**, 1588 (1985).

12. T. Seideman and M. Shapiro, *J. Chem. Phys.* **88**, 5525 (1988); *idem., J. Chem. Phys.* **92**, 2328 (1990).

13. A. D. Bandrauk and O. Atabek, *Adv. Chem. Phys.* **73**(8), 23 (1989); S. Chelkowski and A. D. Bandrauk, *Chem. Phys. Lett.* **186**, 284 (1991); A. D. Bandrauk, J. M. Gauthier, and J. F. McCann, *Chem. Phys. Lett.* **200**, 399 (1992).

14. S.-I. Chu, *Chem. Phys. Lett.* **70**, 205 (1980); T. Ho, C. Laughlin, and S.-I. Chu, *Phys. Rev. A* **32**, 122 (1985); S.-I. Chu and R. Yin, *J. Opt. Soc. Am. B* **4**, 720 (1987).

15. S. O. Williams and D. G. Imre, *J. Phys. Chem.* **92**, 6636 (1988); J. Zhang and D. G. Imre, *Chem. Phys. Lett.* **149**, 233 (1988).

16. V. Engel and H. Metiu, *J. Chem. Phys.* **92**, 2317 (1990); V. Engel, H. Metiu, R. Almeida, R. A. Marcus, and A. H. Zewail, *Chem. Phys. Lett.* **152**, 1 (1988); S. Shi and H. Rabitz, *J. Chem. Phys.* **92**, 364 (1990).

17. B. Hartke, E. Kolba, J. Manz, and H. H. R. Schor, *Ber. Bunsenges. Phys. Chem.* **94**, 1312 (1990); W. Jakubetz, B. Just, J. Manz, and H.-J. Schreines *J. Phys. Chem.* **94**, 2294 (1990).

18. S. Shi and H. Rabitz, *J. Chem. Phys.* **92**, 364 (1990).

19. A. D. Hammerich, R. Kosloff, and M. A. Ratner, *J. Chem. Phys.* **97**, 6410 (1992); A. D. Hammerich, U. Manthe, R. Kosloff, H.-D. Meyer, and L. S. Cederbaum, *J. Chem. Phys.* **101**, 5623 (1994).

20. E. Charron, A. Guisti-Suzor, and F. H. Mies, *Phys. Rev. Lett.* **71**, 692 (1993).

21. A. Bartana, U. Banin, S. Ruhman, and R. Kosloff, *Chem. Phys. Lett.* **219**, 211 (1994).

22. M. Crance and M. Aymar, *J. Phys. B* **13**, L421 (1980).

23. Z. Deng and J. H. Eberly, *Phys. Rev. Lett.* **53**, 1810 (1984).

24. R. Grobe and J. H. Eberly, *Phys. Rev. Lett.* **68**, 2905 (1992); *idem., Phys. Rev. A* **48**, 623 (1993); *idem., Laser Phys.* **3**, 323 (1993); *idem., Phys. Rev. A* **48**, 4664 (1993).

25. K. Rzazewski and R. Grobe, *Phys. Rev. A* **33**, 1855 (1986).

26. M. Dorr and R. Shakeshaft, *Phys. Rev. A* **38**, 543 (1988).

27. K. C. Kulander, *Phys. Rev. A* **35**, 445 (1987); K. J. Schafer and K. C. Kulander, *Phys. Rev. A* **42**, 5794 (1990).

28. R. Blank and M. Shapiro, *Phys. Rev. A* **50**, 3234 (1994); R. Blank and M. Shapiro, *Phys. Rev. A* **51**, 4762 (1995).

29. B. Piraux, R. Bhatt, and P. L. Knight, *Phys. Rev. A* **41**, 6269 (1990).

30. M. Shapiro, *J. Chem. Phys.* **101**, 3844 (1994).

31. A. G. Abrashkevich ad M. Shapiro, *J. Phys. B* **29**, 627 (1996).

32. E. Frishman and M. Shapiro, *Phys. Rev. A* **54**, 3310 (1996).

33. A. Vardi and M. Shapiro, *J. Chem. Phys.* **104**, 5490 (1996).

34. C. Cohen-Tannoudji, *Quantum Mechanics*, Wiley, New York, 1977, Vol. **2**, Chapter XIII.

35. M. Shapiro, *J. Chem. Phys.* **97**, 7396 (1993); *idem., J. Chem. Phys.* **99**, 2453 (1993).

36. P. Brumer and M. Shapiro, *Chem. Phys. Lett.* **126**, 541 (1986); M. Shapiro and P. Brumer, *J. Chem. Soc. Faraday Trans.* 2 **93**, 1263 (1997).

37. E. J. Heller, in *Potential Energy Surfaces; Dynamics Calculations*; D. G. Truhlar, ed., Plenum, New York, 1981; E. J. Heller, *Acc. Chem. Res.* **14**, 368 (1981).

38. M. Shapiro, *J. Phys. Chem.* **102**, 9570 (1998).

39. M. Shapiro, M. J. J. Vrakking, and A. Stolow, *J. Chem. Phys.* **110**, 2465 (1999).

40. See, for example, P. L. Knight, M. A. Lauder, and B. J. Dalton, *Phys. Rep.* **190**, 1, (1990), and references cited therein; O. Faucher, D. Charalambidis, C. Fotakis, J. Zhang, and P. Lambropoulos, *Phys. Rev. Lett.* **70**, 3004 (1993).

41. Z. Chen, M. Shapiro, and P. Brumer, *J. Chem. Phys.* **102**, 5683 (1995); A. Shnitman, I. Sofer, I. I. Golub, A. Yogev, M. Shapiro, Z. Chen, and P. Brumer, *Phys. Rev. Lett.* **76**, 2886 (1996).

42. K. J. Boller, A. Imamoglu, and S. E. Harris, *Phys. Rev. Lett.* **66**, 2593 (1991).

43. A. Abragam, *The Principles of Nuclear Magnetism*, Oxford Univ. Press, London, 1961, pp. 65, 66.

44. D. G. Grischkowski, *Phys. Rev. Lett.* **24**, 866 (1970); D. G. Grischkowski and J. A. Armstrong, *Phys. Rev. A* **6**, 1566 (1972); D. G. Grischkowski, E. Courtens, and J. A. Armstrong, *Phys. Rev. Lett.* **31**, 422 (1973); D. G. Grischkowski, *Phys. Rev. A* **7**, 2096 (1973); D. G. Grischkowski, M. M. T. Loy, and P. F. Liao, *Phys. Rev. A* **12**, 2514 (1975).

45. R. P. Feynmann, F. L. Vernon Jr., and R. W. Hellwarth, *J. Appl. Phys.* **28**, 49 (1957).

46. L. Allen and H. H. Eberly, *Optical Resonance of Two-level Atoms*, Wiley, New York, 1975.

47. M. Takatsuji, *Phys. Rev. A* **4**, 808 (1971).

48. R. G. Brewer and E. L. Hahn, *Phys. Rev. A* **11**, 1641 (1975).

49. R. J. Cook and B. W. Shore, *Phys. Rev. A* **20**, 539 (1979).

50. J. Oreg, F. T. Hioe, and J. H. Eberly, *Phys. Rev. A* **29**, 690 (1984).

51. F. T. Hioe and J. H. Eberly, *Phys. Rev. Lett.* **12**, 838 (1981).

52. U. Gaubatz, P. Rudecki, M. Becker, S. Schiemann, M. Külz, and K. Bergmann, *Chem. Phys. Lett.* **149**, 463 (1988); U. Gaubatz, P. Rudecki, S. Schiemann, and K. Bergmann, *J. Chem. Phys.* **92**, 5363 (1990); J. R. Kuklinski, U. Gaubatz, F. T. Hioe, and K. Bergmann, *Phys. Rev. A* **40**, 6741 (1989); B. W. Shore, K. Bergmann, J. Oreg, and S. Rosenwaks, *Phys. Rev. A* **44**, 7442 (1991); K. Bergmann and B. W. Shore, in *Molecular Dynamic Spectroscopy by Stimulated Emission Pumping*, H. L. Dai and R. W. Field, ed., World Scientific Singapore, 1994.

53. C. E. Carroll and F. T. Hioe, *Phys. Rev. Lett.* **68**, 3523 (1992).

54. T. Nakajima, M. Elk, J. Chang, and P. Lambropoulos, *Phys. Rev. A* **50**, R913 (1994).

55. G. Coulston and K. Bergmann, *J. Chem. Phys.* **96**, 3467 (1992).

56. B. W. Shore, J. Martin, M. P. Fewell, and K. Bergmann, *Phys. Rev. A* **52**, 566 (1995); J. Martin, B. W. Shore, and K. Bergmann, *Phys. Rev. A* **52**, 583 (1995).

57. C. Liedenbaum, S. Stolte, and J. Reuss, *Phys. Rep.* **178**, 1 (1989).

58. B. Broers, H. B. van Linden van den Heuvell, and L. D. Noordam, *Phys. Rev. Lett.* **69**, 2062 (1992).

59. J. S. Melinger, S. R. Gandhi, A. Hariharan, J. X. Tull, and W. S. Warren, *Phys. Rev. Lett.* **68**, 2000 (1992); J. S. Melinger, S. R. Gandhi, A. Hariharan, D. Goswami, and W. S. Warren, *J. Chem. Phys.* **101**, 6349 (1994).

60. J. T. Bahns, W. C. Stwalley, and P. L. Gould, *J. Chem. Phys.* **104**, 9689 (1996).

61. A. Bartana, R. Kosloff, and D. J. Tannor, *J. Chem. Phys.* **99**, 196 (1993).

62. S. Chu, J. E. Bjorkholm, A. Ashkin, and A. Cable, *Phys. Rev. Lett.* **57**, 314 (1986).

63. A. Aspect, C. Cohen-Tannoudji, J. Dalibard, A. Heidemann, and C. Solomom, *Phys. Rev. Lett.* **57**, 1688 (1986).

64. D. Normand, L. A. Lompre, and C. Cornaggia, *J. Phys. B* **25**, L497 (1992); M. Schmidt, D. Normand, and C. Cornaggia, *Phys. Rev. A* **50**, 5037 (1994).

65. P. Dietrich, D. T. Strickland, M. Laberge, and P. B. Corkum, *Phys. Rev. A* **47**, 2305 (1993).

66. E. Charron, A. Giusti-Suzor, and F. H. Mies, *J. Chem. Phys.* **103**, 7359 (1995).

67. A. D. Bandrauk and E. E. Aubanel, *Chem. Phys.* **198**, 159 (1995).

68. T. Seideman, *J. Chem. Phys.* **103**, 7887 (1995).

69. H. Haberland and B. v. Issendorff, *Phys. Rev. Lett.* **76**, 1445 (1996).

70. W. Kim and P. M. Felker, *J. Chem. Phys.* **104**, 1147 (1996).

71. B. Friedrich and D. R. Herschbach, *Phys. Rev. Lett.* **74**, 4623 (1995).

72. H. R. Thorsheim, J. Weiner, and P. S. Julienne, *Phys. Rev. Lett.* **58**, 2420 (1987).

73. Y. B. Band and P. S. Julienne, *Phys. Rev. A* **51**, R4317 (1995).

74. A. Fioretti, D. Comparat, A. Crubellier, O. Dulieu, F. Masnou-Seeuws, and P. Pillet *Phys. Rev. Lett.* **80**, 4402 (1998).

75. A. Vardi, D. Abrashkevich, E. Frishman, and M. Shapiro, *J. Chem. Phys.* **107**, 6166 (1997), The Na-Na potential curves and the relevent electronic dipole moments are from I. Schmidth,—Ph.D. thesis; Kaiserslautern Univ., 1987.

76. B. R. Johnson, *J. Chem. Phys.* **67**, 4086 (1977).

77. R. E. Langer, *Phys. Rev.* **51**, 669 (1937).

78. R. E. Langer, *Trans. Am. Math. Soc.* **34**, 447 (1932; *ibid*, **37**, 937 (1935).

79. R. E. Langer, *Bull. Am. Math. Soc.* **40**, 545 (1934).

80. W. H. Miller, *J. Chem. Phys.* **48**, 464 (1968).

81. W. H. Miller, ed., *Modern Theoretical Chemistry*; Part B, Plenum Press, New York, 1976.

82. M. S. Child, *Semiclassical Mechanics with Molecular Applications*, Clarendon Press, Oxford, UK., 1991, pp. 108–112.

83. For reviews, see A. D. Bandrauk, ed., *Molecules in Laser Fields*, Marcel Dekker, New York, 1994.

84. M. V. Fedorov, O. V. Kudrevatova, V. P. Makarov, and A. A. Samokhin, *Opt. Commun.* **13**, 299 (1975).

85. N. M. Kroll and K. M. Watson, *Phys. Rev. A* **8**, 804 (1973); *idem.*, **13**, 1018 (1976).

86. J. I. Gerstein and M. H. Mittleman, *J. Phys. B* **9**, 383 (1976).

87. J. M. Yuan, T. F. George, and F. J. McLafferty, *Chem. Phys. Lett.* **40**, 163 (1976); J. M. Yuan, J. R. Laing, and T. F. George, *J. Chem. Phys.* **66**, 1107 (1977); T. F. George, J. M. Yuan, and I. H. Zimmermann, *Faraday Disc. Chem. Soc.* **62**, 246 (1977); P. L. DeVries and T. F. George, *Faraday Disc. Chem. Soc.* **67**, 129 (1979); T. F. George, *J. Phys. Chem.* **86**, 10 (1982).

88. A. M. F. Lau and C. K. Rhodes, *Phys. Rev. A* **16**, 2392 (1977); A. M. F. Lau, *Phys. Rev. A* **13**, 139 (1976) *idem.*, **25**, 363 (1981).

89. V. S. Dubov, L. I. Gudzenko, L. V. Gurvich, and S. I. Iakovlenko, *Chem. Phys. Lett.* **45**, 351 (1977).

90. A. D. Bandrauk and M. L. Sink, *Chem. Phys. Lett.* **57**, 569 (1978); *J. Chem. Phys.* **74**, 1110 (1981).

91. A. E. Orel and W. H. Miller, *Chem. Phys. Lett.* **57** 362 (1978); *idem.*, **70**, 4393 (1979); *idem.*, **73**, 241 (1980).

92. J. C. Light and A. Altenberger-Siczek, *J. Chem. Phys.* **70**, 4108 (1979).

93. K. C. Kulander and A. E. Orel, *J. Chem. Phys.* **74**, 6529 (1981).

94. H. J. Foth, J. C. Polanyi, and H. H. Telle, *J. Phys. Chem.* **86**, 5027 (1982).

95. M. Shapiro and Y. Zeiri, *J. Chem. Phys.* **85**, 6449 (1986).

96. T. Seideman and M. Shapiro, *J. Chem. Phys.* **94**, 7910 (1991).

97. T. Seideman, J. L. Krause, and M. Shapiro, *Chem. Phys. Lett.* **173**, 169 (1990); *idem.*, *Faraday Disc. Chem. Soc.* **91**, 271 (1991).

98. A. Zavriyev, P. H. Bucksbaum, H. G. Muller, and D. W. Schumacher, *Phys. Rev. A* **42**, 5500 (1990).

99. A. Guisti-Suzor and F. H. Mies, *Phys. Rev. Lett.* **68**, 3869 (1992).

100. G. Yao, and S.-I. Chu, *Chem. Phys. Lett.* **197**, 413 (1992).

101. E. E. Aubanel and A. D. Bandrauk, *Chem. Phys. Lett.* **197**, 419 (1992); A. D. Bandrauk, E. E. Aubanel, and J. M. Gauthier, *Laser Phys.* **3**, 381 (1993).

102. D. R. Matusek, M. Yu Ivanov, and J. S. Wright, *Chem. Phys. Lett.* **258**, 255 (1996).

103. U. Fano, *Phys. Rev.* **124**, 1866 (1961).

104. D. J. Tannor and S. A. Rice, *J. Chem. Phys.* **83**, 5013 (1985); D. J. Tannor, R. Kosloff, and S. A. Rice, *J. Chem. Phys.* **85**, 5805 (1986); D. J. Tannor, in *Molecules in Laser Fields*, A. D. Bandrauk, ed., Marcel Dekker, New York, 1994, p. 403.

105. T. Seideman, M. Shapiro, and P. Brumer, *J. Chem. Phys.* **90**, 7132 (1989); I. Levy, M. Shapiro, and P. Brumer, *J. Chem. Phys.* **93**, 2493 (1990).

106. A. Vardi, D. Abrashkevich, E. Frishman, and M. Shapiro, *J. Chem. Phys.* **107**, 6166 (1997).

107. C. Eckart, *Phys. Rev.* **35**, 1303 (1930).

108. H. Eyring, J. Walter, and G. E. Kimball, *Quantum Chemistry*, Wiley, New York, 1944.

109. A. Vardi, M. Shapiro, *Phys. Rev. A* **58**, 1352 (1998).

110. B. Liu, *J. Chem. Phys.* **58**, 1925 (1973).

111. D. G. Truhlar and C. J. Horowitz, *J. Chem. phys.* **68**, 2466 (1978).

112. L. L. Chang, L. Esaki, and R. Tsu, *Appl. Phys. Lett.* **24**, 593 (1974).

113. S. C. Kan and A. Yariv, *J. Appl. Phys.* **67**, 1957 (1990).

114. A. Sa'ar, S. C. Kan, and A. Yariv, *J. Appl. Phys.* **67**, 3892 (1990).

115. H. Yamamoto, Y. Kanie, M. Arakawa, and K. Taniguchi, *Appl. Phys. A* **50**, 577 (1990).

116. W. Cai, T. F. Zheng, P. Hu, M. Lax, K. Shun, and R. Alfano, *Phys. Rev. Lett.* **65**, 104 (1990).

117. K. A. Chao, M. Willander, and M. Yu Galperin, *Physica Scripta* **T54**, 119 (1994).

118. J. R. Taylor, *Scattering Theory*, Wiley, New York, 1972, Chapter 11.

119. A. Yariv, *Optical Electronics*; 4th ed., Saunders College Pub., Philadelphia, 1991.

120. I. Vorobeichik, R Lefebvre, and N. Moiseyev, *Europhys. Lett.* **41**, 111 (1998).

121. D. E. Amos, Sandia National Laboratories computer routine library,—slatec/Complex Airy; subsidiary routines.

122. H. Okabe, *Photochemistry of Small Molecules*, Wiley, New York, 1978.

VIBRATIONAL ENERGY FLOW:
A STATE SPACE APPROACH

M. GRUEBELE

Department of Chemistry and Beckman Institute for Advanced Science and Technology, University of Illinois, Urbana, Illinois

CONTENTS

Advances in Chemical Physics, Volume 114, Edited by I. Prigogine and Stuart A. Rice.
ISBN 0-471-39267-7 © 2000 John Wiley & Sons, Inc.

I. INTRODUCTION

In chemical physics, all dynamical questions in the end lead to one: *Where does the energy go*? If we can answer this, we can answer all others about rates, isomerization, product distributions, and control over reactions. The question has fascinated and plagued chemical physicists at the molecular level since the 1930s [1]. It crops up in many areas, from the extraordinarily long-lived vibrational coherence of Si–H stretching vibrations on hydrogen-passivated silicon surfaces [2] to the connection between vibrational energy transfer and macroscopic heat transport [3].

Perhaps the simplest version of the problem asks how vibrational energy is redistributed in an isolated polyatomic molecule prepared in a bound nonstationary vibrational state [4–6]. This has been referred to as *intramolecular vibrational redistribution* or *relaxation* (IVR). A spectral feature carrying oscillator strength acts as the initial state, while the "bath" for energy redistribution mainly has the character of backbone or skeletal vibrations. To discuss the process meaningfully, what constitutes the bath must be defined. In quantum mechanics and in statistical mechanics, it is convenient to distinguish a system (those degrees of freedom of particular interest) from the bath. Quantum mechanically, nonseparability of system and bath degrees of freedom leads to an entangled multidimensional wavefunction. Hence any attempt to average out the environmental degrees of freedom (e.g., via a reduced density matrix) results in dephasing of the system, and usually also in system population relaxation.

Energy flow within isolated molecules can thus be viewed as an either relaxation or dephasing process; if the system consists of the prepared nonstationary vibrational state, and the bath is the orthogonal manifold of states, the population of the system, indeed, relaxes. (At the same time, the system will also undergo pure dephasing.) The overall energy of the molecule is of course conserved in the absence of emission or collisions.* If the molecule as a whole is taken to be the system, and the bath is essentially noninteracting (e.g., a in molecular beam), then intramolecular vibrational energy redistribution can rightly be viewed as a pure dephasing process. It will be convenient in our discussion to adopt either point of view depending on the problem at hand.

---

*Until Section VI, we shall neglect collisions, emission, and "complicated" excitation fields.

To an excellent approximation, nonstationary rovibrational states $|0\rangle$ of a bound isolated molecule are described by

$$|t\rangle = e^{-iH_{vr}t/\hbar}|0\rangle = U(t)|0\rangle = \sum_{s\in F} c_s e^{-i\omega_s t}|s\rangle \qquad (1.1)$$

where $s \in F$ counts the set of eigenstates under the envelope of the spectral feature $|0\rangle$. The problem clearly can be cast into equivalent time-dependent and time-independent formulations; a short pulse can coherently prepare the spectral feature $|0\rangle$, which then dephases according to Eq. (1.1) or a high-resolution laser can scan across the feature, revealing the intensities $I_s = |c_s|^2$ of each contributing eigenstate. Unfortunately, Eq. (1.1) provides only a formal solution. When s grows into the thousands or millions, the knowledge that a finite set of bound states in principle evolves quasiperiodically is a small consolation when none of the c_s and ω_s can be computed! The first viable microscopic IVR theories were developed only in the 1960s [7–9]. Why? Following a better understanding of nonintegrable classical dynamics and its connection to quantum mechanics [10], it became clear from the 1960s onward that the eigenstates of large molecules at high energy can be very complicated objects [11,12]. The superposition (1.1) was therefore almost impossible to construct for a realistic polyatomic rovibrational Hamiltonian H_{vr}. Instead perturbation methods, uniformity assumptions about coupling matrix elements, or a statistical treatment of H_{vr} had to be invoked [4,7,8]. Therefore, the pioneering models could not take into account correlations that exist among the coefficients c_s and angular frequencies ω_s in eq. (1.1). As it turns out, such correlations exist and have subtle but important effects on the dynamics [13–17]; in the energy range from the vibrational flow threshold to beyond the lowest dissociation limit(s), H_{vr} is too strongly coupled for the resulting dynamics to be treated perturbatively at *long* times, yet too weakly coupled to be treated by *global* statistical measures. The latter is due to the fact that most bonds retain their localized nature, even during chemical reactions [18].

Beginning in the 1980s, experiments started to follow the spectrum

$$I(\omega) = \mathscr{F}\mathscr{T}\{\langle 0|t\rangle\} = \sum_{s\in F} I_s\delta(\omega - \omega_s), I_s = |c_s|^2 \qquad (1.2)$$

or equivalently, the survival probability of the initially prepared state $|0\rangle$

$$P(t) = |\langle 0|t\rangle|^2 = \sum_s I_s^2 + 2\sum_{s>u} I_s I_u \cos\omega_{su}t \qquad (1.3)$$

in unprecedented detail [19–37], beyond the level that earlier models were meant to treat. Experiments are limited to states that carry oscillator strength in single or double resonance (e.g., overtone pumping [34] or stimulated emission pumping [26]). Only a small subset of "bright" vibrational features can be prepared among all the potential nonstationary states $|0\rangle$. These features produce clear spectral patterns: IVR even at high energy does not simply lead to a completely random collection of eigenstates with random intensities [38,39]. More recent work is moving beyond studying (1.2) and (1.3) for "bright" states alone; the correlations that survive among the rovibrational eigenstates at high energy are being probed ever more deeply even for states that carry no intrinsic vibrational oscillator strength [40].

In the late 1980s, theory and simulation rose to the renewed experimental challenge; fully quantum models of systems with many degrees of freedom are now computable, thanks to advances in algorithms and computing power [17,41–50]. New analytic treatments have resolved issues that could not be treated by earlier approaches [14,16,51–55]. Correlations among terms in Eq. (1.1) turned out to be significant even when many modes are involved and when the system has already evolved to an 'intermediate' timescale [45,54]. We will abbreviate "correlated intermediate timescale dynamics" as CIT dynamics. It will be the main concern of the state-space treatment given here. A brief overview of the relevant time-scales, and of the assumptions made by models that neglect correlations, is therefore in order.

At very early times, Eq. (1.3) follows a cosine rolloff:

$$\lim_{t \to 0} P(t) = \cos(\omega_{\mathrm{rlf}} t), \quad \omega_{\mathrm{rlf}} = \left[\sum_{s,u} I_s I_s \omega_{su}^2 \right]^{1/2} \tag{1.4}$$

The period $\tau_{\mathrm{rlf}} = 2\pi/\omega_{\mathrm{rlf}}$ becomes very short as the observed density of states ρ_{obs} increases. (It is zero for a Lorentzian spectral envelope even when ρ_{obs} is finite.) Subsequent to this rolloff, a rate k_{IVR} can be defined in terms of the $1/e$ decay time τ_{IVR} of $P(t)$. Eventually, after a time τ_σ, Eq. (1.3) fluctuates about its average value, the dilution factor [21,56]

$$\bar{P} = \sigma = N_{\mathrm{tot}}^{-1} = \sum_s I_s^2 \tag{1.5}$$

The physical interpretation is that an effective maximum number of states N_{tot} participates in the dynamics. When the wavepacket has dephased over all these states, the average population in each, including in $|0\rangle$, will be $\sigma = N_{\mathrm{tot}}^{-1}$ [2].* The value of N_{tot} is constrained by the the total density of

---

*According to Eq. (1.1), $P(t)$ eventually has recurrences to any value < 1 even if the ω_{nm} are not rationally related. As in the famous freshman chemistry movie of the molecules that do not return to the corner of the box unless the movie is run backward, we can safely discount this possibility once $\bar{P} \approx \sigma$, if the molecule is sufficiently large or highly excited. In practice, fluorescence and collisions (Section VI) take care of the problem of recurrences.

states $\bar{\rho}_{tot} = \delta\bar{E}^{-1}$ under the IVR linewidth, as well as by nodal properties of the vibrational wavefunctions; a feature $|0\rangle$ cannot efficiently overlap with all wavefunctions in the eigenbasis $\{|n\rangle\}$ [57]. Even a Lorentzian spectral envelope, which lacks a second moment, has $\sigma > \pi k_{IVR}\hbar\rho_{tot} \tanh(\pi k_{IVR}\hbar\rho_{tot})$. If σ does not reach the smallest value compatible with these constraints, the IVR wavepacket will be termed *nonergodic*. An example would be the presence of a symmetry, which automatically introduces a good quantum number via Nöther's theorem and restricts the number of accessible states (Section V). The concept embodied in σ and N_{tot} is also referred to in the literature as *participation number, fractionation*, or *fragmentation*.

Between τ_{rlf} and τ_σ, the dynamics could be either exponential or CIT. The basic assumption of exponential dynamics can be cast in several forms. As mentioned above, the c_s and ω_s in the eigenbasis must be *independently* and *randomly* distributed; equivalently, application of the propagator $U(\Delta t)$ must not introduce correlations in the IVR wavepacket. [The initial wavepacket $|0\rangle$ may, of course, be a very special function, but $U(\Delta t)$ should not have properties that help preserve this special status.]

Consider a typical exponential model, where the initally excited feature $|0\rangle$ is coupled to a prediagonalized dark manifold $\{|s'\rangle\}$ by matrix elements $V_{0s'}$ with a Gaussian or similar smooth random distribution [7]. Application of first-order time-dependent perturbation theory to the interaction of the feature $|0\rangle$ with a prediagonalized manifold $\{|s'\rangle\}$ of large density of states yields

$$P_{0\to n'}(\Delta t) = 1 - \frac{2\pi}{\hbar}\bar{\rho}_{tot}V_{rms}^2\Delta t = 1 - k_{GR}\Delta t \tag{1.6}$$

Here the individual couplings $V_{0s'}$ from $|0\rangle$ to $\{|n'\rangle\}$ have been replaced by an average value

$$V_{rms} = \lim_{s_{tot}\to\infty}\left(\frac{\sum_{s'}|V_{0s'}|^2}{s_{tot}}\right)^{1/2} \tag{1.7}$$

and the total density of states

$$\rho_{tot} = \sum_{s'=0}\delta(\omega_{s'} - \omega) \tag{1.8}$$

has also been replaced by a window-averaged value $\bar{\rho}_{tot}$. If one further assumes that subsequent population decay steps are not temporally correlated in any way with the previous steps, one obtains

$$\lim_{\Delta t\to 0}P_j(t = j\Delta t) \equiv \lim_{\Delta t\to 0}\left(1 - \frac{\Delta t}{\tau_{GR}}\right)^j = e^{-t/\tau_{GR}} \to (1 - \sigma)e^{-t/\tau_{GR}} + \sigma$$

$$\tag{1.9}$$

The rightmost term is corrected for a finite density of coupled states. The most interesting assumption needed to get to Eq. (1.9) is embodied in the following postulate: There are no correlations within and among ρ_{tot} and $V_{0s'}$ as a function of energy. It follows from the LKL algorithm [58,59] that there are no correlations between c_s and ω_s. Another way of stating this is that ρ_{tot} and $V_{0s'}$ must be sufficiently independent and random functions of the energy. This, in turn [via Eq. (1.1), implies that the successive time propagations in Eq. (1.9) are uncorrelated.

There are obvious instances when such a model must fail. For instance, if the couplings are very strong, peaks in ρ_{tot} are no longer Poisson-distributed. At very low vibrational energy correlations undoubtedly exist, leading to the well-studied quantum beats and other deviations from Eq. (1.9). Also, below a certain threshold energy $k_{IVR} = 0$, even though $\bar{\rho}_{tot}$ and V_{rms} are nonzero [60]. But what about a polyatomic molecule with moderate coupling matrix elements excited to a high energy? $\bar{\rho}_{tot}$ would become so large, and the $V_{0n'}$ would grow in number so rapidly, that in the small energy interval covering the spectral width of a feature $|0\rangle$, random distributions and averaging would seem appropriate. Fortunately for some applications of IVR (Section VI), that is not the case.

Equation (1.9) is based on convenience (e.g., solvability by perturbation theory), not on physical intuition or on a rigorous analysis of H_{vr}. Let us apply some of both to see on what timescale Eq. (1.9) becomes invalid. It has been proved using the Hellman–Feynman theorem that for any spectrum originating from a single feature $|0\rangle$ centered at $\omega_0 = 0$ [61,62].

$$I_s = |c_s|^2 = \frac{1}{2} \frac{\partial \ln \omega_s}{\partial \lambda}\bigg|_{\lambda=1} \qquad (1.10)$$

λ is a unitless factor scaling all couplings from $|0\rangle$ to other states expressed in any basis. The magnitude of c_s is therefore very strongly correlated with ω_s; the postulate fails as a matter of principle. Still, in practice the correlations might by very small. It turns out that they are not, once the IVR wavepacket propagates beyond a critical timescale τ_{loc}. Correlations eventually arise because state space is structured. State space is the collection of all quantum states sampled by the IVR process [63]. It can be expressed in any Hilbert basis, but the constraints that cause the postulate to fail are easier to see in a feature basis (Section IV). A feature basis contains hierarchically sorted orthogonal wavefunctions, such that the smallest possible number of wavefunctions describes the dynamics up to a time t.* In such a basis, $|0\rangle$ is directly coupled to only a small number N_{loc} of states. Until

*A full definition will be given in Section IV.

these states are significantly populated, the IVR wavepacket dynamics are exponential (Section V). After a period τ_{loc}, these states become occupied and the time evolution is correlated and nonexponential. The existence of a feature basis—or equivalently, of a structured state space—is guaranteed for IVR because H_{vr} can always be expressed in a locally coupled form (Section III). In matrix form, such a Hamiltonian has the structure of a hierarchical local random matrix (HLRM) [14,15,17,64,65] whose properties are very different from globally random matrices (e.g., the Gaussian ortogonal ensemble [66]).

Concrete examples of correlations and state space structure, discussed in detail later, are as follows:

1. At short times, no quantum interference effects are possible because no closed loops of paths in state space are populated yet; at longer times, coherent interference effects arise that depend on the nature of the couplings and correlate the eigenstates. This is a purely quantum-mechanical effect.

2. At short times, the IVR wavepacket evolution is not yet confined by the boundaries of state space (i.e., by the fact that there are no states with negative vibrational quantum numbers); at longer times, the IVR wavepacket feels those boundaries, the main reason why $\sigma > 0$. This is analogous to classical trajectories eventually revisiting a phase space volume in a phase space bounded by energy conservation, and can be explained classically.

3. Consider the following thought experiment. A long-chain molecule has two terminal stretching modes with slightly different frequencies and wavefunctions $|v_1 v_2\rangle$. Only mode 1 carries oscillator strength. A high-resolution laser scans over the $|v0\rangle$ vibrational feature [64]. Because of the local nature of bonding, $|v0\rangle$ and $|0v\rangle$ are not significantly coupled whenever they are nondegenerate and the two bonds are far apart (Section III, Fig. 7). Therefore, eigenstates with strong $|0v\rangle$ character may appear very near the center of the $|v0\rangle$ feature, but with negligible intensity. Eigenstates with more $|v0\rangle$ character may light up in the spectrum, but are more likely to appear away from the center of the feature as a result of coupling repulsion. Clearly, the local nature of bonding correlates spectral positions and intensities, showing how failure of the postulate is a consequence of Eq. (1.10). This kind of real-space localization can of course be treated classically. As a general note, a rigorous treatment of state space correlations must include quantum effects, but many of its general features were already presaged by classical approaches because of the close relationship between the state space and action/phase spaces [5].

Example 3 illustrates another important concept that emerges from the state space treatment—the important quantity that controls IVR dynamics is not $\bar{\rho}_{\text{tot}}$, but a local density of states $\bar{\rho}_{\text{loc}}$ closely related to N_{loc} (Section IV) [67,68]. In the example, the $|0n\rangle$ state contributes to $\bar{\rho}_{\text{tot}}$ near the feature $|n0\rangle$, but not to $\bar{\rho}_{\text{loc}}$. Therefore it does not participate in the early-time dynamics. The idea of a local density of states is also reflected in intuitive notions of "tiers" through which the IVR must progress. Such ideas have been in use since the 1970s. That tier models must be rigorously constrained by the structure of the vibrational Hamiltonian was first recognized in the early 1980s [69]. Realistic model tests had to wait for several more years, and the applicability of tiering was rigorously verified only in the 1990s, and emerges naturally from the state-space picture (Section V), [17,70].

The range of times for CIT dynamics is now defined. It begins at τ_{loc} and ends at τ_{σ}. Because $\bar{\rho}_{\text{loc}}$ is a much more slowly increasing function of energy than $\bar{\rho}_{\text{tot}}$, $N_{\text{loc}} \ll N_{\text{tot}}$ and hence $\tau_{\sigma} \gg \tau_{\text{loc}}$. This is *especially* true for large molecules at high energy, where the disparity between $\bar{\rho}_{\text{loc}}$ and ρ_{tot} can be very large. At low energies, it is quite possible that $\tau_{\sigma} \approx \tau_{\text{loc}}$, and the distinction between exponential and correlated dynamics then becomes negligible, especially if quantum beats are present. However, in terms of reactivity, larger molecules and higher energies are the more interesting case. CID governs the vast majority of the IVR process in such cases. For example, consider a feature covering 10^6 eigenstates, but with $N_{\text{loc}} = 100$. Its IVR decay is quasiexponential until $P(t)$ has dropped to 0.01, then becomes correlated until $\bar{P}(t) \approx 10^{-6}$. On a linear plot of $P(t)$ versus t, the decay would appear exponential, yet a log–log plot reveals that most of the dynamics [when $P(t)^{-1}$ grows from 100 to a million participating states] are nonexponential (Fig. 1). The actual strength of the deviations from exponentiality depends on the strength of the surviving correlations. The evidence available so far suggests that the discrepancy can be substantial [50].

Most of the IVR process is thus not characterized by a rate at all [15–17]. More appropriate functional forms for $P(t)$ will be discussed in Section V. CIT dynamics are governed by a dimensional exponent δ instead of a rate [16]. Analytic models [16,54], quantum dynamics simulations [17,44,50, 55,71], classical simulations [71], and analysis of experimental data [50,55] suggest that the asymptotic form

$$P(t) \sim (1 - \sigma)t^{-\delta/2} + \sigma \qquad (1.11)$$

is more appropriate. Of course, as $\delta \to \infty$, this power law becomes indistinguishable from an exponential function [cf. Eq. (1.9) for $j = \delta/2$]. δ will be seen to be related to the number of local resonances allowing population to escape from a locus in state space, namely, to the dimensionality of

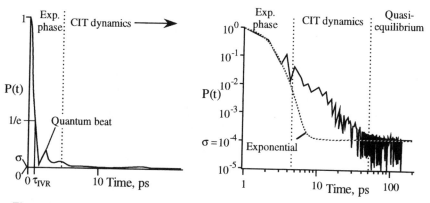

Figure 1. Survival probabilities of an initially prepared state on a linear and logarithmic scale, emphasizing different properties of the IVR process.

the state space manifold on which IVR occurs [55]. The exponent differs among different classes of initial states, but has the same value within a class even though k_{IVR} may differ greatly among states. The main classes to be differentiated in Section IV are edge, partially interior, and interior states. Its lesser dependence on initial-state preparation therefore makes δ a more useful generic descriptor of IVR.

The small values of δ measured and calculated so far even for large (> 50 degrees of freedom) molecules show that CIT dynamics access the avaliable state space much more gradually than does simple exponential dynamics [50]. This, in turn, allows for the possibility of controlling IVR via coherent manipulation [45], thereby generalizing the applicability of coherent control [72–74], which has already been fruitfully applied to smaller molecules or cases where IVR does not occur (Section VI) [75–77]. Vibrational energy redistribution joins a long list of phenomena in chemical physics, including cluster isomerization kinetics [78], protein relaxation and folding [79–81], and reaction kinetics of complex mixtures [82], which can be described by anomalous diffusion and multiscale dynamics [83,84] over all or part of their temporal range.

It is not surprising that the survival probability of IVR initiated in hydrogenic stretching modes is usually well fitted by exponential functions at short times. Because of the longer timescale required, correlated dynamics occurs mostly among the skeletal or backbone vibrations, not among the hydrogenic vibrations. The lower frequency (< 40 THz)* modes account

---

*$1000\,\text{cm}^{-1}$ in vacuo corresponds to $29.9792458\,\text{THz}$; for simplicity, all energy terms and levels will also be expressed in Hz via $\nu = E/h$.

for most of the average character of the eigenstates at high densities of states. Since most chemical reactions involve the making and breaking of bonds between second- and higher-row atoms, such modes are also responsible for transporting energy to and from the reaction coordinate. Backbone IVR is therefore of particular interest, although it has been surprisingly little studied. The main reason is that backbone combination modes are difficult to access experimentally, especially at high energies [22,85,86].

The aim of this review is not to present an encyclopedic medley of all work in the field of IVR. Rather, a specific way of looking at IVR, quantum state space, is proposed as a fruitful approach that can solve all the important aspects of the problem. Although most topics in modern IVR research are at least touched on, the emphasis is on those result most closely related to the state space framework and to IVR among skeletal modes. The purpose of the following sections is to (1) discuss some examples of pertinent experimental data; (2) develop the structure of the (ro)vibrational Hamiltonian and the underlying reasons for that structure; (3) develop the structure of state space in near-optimal representations related to the effective Hamiltonian; (4) examine the dynamical consequences of the Hamiltonian and state space, which can provide a satisfactory explanation for all known experimental data; and (5) to look at some examples of connected research areas to which our detailed understanding of IVR, built on several decades of model building and experimentation, will be applied.

II. THE EXPERIMENTAL TAPESTRY

Vibrational energy flow experiments in the gas phase have been extensively reviewed [27,32,87]. Some examples illustrating the various types of results which must be explained by a successful IVR theory are discussed here, organized by concepts rather than chronology. A few results from collisional energy transfer, and vibrational relaxation in weakly and strongly perturbing solvents will also be mentioned because they point toward future directions in IVR research (Section VI).

Several experiments have verified that the onset of IVR is not smooth; in contrast to Eq. (1.6), there is a threshold for energy flow below which k_{IVR} cannot be defined [39,85,88,89]. High-resolution spectroscopy at low vibrational energy reveals only isolated resonances. As the vibrational energy is increased, bright states become highly fragmented over a relatively narrow range of E_{vib} and $\bar{\rho}_{tot}$. A typical example of this transition is shown in Fig. 2 for propyne [89–92]. Similar observations have been made in time-resolved studies: anthracene, fluorene, and cyclohexylaniline vibrational features decay very slowly with regular quantum beat patterns at low energy, show "intermediate case" irregular decaying beat patterns at higher energy,

Propyne

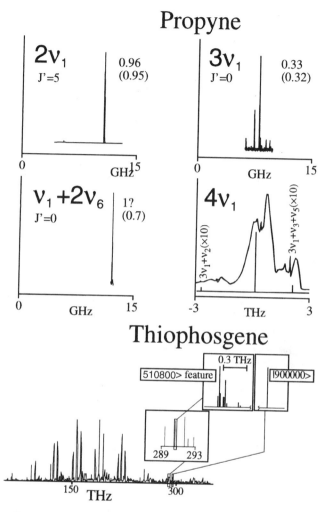

Thiophosgene

Figure 2. Top: overtone and combination bands of propyne from Refs. 36, 89, 91, and 96; the numbers in parentheses are theoretical dilution factors from Ref. 46 below the experimental values. The assignments for the $4\nu_1$ gateway states are also from Ref. 46. Bottom: emission and SEP data from Ref. 37, showing a hierarchical structure for a vibrational spectrum dominated by partially interior skeletal vibrations.

and finally smooth decays at the highest energies (Fig. 3) [22,88,93]. Once the threshold has been reached, the IVR decay rate increases with energy, but not as rapidly as one might expect on the basis of a total density of states [94].

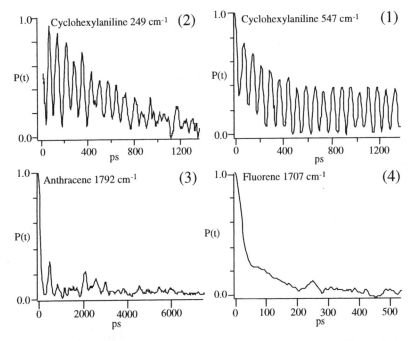

Figure 3. Survival probabilities of three molecules from Refs. 22,88, and 93; panels 1 and 2 illustrate the nonmomotonic energy dependence of k_{IVR}; panel 3 corresponds to more "dissipative" IVR with a quantum beat structure that does not follow the initial exponential decay; panel 4 shows smoother IVR dynamics at high energy; the quantum beats are merging into an envelope that cannot be fitted by an exponential decay.

The pattern of quantum beats in the "intermediate case" can be quite complex (Fig. 3). The beat patterns smooth out as molecular size and energy increase, but never completely go away even at long times. For spectra such as in Fig. 2, the observed long-time fluctuations of $P(t)$ are on the order of $\bar{P} = \sigma$, and indeed, this is expected on very general grounds [56]. At intermediate times, "mini–quantum beats" can be observed, which are larger than the long-time fluctuations. A typical example is given by the $P(t)$ calculated from the overall ground-state SEP spectrum of $SCCl_2$, or the decay patterns observed in anthracene at $E_{vib} \approx 45$–$60\,THz$ in the S_1 state [37,93]. In cases where the experimental data at intermediate times have been analyzed, the decay envelope of such beat patterns fits better to Eq. (1.11) than Eq. (1.9) [50].

The dilution factor σ generally increases with increasing $\bar{\rho}_{tot}$, as verified by time-dependent measurements of a large number of CH overtones of organic compounds [21]. However, the correlation with $\bar{\rho}_{tot}$ is not very good (see Fig. 17). The number of participating states strongly depends on the nature and location of the excited feature, and can vary greatly for similar vibrations excited at nearly the same E_{vib} and $\bar{\rho}_{tot}$. Figure 2 shows that the $3\nu_1$ overtone (acetylenic CH stretch) of the chain molecule propyne is more fragmented than the nearly isoenergetic $2\nu_1 + \nu_6$ combination band (mixed methyl/acetylenic CH stretch) [89]. Another example of strong mode dependence involves the CH stretch/bend polyads of acetylene; states with high *trans*-bend character have much larger dilution factors than do nearly isoenergetic states with more CH stretch character [95]. Statistical analysis of experimental data shows that the distribution of dilution factors is *bimodal*; intermediate degrees of fractionation are rare [68]. Few studies of skeletal IVR dilution factors are available, but $SCCl_2$ offers a dramatic example (Fig. 2). Dilution factors fluctuate almost an order of magnitude among nearly isoenergetic states [37]. Concerning the fate of overtones and combination bands, the situation is reversed for this more compact molecule compared to the chainlike propyne: the $9\nu_1$ (pure CS stretch) is 6 times less fragmented than the $5\nu_1 + 1\nu_2 + 8\nu_4$ state (CS stretch/CCl stretch/out-of-plane bend), which lies $200\,\text{cm}^{-1}$ *lower* in energy.

The same kinds of fluctuations occur in initial hydrogenic IVR rates k_{IVR}, and they are not explainable in terms of the overall density of states. For example, $\bar{\rho}_{tot}$ of $(CF_3)_3CCCH$ exceeds that of $(CH_3)_3CCCH$ by eight orders of magnitude at the CH fundamental, yet its k_{IVR} is only 3 times larger [31]. Studies of cyclohexylaniline and fluorene illustrate that k_{IVR} of features with backbone vibrational character also fluctuates beyond the bounds explainable by $\bar{\rho}_{tot}$ or V_{rms}: initial decay rates are not necessarily monotonic with energy (Fig. 3), implicating local resonance structure [22,88].

The local resonance structure can become experimentally very prominent in the form of gateway states. These are states that light up in the spectrum either outside the central part of the IVR feature, or with particular prominence (Fig. 2). The latter two restrictions must be made for practical reasons; under the spectral envelope of a highly diluted feature, many eigenstates "light up," but these would not be labeled "gateways." Intuitively, gateway states are those most "directly" coupled to the feature. Such states are assigned to the first "tier" of the IVR process. Unfortunately, this depends on the choice of basis. In Sections IV and V the concept of the gateway state will be made more rigorous and basis-independent; it is a state in the optimal feature basis that makes a significant contribution to N_{loc}. There is actually a continuous manifold of states participating in IVR, which ranges from "gatewaylike" at short times to "eigenstatelike" at long times. Gateway states have been observed in many molecules, including propyne [96] and methanol [33].

On a more subtle level than gateway states, there are intensity and level spacing fluctuations in the high-resolution spectra of IVR features. The level spacings follow a Poisson distribution in the weak coupling limit or when too many independent non-Poissonian sequences overlap, and approximately follow the Wigner conjecture in the strong coupling limit when a pure sequence is observed. The Δ spectral rigidity statistic is also frequently used instead of level spacings (Section V) [66]. Acquiring sufficiently good spectra is a challenge because all eigenstates must be observed and no unrelated sequences mixed in. Nevertheless, such statistics have been obtained in a few cases [97], and new techniques are being developed to "unzip" spectra in the search for patterns [98]. The general finding is that the distributions are at most intermediate between Poissonian and Wigner for both purely vibrational couplings and vibronic couplings [99,100]. A complete breakdown of all good quantum numbers appears unlikely at chemical energies: the vibrational Hamiltonian retains structure from the weak coupling limit (Section III).

Structural influences on IVR were already discussed in terms of a chain molecule thought experiment. Chain substitution in halobenzenes results in large changes in the dilution factor [20]. (Only the broadening of the spectrum is relevant; shifts can be accounted for by non-resonant anharmonicity, as discussed in Section IV [101]). Torsional modes are another major influence on IVR dynamics [102]. Chemical timing experiments first showed dramatic increases in k_{IVR} on addition of a methyl group [103]. Later work showed that k_{IVR} is sensitively influenced by the positioning of the rotor with respect to the local mode from which energy is redistributed [35,104]. However, these effects could be due to changes in the density of states, rather than localized higher-order potential couplings [105]. The competition between low-order and high-order contributions from the potential energy will resurface repeatedly, and must be explained by a successful IVR model [102,106]. Rotor effects depend on a *local* density of states, which, in turn, depends on a tradeoff between low- and high-order resonance contributions (Section IV) at energies near the barrier height [65]. Finally, the stretch–bend resonance for CH moieties is a classic example of a structural stretch–bend–bath progression. CH(D)BrClF and other substituted methanes are particularly nice examples, and have been studied by IR multiphoton spectroscopy to reveal the progression of the IVR wavepacket from stretch to bend (with some conservation of symmetry) and finally to the bath [107].

There is experimental evidence that IVR is not an exponential rate process at intermediate times. Most of it comes from experiments that probe combination bands involving backbone modes. The time-resolved fluorescence depletion measurements of backbone combination bands in Figs. 3 (panel 4) and 4 yield nearly smooth IVR decays for fluorene and cyclohexylaniline [22,88]. These decays cannot be fitted by exponentials even if

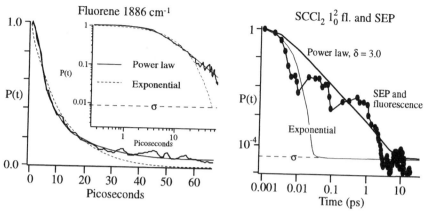

Figure 4. Onset of quantum diffusion governed by a powerlaw. On the left (Ref. 22) the decay of a narrow vibrational feature; on the right, the survival of the hierarchical spectrum in Fig. 2 (Ref. 55).

the dilution factor is taken into account. The $\tilde{X}$ state vibrational band contour of $SCCl_2$, determined by fluorescence and SEP data with $\approx 1:10^5$ dynamic range, is also incompatible with exponential dynamics for decay [37,55]. Figure 2 shows that the skeletal combination bands form a hierarchy with at least four levels of resolution: 300, 30, 3, and 0.3 THz; Fig. 4 shows the corresponding nonexponential survival probability. Nonexponential decays implicating multiple timescales have also been observed in methanol and benzene [108], although the larger dilution factors lead to a reduced timescale over which the IVR can be followed. Infrared–microwave double resonance experiments initiated in hydrogenic modes, but capable of revealing dynamics among bath states with more backbone character, have also shown discrepancies between the initial and "bath" timescales for IVR [109]. Figure 5 compares infrared and double resonance spectra (the latter of one conformer) of 2-fluoroethanol, clearly illustrating the discrepancy in line widths and hence lifetimes (0.3 ns for initial dephasing, >2 ns for "bath" dephasing corresponding to isomerization). These techniques should soon achieve sufficiently good statistical sampling to also study nonexponential dynamics within the bath.

Some gas-phase dissociation and isomerization experiments indirectly reveal contributions of IVR [25,110]. Although the pressure dependence of thermal measurements generally indicates that IVR is fast compared to *threshold* rates, molecular beam experiments on the isomerization of stilbene, and dissociation of acetaldehyde (where reactant and product fluorescence could be monitored independently) [111] reveal deviations from statistical rate theory that can be attributed to IVR. Such non-RRKM

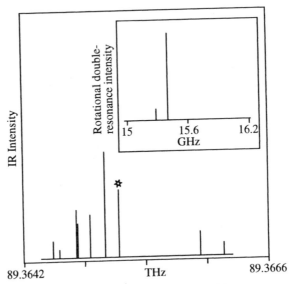

Figure 5. Pure vibrational and tT conformer rotational double-resonance spectra of 2-fluoroethanol reveal slower dynamics among the "dark" states (Ref. 109).

behavior is receiving more theoretical scrutiny [112,113]. Although not the main concern of the review in this chapter, IVR also plays a role in collision processes leading to intermolecular energy transfer. In dense gases, progress has been made toward determining the energy dependence of the vibrational deactivation probability $P(E, \Delta E, n)$ (n represents other quantum numbers of interest, such as total angular momentum) [114]. Very little is known at present about how this probability depends on vibrational state preparation, or about whether some memory of relative phases is retained in the vibrational states populated after collision. Helium clusters have been developed into very weakly coupled quantum baths for small to medium-sized molecules. Rotation is only partially quenched, intramolecular vibrational patterns persists, and individual rovibronic lines are broadened by interaction with the solvent [115]. In more strongly interacting solvents, time-resolved experiments are now measuring vibrational survival probabilities of solutes in unprecedented detail [116,117]. There is evidence for partial survival of intramolecular coherences, and the relative importance of IVR-like and multiphonon mechanisms is under debate. Clearly, such experiments require a marriage of IVR with collisional and intermolecular relaxation models, and it will be very interesting to see to what extent the coherent effects discussed here survive in molecules exposed to a more strongly coupled environment.

III. SCALING OF THE HAMILTONIAN

As discussed in the introduction, the polyatomic rovibrational Hamiltonian H_{vr} has a property that is largely preserved even above the lowest dissociation limits: it has the structure of a hierarchical local matrix (HLM). Such a matrix must satisfy two criteria. It must be sparse, and the size of the nonzero matrix elements must strongly depend on their position within the matrix. On a more intuitive note, this structure arises because covalent bonds are generally highly directional and localized. Only a small number of such bonds (usually 1–6) is connected to any given atom. The reason we can talk about chemical bonds at all is due to the discrepancy between the electron and nuclear masses, and hence the validity of the Born–Oppenheimer approximation. This will be our starting point for the discussion of H_{vr}, and from there we will develop its coupling structure, statistics of matrix elements, and various suitable approximations that are relevant for the discussion of IVR [18,64].

A. Hamiltonian and Coordinate Systems

For the purposes of computing IVR, the molecular Hamiltonian of a singlet state is well approximated by

$$H_{mol} = \frac{1}{2}\sum_e \frac{P_e^2}{m_e} + \frac{1}{2}\sum_n \frac{P_n^2}{m_n} + V(\mathbf{r}_e, \mathbf{r}_n) \qquad (3.1)$$

where the last term is the Coulombic interaction potential among electrons and nuclei. Solution of the electronic part without nuclear kinetic energy yields the effective nuclear potential $V_i(\mathbf{r}_n)$, which is recombined with the nuclear kinetic energy to yield the rovibrational Hamiltonian of electronic state i (see Section VI for comments about the nonadiabatic case). H_{vr} can be transformed to internal coordinates and expanded in a power series

$$H_{vr} = \sum_{n,m} H_{nm} \qquad (3.2)$$

where n counts the vibrational and m the rotational order. Typical terms for $\mathcal{N}$ vibrational degrees of freedom include [118]

$$H_{20} = \frac{1}{2}\sum_{i=1}^{\mathcal{N}} \omega_{ei}\left(a_i^\dagger a_i + \frac{1}{2}\right), \quad H_{02} = \sum_{k=x,y,z} B_k \hat{J}_k^2, \qquad (3.3a)$$

$$H_{30} = \sum_{\mathbf{n},|\mathbf{n}|=3} V_{\mathbf{n}}^{(3)} \prod_{n_i}(a_i^\dagger + a_i)^{n_i}, \quad H_{40} = \sum_{\mathbf{n},|\mathbf{n}|=4} V_{\mathbf{n}}^{(4)} \prod_{n_i}(a_i^\dagger + a_i)^{n_i} + \sum_{k=x,y,z} B_k \hat{\mathscr{P}}_k^2$$

$$(3.3b)$$

In the order written, they account for the harmonic vibrational energy, rigid-rotor energy, cubic anharmonicity, and quartic anharmonicity (with a vibrational angular momentum correction). In the form of Eq. (3.3), the Hamiltonian will be referred to as the "full resonance" or "full effective" Hamiltonian. The symbols are defined in the glossary; the prime on the sum indicates that no redundant permutations of ladder products are to be summed over. The convergence radius of Eq. (3.2) is limited to energies below the first dissociation limit. However, the dissociative branch of the potential can be reparameterized in terms of explicit reaction coordinates such that Eq. (3.3) converges for the remaining purely vibrational coordinates.

Equation (3.3) explicitly formulates Eq. (3.2) in terms of normal coordinates, although any combination of delocalized or localized vibrational coordinates could have been used for the scaling to be described below. The reason is that the main effect of such choices is to repartition kinetic- and potential-energy terms, as expected for any contact (classically: canonical) transformation. Because of the virial theorem, such partitioning does not affect the magnitude of the terms in Eq. (3.2) as long as the transformation is linear to first order, specifically

$$p_i' = f_p(p_j, q_j) = \sum_j C_j^{(pp)} p_j + \sum_j C_j^{(pq)} q_j + \cdots$$

$$q_i' = f_q(p_j, q_j) = \sum_j C_j^{(qp)} p_j + \sum_j C_j^{(qp)} q_j + \cdots \qquad (3.4)$$

to preserve the scaling within a given order. For generalized coordinates and momenta that obey a commutation relation $[q, p] = i\hbar$ to first order, the virial theorem can be applied to each order, and kinetic and potential terms therefore scale the same, no matter how they are redistributed by canonical transformation. With the restriction of Eq. (3.4), the scaling laws derived below are therefore completely general.

B. Locality of Bonding and Scaling of Potential Constants

Equation (3.2) is not a mere convenience; if the adiabatic approximation is valid, it actually sorts the magnitude of terms in the Hamiltonian. Specifically, a typical molecular bond length/dissociation length, mass and dissociation energy in the Born–Oppenheimer approximation are given by

$$\bar{r} \sim 3\frac{\hbar^2}{m_e e^2}, \qquad \bar{\mu} \sim 20 m_H, \qquad \bar{D}(\text{Hz}) \sim 0.2\frac{m_e e^4}{2\hbar^3} \qquad (3.5)$$

The prefactors, of course, vary; in particular, the mass shown above overestimates rotational effects for large molecules, but is appropriate for internal rotor effects that survive even in large molecules. It is remarkable that these quantities depend only on the electron mass, nuclear mass, elementary charge, and Planck's constant. As a result, some manipulation yields

$$\frac{\omega_e}{D} \sim \frac{|V^{(3)}|}{\omega_e} \sim \frac{|V^{(n+1)}|}{|V^{(n)}|} \sim \frac{B}{\omega_e} \sim \kappa^2 = \left(\frac{m_e}{m_H}\right)^{1/2} \tag{3.6}$$

When the coordinates are expressed in terms of unitless operators and the couplings have units of energy (here: of frequency), vibrational couplings of any order decrease exponentially in size with the order of the coupling [17,18,64,119–121].

$$|V_n^{(n)}| \sim |V^{(3)}| a^{3-n} \tag{3.7}$$

with analogous scaling rules for rotation and rotation–vibration terms [118]. The scaling parameter a is of order κ^2 and typically varies from 0.03–0.2, with different scaling for different types of vibrational motions. Figure 6 compares potential constants from a realistic *ab initio* potential surface of $SCCl_2$ to an exponential scaling model, averaged geometrically over each order. The agreement with Eq. (3.7) in this (and many other cases that we have tested) is indeed excellent. The Hamiltonian matrix can therefore be

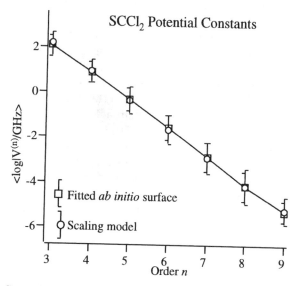

Figure 6. Comparison of average values and variances (the bars are square roots of the variance) of potential constants from a realistic potential surface and a scaling model (Ref. 18).

cast in a form that is sparse. Matrix elements obey a scaling law similar to Eq. (3.7) [18,64],

$$\langle \boldsymbol{u}'|V_{\mathbf{n}}^{(n)}|\boldsymbol{u}\rangle \sim \pm|V^{(3)}|a^{3-n}\bar{v}^{n/2} \qquad (3.8)$$

where $\bar{v}$ is a geometrically averaged quantum number for the two coupled states.* We can therefore write down a vibrational Hamiltonian

$$H_{\mathrm{v}} = \sum_n H'_{n0} \qquad (3.9)$$

where each successive order on average decreases by a (the prime indicates that all rotational terms have been removed; see discussion below for rotations).

C. Correlations in the Hamiltonian

Exponential scaling automatically introduces a positional correlation into the matrix representation of the Hamiltonian. Although matrix elements such as (3.8) have higher-order corrections, to lowest order $n = \sum |v'_i - v_i|$. For any triplet of states there is a triangle rule analogous to the rule for angular momenta [122]:

$$|n_{13} - n_{23}| \le n_{12} \le |n_{13} + n_{23}| \qquad (3.10)$$

If matrix elements are exponentially sorted by n, then this triangle rule applies to the matrix elements as well; triplets of matrix elements of H_{v} have correlated magnitudes. Numeric evaluation using realistic high order ab initio potentials shows that the triangle rule applies to vibrational matrix elements in 90% of cases; this is helped by the fact that the lowest-order corrections to a coupling of order n must be of order $n + 2$. In essence, Eq. (3.9) says that if one of a pair of coupled states is strongly coupled to a third state, the second in the pair is also strongly coupled to the third state.

Equations (3.8) and (3.9) provide a minimal HLM model for the vibrational Hamiltonian. It can be turned into an HLRM model by assuming that the quantum number distribution in each mode is Bose–Einstein, and that the diagonal states are Poisson-distributed. The resulting Bose statistics–triangle rule (BSTR) model is the simplest HLRM that can reproduce some of the experimentally observed correlations not captured by the simplest application of the Fermi "golden rule" [122].

Equations (3.8) and (3.9) capture some correlations among vibrational matrix elements of different orders. Examination of *ab initio* calculations

*States differing by only one or two quanta have lowest orders 3 and 4, respectively; several terms of comparable magnitude contribute in that case.

and experimentally derived potentials also shows strong correlations within a given order. It has been shown that in the limit of infinitely many degrees of freedom, a potential surface with a pronounced local minimum can be expanded in a rapidly converging cumulant series [18,64]:

$$\ln|V_{\mathbf{n}}^{(n)}| = \sum_i n_i \ln c_i^{(1)}| + \sum_{j>i} n_i n_i \ln c_{ij}^{(2)} + \cdots \qquad (3.11)$$

This type of representation has the advantage of being asymptotic: it is capable of giving semiquantitative information about molecular potentials of very large molecules, subject to certain assumptions about the bond connectivity.[†] To first order, the potential constants of Eq. (3.9) can therefore be factorized, or

$$H_{\mathrm{v}} - H_{20} \approx \sum_{n=3}^{\infty} \prod_{i=1}^{\mathcal{N}} \mathrm{sgn}(\mathbf{n}) \mathscr{R}_i^{n_i} (\mathsf{a}_i^\dagger + \mathsf{a}_i)^{n_i} \qquad (3.12)$$

This is not to be confused with the trivial factorization of the product $\prod(\mathsf{a}_i^\dagger + \mathsf{a}_i)^{n_i}$ during evaluation of matrix elements in a product basis. For Eq. (3.12) to satisfy the scaling of Eq. (3.8), the simplest expression for $\mathscr{R}_i$ is [17]

$$\mathscr{R}_i = \left(\frac{V_i^{(3)}}{a_i^3} \right)^{1/n} a_i \qquad (3.13)$$

and a number of related expressions have been used in the literature with the same fundamental factorization in mind. Factorization reduces the number of independent vibrational parameters to $3\mathcal{N}[\omega_{ei}, V_i^{(3)}$ and $a_i]$, introducing further correlations in H_{vr}. It can be a surprisingly good approximation even for a few degrees of freedom: Table I compares representative seventh-order potential constants from a realistic curvilinear potential energy surface (PES) of the SCCl radical to the best factorized set. Although the constants within that order vary over several orders of magnitude, the factorized values are rarely off by a factor of >2.

For longer-chain molecules, the second term in Eq. (3.11) must be introduced. This is again due to the localization of chemical bonding, and best illustrated using the chain molecule example of the introduction [64,65]. Figure 7 illustrates numerically evaluated cubic couplings between

---

[†]Equation (3.11) is unfortunately not valid in the "liquid" limit, where all atoms can be exchanged over small barriers, such as a Lennard-Jones cluster. It works well for covalently bonded systems in which the majority of atoms remains bonded, such as organic thermal and photoreactions.

TABLE I
Representative Seventh-Order Potential Constants from a Curvilinear
Model Surface of SCCl Radical[a]

$V_{[n1n2n3]}$	Full Curvilinear Surface (kHz)	Factorized Surface (kHz)
$V_{[700]}$	18	18
$V_{[610]}$	70	165
$V_{[412]}$	5,540	2,760
$V_{[304]}$	1,890	1,640
$V_{[223]}$	21,600	11,030
$V_{[070]}$	200	200
$V_{[034]}$	4,260	4,650
$V_{[007]}$	660	610

[a]Although not of experimental accuracy, the surface correctly represents global bending and single-bond dissociation, and is expected to show the same correlations as an experimentally determined potential. The second column shows the set of completely factorized potential constants. (A full list is presented in Ref. 18.)

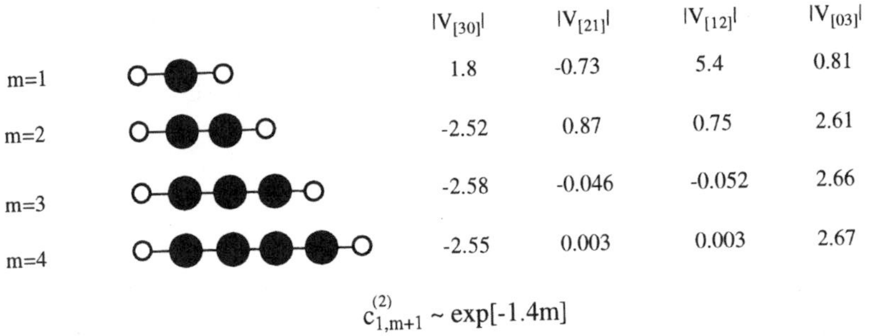

		$|V_{[30]}|$	$|V_{[21]}|$	$|V_{[12]}|$	$|V_{[03]}|$
m=1	O-●-O	1.8	-0.73	5.4	0.81
m=2	O-●●-O	-2.52	0.87	0.75	2.61
m=3	O-●●●-O	-2.58	-0.046	-0.052	2.66
m=4	O-●●●●-O	-2.55	0.003	0.003	2.67

$$c_{1,m+1}^{(2)} \sim \exp[-1.4m]$$

Figure 7. Effect of intervening bonds on the coupling constants between two nearly degenerate vibrational modes. Masses and Morse potentials were chosen to correspond to a substituted hydrocarbon. the potential constants are in THz (Ref. 64).

two nearly degenerate terminal CH-like stretches of a Morse chain. The parameters are typical of a substituted alkane. As chain length increases, the cross-couplings diminish exponentially with the number of intervening bonds and the CH stretching modes become localized. This behavior cannot be reproduced by Eq. (3.13), and requires $c_{ij}^{(2)} < 1$ in Eq. (3.11). A distribution of $c_{ij}^{(2)} < 1$ obtained by analyzing curvilinear model potentials for a number of small organic molecules is shown in Ref. 64.

The rovibrational matrix is thus local in two ways. *Adiabatic correlations* (eq. (3.10) arise because of the validity of the Born–Oppenheimer approximation and the resulting smooth potential surface; wavefunctions with very different nodal structure are weakly coupled (Eqs. (3.8) and (3.9)). *Spatial correlations* (Fig. 7) lead to localized modes in sufficiently large molecules of low symmetry: couplings among modes drop off exponentially with the number of bonds intervening between the atoms mainly involved in the modes [Eq. (3.11), second term]. Both of these effects are ultimately due to the locality of chemical bonds, distinguishing H_{rv} from the nucleon Hamiltonian or other "strongly" interacting systems. It will be seen in sections IV and V how this leads to highly anisotropic spreading of the IVR wavepacket in state space. An interesting consequence is that large molecules with spatially segregated vibrational excitation can behave like smaller molecules at lower energy. An experimental example from Section II is the enhanced stability of the propyne $2v_1 + v_6$ CH stretching combination compared to $3v_1$ (an "extreme motion" or "edge" state in Section IV); the spectral fragmentation of $2v_1 + v_6$ is much closer to $2v_1$ because the methyl and acetylenic CH stretches behave like two separate entities, each at lower local vibrational energy. As seen in Section IV, localization means that most interesting properties of IVR already emerge in low dimensional model systems.

D. Matrix Element Phases and Coupling Chains

So far, we have neglected the function $\text{sgn}(\mathbf{n})$ in Eq. (3.12). The purely vibrational Hamiltonian can be represented by a real-symmetric matrix; $\text{sgn}(\mathbf{n}) = \pm 1$ represents the sign associated with each vibrational matrix element. For direct couplings, the sign of the matrix element is immaterial. However, calculations have shown that coupling chains (e.g., in the form of superexchange) can make substantial contributions to the IVR rate [17,123–125]. The importance of direct high-order versus sequential low-order couplings depends crucially on the distribution of matrix element signs. If coupling chains are suppressed, then high-order couplings will dominate over low-order couplings at sufficiently high energy, despite the larger potential constants associated with low-order couplings [64]. If all signs were equal, coupling chains would dominate over high-order direct resonances. The reason is that for every direct coupling of order n, between two states, there are $r = \mathcal{O}\!\left(\binom{n}{n/2} N^{\Delta n}\right)$ indirect coupling chains of order $n + \Delta n$. If the signs are evenly distributed, the situation becomes more subtle.

Figure 8 shows the distribution of $\text{sgn}(\mathbf{n})$ for the normal-mode surface of CDBrClF derived from a realistic curvilinear model potential [64]. The signs balance within statistical error for this and other polyatomic test cases, except perhaps for the third order. Coupling chains therefore extensively cancel one another due to quantum interference, as illustrated in Fig. 9; the upper and lower states in 9a are strongly mixed, whereas in Fig. 9b changing the sign of a single matrix element leaves these states entirely unmixed. Because of the cancellation, the contribution of r sequential couplings grows

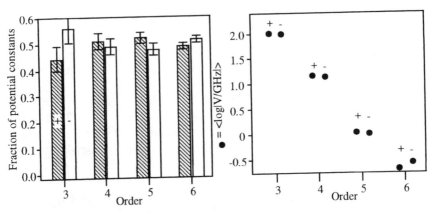

Figure 8. Sign distribution of normal-mode vibrational matrix elements from a curvilinear CDBrClF surface (Ref. 64). Left: fraction of potential constants with a given sign at a given coupling order. Right: average size of potential constants of a given order with a given sign.

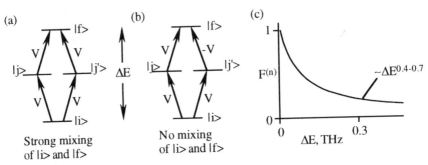

Figure 9. (a, b) Cancelation of a coupling chain between $|i\rangle$ and $|f\rangle$ because of a matrix element sign. (c) Fraction of the effective coupling between two states provided by direct couplings of order n versus coupling chains of order $n + 2$ (chains of order n make a negligible contribution for large molecules; Ref. 64).

as $r^{1/2}$ rather than r. Higher order direct couplings and coupling chains therefore make comparable contributions to the coupling structure. The excat relative importance of high order couplings versus chains becomes a function of the energy or time window being considered (Fig. 9c) [64]. At short times, coupling chains (or low order couplings at low energies) dominate the dynamics; at intermediate times (as defined in the introduction), direct couplings can dominate, particularly if $V_i^{(3)}$ and a_i in eq. (3.13) are small. Direct couplings are therefore important for a quantitative understanding of the "filling in" of an IVR lineshape. The same observation has been made purely on the basis of semiclassical arguments [52]; at high

$\bar{\rho}_{tot}$ (corresponding to observation at long times), dynamical tunneling among states is mediated by high-order resonances. This is a nonclassical effect because quantum states have a finite volume in state space, and will contaminate one another even if only weak resonances are available. A dramatic example is given by the broadened rovibrational lines of organic molecules dissolved in cold He clusters [115]. As a final note, if $V_i^{(3)}$ and a_i are large, then Fig. 9c decays very slowly, as a power law with an exponent of approximately $\frac{1}{2}$. Calculations of k_{IVR} or other "early time" quantities will then be in error by a factor of 2–3 if high order couplings are neglected, but no order-of-magnitude errors are introduced by the neglect.

E.   Overall and Internal Rotation

Rotational effects can play an important role in IVR by enhancing the local density of states [105], particularly if the H_{21} term in Eq. (3.2) destroys good angular momentum quantum numbers (other than total angular momentum) [91,106,126–128]. The largest effects are observed in linear molecules or molecules with a very large A-axis rotational constant (e.g., propyne). As molecular size grows, the effect of overall rotation diminishes the prefactor of $\bar{\mu}$ in Eq. (3.5) increases rapidly with molecular size, while the local cubic coupling constant $V^{(3)}$ remains unaffected. Therefore $|V_n^{(3)}|/\omega_e \ll B/\omega_e$ in large molecules (all $B_i \leq 3\,\text{GHz}$), even though they nominally have the same Born–Oppenheimer scaling. H_{21} scales as $(T_{rot}B)^{1/2}$, and the Coriolis timescale for large molecules in a jet reaches ns for large molecules, longer than even the CIT timescale of interest here [45,129]. When rotations need to be included, this can be done explicitly in HLRM or RM models, and has been done in the latter [106]. If the prefactor $\bar{\mu}$ is adjusted explicitly, the same scaling as in Eq. (3.3) can be used for rotational terms, with a $+2$ increase in the scaling exponent for each factor $|\hat{J}^2|$ or $\mathscr{P}^2|$.

Internal angular momentum contributions to IVR persist for any molecular size. These are of two types: internal rotations [102–105,130] and other vibrational angular momentum couplings (e.g., due to degenerate bending modes) [131]. The latter are most important in smaller molecules of high symmetry, while the former play a role even in the largest molecules. In fact, most large molecules contain internal rotors such as methyl groups or XY single bonds (where X and Y exclude halogens or hydrogens). As discussed in Section IV the local density of states is influenced by both the mass of the rotor, as well as relaxed selection rules for matrix elements under certain circumstances [65].

F.   Examples and Summary

The 1990s saw an explosion of methods to calculate realistic Hamiltonians for covalently bonded molecules with ≥ 6 degrees of freedom. They range from HLRMs including the most basic correlations [122], to semiquantitative scaling models [18,64,121], and finally to full *ab initio* [18,41,132–135] or experimentally adjusted *ab initio* models [37,132]. Techniques for calculating energy levels and states have also been highly developed. A few of the many available examples include benzene [41,136], $SCCl_2$ [37], HFCO [137], and fluorene [50] treated by variational techniques employing Lanczos iteration, usually with basis set and/or Hamiltonian filtering; and formaldehyde, treated by high order perturbation and Lanczos techniques [132,138]. Finally, $P(t)$ and other dynamical information can be obtained directly by applying Chebyshev, symplectic, and other propagators to the IVR wavepacket [139–141]. Promising new techniques include vibrational SCF[142] and Chebyshev iteration for window diagonalization (based on Neuhauser's method) [143]. The 1990s saw simulation techniques come of age, and realistic systems with 6–60 degrees of freedom can now be treated semiquantitatively to quantitatively.

In summary, the vibrational Hamiltonian has the structure of a sparse matrix with correlated magnitudes of the matrix elements (HLRM). This structure leads to the correlation among the eigenstates described in the introduction. Most of the discussion in this section was given in terms of a rectilinear coordinate expansion (applicable to normal, rectilinear local, and similar modes). This is sufficient because H_{rv} has the HLRM structure as long as Eq. (3.7) holds in *some* representation. The GOE Hamiltonian (with Gaussian distributed elements) illustrates that this property is nontrivial and not shared by global random matrices: A GOE matrix can of course be brought into sparse (or even diagonal!) form; but no representation of a GOE Hamiltonian can have correlations of the type given by Eqs. (3.10) and (3.11) because its couplings are completely nonlocal.

The relatively small value of $a = 0.03 - 0.2$, and of many $c_{ij}^{(2)}$ means that the molecular Hamiltonian retains some memory of the regular types of motion that would occur in the uncoupled case (i.e. when all mixed products of ladder operators are removed from Eq. (3.10). a ensures that the resonances are weak, and $c_{ij}^{(2)}$ ensures that not too many of them act at once on any given state. As a result, vibrational transitions do not occur all over the place, but are grouped into features [27]. "Almost good" quantum numbers can be assigned to such features, even though the feature states can be fragmented into many eigenstates. When considering optical excitation experiments, it turns out that a quantum state space constructed in a basis of wavefunctions resembling such features is the most natural to represent the vibrational energy flow, in the sense that the correlations of H_{rv} are easiest to formulate in such a basis. We therefore turn to the state space next.

IV. STATE SPACE

Intuition based on classical ball-and-spring models of molecules, leads to IVR models in terms of internal coordinates such as bond distances and bond angles, and as a function of time. An initial displacement along a given bond eventually rattles all the atoms, causing complex temporal displacement patterns of the bond distances, bond angles, and dihedral angles. Internal coordinates or cartesian coordinates are also useful in large-scale numerical simulations of IVR (whether classical [144], semiclassical [145], or quantum [146]) because the Hamiltonian is easily formulated in these coordinates.

This section espouses a state space formalism instead. The reason is that internal or Cartesian coordinates do not allow the most parsimonious formulation of IVR. Because of the complex nature of vibrational states at high energy, parsimony is of the essence if one is to understand, rather than just simulate, the dynamics [147]. Spectroscopists and (semi)classical dynamicists who study IVR are well aware of this need for simplification; the ideas of action variables, of an effective reference Hamiltonian, and of hierarchical analysis [5,15,148] underlying the state space concept are well known to them.

For a molecule with $\mathcal{N}$ internal degrees of freedom, state space is simply the $\mathcal{N}$-dimensional Hilbert space containing a complete set of basis states needed to represent an arbitrary IVR wavepacket $|t\rangle$. Infinitely many bases are at our disposal: the trick is to find a (near-)optimal representation. Since we are interested in a simple representation of IVR dynamics, optimal means that the number of states required to describe the dynamics should grow as slowly as possible with time. Examination of this small number can then provide insights into the CIT dynamics that gives rise to nonexponential survivals. One can also view the problem from the perspective of the Hamiltonian; the eigenstates of the optimal reference Hamiltonian will enter into the IVR dynamics in a temporally sorted hierarchy. In Section III it was shown that H_v is relatively weakly perturbed ($|V_i^{(3)}|/\omega_{ei} \sim a_i, a_i < 0.1$) and that the coupling structure is local. This is reflected in the best state-space bases discussed below.

A. State Space Representations

Figure 10 shows a hypothetical vibrational spectrum at low and high resolution. Let the low resolution feature be assigned nominally to a $|1, 1, \cdots\rangle$ combination state (only two of potentially many dimensions are shown). The high resolution spectrum might reveal fragmentation of this feature into many lines. The corresponding eigenstates typically have the complicated appearance of the one shown. With a pulsed laser, one can

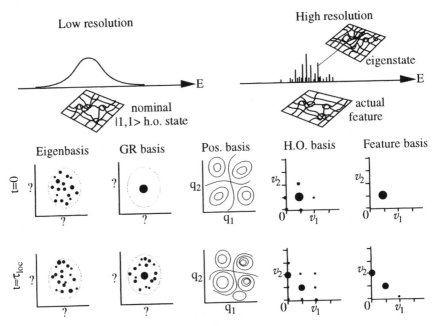

Figure 10. Low-resolution and high-resolution structures of a feature. State space is represented in five ways, from left to right.

excite the entire feature. Its examination at $t \approx 0$ would reveals a much simpler wavefunction, the actual 'bright' feature state.*

The bottom of Fig. 10 illustrates how state populations evolve in various representations. To differentiate the representations, we now introduce a criterion for parsimony, the *dispersion* of the IVR wavepacket, defined at $t > 0$ as [55]

$$\mathscr{D}(t) \equiv \sigma^{-1}(t) \equiv \frac{(1 - |c_i(0)|^2)}{\sum_{i \neq 0} |c_i(t)|^4} + 1 \qquad (4.1)$$

In this expression, $c_i(t)$ is the overlap between the dephased feature $|t\rangle$ and the basis state $|i\rangle$ in whatever representation has been chosen. The sum in the

---

*Actually, one would have to use a phase-chirped pulse. The eigenvector components of the Hamiltonian matrix can have positive and negative signs, a Fourier transform-limited pulse would excite all eigenstate components of the feature with the same phase. However, the survival amplitude $\langle 0|t\rangle$ and all other dynamically interesting conclusions are not affected by this, because the phase factors in $\langle 0|$ and $|t\rangle$ exactly cancel; so for all practical purposes one can think of the $P(t)$ of the feature as obtained from a Fourier transform–limited pulse.

denominator measures how much the wavepacket is dispersed across the basis $\{|i\rangle\}$, with the exception of a state $|i = 0\rangle$, which has the greatest overlap with the feature state $|t = 0\rangle$. The ratio in Eq. (4.1) therefore counts the number of states participating in the dynamics at time t, excluding the state $|i = 0\rangle$. The $+1$ adds $|i = 0\rangle$ back in, but weighted equally with the other states. This procedure ensures that at $t \approx 0$ the states that are just beginning to participate in the dynamics are not overshadowed by the initially prepared state to which they are coupled. For example, a prepared state from which N_{loc} basis states are evenly populated will have $\mathscr{D}(t = 0^+) = N_{loc} + 1$, not $\mathscr{D}(t = 0^+) \approx 1$. Note that $\mathscr{D}$ is ill-defined if $|0\rangle$ is time-independent, but can be defined at $t = 0^+$ if $|0\rangle$ is even slightly nonstationary. Figure 11 illustrates this: $P(t) = \frac{1}{2}$ in both cases, but $\sigma(t)$ is much smaller in the bottom case, where many states immediately participate in the dynamics.

Different representations behave as follows:

1. In the eigenbasis, the $|c_i(t)|^2 \hat{=} |c_s|^2$ are constants [see eq. (1.1)]. $\mathscr{D} = N_{tot}$ is time-independent and equals the inverse of the dilution factor for $t \geq 0$. Unfortunately, this means that the maximum number of states must be included in the dynamics at all times and any correlations between different times are difficult to discern. As far as $P(t)$ is concerned, the eigenbasis merely restates the experimental facts since its parameters $|c_s|^2 = I_s$ and ω_s can be read off directly from the spectrum. No insight into the origin of the

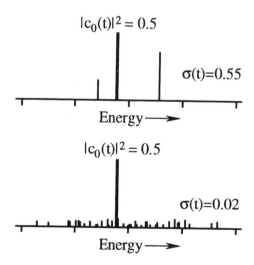

Figure 11. Two extreme cases an IVR wavepacket can spread in state space; in both cases $P(t)$ is the same. Top: optimal feature basis; little spreading has occured. Bottom: GR basis; spreading into all states occurs simultaneously (Ref. 55).

dynamics is obtained unless the individual eigenstates themselves can be carefully analyzed.

2. The "golden rule" (GR) basis can be obtained computationally from any basis as follows. The "bright" state $|0\rangle$ which carries oscillator strength and its direct couplings are projected from the Hamiltonian. The resulting matrix of dimension 1 smaller is diagonalized. The bright state couplings are transformed to the new basis using the diagonalizing matrix, and the complete Hamiltonian matrix is reassembled in a form that contains $|0\rangle$, a prediagonalized dark manifold, and the couplings from $|0\rangle$ to that manifold. In this picture, $\mathscr{D} = 1$ at $t = 0$, an improvement over the eigenbasis. But for any time $t \geq 0^+$, $\mathscr{D} = N_{tot} = \sigma^{-1}$ again, and all states must be included in the dynamics. The GR representation of H_v can be obtained directly from the experimentally measured $I_s = |c_s|^2$ and ω_s, via the LKL algorithm [58,59]. The $2N_{tot} - 2$ relative frequencies and relative intensities are exactly equivalent to the $2N_{tot} - 2$ diagonal and off-diagonal elements of the GR matrix $[E_0^{(0)} = 0]$. The GR basis is therefore essentially an eigenbasis $\{|s'\rangle\}$. This has invited a tautological treatment of experimental IVR data in the literature. Experiments directly measure k_{IVR} via the lineshape or survival probability. In addition, V_{rms} [Eq. (1.7)] can be obtained from experiment by a GR expression.* Unfortunately, V_{rms} represents the average coupling of the bright state to a manifold of quasieigenstates uncharacterized except for their energies. V_{rms} cannot be related to known molecular parameters except by extensive computation, and contains no more information than k_{IVR} itself (Section V). Hence there is a question mark on the eigenbasis and GR coordinate axes in Fig. 10. In contrast, measurement of $\rho_{tot}(\omega)$ and I_s without averaging is a fruitful exercise, if the proper analysis is applied to reveal statistical correlations.

3. In a position basis, $\Psi(q_1, q_2, t)$ is labeled by the continuous coordinate parameters. In this representation, full information about the wavefunctions is, in principle, available [95,149]. For example, inspection of the IVR wavepacket at time τ_d clearly shows that a state with approximate character $|0, 2\rangle$ has been mixed into the initial feature $\Psi(q_1, q_2, t = 0) = \langle q_1, q_2, |0\rangle$. This provides a clue that H_{40} contains a 1 : 1 resonant term. The drawback of a coordinate representation lies in the fact that a continuous representation is used for a discrete system. $\bar{\rho}_{tot}$ is finite, but even if it is very large, the partially evolved wavefunction should be intuitively representable by about two simple basis functions in Fig. 10. The coordinate representation may require many more grid points than that. In principle, $\mathscr{D}$ equals the number

---

*In a high-resolution measurement, $V_{0s'}V_{0s'}$ can be determined using the LKL inversion, and Eq. (1.7) yields V_{rms}. In time-dependent measurements $\bar{\rho}_{tot}$ is often estimated from the Whitten–Rabinovitch density of states and inserted into Eq. (1.6).

of gridpoints required; in practice, visual inspection of low-dimensional cuts corresponds to a lower $\mathscr{D}$ because important structures (e.g., nodal lines) can be made out at much lower resolution [5,150]. Advances in discretized path integrals and grid representations may make this representation suitable for simulations in higher dimensions [151], but not for gaining insight into high-dimensional systems unless the coordinates q_i are very carefully chosen. There is one exception—if IVR is studied above the dissociation limit, it will be useful to parameterize the reaction coordinates continuously, and treat the remaining modes as in representation 4 or 5 in a hybrid approach. Whatever its shortcomings, representation 3 is clearly far superior to representation 1 or 2 for understanding the dynamics, and has been widely used in spectroscopic and hierarchical analysis of low-dimensional molecular models [95,149].

4. In this particular case, the normal mode basis allows a fairly parsimonious representation of the dynamics. Its $|1,1\rangle$ state closely resembles the actual feature, as seen by its highly weighted contribution at $t = 0$. As the IVR wavepacket evolves, population appears on the lattice sites of the directly coupled states, immediately revealing a resonant interaction with the $|0,2\rangle$ basis state. The coupling structure is relatively local, so the IVR wavepacket, although expanding on the grid, remains localized for a long time. The discrete lattice efficiently represents the IVR process. For semirigid vibrations, the $\mathscr{N}$-dimensional lattice of normal-mode quantum numbers is thus a good approximation of the feature state space. $\mathscr{D} \approx 4 - 5$ at $t = 0^+$ in the example, and rises very slowly with time.

5. The feature state space provides the most parsimonious representation. The bright state $|0\rangle$ (in this case $|1,1\rangle$) is one of the basis functions, as in representation 2. Because features conserve quantum numbers at short times, the lattice can again be labeled by vibrational quantum numbers. However, these quantum numbers may represent motions other than normal modes (e.g., local modes or precessional modes [152]); there may even be a transition in the qualitative nature of a quantum number due to a resonance that "locks in" above a certain energy. In Fig. 10, $\mathscr{D}(t = 0^+) \approx N_{\mathrm{loc}} \approx 2$ at $t = 0^+$, and rises to ~ 4 at τ_d. As in representation 3 and 4, a 1 : 1 resonance between $|1,1\rangle$ and $|0,2\rangle$ can be recognized immediately as the cause of the early local dynamics. State-space furthermore facilitates conversion between time-dependent and frequency-dependent views, which are related by Fourier transform. In the time-dependent view, the IVR wavepacket spreads locally and anisotropically in state space, as the population hops from lattice point to lattice point. The spreading is local because of the exponential decay of couplings in Eq. (3.7). In the frequency domain, the anisotropy is recognized as having two sources: energy conservation if excitation is

quasimicrocanonical, and (ro)vibrational resonances that facilitate certain paths through state space (e.g., from $|1,1\rangle$ to $|0,2\rangle$ but not to $|2,0\rangle$).

B. Optimal Representation and Hierarchical Analysis

Using Eq. (4.1) one can give a simple variational definition of the ideal feature basis. It should contain $|0\rangle$ and maximize the integral [55]

$$\lim_{t\to\infty} \int_0^t dt'\,\sigma(t') \tag{4.2}$$

subject to the constraint that the basis functions used to construct $\sigma(t)$ can be assigned a full set of quantum numbers. Equation (4.2) puts a precise meaning on a minimum number of functions to describe the dynamics for a maximum length of time. A close analogy to coordinate formulations of type 3 also exists: the optimal features corrrespond to "natural orbitals" of a nodal analysis of eigenstates in terms of harmonic oscillator bases [11,153, 154].

Figure 12 shows $\sigma(t)$ for the $|233222\rangle$ combination state of $SCCl_2$ in a harmonic state space-basis; it decays by an anomalous diffusion law $t^{-\delta_\sigma/2}$ with an exponent [Eq. (3.11) of $\delta_\sigma \approx 3$ and an initial value of $\mathscr{D}(t = 0^+) \approx 5$. τ_{loc} is approximately 30 fs and τ_σ is approximately 1 ps, so CIT dynamics occur over nearly three orders of magnitude in time. On the other hand, the

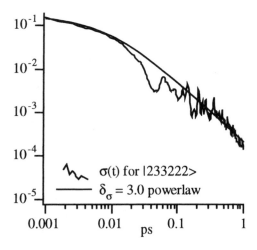

Figure 12. Inverse dispersion for an interior state of $SCCl_2$ in a harmonic basis, showing a slow power-law decay.

first 30 fs of the dynamics are well approximated by an exponential decay [55]. A normal-mode basis of the type shown in Fig. 10 is a reasonable approximation to the optimal feature basis in this case.

Equation (4.2) can also be related to hierarchical analysis [15,16,155, 156], although the wavefunctions in the $t \to \infty$ limit of a hierarchical analysis are not necessarily assignable to quantum numbers any longer. The idea in spectral hierarchical analysis is to convolve the eigenspectrum with a resolution function. At low resolution there is only one peak in the spectrum, and as the resolution is improved, local maxima appear and can be grouped into a tree of resolution versus frequency. Smoothed states correspond to junctions of the tree, with the overall feature state corresponding to the first junction (at lowest resolution). The resolution axis can be mapped naturally into a time axis with t inversely proportional to resolution, so that the feature state corresponds to the shortest time and the nearly unsmoothed eigenstates correspond to the longest times. Hierarchical analysis has been realized in real space (representation 3 in Section IV. A) [149]. In state space, a similar hierarchy exists as defined by Eq. (4.2), but has been discretized onto a lattice [16].

C. Near-Optimal Representations

Although it is difficult to prove that (4.2) has a unique solution in the most general case, it defines a unique basis if the potential surface has one minimum and t is taken to be large but not infinite. In that sense, the optimal feature state space is as unique as the eigenbasis of H_v. In practice, solving Eq. (4.2) can be a daunting problem. Fortunately a near-optimal state space, such as representation 4 in Section IV A, can be constructed much more easily. Take, for example, a scenario in which a semirigid vibration is coupled to an internal rotor of C_s symmetry, $H^{(0)} = \omega(n_1 + \frac{1}{2} + BJ_2^2 + b(1 - \cos\theta_2)/2$. One could pick a product normal-mode or a free rotor/ normal-mode mixed representation. Even better, one can switch representations smoothly at the barrier by introducing a quantum number ℓ such that

$$v_2 = 2\left|\ell_2 + \tfrac{1}{4}\right| - \tfrac{1}{2}\left(E - \omega\left(n_1 + \tfrac{1}{2}\right)\right) < b)$$
$$J_2 = \ell_2\left(E - \omega\left(n_1 + \tfrac{1}{2}\right) > b\right) \tag{4.3}$$

which allows one to interpolate from a vibrational to a rotor state space along a transition surface in state space defined by the barrier height. [In Eq. (4.3) the "surface" is a line embedded in the 2D state space.) When the switching occurs at $E - \omega\left(n_1 + \tfrac{1}{2}\right) \approx b$, matrix elements can be anomalously enhanced, leading to a larger local density of states in Eq. (4.8) below [65]. For hindered rotor states that lie above or below the barrier, the same exponential scaling described in Section III is obtained, and the contribution of internal rotation to the local density of states is similar to that of any other vibrational mode.

For a normal/local mode transition, the switching of quantum numbers involves a rotational shearing of state space [157], and other simple transformations account for other resonance-induced changes in state space structure. As a rule, couplings in such transition zones have relaxed selection rules when the energy in the relevant mode lies near the transition energy; couplings within zones and across zones obey the simple exponential scaling of Section III, albeit with different constants $V_i^{(3)}$. This explains why spectral progressions often go from a simple to a complicated and back to a simple structure as total energy is increased [95].

Most importantly, any of these near-optimal state-space representations, from pure normal modes to combinations of energy dependent normal/rotor/local modes, are virtually identical to one another when compared to an eigen- or GR basis. It may well be that a particular feature $|0\rangle$ requires five normal-mode functions and only one local-mode function, or eight normal-mode instead of two rotor functions, but these are all major improvements compared to N_{tot} eigenfunctions or GR functions for a highly diluted feature! A suboptimal but adequate feature basis can therefore be constructed from simple oscillator and rotor bases. For a large molecule, the recipe of "rotors for torsions, (an)harmonic normal or local oscillators for all semirigid vibrations" is sufficient to extract the average properties of interest, such as the dimensional exponent δ, and to understand their origin [50,68]. Carefully optimized formulations have been succesfully constructed for small model systems and can be added to the repertoire once their scaling behavior has been well characterized [152].

D. Interior and Edge States

Figure 13 illustrates a three-dimensional state space in a near-optimal representation. IVR experiments are often quasimicrocanonical; they are carried out on a well-defined energy shell because the width of a feature is much smaller than the vibrational energy. In a harmonic basis, the energy shell is flat; otherwise, it is curved as discussed in the next subsection. The most obvious structure in state space is provided by the boundaries, and allows different types of states to be distinguished. Interior states (shown in black) have all quantum numbers nonzero. At high vibrational energy and for large $\mathcal{N}$, such states dominate the CIT and late dynamics because they make up most of the density of states. At the other extreme are the edge states that sit on the coordinate axes or "edges" of state space [17]. Such states are often prepared in experiments as initial states ("overtone excitation") [89]. Between these extremes lie partially interior states or surface

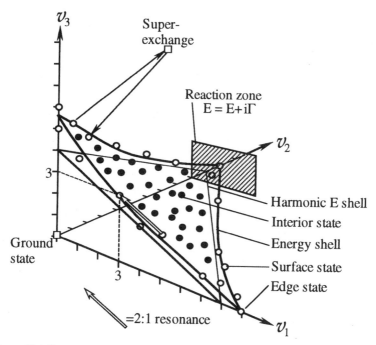

Figure 13. State space, illustrating harmonic and anharmonic energy shells, and interior, surface and edge states. A 2 : 1 Fermi-resonant coupling between two states is shown; the $|420\rangle$ does not couple by an analogous mechanism because its partner is too far off the energy shell; for strong couplings, superexchange can become significant, as shown. The shaded area at $\nu_2 \geq 9$ corresponds to reactive lattice points in state space which have a finite lifetime. Well above threshold there may be no such 'preferred' area, and all points on the energy shell are reactive: statistical reaction occurs if IVR redistributes population among lattice points sufficiently fast.

states of the energy shell* which have some, but not all quantum numbers, nonzero. Such states correspond to combination bands with zero quanta in some of the modes, and can also be prepared in some cases. Any IVR wavepacket will eventually feel the boundaries of state space shown in Fig. 13 and will be limited to a dilution factor inversely proportional to the number of all interior, surface, and edge states near the energy shell.

Interior and edge states differ qualitatively in several ways, and the existence of a distinction was recognized early on [5]. The most obvious

The "surface" in Fig 13 looks rather like an edge in three dimensions! It would be a surface in four dimensions for states with one zero quantum number. The choice of interior–surface– edge nomenclature instead of interior–edge–apex is somewhat arbitrary, but serves as a reminder that 3D illustrations should be taken with a grain of salt.

difference is their number. If d is a vector in state space measuring the distance from the ground state to the energy shell, then the number of interior states scales as $\sim d^{\mathcal{N}-1}$, the number of edge states scales as $\mathcal{O}(\mathcal{N}d^0)$, and the partially interior surface states scale with powers from 1 to $\mathcal{N} - 2$. Interior states are not only more numerous but also surrounded by more potential coupling partners. Because there are no negative vibrational quantum numbers in state space, every zero quantum number cuts the number of potential coupling partners in half, every unit quantum number cuts the number of cubic coupling partners by one quarter and of higher than cubic coupling partners by one half, and so on.

E. Effective Hamiltonian

There is also structure in state space independent of its boundaries. In terms of energy dependence, much of this structure arises because of resonances, which brings us to the effective Hamiltonian. In Fig. 13, the Fermi resonant term $V_{[120]}^{(3)}$ in H_{30} may couple states $|n, m, p\rangle$ and $|n - 2, m+ 1, p\rangle$, which both lie near the energy shell if $\omega_2 \approx 2\omega_1$. Strong resonances can destroy good quantum numbers or create a different set of good quantum numbers, as in the case of the normal to local mode transition of Darling–Dennison resonant stretching vibrations [157]. In the spectrum, strong low order resonances can lead to the appearance of gateway states (Section II). Additional structure can arise due to quantum interference effects, as illustrated by the coupling chain in Fig. 9, where mixing of $|i\rangle$ and $|f\rangle$ is canceled despite strong direct resonances via $|j\rangle$ and $|j'\rangle$. Higherorder resonances bridging small energy gaps can lead to slight broadening of otherwise stable spectral features at high $\bar{\rho}_{tot}$.

The off-resonant and resonant effects for semirigid vibrations can be represented by an effective Hamiltonian of the form [158]

$$H_{eff} = \sum_i \omega_{ei}\left(v_i + \tfrac{1}{2}\right) + \sum_{i,j} \chi_{ij}\left(v_i + \tfrac{1}{2}\right)\left(v_j + \tfrac{1}{2}\right) + \sum_{\mathbf{n}} V_{\mathbf{n}}^{(n)\prime} \prod_i (a_i^\dagger + a_i)^{n_i}$$

$$= E^{(0)} + H_{res} \tag{4.4}$$

The first two terms include the harmonic and off-resonant energy shifts (sometimes defined without cross-terms), while the last term collects the most important resonant couplings with the resonant part removed. This notation is ideally suited to the state space picture (Fig. 13). Eigenfunctions of the nonresonant $E^{(0)}$ of Eq. (4.4) correspond to feature states, and are subject to an anharmonic energy shift. The resonant coupling term causes transport of probability in state space among feature states, corresponding to diffusion of the IVR wavepacket probability distribution. The resonance condition for Eq. (4.4) is given by [15]

$$\mathbf{n} \cdot \boldsymbol{\omega} = 0, \qquad \omega_i = \frac{\partial E^{(0)}}{\partial n_i} \qquad (4.5)$$

This Hamiltonian still allows states off the energy shell to contribute to the dynamics by superexchange (Fig. 13) [123], although some of the anharmonic energy shifts of feature states have been explicitly incorporated in the second term.

The first term in Eq. (4.3) leads to a planar energy shell. Because most anharmonic constants are negative, the second term in Eq. (4.3) causes a concave distortion of the energy shell if an anharmonic feature basis is used, allowing certain pairs of states to be in resonance while analogous pairs are not. In the figure, the $|330\rangle$ and $|520\rangle$ feature states are in $2:1$ resonance but the $|420\rangle$ and $|230\rangle$ states are not, because $|230\rangle$ is not close enough to the energy shell to be coupled. The deviation of the energy shell from the harmonic planar shell is generally smallest for the interior states, and grows for the states at the edge of state space (Fig. 12). The curvature of the energy shell causes states to tune in and out of resonance as their quantum numbers are systematically varied. It is even possible for a pair of states to be in resonance, while an analogous higher energy pair is out of resonance.* Such cases have been observed experimentally [33,38].

As discussed in the previous subsection, interior and edge states differ in the number of coupling partners. In addition, interior states of a set of coupled anharmonic oscillators are generally more nearly separable than edge states because each mode is on average excited to lower energy, so the coupling matrix elements in Eq. (4.4) are smaller. These differences can lead to qualitatively different dynamics of nearly isoenergetic edge and interior states [5,17,89]. If the higher coupling strength near the edge is more important, edge states will decay sooner; if the number of coupling partners dominates, then interior states will decay sooner. Experimentally, features corresponding to edge states (particularly near the apices) can often be realized by high-resolution scanning or pulsed excitation. Interior states are less accessible because of unfavorable Franck–Condon factors. Experimental studies thus primarily probe edge-to-interior evolution, although double-resonance methods [40,109] and SEP spectra probing low-frequency modes [37,86] allow a look deeper into the interior and skeletal IVR regions of state space.

Effective Hamiltonians such as those in Eq. (4.3) have been determined from spectroscopic measurements or symmetry considerations for a number

---

* "Analogous" here means pairs $|v_1, v_2\rangle, |v_1 + n_1, v_2 + n_2\rangle$ with different v_1 and v_2.

of small molecules [30,38,159]. The main difficulty is that the effective molecular constants are often difficult to relate to the parameters given in Section III, and hence to specific feature basis sets. Much progress was made in this regard in the 1990s. As discussed in Section III, potential surfaces for several molecules have been highly sampled by ab initio calculations, leading to curvilinear or normal coordinate force fields. In some cases, the potential constants have been adjusted such that (ro)vibrational eigenvalues agree with experiment, through either scaling [18,132], or full variational calculations [37]. An effective Hamiltonian can then be obtained by converting H_v to a harmonic or diagonal anharmonic representation and keeping the most important resonance terms [55].

F. Dimension of IVR Flow Manifold and Local Measures

For large molecules, it may be more fruitful to first understand the average properties of the resonance structures in state space before jumping in and analyzing them in detail. A question that naturally arises pertains to the dimensionality of the manifold on which the IVR wavepacket flows [16,54].

One can envision two extremes. In one case, the initial state is strongly coupled to all surrounding states. It expands as a "hyperglobe" of dimension $\mathcal{N}$ in state space (a "hyperpancake" of dimension $\mathcal{N} - 1$ if the preparation step is quasimicrocanonical). The IVR flow manifold covers all the surrounding state space. In the other extreme only one resonance direction is active and the IVR manifold has dimension 1. In the weak coupling limit, it is, of course, possible that no significant flow occurs, and the dimension is then less than 1. This indicates that the IVR threshold has not been reached. In general, the IVR flow manifold lies between these extremes, and is embedded in the full state space with dimension $0 \leq D_v \leq \mathcal{N}$.

The dimensionality of the IVR flow manifold is related to the goodness of approximately conserved quantum numbers: [15,63,160]. If all quantum numbers are bad, the flow will be $\mathcal{N} - 1$ to $\mathcal{N}$-dimensional; if some quantum numbers are less well conserved than others, the flow will expand in the corresponding direction in state space more. The sequential loss of approximate quantum numbers is therefore directly linked to an anisotropic and nonexponential diffusion of the IVR wavepacket in state space. The principal axes of this anisotropy do not generally lie along the **n** axes of the zeroth-order Hamiltonian in Eq. (4.4); as seen in Fig. 14 and discussed in detail below, the actual principal axes of the flow manifold correspond to resonance conditions such as Eq. (4.5), subject to constraints forbidding negative quantum numbers if the initial state $|0\rangle$ lies near the edge. (For internal rotor bases, negative quantum numbers are allowed, or can be replaced by a $\pm$ phase label for each angular momentum state.)

Motivated by general scaling arguments applied to state space [16], it has been shown that the dimensionality of the IVR manifold can be approximated directly from the Hamiltonian if certain kinds of superexchange are neglected [55]:

$$D_v(n) \approx \frac{\partial \ln \int_0^n dn' N_{loc}(n')_{n_0}}{\partial \ln n} \Bigg|_{n_0} \approx \frac{\ln N_{loc}^{(n)}}{\ln[n] - \ln[n-2]} \Bigg|_{n_0} \qquad (4.6)$$

In essence, this equation is the logarithmic derivative of the integrated local number of coupled states with respect to distance in state space. Because state space is discrete, this has to be realized as a finite difference; $n = 4$ gives the dimensionality for quartic couplings, $n = 3$ for cubic couplings.

The quantity $\mathscr{L}_{i0}$ is the magnitude of the local coupling amplitude between features $|0\rangle$ and $|i\rangle$. It has been shown that this is well approximated by [17,45]

$$\mathscr{L}_{i0} = \left[1 + \left(\frac{\Delta E_{0i}}{\langle 0|H_v|i\rangle} \right)^2 \right]^{-1/2}. \qquad (4.7)$$

This equation yields the correct mixing coefficient of states $|0\rangle$ and $|i\rangle$ both in the limits $\Delta E_{0i} \to 0$ (resonance) and $\Delta E_{0i} \to \infty$. Especially for an effective Hamiltonian, or in the form of Eq. (3.9) or (3.12), Eq. (4.7) can be evaluated efficiently.

Equation (4.7) can be used to evaluate the local quantities mentioned in Section I. Replacing H_v by H_{n0} in Eq. (4.7) and summing, one obtains

$$A_{loc}^{(n)} = \sum_i \mathscr{L}_{i0} \Bigg|_{n=const}$$

$$N_{loc}^{(n)} = \sum_i \mathscr{L}_{i0}^2 \Bigg|_{n=const} \qquad (4.8)$$

$$\rho_{loc}^{(n)}(\Delta E) = \frac{1}{\delta E} \sum_i \mathscr{L}_{i0}^2 \Bigg|_{n=const,[E+\Delta E-\delta E/2,E+\Delta E+\delta E/2]}$$

Here $A_{loc}^{(n)}$ is the local mixing amplitude of states directly coupled to $|0\rangle$ by nth-order couplings, $N_{loc}^{(n)}$ is the corresponding effective number of coupled states, and $\rho_{loc}^{(n)}(\Delta E)$ is the local density of states at energy ΔE from state $|0\rangle$. By summing these quantities over all n, A_{loc}, N_{loc} and ρ_{loc} are obtained. Because the local density of states coupled by H_{n0} decreases rapidly as $\Delta E_{0i} \to \infty$, the amplitude and local state number are actually guaranteed to be finite even if the summation is extended to all states off the energy shell [64]. The local density of states grows much more slowly with energy than does the total density of states with energy. Useful analytic approximations to Eq. (4.8) have been derived for both interior and edge states [14,64, 67,161].

Figure 14 shows two IVR wavepackets from 6D quantum dynamics simulations of $SCCl_2$ on a potential surface of near-experimental accuracy (the $|233222\rangle$ packet corresponds to the $\sigma(t)$ in Fig. 12). Two different initial states are shown, the $7v_1$ CS stretching overtone and a nearby combination band. The axes correpond to the three most important resonances in the full 6D state space, $\mathbf{n} = [0, -1, 0, 0, 1, -1]$, $\mathbf{n} = [-1, -1, 0, 0, 2, 0]$, and $\mathbf{n} = [1, 0, 0, 0, -2, 0]$, two cubic and one quartic. The energy flow is clearly not isotropic, but strongly guided by state-space structure. Visual inspection immediately indicates a dimension ≈ 1 for the overtone and 2–3 for the combination band. The analytic result in Eq. (4.6) applied to the same Hamiltonian yields $D_v = 0.8$ and 2.8, respectively. [Equation (4.6) can, of course, be evaluated rapidly, while the 6D quantum dynamics required about a day of IBM SP2 CPU time.]

The difference between the eigenstate and feature state representations is that in the latter, the IVR manifold is represented as compactly as possible. Its dimension can be read off visually from Fig. 14, whereas in the eigenbasis, the distribution of the population approaches a state space-filling fractal with any smooth choice of "coordinates." However, the dimension D_v is approximately basis-independent. The main uncertainty in D_v as given

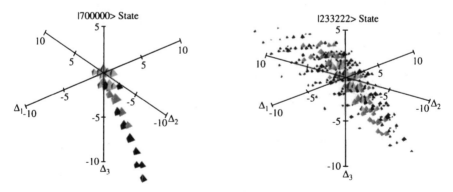

Figure 14. The three most important resonances of $SCCl_2$ are used to construct a 3D projection of the state space and of the IVR flow manifold (Ref. 55). The initial state lies at the origin. Energy flow of the edge state (left) is on a low-dimensional manifold; the interior state (right), flows on a nearly three-dimensional manifold (embedded in six dimensions). The edge state flows out in only one direction because negative vibrational quantum numbers are not allowed.

by Eq. (4.6) arises from superexchange. When superexchange is important (Fig. 13), (4.7) probably underestimates the actual dimensionality of the IVR manifold [123]. As discussed in Section III, neither direct nor indirect coupling mechanisms are likely to gain complete upper hand because of the phase cancellation of multiple paths through state space (Fig. 9) [64]. D_v could be greater by about $+2$ in cases where coupling chains account for 90% of the dynamics.

It is no accident that D_v almost equals the power law coefficient of the IVR wavepacket dispersion in Fig. 12. If IVR is indeed a quantum diffusion process, than the exponent of diffusion should equal the dimensionality of the IVR manifold. The exponent δ_σ in Fig. 12 is strongly basis-set-dependent but $\delta_\sigma \approx D_v$ in a feature basis. In Section V we will compare D_v to δ, the exponent of the survival probability. All three quantities will be seen to track closely in the examples studied so far.

Figure 14 illustrates another important concept in IVR, that of polyad quantum numbers [27,38]. For the $|700000\rangle$ state of $SCCl_2$, Δ_1 and a linear combination of Δ_2 and Δ_3 is are conserved at short times (and so are two more quantum numbers and energy); only one quantum number really breaks down, given by the linear combination of Δ_2 and Δ_3 at $45°$ between the axes. Even for the $|233222\rangle$ state, the three axes not shown in Fig. 14 participate on a much longer timescale than do the three resoances shown, so < 3 quantum numbers have broken down. One therefore expects hierarchical structure in the spectrum, as indeed experimentally found for $SCCl_2$ in Fig. 2. The number N_P of good polyad quantum numbers at *short* times is related to the local dimension D_v by $N_P = \mathcal{N} - D_v$ (including energy in the case of microcanonical excitation). The best-known case is undoubtedly that of acetylene [28,38,95].

Further examples in Section V show that a small D_v obtains even in much larger molecules; the dimension of the state-space IVR manifold is much smaller than $\mathcal{N}$ and corresponds to the number of active resonances. This fact bodes well for the development of low-dimensional resonance Hamiltonians [71,152,162] even for large molecules. There is, however, a practical problem to be overcome—a parsimonious effective Hamiltonian has to be built. For a small molecule like $SCCl_2$, it can be derived directly from the full potential surface. For large molecules, the scaling techniques described in Section III have to be used, and have been successfully applied [50,68].

G.   Connection of State Space to Tiers and to Chemical Reactions

The tier picture of IVR is now revealed as a 2D projection of the full state space. One projection axis measures energy from the bright state, the other a quantum number distance in state space. In Fig. 15, the energy axis through $|0\rangle$ and two polygons of constant n are shown. This groups the states by

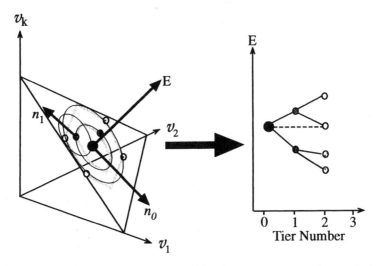

Figure 15. Relation between state space and tier pictures. n_0 measures the magnitude (but not direction) of the quantum number difference to the bright state. Tiers can be referenced to n_0, or re-referenced to the previous tier, as shown for n_1 measuring the distance from a first tier state. On the left, this has been projected into a tier diagram, which loses the directional ($\mathbf{n_0}$) information. Higher-order couplings (dashed) can weaken the hierarchical sorting of the tree.

energy and coupling distance, which is in essence the tier picture because the grouping can be represented by a tree as shown. Rigorous mapping of polygons about $|0\rangle$ to tiers corresponds to a sorting of tiers strictly by coupling distance from the bright state, averaging over any anisotropy. Anisotropy is maintained indirectly because coupling constants vary. Alternatively, criterion (4.7) in combination with a cutoff has been used for tier classification of states. This builds in anisotropy by mixing different order couplings across tiers, corresponding to an irregular diameter polygon in Fig. 13, and building the anisotropy directly into the tiers themselves. Finally, Fig. 15 also shows how polygons can be re-referenced to states in the next tier (n_1) to define further tiers.

State space can also account for "leaks" due to chemical reactions [113] Figure 13 shows a reaction zone, in which each state energy can be assigned a complex value. This value can be obtained independently of any IVR calculations by various approaches used in reaction dynamics. In the example given, the reaction zone requires that v_2 exceed a threshold value (e.g., the last bound state of an anharmonic stretching oscillator). Diffusion of the IVR wavepacket into this region results in a loss of population to reaction products. This type of model is very conducive to studying the breakdown of statistical rate theories due to slow IVR [113]. If the quantum transport is governed by a power-law exponent δ, the possibility of depleting states in the reaction zone and thereby preventing the IVR packet from reaching equilibrium [in the sense that $\sigma(t) \approx \sigma$] must be reckoned with (Section VI). A more rigorous, but computationally intensive, alternative would be to treat the reaction coordinate continuously and solve the IVR/ dissociation problem simultaneously.

H. (Semi)classical Connection

The feature state space can also be motivated via classical or semiclassical dynamics. In classical dynamics, IVR is studied in the $2\mathcal{N}$-dimensional phase space (q_i, p_i). Since the beginning of classical IVR studies, it was recognized that action-angle variables lead to the simplest formulation of the problem [5,63]. The reason is that according to Eq. (1.1), the overall motion, as complicated as it may be, can be generated by a superposition of periodic motions with frequencies ω_s. Action variables, which for a separable system (such as H_{20}) are defined in units of $\hbar$ by the integral

$$I_i = \frac{1}{\hbar} \oint p_i \cdot dq_i \qquad (4.9)$$

around a full period of the motion, are natural variables for quasiperiodic systems. In particular, the canonical transformation $(p, q) \rightarrow (I, \theta)$ from internal to action-angle variables for the simplest reference systems pertinent to the present discussion, the harmonic oscillator and the rigid rotor, leads to the Hamiltonian functions

$$H_{\text{ho}} = \omega I_v \quad \text{and} \quad H_{\text{rot}} = B I_r^2 \qquad (4.10)$$

In this format, action variables are clearly recognizable as classical analogs of the quantum numbers v and J.* Quantum-mechanically, only $\mathcal{N}$ coordinates or actions are required to fully describe the wavefunction $\Psi(\mathbf{x}) = \langle \mathbf{x} | n_i \rangle$; the remainder are obtainable by Fourier transform. Hence state space or quantum-number space is analogous to the classical action space, not to the full phase space. A large body of literature has been developed to deal with quantization and semiclassical treatments of polyatomic molecules in cases where the classical phase space is partially chaotic [163–165]. Although the treatment given here is purely quantum-mechanical, close analogies exist [162]; an example from section 3 is the

---

*Because the vibrational motion has turning points, a Maslov index of $\frac{1}{2}$ must be added onto I_v when quantizing, i.e., $I_v \rightarrow n + \frac{1}{2}$; rotational motion leads to a continuous increase in the angle, therefore $I_r \rightarrow J$ without modification.

comparison between high order ladder operator terms in H_v and semiclassical resonance-mediated tunneling leading to line broadening [52]. As discussed in the previous section, the validity of the Born–Oppenheimer approximation assures that the couplings in H_v are sufficiently weak so that some quantum numbers are conserved or at least temporarily conserved (analogous to a partially chaotic phase space) even at energies of chemical interest.

V. SCALING OF THE DYNAMICS

We now discuss numeric simulations and analytic predictions of IVR dynamics in the light of the state-space model, with an emphasis on recent results related to CIT dynamics not covered in earlier reviews [5,6]. As in section II, aspects such as threshold behavior, k_{IVR}, σ and its statistics, level spacing distributions, or power-law decay of the initial population will be grouped by topic, rather than by the history of the subject.

A. Energy Threshold

Equation (1.9) must fail at low energies. If $D_v < 1$ because the energy shell is only sparsely populated and the couplings (which scale as a polynomial in E) are small, the IVR flow manifold in state space remains underconnected and no long-range population transport can occur in state space. The threshold behavior has been worked out in detail via a diagrammatic Green function approach using a low-order truncation of H_v in the form of Eqs. (3.3) and (3.9) [51]. It yields a self-consistent set of equations for the average frequency shifts and widths associated with different modes as a function of energy. The most interesting result is that the averaged width can be nonmonotonic with vibrational energy, due to tuning in and out of resonance. For example, CF_3I is found to have islands of free energy flow as a function of energy, above which no IVR occurs [51]. From Fig. 12 it is clear how differences in the curvature of the energy shell due to anharmonicity can allow energy flow among a pair of states whose higher energy analog still lies below threshold. The model directly relates the onset of IVR to $V^{(3)}$ from Section III, but does not take into account high order perturbations explicitly.

A more microscopic study of this problem in terms of a Caley tree tier structure has been undertaken by making use of the analogy between IVR mediated by low-order couplings, and Anderson localization (AL) [14]. Anderson localization was originally developed to explain the cutoff of energy transport in disordered solids below a certain coupling strength. Although a Cayley tree structure does not allow for coupling loops such as in Fig. 9, it does allow for interference effects due to "reflection" at the nodes. The model makes detailed predictions about the threshold for vibrational energy flow as a function of $V^{(3)}$. Even more remarkably, the model predicts a linear increase of $k_{IVR}/\bar{\rho}_{tot}$ with $V^{(3)}$ above threshold, instead of the GR parabolic dependence. This does not mean that the GR will not yield the usual quadratic dependence on V_{rms} when used as described in Section I or section IV.B; rather, the cubic coupling strength and V_{rms} are not linearly related; thus an experimental determination of V_{rms} alone is not directly connected to the usual spectroscopic parameters. A similar analytic approach has also proved the longstanding conjecture that a dimensionless parameter $T \sim \rho V$ governs the transition from "restricted" to "full" IVR [67].

Quantum-dynamics simulations have shown similar behavior [122,141]. Figure 16 shows a HLRM simulation of $SCCl_2$ IVR lineshapes and $k_{IVR}/\bar{\rho}_{tot}$ as a function of $V^{(3)}$ at $E_{vib} \approx 360\,THz$, using the BSTR model of Section III. The transition from quantum beats to free flow is sharp, and occurs at $V^{(3)} \approx 10\,GHz$. Subsequent to that, the IVR rate increases as $[V^{(3)}]^4$ up to $\approx 35\,GHz$. This is probably due to a two-step superexchange into the dark manifold, corresponding to an effective sequence of two GR steps. Finally, the relationship in Fig. 16 becomes linear. As discussed later in detail, this happens despite the fact that $D_v \approx 3$, less than the possible maximum of 5–6: the IVR is not covering all of state space uniformly and is certainly not "chaotic." One might eventually expect D_v to jump from 3 to 5 or 6, and an even steeper slope in Fig. 16 above a coupling of $180\,GHz$: the $k_{IVR}/\bar{\rho}_{tot} \cdot V^{(3)}$ plot would then be piecewise linear. However, the actual molecular coupling is only $V^{(3)} \approx 75\,GHz$: vibrational anharmonicity is sufficiently small so that even at vibrational energies comparable to dissociation energies, molecules do not undergo completely unstructured energy flow. Indeed, our recent unpublished work on $SCCl_2$ shows regular progressions of skeletal vibrations even above the first dissociation limit.

B. Initial Decay

Initial IVR decays and k_{IVR} are now accessible to high-level PES/quantum dynamics calculations with good accuracy [41,45,47,137,166]. The results obtained from scaling models are also encouraging: lifetimes can be computed within a factor of 2–3 of experimental values [17,50]. The importance of the local density of states has been verified for $(CX_3)_3YCCH$ using a low order resonance Hamiltonian; as observed experimentally, the rate does not track $\bar{\rho}_{tot}$ when the masses of X (H or halogen) and Y (C or Si) are varied; rather, they depend on the local resonance structure [123]. Calculations on a filtered state space also reveal fluctuations in rates among nearby states of a single molecular species, even though $\bar{\rho}_{tot}$ is essentially constant across the energy window [37].

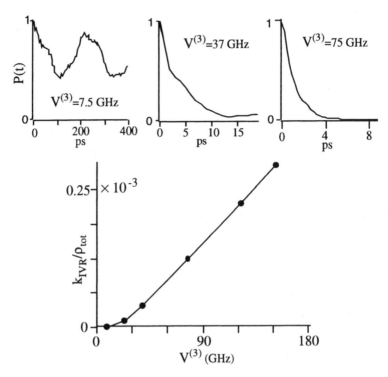

Figure 16. BSTR model decays for $SCCl_2$ at $\approx 360\,THz$ vibrational energy (Ref. 122). As the anharmponic coupling is tuned, quantum beats give way to an initially exponential, later power-law decay. Bottom: initial rate versus cubic coupling magnitude. The threshold lies at $V^{(3)} \approx 10\,GHz$. $D_v \approx 3$ (see text) in the above-threshold range, and the rate is linearly related to the cubic coupling magnitude. If D_v increases at very strong coupling couplings, the slope may rise further.

The simple application of the GR to a bright state $|0\rangle$ cannot take into account such rate fluctuations a priori if the full density of states $\bar{\rho}_{tot}$ is used in Eq. (1.6). The reason is that the rate fluctuations must be absorbed by V_{rms}, which is not straightforwardly related to known molecular parameters. Nonetheless, the GR is a viable approach to early time dynamics. In any reasonable representation, the product ρV^2 will be proportional to the rate. However, better choices for ρ and V than $\bar{\rho}_{tot}$ and V_{rms} exist because the latter can be obtained only if the rate is already known.

This is best seen by considering a somewhat tautological application of the LKL algorithm [58,59]. Let us say that a spectrum of width k_{IVR} has been computed or measured at high signal-to-noise ratio (SNR), so all eigenstates are observed. Using the LKL algorithm, the energies of all basis functions in the GR basis (Section IV) and all couplings $V_{0n'}$ from the bright state to the GR basis $\{|n'\rangle\}$ can be determined. This allows $\bar{\rho}_{tot}$ and V_{rms} to be computed rigorously. Inserting into Eq. (1.6), the resulting rate k_{GR} agrees with the experimental rate k_{IVR}. Now let us say that the SNR is lowered so only half the energy levels are known. The observed rate is of course still approximately k_{IVR}: removing some grass from the spectrum hardly changes the spectral envelope. Application of the LKL algorithm now results in half the total density of states, and a $\sqrt{2}$ larger V_{rms}; k_{GR} still equals k_{IVR}. This is still true if we use the local density of states $\rho_{loc}^{(n)}$ and the local coupling strength, which, unlike V_{rms}, is directly proportional to $V^{(n)}$. Therefore the GR can be applied directly to $\rho_{loc}^{(n)} \ll \bar{\rho}_{tot}$ and $V^{(n)} > V_{rms}$ to compute the rate. The first tier in Fig. 15 is sufficient for a rate calculation [167] if high order couplings are absent. If superexchange plays a role, then effective couplings for coupling chains and the resulting densities of states must be used [64], and two to three tiers are needed for rate computations, still a very small number. Experimentally, as already discussed in Section IV, the practice of evaluating V_{rms} is not very helpful. Rather, local effects such as gateway states or low-order couplings due to resonances should be observed to help in the construction of the proper hierarchical vibrational Hamiltonian.

C. Dilution Factors

Figure 16 is concerned mainly with the early dynamics before $t \approx \tau_{loc}$, but also illustrates the tuning of the very long time dynamics: the dilution factor is seen to decrease rapidly with increased coupling strength for the particular state examined. Experiments have shown that dilution factors for CH stretches with a similar total density of states can fluctuate over orders of magnitude [21]; σ for nearly isoenergetic skeletal combination bands also fluctuate over nearly an order of magnitude [37]. One would like to have models that can describe the overall statistics of σ, as well as the σ of specific features quantitatively. Such models now exist within the state-space formalism. In order to disucss them, consider the transformation

$$u = \frac{1 - \sigma}{\sigma} \tag{5.1}$$

$\sigma = 1$ is mapped to $u = 0$; $\sigma = 0$ is mapped to $u = \infty$. In that respect, u behaves like the first term in Eq. (4.1). An analytic model for σ has been developed by analyzing an HLRM based on the hierarchical vibrational Hamiltonian of Eq. (3.12) [161]. This model assumes that matrix elements $\langle V^{(n)} \rangle$ can be treated as random variables with a well-defined averages at each coupling order. When the local coupling criterion

$T \sim \sum_n \rho_{\text{loc}}^{(n)}(0)\langle V^{(n)}\rangle$ exceeds unity, the probability distribution for σ becomes

$$P(u) = u^{1/2}e^{-a/u} \qquad (5.2)$$

which turns out to be a gamma distribution of order $\frac{1}{2}$ whose integrated probability density is an error function of $u^{-1/2}$, a beautifully simple result. This distribution is bimodal in σ at intermediate coupling strengths or local state densities. In the weakly coupled regime, $P(\sigma)$ is peaked near 1; as the coupling strength increases, $P(\sigma)$ becomes peaked near 0 and 1; finally, it is peaked near 0 only. Thus, IVR dynamics avoid intermediate dilution factors. The dynamics either spreads the IVR wavepacket very little or very extensively.

Numerical simulations using the BSTR model (section 3C) and a nonrandom scaled Hamiltonian [Eq. (3.8) for SCCl$_2$ [122] and many organic molecules [46,68] show a similarly bimodal distribution. In these cases, N_{loc} as defined by summing Eq. (4.8) over all orders n is used instead of T. Figure 17 shows plots of σ as a function of analytical or computed local state number for many organic molecules. The transition from no energy flow ($D_v < 1$) to partially free energy flow is much sharper when plotted against a local state number (or density of states) than against the total density of states. The experimental as well as calculated $P(\sigma)$ from Fig. 17 are bimodal near $N_{\text{loc}} \approx 1$, as predicted. Dilution factors can also be computed for individual features if an accurate Hamiltonian is available. In Fig. 2 (top),

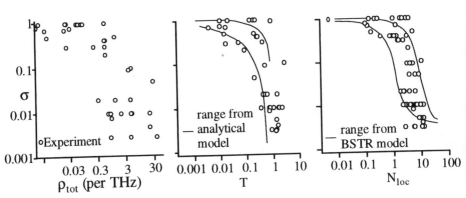

Figure 17. Dilution factors measured for hydrocarbon CH fundamentals (Ref. 21) as a function of total density of states; the correlation is weak. Middle: dilution factors versus an analytical local number of states, and range predicted by the analytical model. Right: dilution factors versus a numerically calculated local number of states, and range predicted by the BSTR model. A sharper onset of IVR as a function of local number of coupled states is evident (Ref. 68).

the numbers in parenthesis are dilution factors calculated by time-propagating an IVR wavepacket represented by $\approx$30,000 harmonic basis states using a fifth-order normal-mode Hamiltonian. It shows quantitative agreement for fluctuations of σ among nearly degenerate feature states.

D. Level Statistics

Level statistics are a useful experimental descriptor of state mixing (Section II). The level spacing distribution $P_\beta(s)$ itself is the simplest statistic, and the spectral rigidity $\Delta(L)$, which gives the least-squares deviation of the integrated level density $W_\beta(s) = \int^s ds' P(s')ds'$ from a straight line, is also often used. According to the Brody conjecture, the distribution function is approximately given by [66]

$$P_\beta(s) = (1+\beta)\left(\frac{s}{\bar{s}}\right)^\beta \Gamma\left(\frac{\beta+2}{\beta+1}\right)^{(1+\beta)} \exp\left[-\left[\Gamma\left(\frac{\beta+2}{\beta+1}\right)\frac{s}{\bar{s}}\right]^{(1+\beta)}\right]. \quad (5.3a)$$

$$W_\beta(S) = 1 - \exp\left[-\left[\Gamma\left(\frac{\beta+2}{\beta+1}\right)\frac{s}{\bar{s}}\right]^{(1+\beta)}\right] \quad (5.3b)$$

with $\beta = 0$ for a regular spectrum, and $\beta = 1$ in the strong coupling limit corresponding to classical chaos. Intermediate cases are possible if only some of the quantum numbers are conserved, although the Poisson limit is rapidly approached. Unlike experiments, model calculations can easily generate uncontaminated sequences of all eigenstates. Figure 18 shows $P_\beta(s)$ for the SCCl$_2$ calculations in Fig. 16, At $V^{(3)} = 75$ GHz, the average coupling strength compatible with experimental and ab initio data [18,37]. $\beta = 0.6$ provides the best fit, indicating conservation of 1–2 quantum numbers in this case. This number of conserved quantum numbers is

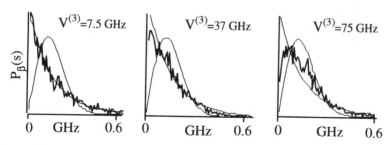

Figure 18. Level distribution functions for the same calculations as in Fig. 16. A transition toward Wigner statistics is incipient but incomplete at realistic coupling strengths (Ref. 122).

compatible with quasimicrocanonical excitation, $\mathcal{N} = 6$ and the computed $D_v \approx 3$ discussed below. The conclusion from simulations, as from experiments, is that at chemical energies, IVR is usually not fully chaotic, but preserves some good quantum numbers. In large molecules, "some" means the majority, as will be evident from the discussion of δ and D_v below. This should not be too surprising. With a typical dissociation energy of 1000 THz and 50 degrees of freedom for a medium-size organic molecule, only 20 THz per mode remain at chemically relevant energies, corresponding to a low-frequency vibrational energy. From a time-dependent point of view, intermediate statistics such as Fig. 18 may actually indicate a sweep from irregular toward more regular dynamics, as the energy spreads from a localized feature $|0\rangle$ throughout a molecule, and as the average excitation in all the modes begins to balance out. As a result, some caution has to be applied to measurements of $\bar{\rho}_{\mathrm{tot}}$ from spectra. The fluctuations in partially regular spectra can be much larger than the Porter–Thomas fluctuations expected for well-mixed spectra. In a very large molecule, the majority of molecular states are not well coupled to $|0\rangle$ even by coupling chains, and hence simply not represented in the spectrum at all at achievable SNRs.

E. Interior and Edge States

As discussed in Section IV, features in state space can be classified as interior or edge states. For large $\mathcal{N}$, the distinction is rather gradual depending on the number of nonzero quanta [5,17,89]. The difference is best illustrated by considering the two extremes, say, a $|700000\rangle$ tetratomic overtone state and a $|233222\rangle$ interior interior state that are nearly isoenergetic. Figure 14 illustrated the difference in time evolution between just these two states of $SCCl_2$. The interior state tends to spread through state space much more effectively, and its dynamics are closer to being ergodic. In Fig. 2, this trend is also seen by comparing the $SCCl_2$ combination bands and the higher-lying $|900000\rangle$ state. Figure 2 also illustrates the opposite behavior for propyne; the combination band consists of a single line, while the overtone has $\sigma \lesssim 0.3$.

Two main effects compete to decide which relaxes faster, an interior state or an edge state. Interior states are fully surrounded by other states; in the limit of $\mathcal{N} \gg \bar{n}$ [64] and strong couplings,

$$\rho_{\mathrm{loc}}^{(n)} \sim \frac{N^n}{\left(\dfrac{n}{2}\right)!^2}. \tag{5.4}$$

On the other hand, a state with k quanta within less than n of the edge of state space has its coupling partners reduced by 2^k. This can be a large number for a near-apex state in a large molecule. As a result, edge states tend to relax more slowly, all other things being equal. All other things are not equal, however. As seen in Fig. 13, edge states lie in regions of higher curvature of the energy shell in an anharmonic basis, indicating stronger off-resonant and resonant couplings. The latter scale as products of $v_i^{n/2}$ for each coupled mode. Edge states therefore tend to be more strongly coupled to their fewer neighbors, potentially outweighing the decreased number of coupling partners. State space calculations with either scaled or experimentally refined Hamiltonians predict exactly the observed trend in propyne, $SCCl_2$, and other molecules [17,37,45,55]. Recently, differences between initial CH overtone IVR rates (edge state), and subsequent "bath" rates within the manifold of dark states (more interior states), have been measured starting with the *same* initial state using vibration/rotation double-resonance experiments (Fig. 5). They have been interpreted in terms of a simple two-tier picture involving the bright state and two conformational populations [40]. Again, this approach finds bath rates to be slower in some cases, faster in other cases. Excellent agreement with experiment is obtained for the interior states using the HLRM state-space treatment [168]. Although random matrix models average over the energy shell and therefore weight interior states more, they can be suitably modified to account for edge states.

Qualitatively, the picture can be summarized as follows in terms of the $c_{ij}^{(2)}$. If all mode couplings are relatively uniform, interior states will decay faster than edge states; if the mode couplings are highly nonuniform, most of the neighbors in state space are not available, leading to slower interior state decays. In the $SCCl_2$ example, several modes interact with v_1. In the propyne example, the coupling between the acetylenic and methyl CH stretches is weak because of the chainlike geometry of the molecule (Fig. 7). Exciting the $2v_1 + v_6$ band is almost like exciting two separate molecules to only $\frac{2}{3}$ and $\frac{1}{3}$ of the total 290-THz vibrational energy; hence the $2v_1 + v_6$ IVR resembles more the $2v_1$ than the $3v_1$ IVR.

F. Quantum Diffusion on the IVR Manifold

In strict tier models, closed-coupling loops among states are not allowed. Although such loops are important when considering coherence effects (Section VI), states can be meaningfully grouped into tiers even if couplings within and across more than 1 tier are included. For example, model calculations for the fourth CD stretching overtone of CDBrClF have been carried out with a complete state space and no tiering. When the states are assigned to tiers by the criterion in Eq. (4.7), with a cutoff of 0.01, successive tiers from 1 to 4 reach maximal populations at 0.46, 0.63, 0.91 and 1.2 ps [17].

The fact that tiers can be defined succesfully indicates that IVR is a multitimescale process that cannot be identified with a single rate. One reason for this has aleady been discussed: as IVR progresses, the nature of the features changes from edge to interior, which can lead to an increase or decrease in the time-dependent rate $k(t)$, depending on the character of the interior states. An experimental example of this was shown in Fig. 5. What about the case where IVR is initiated in an interior state, or at least in a combination band already involving substantial skeletal activity? As it turns out, one still cannot define a unique rate constant: not only do bath–bath rates differ from system-bath rates, but a multitude of bath–bath timescales exist, leading to CIT dynamics. The reason is the HLRM scaling behavior of the vibrational Hamiltonian, which translates into a scalable structure of state space [16].

This has been described by a simple analytic model that employs rescaling of state space to conclude that IVR is a quantum diffusion process [16], with a survival probability for transfer among interior states given by

$$P(t) \sim t^{-\delta/2} \qquad (5.5)$$

where δ is a dimensional power that can vary between 0 and $\mathcal{N}$ (or $\mathcal{N}-1$ for microcanonical excitation). Even when flow in state space is anisotropic (Section IV), the time decay of the survival of an interior state $|0\rangle$ retains the simple form of Eq. (5.5) [54]. In that case, not all directions in state space make the same contribution to the overall value of δ. Expressions have been given that extend the power law to $t = 0$ to provide exact short-time or long-time behavior [50]. A simple approximation is given by the exponential limit polynomial of Eq. (1.9)

$$P(t) = \frac{(1-\sigma)}{[1 + (2t/\delta\tau)]^{\delta/2}} + \sigma \qquad (5.6)$$

which, however, does not reproduce the early-time rolloff correctly. Functions such as demonstrated in this equation have been found to give satisfactory fits to the average global dynamics (Figs. 12 and 19), although the survival in the exponential regime $t < \tau_{\text{loc}}$ is usually overestimated.

If δ were rather large ($\approx \mathcal{N}$), then the power law would be indistinguishable from an exponential decay. Simulations for a variety of molecules with realistic model Hamiltonians show that δ is always substantially smaller than $\mathcal{N}$ [17,50,55]. Figure 12 has illustrated this for the decay of $\sigma(t)$ of $SCCl_2$, which, however, is basis-set-dependent. Figure 19 shows a simulation of $P(t)$ using an experimentally fitted vibrational Hamiltonian for

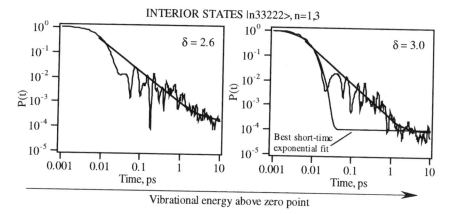

Figure 19. Power-law decays in the CIT regime, indicating that IVR is a quantum diffusion process at intermediate times. The early dynamics are better fitted by an exponential decays, but the system requires at least 1.5 orders of magnitude longer in time to reach quasiequilibrium (Ref. 55). The power-law exponent δ increases slowly with energy.

the $|n33222\rangle$ interior states of $SCCl_2$ lying above and below the interior state in Fig. 14, but otherwise analogous. After an initial exponential phase to $\tau_{loc} \approx 30$ fs, the average decay follows a power law with $\delta = 2.6–3.0$. The maximum possible value is ~ 5, and despite an energy difference $>2000 \, cm^{-1}$, the two decays are very similar at intermediate times. The case is even more dramatic for larger polyatomic molecules. An analysis [50,55] of experiments on combination bands of several polyatomic molecules (fluorene, methanol, cyclohexylaniline, $SCCl_2$) [22,33,68,88] agrees with the simulations, yielding power law coefficients of <5 in all cases. This is to be contrasted with the number of vibrational degrees of freedom of some of these molecules; for example, $\mathcal{N} = 72$ for fluorene, which could easily fall in the exponential limit but does not.

The reason why CIT dynamics is far more correlated (i.e., $\delta \ll \mathcal{N}$) than it needs to be is given by Eq. (4.6). The dimension D_v of the IVR coupling manifold controls the expansion of the IVR packet, and D_v is much smaller than $\mathcal{N}$. Indeed, for the $SCCl_2$ example, the dynamical quantity δ is essentially equal to D_v derived only from the coupling structure of H_v in Eqs. (4.6)–(4.8). Thus an IVR wavepacket cannot simply hop anywhere in state space, but rather has to wind its way through the coupling manifold as illustrated in Fig. 20. The identity of D_v and δ confirms that IVR is a quantum diffusion process, and not a rate process at intermediate to long times. However, the two need not be exactly identical; D_v as defined by Eq. (4.6) includes only local coupling effects, while δ characterizes the overall dynamics. In cases where superexchange plays an important role, δ could be

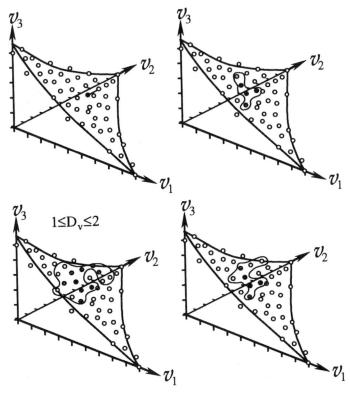

Figure 20. Clockwise from top left: schematic expansion of an IVR wavepacket with quasimicrocanonical excitation. The initial state is shown in black, and maximal populations reached at later times are indicated by shading. The initial state cannot directly couple to the neighbor shown by a connecting line in the top left panel; rather, the population has to wind its way through the IVR manifold with $D_v \approx 1.5$.

larger than D_v by 1–2, as discussed in Section IV. Comparison of Figs. 19 and 12 shows that δ and δ_σ obey similar power laws, when the latter is calculated in a near-optimal feature basis. This will be of interest for IVR control as discussed in section VI.

In order to probe the structure of the IVR manifold, the IVR wavepacket has to populate at least the locally coupled states $|\ell\rangle$ such that

$$\sum_{\ell=0} \langle \tau_{\mathrm{loc}} | \ell \rangle \tag{5.7}$$

is maximized. This defines the timescale for the onset of CIT dynamics from interior states. For edge states, this timescale can become even longer because the transition from edge to interior states can instead control the dynamics. This is the reason why rotational double-resonance measurements initiated in overtones, although capable of distinguishing edge and interior timescales [109], cannot yet probe the fact that the interior evolution in fact has no timescale; instead, one has to look at the deviations of lineshapes from Lorentzians and of survival probabilities from exponentials measured in overtone or preferably pump–dump combination band experiments [22,37]. In the future, as statistics improve so more than a linewidth can be assigned to rotational double resonance spectra, and experiments are initiated in states closer to the interior of state space, vibration–rotation double-resonance measurements are expected to become another excellent probe of multiscale IVR among bath modes.

One issue that has not been fully addressed by experiment or theory, is the question of how D_v varies with local position in state space. One can imagine three scenarios: (1) the resonances responsible for diffusion are the same everywhere, and D_v remains fixed; (2) the resonances responsible for diffusion change locally, but the average number of such resonances remains nearly constant, so the overall D_v remains constant; and (3) the number of resonances differs drastically among different parts of state space, and D_v could change value as the IVR wavepacket spreads. The first of these is highly unlikely, except as an idealized model for small molecules. The second case may well correspond to many actual cases of IVR. It is important to note that in case 2, most or all quantum numbers can still break down at long time, even though D_v remains much smaller than $\mathcal{N}$: D_v is a *local*, not a global measure of the IVR manifold. (For example, in Fig. 20 D_v may be 1.5 everywhere locally, but the resonances point in different directions in different places, so no globally good quantum numbers exist).* The third case could result in a change of D_v in the course of the dynamics. This is not borne out so far by measurements of *averaged* IVR decays of large molecules (Fig. 19). However, it does provide another way of looking at intermediate regime quantum beats; these can be ascribed to fluctuations in D_v as sampled by an increasingly delocalized IVR wavepacket. If so, one would expect them to decrease in magnitude with time, and this is, indeed, what is observed experimentally. As mentioned in Section I, at long times the beat amplitudes are proportional to σ.

In small molecules, the value of D_v depends largely on accidental degeneracies and coupling constants, and can be estimated from the effective resonance Hamiltonian. Figure 19 shows that it is a slowly increasing function of energy, and data in Ref. 50 indicate that it is also a slowly, if at

---

*Classically, this would present a problem; quantum-mechanically, there is no problem because state-space volumes are no longer independent below a scale $h^{\mathcal{N}}$.

all, increasing function of molecular size. Although this has not yet been studied systematically, it is plausible that the size dependence of D_v saturates in large molecules. The likely reason brings us full circle to the beginning of Section III. Chemical bonding is local and highly directional due to the small value of κ. In large, low symmetry molecules, most $c_{ij}^{(2)}$ are very small once ≥ 2 bonds intervene between atoms (Fig. 7). The major amplitude of medium to high-frequency vibrational modes is therefore inherently localized to relatively small groups of 2–6 atoms. Even low-frequency modes of large molecules involve a relative displacement of subunits that are quasirigid. Either way, the number of active coordinates is usually less than 6–18.[†] Once vibrational energy spreads over a sufficiently large part of a molecule, the average local dimensionality of the problem is "frozen" in and the energy merely diluted, resulting in the characteristic slowing of IVR during CIT dynamics. Localization wins.

VI. FUTURE DIRECTIONS

The introduction began by extolling the virtues of energy flow in a large number of chemical fields. Although each area to which our knowledge of IVR processes could be applied deserves a review of its own, this review would be incomplete without a brief discussion of some of these applications, which will undoubtedly occupy IVR researchers for years to come.

A. Control of Molecular Reactivity

If the "typical" value of D_v in Section V stands the test of time, prospects are rosy for completely general coherent control of molecules. Since the late 1970s, investigators have suspected that ideas from IVR research can be applied to molecular control [5]. The principles of coherent control were developed in the mid-1980s after it became clear that multiphoton or Franck–Condon-specific excitation must be limited in the reactive control they can exert [169,170]. That is not to say that Franck–Condon control of reactivity has not been successful. Into this category fall experiments from laser isotope separation [171], to control of unimolecular dissociation on dissociative excited states [172], on to bimolecular reaction probabilities of local mode excited molecules [173]. However, molecules are fundamentally quantum mechanical objects, and the addition of coherence and interference effects to the repertoire makes possible the most general kind of

---

[†]Some nonlocal transport can occur through high-order resonances between nearly degenerate states, as discussed in Section III, but this is likely to occur on a much slower timesale than locally mediated transport. Certainly in the macroscopic limit of IVR, which is heat conductivity, there is no evidence for dominant nonlocal transport.

manipulation of the wavefunction. Controlling the amplitude and phase of the molecular wavefunction in principle has two great advantages—one has a new and sensitive tool to study reaction dynamics, and one can control the specificity of reactions.

This is the principle. What about the practice? With the equivalence of time-dependent and time-independent approaches proved [73], and optimal control theory providing a framework for iterative improvement that can be applied to coherent control [174], the general theory is well established. Experiments have had great successes in demonstrating the principle of coherent control in atoms, [175], diatomic molecules [176], XH_2 molecules [76] and even a large molecule with light substituents coupled weakly to a massive central atom [75]. Control of a strongly coupled organic polyatomic molecule on a bound surface has not yet been achieved. Probably the main reason for this is IVR; it would seem that when a very large number of states in many dimensions is involved, the control problem can become very complicated and also lose robustness, in the sense that small errors can lead to a complete loss of control.

There is certainly engineering precedent that coupled with feedback, this need not be a problem [177]. For instance, spacecraft can be moved into position on highly irregular orbits which allow minimal energy expenditure by coupling knowledge of the equations of motion (Hamiltonian) with position feedback and occasional corrections from a small engine. The problem with this analogy is that in the molecular context, the Hamiltonian is not well known and the feedback information is extremely limited. Although feedback certainly can be conducted with a 'blackbox Hamiltonian', engineers for good reason generally try to avoid this purist approach and take all the help they can get. To alleviate these problems, we can have recourse to another engineering analogy. Maintaining a spacecraft in a simple orbit is also a control problem, but a much simpler one; the required positional feedback and corrections are minimal and the equations of motion have a structure that almost automatically maintains the orbit and causes the spacecraft to explore only a small, familiar corner of phase space. This maintenance problem is therefore more robust and straightforward than the full control problem.

The maintenance problem resembles the situation of an initially prepared feature state; which lies in a small, optically accessible region of state space. The local couplings can be determined spectroscopically, so the local Hamiltonian can be characterized. The hierarchical structure of the Hamiltonian causes unusual stability of features; in the CIT regime they decay via a power law $t^{-\delta/2}$. In addition, features with small initial decay constants k_{IVR} can usually be found. If features could be linked to selective reactivity, then the most general control could be reduced to two steps:

(1) prepare and maintain a feature; then (2) wait for the selective chemistry to happen because of the Franck–Condon properties of the feature.

There is indeed a link between features and reactivity, and it has been experimentally demonstrated [173,178–180]. For example, excitation of HOD local modes which have a long lifetime even on a collisional timescale leads to selective bimolecular reaction of H + HOD, purely via Franck–Condon control. This should not be surprising; wavefunctions with a simple structure will generally have very different overlaps with different product channels, whether in unimolecular dissociation, predissociation, or direct dissociation from a continuum resonance. The difference between HOD and a large organic molecule at chemical energies is that the latter's features are much shorter-lived, and therefore require maintenance.

The proposed scheme thus combines coherent and Franck–Condon control: a feature state near the desired reaction channel is maintained by coherent control. The molecular Franck–Condon structure then leads to the desired unimolecular or bimolecular reaction (Fig. 21) This will be called "static coherent control." It is not as optimal as full dynamical coherent control, but trades efficiency for robustness. The robustness of the first half of this scheme has been verified computationally using a realistic 6D vibrational Hamiltonian for $SCCl_2$. Modest shaped pulses with < 100 phase/ amplitude parameters should be able to create a "frozen" feature that survives 10–100 times longer than on simple pulsed excitation [45]. Variations of the Hamiltonian within the bounds allowed by spectroscopic measurement still lead to 3–10 enhancements in feature lifetime. This is important: by working exclusively in an optically explorable part of state space by maintaining a feature in its original state, one automatically knows something about the "blackbox."

Freezing of feature states is possible because of the small dimension D_v of the IVR manifold embedded in state space, and the resulting small power exponent δ. As discussed in Section IV, there are a minimal number of states required to represent the dynamics, given by the optimal dispersion $\mathscr{D}(t)$ in Eq. (4.1). These are the states that minimally must be controlled. Each such feature requires an amplitude and a phase to be controlled by the radiation field, so the number of control parameters C increases very slowly as

$$C = 2\mathscr{D}(t) \sim t^{-\delta/2} \qquad (6.1)$$

in the CIT regime. Figure 12 shows a close upper bound to this number for the $|233222\rangle$ interior state of $SCCl_2$ at $\approx 238\,THz$. The initial exponential decay would reach the $\mathscr{D} = 10^4$ baseline in approximately 70 fs in this example; the power law requires 3 ps, a factor 40 improvement. It is not very likely that $2 \cdot 10^4$ control channels will be achieved very soon, so the initial

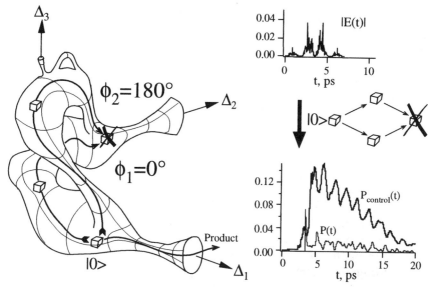

Figure 21.   Schematic of control on an IVR manifold: Franck–Condon preparation creates a feature $|0\rangle$ near the Δ_1 product channel. Because of the presence of a bottleneck (transition state) in channel Δ_1, this could rapidly dephase to channel Δ_2, for example, by the two paths on the IVR manifold shown. Application of an electric field (actual control field from Ref. 45 shown) can cancel undesirable paths in state space and maintain the feature. The bottom right panel shows $P(t)$ of the $\nu_1 = 8$ feature of $SCCl_2$ with a weak-field FT-limited pulse (weak line), and a strong-field shaped pulse (shown above). In terms of Fig. 4, let all lattice points on the energy shell be reactive, but to different product channels depending on their vibrational structure. Rapid IVR would yield an average statistical product formation. 'Freezing' a lattice point with a control field will yield only the product(s) characteristic of that lattice point.

exponential phase still presents a problem. Fortunately, examination of IVR in several organic molecules and for other states of $SCCl_2$ has shown that the exponential phase can be relatively short for some states [50,55], allowing CIT dynamics to take over when $P(t) = 10^{-1} - 10^{-2}$. Ten to a thousand control channels are easily available with present-day equipment [181,182].

The goal of static control differs from full dynamical coherent control in another interesting way. Given a reaction timescale τ_{rx}, there is no need for IVR wavepacket maintenance to extend much beyond τ_{rx} for the Franck–Condon effect to perform the second step. The overall energy spread of the multiple coherent control paths must therefore satisfy only

$$\Delta E \lesssim \frac{1}{\tau_{rx}} \tag{6.2}$$

whereas full control requires complete degeneracy, so the coherence can manifest itself at infinite time for the products. Equation (6.2) provides a more general formulation of the usual degeneracy criterion for coherent control paths.

The question addressed here is closely related to the applicability of RRKM theory. Recently, there has been much progress in measuring accurate rates for isomerization and unimolecular dissociation reactions [87,110]. RRKM theory [183] and its modern variants [133] provide an excellent description of such reactions over most of the pressure range in the thermal case. RRKM theory has also been applied with much success to isomerization/dissociation reactions in molecular beams above threshold [184,185], indicating that IVR is sufficiently fast to compete favorably with the higher dissociation rates possible in photochemical experiments. Nonetheless, the $t^{-\delta/2}$ dependence during CIT dynamics leads one to suspect that vibrational randomization, although sufficiently complete for RRKM theory to hold in many cases, is not necessarily extensive. In particular, for photochemical reactions well above threshold, τ_{rx} rapidly decreases, while the IVR dynamics speed up only slowly with energy. There must be an above-threshold energy where $\tau_{IVR} \approx \tau_{rx}$ and RRKM theory just barely holds. In addition, reactions can have an instantaneous time component [186]. There are beginning to be experimental indications that IVR is not necessarily complete on the τ_{rx} timescale for photochemical reactions. One example is the isomerization of partially deuterated stilbenes [110], where the d^8 ring-deuterated (higher total density of states) and d^2 ethenyl-deuterated rates are inverted from the RRKM prediction. A natural explanation would be that the rate above threshold becomes increasingly limited by IVR, and that ethenyl deuteration is more effective at increasing the local density of states of modes near the reaction coordinate, thereby increasing the rate more than d^8-ring deuteration [112,113,168]. If so, then slowing the IVR rate further by freezing the $t^{-\delta/2}$ decay of a feature state could indeed be very effective in photochemical control.

Although feature freezing has been computationally modeled, and the Franck–Condon step has been experimentally verified in small molecules, static control has yet to be applied experimentally. However, the results discussed here and in Sections III–V are cause for cautious optimism that robust coherent control of organic molecules will soon be a reality.

B. Other Areas

Several other areas of application are equally exciting, as illustrated by the following four examples:

1. Vibrational relaxation in dense gases and liquids is now subject to a new generation of experiments which provide an unprecedentedly detailed

look at quantities such as the collisional deactivation probability $P(E, E', J, J')$ or mode-resolved relaxation of energy into the solvent bath [114–116]. Intramolecular energy redistribution is tightly coupled to system–bath energy exchange and phase randomization in such cases. A question of particular interest is: On what timescale are relative internal phases of vibrationally excited molecules randomized by the solvent? In a strongly coupled solvent, multiphonon processes in the solvent may compete with smaller quantum number steps taken by the intramolecular process. Superfluid helium clusters, which provide an extremely weakly interacting solvent, will be excellent laboratories for studying the coupling between IVR and solvent modes [115].

2. Nonadiabatic interactions often cannot be neglected in photochemical reactions [187]. The treatment of IVR given here assumed the Born–Oppenheimer approximation. Some of these assumptions must be relaxed near conical intersections and avoided crossings. At the same time the influence of crossings and intersections on photochemistry is becoming better understood, and exceptions to the Woodward–Hoffman rules are the rule [188]. So far, there are relatively few studies that combine IVR and vibronic dynamics at the state-to-state level [189]. The interaction between vibrational energy flow and electronic energy flow is likely to reveal many interesting principles, which will have direct applicability to photochemical coherent control.

3. IVR studies need to be extended to mesoscopic systems. It is now feasible to perform calorimetric measurements on very small structures at the one-phonon level, where macroscopic heat flow and molecular IVR meet [190]. Interesting phenomena, such as a size-dependent turnover regime for the thermal conductivity have been proposed [3]. At very low temperatures, dephasing in mesoscopic systems may also eventually answer questions about how quantum interactions turn into measurements as one moves from the microscopic to the macroscopic scales.

4. Closely connected to this, IVR is an excellent laboratory for questions about dephasing, or "decoherence," as it is fashionably called when a subsystem is split off from a "bath" and its dephasing examined. Molecular IVR allows tuning through a wide range of system and bath sizes. Also, the vibrational Hamiltonian in Eqs. (3.11) and (3.12) is not likely to be the only hierarchically scaled Hamiltonian. On the contrary, the general conditions of exponential decay of matrix elements and correlations due to connectivity (whether in real space or in state space) are likely to be satisfied by a variety of systems, such as coupled spins in a magnetic field or solute–solvent couplings [191]. It may well turn out that following an initial exponential phase, power laws such as Eq. (5.4) are the rule, not the exception, in complex coupled systems.

ACKNOWLEDGMENTS

Major support for the author's work on IVR dynamics and dephasing was provided by the National Science Foundation and the David and Lucile Packard Foundation.

GLOSSARY

Non-Roman symbols are sorted phonetically; not all symbols used only once in the text are included.

$a_i^\dagger, a_i$	Unitless vibrational ladder operators
a_i	Born–Oppenheimer scaling parameter for mode i
B_k	Rotational constant in hertz
BSTR	Bose–Einstein statistics triangle rule model
CIT dynamics	Correlated intermediate timescale dynamics
c_i, c_s	Overlap amplitude of states $|i\rangle$ or $|s\rangle$ with feature $|0\rangle$; constant if $|s\rangle$ is an eigenstate
$c_{ij}^{(2)}$	Coupling correlation coefficient: if $c_{ij}^{(2)} \ll 1$, potential constants between modes i/j are small
$\mathscr{D}(t)$	Dispersion of IVR wavepacket (number of states required to represent the wavepacket at time t); inverse of $\sigma(t)$
δ, δ_σ	Power-law exponent of $P(t)$ [subscript σ indicates exponent of $\sigma(t)$]
E_{vib}	Total vibrational energy
$E^{(0)}$	Diagonal contribution to the effective resonance Hamiltonian
GR	"Golden rule", the (Fermi)
HLM	Hierarchical local matrix
HLRM	Hierarchical local random matrix
H_{vr}	Rovibrational Hamiltonian
H_{v}	Vibrational Hamiltonian
H_{nm}	nth-order vibrational/mth-order rotational term in a power series expansion of H_{rv}
I_s	Intensity of an eigenstate in the spectrum of feature $|0\rangle$
IVR	Intramolecular vibrational redistribution (or relaxation, depending on context)
$\hat{J}_k$	Unitless total angular momentum operator
k_{GR}	GR initial decay rate of $P(t)$
k_{IVR}	Initial IVR decay rate defined by $P(1/k_{\mathrm{IVR}}) = \mathrm{e}^{-1}$
n	Overall order of a coupling, given by the 1-norm $n = |\mathbf{n}|_1 = \Sigma n_i = \Sigma|v_i' - v_i|$

n_i	Partial order of a coupling in mode i
$\mathbf{n}$	Array of mode coupling orders $[n_2, n_2, \ldots, n_{3N-6}]$
$N_{\text{loc}}^{(n)}$	Effective number of locally coupled states [directly to $|0\rangle$ by $V^{(n)}$]
N_{loc}	Number of states locally coupled to $|0\rangle$ to any order
N_{tot}	Effective total number of states composing an IVR wavepacket
$\mathcal{N}$	Number of vibrational degrees of freedom
ω_s	Eigenstate or other frequency (in rad/s unless "e" appears in the subscript)
ω_{ei}	Harmonic normal mode vibrational frequencies (in Hz, not rad/s)
ω_{rlf}	Rolloff frequency of $P(t)$ defined in Eq. (1.4)
$p_i = i(a_i^\dagger - a_i)$	Unitless momentum operators
$P(t)$	Survival probability of an initially prepared state
P_β	Brody distribution of line spacings ($\beta = 0$ is Poisson, $1 = $ Wigner)
$\mathcal{P}_i$	Unitless vibrational angular momentum operator
$q_i = (a_i^\dagger + a_i)$	Unitless normal coordinate operator
ρ_{tot}	State density distribution function
$\bar{\rho}_{\text{tot}}$	Window-averaged total density of states (vibrational or rovibrational depending on context)
$\rho_{\text{loc}}^{(n)}$	Local density of states coupled to $|0\rangle$ to order n
SNR	Signal-to-noise ratio
σ	Dilution factor ($1/N_{\text{eff}}$, the participating number of states)
$\sigma(t)$	Inverse of the dispersion, goes approximately to the dilution factor as $t \to \infty$
τ_{loc}	Time required for the IVR wavepacket to explore states directly coupled to $|0\rangle$
τ_{rlf}	Initial rolloff time of $P(t)$ at finite density of states
τ_σ	Time required for $P(t)$ to fluctuate about its average value $\overline{P(t)}$ with fluctuations of order $\overline{P(t)}$
τ_{rx}	Reaction lifetime
RM	Random matrix
$v, \boldsymbol{u}$	Vibrational quantum number (array if bold)
$V_{0s'}$	Coupling from the bright state to a prediagonalized GR manifold of states $\{|s'\rangle\}$
$V_{\mathbf{n}}^{(n)}$	Potential constants including all permutation factors; added prime indicates that off-resonant contributions have been removed.
$V^{(n)}$	Potential constants geometrically averaged over a fixed order n

M. GRUEBELE

$V_i^{(3)}$ Average cubic potential constant for mode i

V_{rms} Root-mean-square coupling constant from a bright state to the full dark state manifold

W_β Integral of P_β

References

1. G. Kortüm and B. Finckh, *Z. Phys. Chem.* **B52**, 263 (1942).

2. P. Guyot-Sionnest, P. H. Lin, and E. M. Miller, *J. Chem. Phys.* **102**, 4269 (1995).

3. D. M. Leitner and P. G. Wolynes, *Phys. Rev. E* **61**, 2902 (1999).

4. K. F. Freed and A. Nitzan, *J. Chem. Phys.* **73**, 4765 (1980).

5. S. A. Rice, *Adv. Chem. Phys.* **47**, 117 (1981).

6. T. Uzer, *Phys. Rep.* **199**, 73 (1991).

7. M. Bixon and J. Jortner, *J. Chem. Phys.* **48**, 715 (1968).

8. W. M. Gelbart, S. A. Rice, and K. F. Freed, *J. Chem. Phys.* **57**, 4699 (1972).

9. W. M. Gelbart, S. A. Rice, and K. F. Freed, *J. Chem. Phys.* **52**, 5718 (1970).

10. A. M. O. D. Almeida, *Hamiltonian Systems: Chaos and Quantization*, Cambridge Univ. Press, Cambridge, UK, 1988.

11. K. S. J. Nordholm and S. A. Rice, *J. Chem. Phys.* **61**, 203 (1974).

12. S. Nordholm and S. A. Rice, *J. Chem. Phys.* **62**, 157 (1975).

13. A. A. Stuchebrukhov, *Sov. Phys. JETP* **64**, 1195 (1986).

14. D. E. Logan and P. G. Wolynes, *J. Chem. Phys.* **93**, 4994 (1990).

15. F. Remacle and R. D. Levine, *J. Chem. Phys.* **98**, 2144 (1993).

16. S. Schofield and P. G. Wolynes, *J. Chem. Phys.* **98**, 1123 (1993).

17. R. Bigwood and M. Gruebele, *Chem. Phys. Lett.* **235**, 604 (1995).

18. D. Madsen, R. Pearman, and M. Gruebele, *J. Chem. Phys.* **106**, 5874 (1997).

19. R. E. Smalley, *J. Phys. Chem.* **86**, 3504 (1982).

20. D. E. Powers, J. B. Hopkins, and R. E. Smalley, *J. Chem. Phys.* **72**, 5721 (1980).

21. G. M. Stewart and J. D. McDonald, *J. Chem. Phys.* **78**, 3907 (1983).

22. J. F. Kauffman, M. J. Coté, P. G. Smith, and J. D. McDonald, *J. Chem. Phys.* **90**, 2874 (1989).

23. R. A. Coveleskie, D. A. Dolson, and C. S. Parmenter, *J. Phys. Chem.* **89**, 655 (1985).

24. J. S. Baskin, M. Dantus, and A. H. Zewail, *Chem. Phys. Lett.* **130**, 473 (1986).

25. P. M. Felker, W. R. Lambert, and A. H. Zewail, *J. Chem. Phys.* **82**, 3003 (1985).

26. C. E. Hamilton, J. L. Kinsey, and R. W. Field, *Annu. Rev. Phys. Chem.* **37**, 493 (1986).

27. D. J. Nesbitt and R. W. Field, *J. Phys. Chem* **100**, 12735 (1996).

28. D. M. Jonas, S. A. B. Solina, B. Rajaram, R. J. Silbey, R. W. Field, K. Yamanouchi, and S. Tsuchiya, *J. Chem. Phys.* **99**, 7350 (1993).

29. K. Yamanouchi, N. Ikeda, S. Tsuchiya, D. M. Jonas, J. K. Lundberg, G. W. Adamson, and R. W. Field, *J. Chem. Phys.* **95**, 6330 (1991).

30. A. Beil, D. Luckhaus and M. Quack, *Ber. Bunsenges. Phys. Chem.* **100**, 1853 (1997).

31. J. E. Gambogi, K. K. Lehmann, B. H. Pate, G. Scoles, and X. Yang, *J. Chem. Phys.* **98**, 1748 (1993).

32. K. K. Lehmann, G. Scoles, and B. H. Pate, *Annu. Rev. Phys. Chem.* **45**, 241 (1994).

33. O. V. Boyarkin, L. Lubich, R. D. F. Settle, D. S. Perry, and T. R. Rizzo, *J. Chem. Phys.* **107**, 8409 (1997).

34. R. D. F. Settle and T. R. Rizzo, *J. Chem. Phys.* **97**, 2823 (1992).

35. G. A. Bethardy, X. Wang, and D. S. Perry, *Can. J. Chem.* **72**, 652 (1994).

36. J. Go, T. J. Cronin, and D. S. Perry, *Chem. Phys.* **175**, 127 (1993).

37. R. Bigwood, B. Milam, and M. Gruebele, *Chem. Phys. Lett.* **287**, 333 (1998).

38. S. A. B. Solina, J. P. O'Brien, R. W. Field, and W. F. Polik, *Ber. Bunsenges. Phys. Chem.* **99**, 555 (1995).

39. A. Geers, J. Kappert, F. Temps, and J. W. Wiebrecht, *J. Chem. Phys.* **101**, 3618 (1994).

40. B. H. Pate, *J. Chem. Phys.* **110**, 1990 (1999).

41. R. E. Wyatt, *J. Chem. Phys.* **109**, 10732 (1998).

42. R. E. Wyatt, *Phys. Rev. E* **51**, 3643 (1995).

43. R. E. Wyatt, *Adv. Chem. Phys.* **73**, 231 (1989).

44. S. A. Schofield, R. E. Wyatt, and P. G. Wolynes, *Phys. Rev. Lett.* **74**, 3720 (1995).

45. M. Gruebele and R. Bigwood, *Int. Rev. Phys. Chem.* **17**, 91 (1998).

46. R. Bigwood and M. Gruebele, *ACH Models Chem.* **134**, (1997).

47. C. Iung and C. Leforestier, *J. Chem. Phys.* **97**, 2481 (1992).

48. E. L. Sibert, *Int. Rev. Phys. Chem.* **9**, 1 (1990).

49. E. L. Sibert, W. P. Reinhardt, and J. T. Hynes, *J. Phys. Chem.* **81**, 1115 (1984).

50. M. Gruebele, *Proc. Natl. Acad. Sci.* (USA) **95**, 5965 (1998).

51. A. A. Stuchebrukhov, M. V. Kuzmin, V. N. Bagratashvili, and V. S. Lethokov, *Chem. Phys.* **107**, 429 (1986).

52. E. J. Heller, *J. Phys. Chem.* **99**, 2625 (1995).

53. E. J. Heller, *J. Chem. Phys.* **92**, 1718 (1990).

54. S. A. Schofield and P. G. Wolynes, *J. Phys. Chem.* **99**, 2753 (1995).

55. V. Wong and M. Gruebele, *J. Phys. Chem.* **103**, 10083 (1999).

56. P. Pechukas, *Chem. Phys. Lett.* **86**, 553 (1982).

57. E. B. Stechel and E. J. Heller, *Annu. Rev. Phys. Chem.* **35**, 563 (1984).

58. W. D. Lawrance and A. E. W. Knight, *J. Phys. Chem.* **89**, 917 (1985).

59. K. K. Lehmann, *J. Phys. Chem.* **95**, 7556 (1991).

60. D. E. Logan and P. G. Wolynes, *Phys. Rev. B* **29**, 6560 (1984).

61. M. Gruebele, *J. Chem. Phys.* **104**, 2453 (1996).

62. M. Gruebele, *J. Phys. Chem.* **100**, 12178 (1996).

63. C. C. Martens, *J. Stat. Phys.* **68**, 207 (1992).

64. R. Pearman and M. Gruebele, *J. Chem. Phys.* **108**, 6561 (1998).

65. R. Pearman and M. Gruebele, *Zeitschr. für Phys. Chem.*, **in press** (2000).

66. M. L. Mehta, *Random Matrices*, Academic Press, San Diego, 1991.

67. D. M. Leitner and P. G. Wolynes, *J. Chem. Phys.* **105**, 11226 (1996).

68. R. Bigwood, M. Gruebele, D. M. Leitner, and P. G. Wolynes, *Proc. Natl. Acad. Sci.* (USA) **95**, 5960 (1998).

69. M. Davis and E. J. Heller, *J. Chem. Phys.* **75**, 246 (1981).

70. A. A. Stuchebrukhov, A. Mehta, and R. A. Marcus, *J. Phys. Chem.* **97**, 12491 (1993).

71. S. Keshavamurthy, *Chem. Phys. Lett.* **300**, 281 (1998).

72. D. J. Tannor, R. Kosloff, and S. A. Rice, *J. Chem. Phys.* **85**, 5805 (1986).

73. M. N. Kobrak and S. A. Rice, *J. Chem. Phys.* **109**, 1(1998).

74. P. Brumer and M. Shapiro, *Acc. Chem. Res.* **22**, 407 (1989).

75. A. Assion, T. Baumert, M. Bergt, T. Brixner, B. Kiefer, V. Seyfried, M. Strehle, and G. Gerber, *Science* **282**, 919 (1998).

76. V. D. Kleiman, L. Zhu, X. Li, and R. J. Gordon, *J. Chem. Phys.* **102**, 5863 (1995).

77. C. J. Bardeen, J. Che, K. R. Wilson, V. V. Yakovlev, V. A. Apkarian, C. C. Martens, R. Zadoyan, B. Kohler, and M. Messina, *J. Chem. Phys.* **106**, 8486 (1997).

78. R. S. Berry, *Int. J. Quantum Chem.* **58**, 657 (1996).

79. R. H. Austin, K. W. Beeson, L. Eisenstein, H. Frauenfelder, and I. C. Gunsalus, *Biochemistry* **14**, 5355 (1975).

80. S. J. Hagen and W. A. Eaton, *J. Chem. Phys.* **104**, 3395 (1996).

81. J. Sabelko, J. Ervin, and M. Gruebele, *Proc. Natl. Acad. Sci.* (USA) **96**, 6031 (1999).

82. M. J. Davis and R. T. Skodje, *J. Chem. Phys.* **111**, 859 (1999).

83. A. Blumen, J. Klafter, and G. Zumofen, *J. Phys. A* **19**, L77 (1986).

84. E. Shalev, J. Klafter, D. F. Plusquellic, and D. W. Pratt, *Physica A* **191**, 186 (1992).

85. P. M. Felker and A. H. Zewail, *Chem. Phys. Lett.* **102**, 113 (1983).

86. J. C. Crane, H. Nam, H. Clauberg, H. P. Beal, I. J. Kalinovski, R. G. Shu, and C. B. Moore, *J. Chem. Phys.* **102**, 9433 (1998).

87. A. H. Zewail, *J. Phys. Chem.* **100**, (1996).

88. P. G. Smith and J. D. McDonald, *J. Chem. Phys.* **92**, 1004 (1990).

89. J. E. Gambogi, J. H. Timmermans, K. K. Lehmann, and G. Scoles, *J. Chem. Phys.* **99**, 9314 (1993).

90. E. R. T. Kerstel, K. K. Lehmann, B. H. Pate, and G. Scoles, *J. Chem. Phys.* **100**, 2588 (1994).

91. J. E. Gambogi, E. R. T. Kerstel, K. K. Lehmann, and G. Scoles, *J. Chem. Phys.* **100**, 2612 (1993).

92. A. McIlroy, D. J. Nesbitt, E. R. T. Kerstel, B. H. Pate, K. K. Lehmann, and G. Scoles, *J. Chem. Phys.* **100**, 2596 (1994).

93. P. M. Felker and A. H. Zewail, *Chem. Phys. Lett.* **108**, 303 (1984).

94. P. M. Felker and A. H. Zewail, *J. Chem. Phys.* **82**, 2975 (1985).

95. M. P. Jacobson, J. P. O'Brien, R. J. Silbey, and R. W. Field, *J. Chem. Phys.* **109**, 121 (1999).

96. A. McIlroy and D. J. Nesbitt, *J. Chem. Phys.* **91**, 104 (1989).

97. A. Callegari, *Intramolecular Vibrational Energy Redistribution in Aromatic Molecules* Ph.D. thesis, Princeton Univ., 1998.

98. J. P. O'Brien, M. P. Jacobson, J. J. Sokol, S. L. Coy, and R. W. Field, *J. Chem. Phys.* **108**, 7100 (1998).

99. S. Choi, A. Callegari, H. K. Srivastava, P. Engels, K. K. Lehmann, and G. Scoles, *52nd Int. Sympo. Molecular Spectroscopy* (**RG09**), (1997).

100. R. Georges, A. Delon, and R. Jost, *J. Chem. Phys.* **103**, 1732 (1995).

101. D. Gruner and P. Brumer, *J. Chem. Phys.* **94**, 2862 (1991).

102. W. M. Gelbart, K. F. Freed, and S. A. Rice, *J. Chem. Phys.* **52**, 2460 (1970).

103. P. J. Timbers, C. S. Parmenter, and D. B. Moss, *J. Chem. Phys.* **100**, 1028 (1994).

104. D. S. Perry, G. A. Bethardy, and X. Wang, *Ber. Bunsenges. Phys. Bhem.* **99**, 530 (1995).

105. J. E. Gambogi, R. P. L'Esperance, K. K. Lehmann, B. H. Pate, and G. Scoles, *J. Chem. Phys.* **98**, 1116 (1993).

106. J. Go and D. S. Perry, *J. Chem. Phys.* **103**, 5194 (1995).

107. A. Beil, D. Luckhaus, M. Quack, and J. Stohner, *Ber, Bunsenges. Phys. Chem.* **101**, 311 (1997).

108. D. Perry and G. Scoles, private communication.

109. D. A. McWhorter, E. Hudspeth, and B. H. Pate, *J. Chem. Phys.* **110**, 2000 (1999).

110. S. H. Courtney, M. W. Balk, L. A. Philips, and G. R. Fleming, *J. Chem. Phys.* **89**, 6697 (1988).

111. G. H. Leu, C. L. Huang, S. H. Lee, Y. C. Lee, and I. C. Chen, *J. Chem. Phys.* **109**, 9340 (1998).

112. D. V. Shalashilin and D. L. Thompson, *J. Chem. Phys.* **107**, 6204 (1997).

113. D. M. Leitner and P. G. Wolynes, *Chem. Phys. Lett.* **280**, 411 (1997).

114. A. S. Mullin, C. A. Michaels, and G. W. Flynn, *J. Chem. Phys.* **102**, 6032 (1995).

115. K. K. Lehmann and G. Scoles, *Science* **279**, 2065 (1998).

116. J. C. Deak, L. K. Iwaki, and D. D. Dlott, *J. Phys. Chem.* **102**, 8193 (1999).

117. R. Laenen, C. Rauscher, and A. Laubereau, *Chem. Phys. Lett.* **283**, 7 (1998).

118. D. Papousek and M. R. Aliev, *Molecular Vibrational-Rottaional Sepctra*, Elsevier, Amsterdam, 1982.

119. T. Oka, *J. Chem. Phys.* **47**, 5410 (1967).

120. S. Rashev, *Chem. Phys.* **147**, 221 (1990).

121. W. J. Bullock, D. K. Adams, and W. D. Lawrance, *J. Chem. Phys.* **93**, 3085 (1990).

122. M. Gruebele, *J. Phys. Chem.* **100**, 12183 (1996).

123. A. A. Stuchebrukhov and R. A. Marcus, *J. Chem. Phys.* **98**, 6044 (1993).

124. A. Mehta, A. A. Stuchebrukhov, and R. A. Marcus, *J. Phys. Chem.* **99**, 2677 (1995).

125. D. M. Leitner and P. G. Wolynes, *Phys. Rev. Lett.* **76**, 216 (1996).

126. E. L. Sibert, *J. Chem. Phys.* **90**, 2672 (1989).

127. M. Gruebele, J. W. C. Johns, and L. Nemes, *J. Mol. Spectrosc.* **198**, 376 (1999).

128. C. C. Martens and W. P. Reinhardt, *J. Chem. Phys.* **93**, 5621 (1990).

129. D. S. Perry, G. A. Bethardy, M. J. Davis, and J. Go, *Faraday Disc.* **102**, 215 (1995).

130. R. Wallace and J. P. Leroy, *Chem. Phys.* **144**, 371 (1990).

131. D. M. Jonas, S. A. B. Solina, B. Rajaram, R. J. Silbey, R. W. Field, K. Yamanouchi, and S. Tsuchiya, *J. Chem. Phys.* **97**, 2813 (1992).

132. A. L. L. East, W. D. Allen, and S. J. Klippenstein, *J. Chem. Phys.* **102**, 8506 (1995).

133. S. J. Klippenstein and W. D. Allen, *J. Chem. Phys.* **105**, 118 (1996).

134. T. G. Wei and R. E. Wyatt, *J. Phys. Chem.* **97**, 13580 (1993).

135. M. Quack and M. Willeke, *J. Chem. Phys.* **110**, 11958 (1999).

136. S. Rashev, M. Stamova, and L. Kancheva, *J. Chem. Phys.* **109**, 585 (1998).

137. A. Viel and C. Leforestier, *J. Chem. Phys.* **112**, 1212 (1999).

138. A. McNichols and J. T. Carrington, *Chem. Phys. Lett.* **202**, 464 (1993).

139. H. Tal-Ezer and R. Kosloff, *J. Chem. Phys.* **81**, 3967 (1984).

140. S. K. Gray and D. E. Manolopoulos, *J. Chem. Phys.* **104**, 7099 (1996).

141. R. Bigwood and M. Gruebele, *Chem. Phys. Lett.* **233**, 383 (1995).

142. G. M. Chaban, J. O. Jung, and R. B. Gerber, *J. Chem. Phys.* **111**, 1823 (1999).

143. V. A. Mandelshtam and H. S. Taylor, *J. Chem. Phys.* **106**, 5085 (1997).

144. D. Lu and W. Hase, *J. Phys. Chem.* **92**, 3217 (1988).

145. K. Thompson and N. Makri, *J. Chem. Phys.* **110**, 1343 (1999).

146. M. Topaler and N. Makri, *J. Chem. Phys.* **97**, 9001 (1992).

147. K. G. Kay and S. A. Rice, *J. Chem. Phys.* **38**, 4852 (1973).

148. M. J. Davis, *J. Chem. Phys.* **98**, 2614 (1993).

149. M. J. Davis, *Int. Rev. Phys. Chem.* **14**, 15 (1995).

150. M. J. Davis, *J. Chem. Phys.* **107**, 1 (1997).

151. K. Thompson and N. Makri, *Phys. Rev. E* **59**, R4729 (1999).

152. J. P. Rose and M. E. Kellman, *J. Chem. Phys.* **105**, 10743 (1996).

153. S. Nordholm and S. A. Rice, *J. Chem. Phys.* **61**, 768 (1974).

154. R. M. Stratt, N. C. Handy, and W. H. Miller, *J. Chem. Phys.* **71**, 3311 (1979).

155. M. J. Davis, *Chem. Phys. Lett.* **192**, 479 (1992).

156. M. A. Sepulvéda and E. J. Heller, *J. Chem. Phys.* **101**, 8016 (1994).

157. M. S. Child and L. Halonen, *Adv. Chem. Phys.* **57**, 1 (1984).

158. G. Herzberg, *Molecular Spectra and Molecular Structure II* orig. ed. 1945, Krieger, Malabar, 1991.

159. M. E. Kellman, *Annu. Rev. Phys. Chem.* **46**, 395 (1995).

160. R. W. Field, S. L. Coy, and S. A. B. Solina, *Prog. Theor. Phy.* **116** (Suppl.), 143 (1994).

161. D. M. Leitner and P. G. Wolynes, *Chem. Phys. Lett.* **258**, 18 (1996).

162. J. Svitak, Z. Li, J. Rose, and M. E. Kellman, *J. Chem. Phys.* **102**, 4340 (1995).

163. G. S. Ezra, C. C. Martens, and L. E. Fried, *J. Phys. Chem.* **91**, 3721 (1987).

164. M. V. Berry and M. Robnik, *J. Phys. A* **17**, 2413 (1984).

165. W. P. Reinhardt, *J. Phys. Chem.* **86**, 2158 (1982).

166. C. Iung and R. E. Wyatt, *J. Chem. Phys.* **99**, 2261 (1993).

167. D. M. Leitner and P. G. Wolynes, *J. Phys. Chem. A* **101**, 541 (1997).

168. D. M. Leitner, *Int. J. Quantum Chem.* **75**, 523 (1999).

169. D. J. Tannor and S. A. Rice, *J. Chem. Phys.* **83**, 5013 (1985).

170. M. Shapiro and P. Brumer, *J. Chem. Phys.* **84**, 4103 (1986).

171. G. N. Makarov, D. E. Malinovsky, and D. D. Ogurok, *Laser Chem.* **17**, 205 (1997).

172. F. F. Crim, *Annu. Rev. Phys. Chem.* **44**, 397 (1993).

173. M. J. Bronikowski, W. R. Simpson, B. Girard, and R. N. Zare, *J. Chem. Phys.* **95**, 8647 (1991).

174. R. S. Judson and H. Rabitz, *Phys. Rev. Lett.* **98**, 1500 (1992).

175. W. S. Warren, H. S. Rabitz, and M. Dahleh, *Science* **259**, 1581 (1993).

176. C. J. Bardeen, J. Che, K. R. Wilson, V. V. Yakovlev, P. Cong, B. Kohler, J. L. Krause, and M. Messina, *J. Phys. Chem.* **101**, 3815 (1996).

177. L. C. Young, *Calculus of Variations and Optimal Control Theory* Chelsea, New York, 1980.

178. R. L. Vander-Wal, J. L. Scott, and F. F. Crim, *J. Chem. Phys.* **92**, 803 (1990).

179. A. Sinha, M. C. Hsiao, and F. F. Crim, *J. Chem. Phys.* **92**, 6333 (1990).

180. T. M. Ticich, T. R. Rizzo, H.-R. Dübal, and F. F. Crim, *J. Chem. Phys.* **84**, 1508 (1985).

181. M. A. Dugan, J. X. Tull, and W. S. Warren, *J. Opt. Soc. Am. B* **14**, 2348 (1997).

182. H. Kawashima, M. M. Wefers, and K. A. Nelson, *Annu. Rev. Phys. Chem.* **46**, 627 (1995).

183. R. A. Marcus, *J. Chem. Phys.* **20**, 359 (1952).

184. E. A. Wade, A. Mellinger, M. A. Hall, and C. B. Moore, *J. Phys. Chem. A* **101**, 6568 (1997).

185. E. D. Potter, M. Gruebele, L. R. Khundkar, and A. H. Zewail, *Chem. Phys. Lett.* **164**, 463 (1989).

186. F. Remacle and R. D. Levine, *J. Phys. Chem.* **100**, 7962 (1996).

187. L. J. Butler, *Annu. Rev. Phys. Chem.* **49**, 125 (1998).

188. W. Fuss, S. Lochbrunner, A. M. Muller, T. Schikarski, W. E. Schmid, and S. A. Trushin, *Chem. Phys.* **232**, 161 (1998).

189. J. Nygard, A. Delon, and R. Jost, *ACH Models Chem.* **134**, 541 (1997).

190. T. S. Tinghe, J. M. Worlock, and M. L. Roukes, *Appl. Phys. Lett.* **70**, 2687 (1997).

191. V. Wong and M. Gruebele, *Phys. Rev. B* submitted (2000).

DISCRETE-VARIABLE REPRESENTATIONS AND THEIR UTILIZATION

JOHN C. LIGHT

Department of Chemistry and The James Franck Institute, The University of Chicago, Chicago, Illinois

TUCKER CARRINGTON JR.

Département de chimie, Université de Montréal, Montreal, Canada

CONTENTS

Advances in Chemical Physics, Volume 114, Edited by I. Prigogine and Stuart A. Rice.
ISBN 0-471-39267-7 © 2000 John Wiley & Sons, Inc.

I. INTRODUCTION AND HISTORY

A. Scope

The title of this chapter, "Discrete-Variable Representations" or (DVRs), suggests an oxymoron—spatial coordinate variables are not discrete, and, if discrete values are used, are not exact representations. A better description would be: that a DVR is a representation; whose associated basis functions are localized about discrete values of the variables. The term DVR also implies the use of an approximation; coordinate operators are assumed diagonal in this representation and are approximated by their values at the DVR points. DVRs have enjoyed great success as highly accurate representations for the solution of a variety of problems in molecular vibration–rotation spectroscopy and molecular quantum dynamics.

DVRs are highly advantageous for most of these problems for two reasons. First, they greatly simplify the evaluation of the Hamiltonian matrix; kinetic energy matrix elements are calculated simply, and potential matrix elements are merely the value of the potential at the DVR points, (i.e., no integral evaluations are required). Second, for direct product DVRs in multidimensional systems, the Hamiltonian is sparse and the operation of the Hamiltonian on a vector is always fast. DVRs provide simple and well defined representations that permit efficient and accurate numerical solutions to quantum-dynamical problems of interest.

In this chapter we focus on three aspects of DVRs: what they are and how they compare with other "pointwise representations"; how they simplify solutions of multidimensional quantum-dynamical problems, in both time-dependent and time-independent frameworks; and their limitations, in terms of both their applicability to various coordinate systems and operators and the accuracy of the solutions of the quantum problem at hand.

The objectives of this chapter are then to (1) review various definitions of DVRs, the differences between them, and their mathematical and physical foundations; (2) note the basic advantages of DVRs in the solution of quantum-dynamical problems in chemical physics; (3) indicate the modes of solution for such problems; and (4) note the circumstances in which DVRs are *not* advantageous. We hope this will remove confusion about DVRs and make their use both better understood and simpler.

We should note at the outset that the bias of this chapter is toward the solution of the Schrödinger equation, a *linear* second-order partial-differential equation with specified boundary conditions, primarily for nuclear dynamics, not electronic structure. This is a broad area encompassing theoretical molecular spectroscopy, dynamics, chemical reactions, and related fields. Although applications have been made to other nonlinear systems, this chapter does not include these areas.

A fundamental problem in chemical physics is to solve the Schrödinger equation:

$$H\Psi = E\Psi \tag{1.1}$$

or

$$i\hbar \frac{\partial \Psi}{\partial t} = H\Psi \tag{1.2}$$

which governs the properties and dynamics of matter at the atomic and molecular level. We will focus on nuclear motion on potential-energy surfaces (PESs), and thus not be concerned with spin operators, and similar, although nonadiabatic coupling between PESs may be included.

The Hamiltonian, H [Eqs. (1.1) and (1.2)], contains a second-order differential (kinetic energy) operator in coordinate space, $K(\{q\})$; and a potential-energy operator, $V(\{q\}, t)$, which depends on coordinates and perhaps the time. $\{q\}$ stands for the set of coordinates describing the system, and the Hamiltonian is

$$H(\{q\}) = K(\{q\}) + V(\{q\}, t) \tag{1.3}$$

We are interested in solutions both of molecular bound state and of scattering problems. In the bound-state problems, the solutions, $\Psi(\{q\})$, of Eq. (1.1) are localized, and therefore are a discrete set of square integrable (L^2) eigenfunctions with discrete eigenvalues, $E = \epsilon_i$. We usually take these eigenfunctions to be normalized over the coordinate range. Scattering solutions of the time independent Schrödinger equation are not square-integrable, but the asymtotic forms of the solutions are known, and appropriate scattering boundary conditions may be applied. There is usually some finite range in which a numeric vs. analytic solution is required, and this solution is then matched in some fashion to the known asymptotic form. Thus in both bound and scattering problems we may consider the numerical solution of the Schrödinger equation in only a finite coordinate range. For scattering solutions of the time-dependent Schrödinger equation, the use of absorbing potentials at finite ranges (or other techniques) also limits the range over which numerical solutions are required.

This has an enormously important consequence for the mathematical representation of Eqs. (1.1) and (1.2) We need represent the solutions numerically in only a finite coordinate range, with boundary conditions either zero for bound states or permitting appropriate asymptotic matching for scattering states. An important consequence of this is that accurate dynamics, up to a specified energy, E, can be represented in a *finite* basis of N functions of the coordinates, $\{\theta_n(\{q\})\}_N$, (where, of course, N may be very large depending on the accuracy desired).

After a brief presentation of the historical context the chapter is divided into the following sections:

II. DVRs in one dimension: various types and comparisons

III. Properties of multidimensional DVRs and methods used to solve the Schrödinger equation equation in DVRs

IV. Caveats: problems and resolutions; numerical quadratures, etc;

V. Conclusions

B. Historical Context

Unfortunately, this description of the historical context of DVRs must be limited by the authors' knowledge and understanding. Since we are neither applied mathematicians nor historians there may be large gaps in our presentation, particularly with respect to the numerical analysis literature. We have, however, tried to be accurate as to the hitorical roots of these approaches in chemical physics.

The numeric solution of differential equations has been studied for centuries, presumably since the time of Newton. However, the restriction to linear equations such as the Schrödinger equation permits a variety of powerful approaches based on linear algebra, and the spectral properties of the Hamiltonians permit variational solutions. Thus, for the Schrödinger equation, a solution may be represented by an exact analytic functional form, $\Psi_l(\{q\})$; by values on a grid in coordinate (or momentum) space, $(\Psi_l)_i = \Psi_l(q_i)$; or by a representation in a complete basis

$$\Psi_l(\{q\}) = \sum_j c_{jl}\phi_j(\{q\}) \tag{1.4}$$

Basis-set representations started, of course, with Fourier series. More general Hilbert spaces and representations in terms of orthogonal basis expansions (due to Lord Rayleigh, 1842–1919) were not developed until about 1900 [1].

Discrete-variable representations are representations in bases of continuous functions that are in some sense localized "on a grid" in coordinate space. In particular, to construct a DVR, a finite basis of "global" orthnormal functions is transformed to another orthnormal basis set (the DVR) in which each basis function is "localized" about one point of a coordinate space grid. The relationship is most clear for basis sets of orthogonal polynomials (with weight functions) and their associated grids of Gaussian quadrature points. This relationship may have been known earlier, but was pointed out for Chebyshev polynomials in the last section of Lanczos' book on applied analysis [1] published in 1956. Somewhat later (1966), Fox [2,3], also noted that sets of orthogonal (Chebyshev) polynomials evaluated at the appropriate set of Gaussian quadrature points, formed an orthogonal transformation between the original polynomial representation and a more localized representation.

In chemical physics, Harris et al. [4] in 1965 generated the appropriate transformation via diagonalization of the coordinate operator. Their purpose was to evaluate approximately the matrix elements of coordinate (potential) functions in the basis by transformation of the diagonal matrix of the potential evaluated at the coordinate eigenvalues. In their specific case the basis functions were harmonic oscillator functions. In 1968 Dickinson and Certain [5] noted that the eigenvalues of the coordinate operator so obtained were, in fact, the Gaussian quadrature points for the (Hermite) polynomial used. This lent support to the accuracy of the matrix elements evaluated in this fashion.

In 1982 Lill et al. [6,7] and then Heather and Light [8] first explicitly used the transformed representation in which the coordinate operator is diagonal, the discrete-variable representation, as a basis representation for quantum problems rather than only a means for evaluation of matrix elements. The approach was introduced independently by Blackmore and Shizgal [9,10] in 1983 and 1984 under the name "discrete-ordinate method." In one dimensional problems, the use of a DVR may offer only a slight advantage. In higher-dimensional problems with direct product basis sets, however, it becomes highly advantageous. This was noted when using a DVR in conjunction with a distributed Gaussian basis (DGB) by Bacic, Light, and others [11–14] and for multidimensional DVRs by Whitnell and Light [15] and Light et al. [16].

About 1990, discrete variable representations became increasingly widely used, and a number of variations and different prescriptions for their definition were presented by Manolopoulos and Wyatt [17], Echave and Clary [18], Wei and Carrington [19], Muckerman [20], and Colbert and Miller [21]. More recently Szalay [22] has proposed a generalized multidimensional DVR. Several earlier review articles on the definitions and uses of DVRs are also available [14,16,23,24].

II. "POINTWISE" REPRESENTATIONS IN ONE DIMENSION

A. Introduction to Types of Discrete-Variable Representations (DVRs)

In the general variational approach to quantum problems, the unknown solution is represented exactly in a (usually infinite) basis, or, more properly, a Hilbert space. In practice, of course, the infinite basis is truncated, and the approximate solutions in this truncated representation are variational; that is, the energy eigenvalues of the Hamiltonian in this truncated basis are all larger than or equal to the corresponding exact eigenvalues. We call the representation in which all Hamiltonian matrix elements are evaluated exactly the *variational basis representation* or (VBR). In the VBR errors are due *only* to the truncation of the basis.

In general terms, DVRs are representations in terms of localized functions that are usually obtained by transformation from a truncated "global" basis. DVRs are generally used with the *approximation* that in the DVR the matrix representation of functions of the coordinate are diagonal and the diagonal matrix elements are values of the function at the DVR points. The term DVR in this chapter implies that this approximation is made. It should be noted, however, that although the DVR functions are "focused" on the grid points, they are *not* perfectly localized. Each function extends throughout the range of the original basis. For example the infinite set of sinc functions $f_n(x) = \sin[\pi(x - x_n)/\Delta x]/\pi(x - x_n)$ with $n = 0, \pm 1, \pm 2, \ldots$, and $x_n = n\,\Delta x$, is an infinite DVR basis with DVR points at x_n [21]. Each function f_n is unity at $x = x_n$ and zero at all other DVR points, but it is not zero for values of x between the DVR points. The sinc functions are symmetric about their DVR points but are infinite in extent, having an oscillatory shape typical of the "localized" DVR functions. The "focal points" (DVR points) of the DVR basis functions form a grid of points, and the Hamiltonian is then approximated as a matrix labeled by the grid points because DVR functions can be labeled by the grid point about which they are localized.

There are several ways to construct DVRs: one may establish a connection between a set of basis functions and an appropriate numerical quadrature; one may diagonalize a function of the coorodinate operator in a basis set; or one may choose a basis of localized functions initially. In the approach based on numeric quadrature, the relationship between basis size and quadrature accuracy is direct. In the first two approaches, the transformation between DVR and global basis representation is known. In cases where this is a unitary or orthogonal transformation, operators in the global basis obtained by transformation from the DVR contain exactly the same approximations as the DVR. To distinguish this approximation from the VBR, we call it a *finite basis representation* (FBR). Thus the VBR is an exact representation in terms of global basis functions, while the DVR and FBR are "local" and corresponding "global" representations in which an approximation has been made. We first will examine the different one dimensional DVRs and their relations to the FBR and VBR from the viewpoint of Gaussian quadrature, and then examine DVRs from the "product" point of view which is useful for complex kinetic energy operators.

B. Orthogonal Polynomial Bases and Gaussian Quadrature DVRs

The "standard" DVRs are defined in terms of classical orthogonal polynomials, weight functions, and their related Gaussian quadratures [6,25]. The approximations in these DVRs (and the corresponding FBRs) are related directly to the approximations of the Gaussian quadratures associated with the polynomials used. Classical orthogonal polynomials form many common one-dimensional basis sets such as particle-in-a-box functions (Chebyshev polynomials), harmonic oscillator functions (Hermite polynomials), Legendre polynomials, and Laguerre polynomials. An excellent description of these polynomials is given in Dennery and Krzywicki [26], and concise definitions are given in Abramowitz and Stegun [27]. The basic property of classical orthogonal polynomials is that they are (normed) polynomials orthogonal with respect to integration over their range with a specific weight function, $\omega(x)$:

$$\int_{x=a}^{x=b} \omega(x) C_l(x) C_n(x) dx = \delta_{ln} \qquad (2.1)$$

where $C_l(x)$ is a polynomial in x of lth degree.

There are three general types of classical orthogonal polynomials based on the range of their argument, infinite, semiinfinite, or finite. These are summarized in Table I. Note that the arguments can be scaled and shifted so the mathematical ranges below correspond to the appropriate physical coordinates and ranges as desired.

One simple way to introduce the DVR is to use the well known properties of Gaussian quadratures [28] to generate the transformation between the FBR and the DVR. We show that the same transformation is obtained, following Harris et al. [4], by diagonalizing the argument of the orthogonal polynomials. The argument might be the coordinate operator itself of a function of the coordinate operator. We emphasize that in this fashion DVRs can be easily constructed for most standard basis functions without explicit consideration of the properties (or points and weights) of Gaussian

TABLE I
Ranges and Weight Functions for Classical Polynomials

Range	Weight function	Name of Polynomials
$(-\infty, \infty)$	$\exp(-x^2)$	Hermite, $H_n(x)$ (harmonic oscillator functions)
$(0, \infty)$	$x^\nu \exp(-x)$ $\nu > -1$	Laguerre, $L_n^\nu(x)$
$[-1, 1]$	$(1 - x)^\alpha (1 + x)^\beta$ $\alpha, \beta > -1$	Jacobi, $P_n^{\alpha,\beta}$ (Legendre and Chebyshev polynomials)

quadratures. The primary purpose of this section is to show the relations that exist between the DVR and Gaussian quadrature and thus demonstrate the basis for the high accuracy of DVRs.

The basic property of Gaussian quadratures for these polynomials is that for each type of polynomial a quadrature defined on N specified points $\{x_\alpha\}_N$ with particular weights $\{\omega_\alpha\}_N$ is exact for integrals of the weight function $\omega(x)$ times polynomials up to order $2N - 1$. Thus if our basis functions are normalized orthogonal polynomials times the square root of the weight functions, $\omega(x)$

$$\phi_n(x) = \sqrt{\omega(x)} C_n(x) \tag{2.2}$$

then the orthonormality relations are given exactly by the quadrature

$$\delta_{ln} = \int_{x=a}^{x=b} \phi_l^*(x)\phi_n(x)dx$$
$$= \sum_{\alpha=1}^{N} \frac{\omega_\alpha}{\omega(x_\alpha)} \phi_l^*(x_\alpha)\phi_n(x_\alpha) \tag{2.3}$$

and the matrix elements of the coordinate x are also given exactly by

$$\mathbf{X}_{ln} = \int_{x=a}^{x=b} \phi_l^*(x)x\phi_n(x)dx$$
$$= \sum_{\alpha=1}^{N} \frac{\omega_\alpha}{\omega(x_\alpha)} \phi_l^*(x_\alpha)\phi_n(x_\alpha)x_\alpha \tag{2.4}$$

Both relations are exact for $0 \le l, n \le N - 1$ since the N basis functions contain only powers of x from x^0 to x^{N-1}. The polynomial basis functions, $\phi_n(x)$, are frequently chosen so that they are eigenfunctions of an operator H_0, where $H = H_0 + V$, with $H_0 = K + V_0$, where K is the kinetic energy operator and V_0 is the part of the potential (if any) included in the definition of the basis functions.

We may now write these equations in matrix form if we identify the elements of the transformation matrix, $\mathbf{T}$, as

$$\mathbf{T}_{\alpha j} = \sqrt{\frac{\omega_\alpha}{\omega(x_\alpha)}}\, \phi_j(x_\alpha) \tag{2.5}$$

Note that $\mathbf{T}$ is a *square* matrix; there are the same number of points as basis functions. Then Eqs. (2.3) and (2.4) are equivalent to the matrix relations

$$\mathbf{I} = \mathbf{T}^\dagger \mathbf{I}^{DVR}\, \mathbf{T} \tag{2.6}$$

$$\mathbf{X} = \mathbf{T}^\dagger \mathbf{X}^{DVR}\, \mathbf{T} \tag{2.7}$$

where $\mathbf{T}^\dagger$ is the Hermitian transpose of $\mathbf{T}$, $\mathbf{I}^{DVR}$ is a unit matrix labeled by the quadrature points, and $\mathbf{X}^{DVR}$ is the *diagonal* matrix of values of x at the quadrature points $\{x_\alpha\}$, namely, the DVR points.

Since Eq. (2.6) demonstrates that $\mathbf{T}$ is orthogonal (or unitary), we can multiply Eq. (2.7) by $\mathbf{T}$ on the left and by $\mathbf{T}^\dagger$ on the right and find that $\mathbf{T}^\dagger$ is the matrix that diagonalizes the exact coordinate matrix, $\mathbf{X}$:

$$\mathbf{T}\mathbf{X}\mathbf{T}^\dagger = \mathbf{X}^{DVR} \text{ (diagonal)} \tag{2.8}$$

Thus we see that the diagonalization of $\mathbf{X}$ generates the Gaussian quadrature points as the eigenvalues, and the transformation matrix related to the Gaussian weights and points given by Eq (2.5). Thus the Gaussian quadrature points are the DVR "points" and $\mathbf{T}$ is the DVR–FBR transformation. Obviously it can be generated either from the polynomials themselves via Eq. (2.5) or by diagonalization of $\mathbf{X}$. The tridiagonal coordinate matrix $\mathbf{X}$ is easily generated from the three-term recursion relations satisfied by all classical orthogonal polynomials [27,26]. DVRs are most commonly and most easily determined by diagonalization of the exact coordinate matrix, and no explicit reference to quadrature points and weights is then required.

If we label the DVR eigenvalues using Greek letters, then the basis function localized at x_α is

$$\theta_\alpha(x) = \sum_j \mathbf{T}_{\alpha j}^* \phi_j(x) \tag{2.9}$$

An important property of these functions is that each function is nonzero at the DVR point about which it is "localized," but is zero at the remaining

$N - 1$ points. This is easily seen by evaluating Eq. (2.9) at a DVR point:

$$
\begin{aligned}
\theta_\alpha^{DVR}(x_\beta) &= \sum_j T_{\alpha j}^* \phi_j(x_\beta) \\
&= \sum_j \sqrt{\frac{\omega_\alpha}{\omega(x_\alpha)}} \phi_j^*(x_\alpha)\phi_j(x_\beta) \\
&= \sqrt{\frac{\omega(x_\beta)}{\omega_\beta}}(TT^\dagger)_{\alpha\beta} \\
&= \sqrt{\frac{\omega(x_\beta)}{\omega_\beta}}\delta_{\alpha\beta}
\end{aligned}
\tag{2.10}
$$

Note that this also yields the normalization of the DVR functions over the quadrature:

$$
\begin{aligned}
\int_a^b \theta_\alpha^2(x)\,dx &= \sum_\beta \theta_\alpha^2(x_\beta)\frac{\omega_\beta}{\omega(x_\beta)} \\
&= \sum_\beta \delta_{\alpha\beta} \\
&= 1
\end{aligned}
\tag{2.11}
$$

The equations $TT^\dagger = I$ and $T^\dagger T = I$ are the discrete orthonormality relations of DVRs.

In practice the simplest procedure to define a Gaussian quadrature DVR is to generate the exact tridiagonal coordinate matrix implicit in the three-term recursion relation satisfied by the appropriate orthogonal polynomials and then to diagonalize this matrix. The only care required is that the normalization of the polynomials must generate an orthonormal basis from Eq. (2.2). After determining the Gaussian quadrature points, they must be scaled to the appropriate physical range via a linear transformation.

Viewed from the Gaussian quadrature viewpoint, it is clear that a Gaussian DVR is as accurate as an FBR with potential matrix elements evaluated by a Gaussian quadrature with an equal number of DVR quadrature points and basis functions. An alternate view is to view the DVR as a "product approximation" described below.

C. Product Approximation and Potential Optimized DVRs (PODVRs)

We have discussed three representations: the VBR, the FBR, and the DVR. In a VBR all matrix elements are computed exactly; errors with a VBR occur only because the VBR basis is finite. The FBR and the DVR are unitarily equivalent and an FBR may be regarded in two ways. If the underlying basis for the DVR is a classical orthogonal polynomial basis, then a "quadrature" FBR is the representation obtained by evaluating the residual potential matrix elements by Gaussian quadrature and taking exact, analytic kinetic energy (or H_0) matrix elements. (If it is not possible to write down analytic kinetic energy matrix elements an FBR is constructed as explained below in the section on complicated kinetic energy operators). However, a "product" FBR can be generated from any representation in which the kinetic energy matrix (or H_0) and the transformation to the representation in which the coordinate matrix is diagonal are known. The FBR for such a *product approximation* is obtained for a Hamiltonian by using the exact, analytic kinetic energy matrix and replacing the matrix representations of products of the coordinate in the potential with products of the matrix representation of the coordinate.

Thus the matrix representation of a function of x is approximated as the function of the matrix representation of x (assuming the function has a Taylor series expansion in x):

$$[\mathbf{V}(x)]_{ij} \approx [\mathbf{V}(\mathbf{X})]_{ij} \tag{2.12}$$

However, for a product of operators to be evaluated exactly as a product of matrices, a complete basis must inserted between the operators:

$$[\mathbf{AB}]_{ij} = \sum_{n}^{\infty} A_{in}B_{nj} \tag{2.13}$$

Thus the preceding expression [Eq. (2.12)] containing products of the operator, x, is only an approximate evaluation of the potential matrix in a finite representation. However, as the dimension, N, of the basis increases, matrix elements for a given $\{i,j\}$ become more exact.

Errors (if any) in results computed with a quadrature FBR (or its corresponding DVR) occur because the VBR (from which the FBR is obtained) is finite and because some potential matrix elements are not computed perfectly by quadrature. Errors (if any) in results computed with a product FBR occur because the VBR is finite and because the product approximation is not perfect. If the matrix representations of the factors of the products are banded matrices, the error introduced by the product approximation is restricted to the bottom righthand corner of the matrix [19].

If the VBR basis functions are orthogonal polynomials as discussed above and one chooses as many quadrature points as basis functions, the quadrature FBR and the product FBR are identical (the matrix element errors are also indentical). Since a DVR matrix is unitarily equivalent to its FBR counterpart, errors (if any) in results computed with a DVR are the same as the errors that would be obtained with the corresponding FBR. They may equally well be considered as being due to the finiteness of the VBR basis and either (1) the quadrature error or (2) the product approximation error.

Obviously one wants to make the errors as small as possible. If one uses a Gaussian quadrature FBR, errors are minimized by choosing good VBR basis functions (basis functions that represent wavefunctions compactly) and using many functions such that the quadrature error for the residual potential is minimal.

If one uses a basis not associated with a Gaussian quadrature, then product FBR errors would be minimized by choosing VBR basis functions that both represent wavefunctions accurately and compactly (i.e., decrease the size of off-diagonal matrix elements) *and* improve the accuracy of the product approximation. However, it is often (but not always) the case that basis functions that render the Hamiltonian matrix more diagonal (good VBR basis functions) will decrease the accuracy of the product approximation since the coordinate matrix, from which the product approximation FBR is built, will be far from diagonal. If one is willing to accept this degradation of the accuracy of the product approximation one can optimize the VBR. This leads to the potential optimized DVR (PODVR).

In 1992 Echave and Clary [18] and Wei and Carrington [19] introduced a means of generating a DVR that concentrates the DVR points in the region of space where they do the most good and which minimizes the residual potential. The approach is useful only in multidimensional problems, where it may greatly reduce the size of the basis required for each coordinate, but is easily described for one coordinate.

If one wants a *small* DVR for an arbitrary one dimensional bound potential, then an advantageous DVR (with a good distribution of points in x and a smaller residual potential) may be obtained by using a DVR determined from eigenfunctions of a model potential in x. The procedure is simply that of Harris et al. [4] outlined above, except that eigenfunctions of a model Hamiltonian are used as the basis.

A PODVR with N points is obtained by first defining a good model potential for the coordinate in question, and then solving accurately for N eigenfunctions. (This may require a basis size considerably larger than N regardless of whether a DVR is used.) The coordinate matrix is then diagonalized in the N eigenfunctions to yield the DVR–FBR transformation and the DVR points. In constructing the Hamiltonian in the PODVR (e.g., for multidimensional problems), the N eigenvectors are transformed to the DVR and the *residual* potential is evaluated in the DVR. This *product approximation* is found to be very useful in multidimensional problems, permitting a much smaller DVR in the dimension for which the PODVR is constructed.

In a multidimensional problem one defines a PODVR basis for each coordinate in a similar fashion using an appropriate model potential. Then products of the the PODVR functions for each coordinate are used as direct product basis functions for the full problem. Although DVR points for a single degree of freedom will be strewn between the limiting values of the coordinate, PODVR points are concentrated in regoions where the amplitude of the wavefunction is significant.

PODVRs are most advantageous when coupling between degrees of freedom is weak. If there were no coupling, the products of solutions of one-dimensional Hamiltonians would be solutions of the Schrödinger equation for the multidimensional Hamiltonian. When the coupling is weak, such products (and the associated PODVR functions) are very good basis functions. When, however, the coupling is strong, PODVR functions will be less useful.

PODVRs share some of the properties of Gaussian DVRs. The transformation matrix **T** is composed of the eigenvectors of the coordinate function diagonalized and is orthogonal, but because PODVRs are not derived from a Gauss quadrature scheme, it is not possible to write the PODVR transformation matrix as Eq. (2.9). PODVR functions will, therefore, not exactly satisfy Eq. (2.10) but this does not necessarily limit the quality of the results obtained with a PODVR: the accuracy of PODVR results is determined by the accuracy of the associated product approximation and the minimization of the residual potential.

D. Hamiltonian Evaluation

We now look briefly at the consequences of approximating the Hamiltonian in the DVR. The choice of the underlying basis for the DVR is based on the Hamiltonian, the boundary conditions of the wavefunctions, and the domains of the coordinates. It would be ideal to have the spacing of the quadratures points proportional to the nodal spacing of the highest eigenfunctions of interest, and to have the basis satisfy the appropriate boundary conditions. Chebyshev polynomials (particle in the box functions) yield DVRs with equally spaced points in $x = \cos^{-1}\theta$; Legendre and associated Legendre polynomials [with $x = \cos(\theta)$] are appropriate for angular functions on $\theta = (0, \pi)$ satisfying boundary conditions of unity and zero respectively, and so on. Using the Jacobi polynomials with differing α and β shifts the DVR points away from $x = 1$ and -1.

Given an exact evaluation of the Hamiltonian matrix in the VBR, the Hamiltonian could be evaluated exactly in the DVR by transformation from the exact VBR Hamiltonian:

$$\tilde{\mathbf{H}}^{DVR} = \mathbf{T}\mathbf{H}^{VBR}\mathbf{T}^{\dagger} \qquad (2.14)$$

where

$$\mathbf{H}^{VBR} = \mathbf{H}_0^{VBR} + \mathbf{V}^{VBR} \qquad (2.15)$$

Here $\mathbf{V}^{VBR}$ is the exact representation of the *residual* potential, the portion of the potential not included in $\mathbf{H}_0$. $\mathbf{H}_0^{VBR}$ is transformed from the VBR, where it is usually simple to determine. Szalay [29] has given analytic formulas for the differential operators, d/dq and the second-order differentials, (d^2/dq^2) for all orthogonal polynomial DVRs of Gaussian quadrature accuracy. It is therefore often easy to evaluate $\mathbf{T}\mathbf{H}_0^{VBR}\mathbf{T}^{\dagger}$ (see Section III. C for alternative methods). However, to obtain $\tilde{\mathbf{H}}^{DVR}$ the residual potential matrix, $\mathbf{V}^{VBR}$, would have to be evaluated exactly in the VBR beforehand, and this eliminates any advantage of the DVR. The potential matrix evaluated exactly as above in the DVR is diagonally dominant, but is not exactly diagonal.

The power and simplicity of the DVR come from approximating the residual potential matrix in the DVR. The matrix representation of $\mathbf{V}$, the residual potential energy, is, however, *approximated* by the quadrature of the DVR itself. Thus

$$H^{DVR} = H_0^{DVR} + V^{DVR} \qquad (2.16)$$

where

$$\mathbf{H}_0^{DVR} = \mathbf{T}\mathbf{H}_0^{VBR}\mathbf{T}^{\dagger} \qquad (2.17)$$

and we approximate the residual potential as

$$\mathbf{V}_{\alpha,\beta}^{DVR} = V(x_\alpha)\delta_{\alpha\beta} \qquad (2.18)$$

Thus the DVR approximation is equivalent to approximating the exact residual potential matrix, $\mathbf{V}$ as

$$\mathbf{V} \approx \mathbf{T}^{\dagger}\mathbf{V}^{DVR}\mathbf{T} \equiv \mathbf{V}^{FBR} \qquad (2.19)$$

The DVR with an underlying orthogonal polynomial basis is exactly equivalent to evaluating potential matrix elements using the Gaussian quadrature appropriate to the basis, with N quadrature points. The disadvantage of this is that the eigenvalues resulting from evaluating the Hamiltonian in the DVR are *not* variational; the quadrature error may cause some eigenvalues to be below their true value. (For a recent discussion, see Wei [30]). However, the advantage of DVRs based on the classical orthogonal polynomials is that the convergence of the Gaussian quadratures to the exact integrals is excellent. For smooth potentials as the basis size is increased (and the number of DVR points is increased), the quadrature error quickly disappears at least for lower levels. The convergence of these eigenvalues is then limited by the basis set in a variational fashion.

E. Other DVRs

Alternate grid representations have been proposed as DVRs by Muckerman [20], Manolopoulos and Wayatt [17], Colbert and Miller [21], and others. These all share the characteristic that in representing the Hamiltonian, coordinate functions are represented by their values at grid points. However, all differ from Gaussian DVRs in that not all of the equations of orthonormality, exact quadrature for the coordinate and the definition of the DVR–FBR transformation in terms of the basis functions and quadrature points hold exactly [Eqs. (2.3)–(2.5)]. All three approaches define the DVR functions directly as Lagrange interpolating polynomials or by analytic transformation and summation. Of these DVRs we will discuss only the Lobatto quadrature DVR and the "sinc" DVR of Colbert and Miller as these have seen the most frequent application.

1. Lobatto DVR

The Lobatto quadrature/function DVR was introduced by Manolopoulos and Wyatt [17] for scattering problems in which simple evaluation of non-zero boundary conditions was desired. The Lobatto shape functions are Lagrange interpolating polynomials over the Lobatto quadrature points. These *are* the DVR functions. Lobatto quadratures are well known [27]. Lobatto shape functions satisfy the pointwise orthogonality relations on the quadrature points as in Eq. (2.10) and are orthogonal over the quadrature, but their squares are not themselves integrated exactly by the quadrature. The kinetic energy operator can be determined exactly in this DVR.

Manolopoulos et al. demonstrated that their DVR provided a simple L^2 basis for scattering when L^2 methods requiring log derivative evaluation at the boundary are used. Since most basis sets fix the log derivative at the endpoints, they do not have sufficient flexibility to be used. Although Jacobi polynomials could be used, the Lobatto functions have the advantage that the log derivative is simple to evaluate in this basis as only one function is non-zero (unity) at the boundary.

2. Sinc DVR

In 1992 Colbert and Miller [21] introduced a "novel universal" DVR in which they evaluated the kinetic energy operators by infinite order finite differences on infinite uniform grids. The kinetic energy representation on the grid was derived using particle-in-a-box functions, which are, of course, Chebyshev polynomials, and the number of functions was permitted to go to infinity as the range became infinite, with $\Delta x = L/N$ finite. (In the finite and semiinfinite range cases discussed Ref. 21, the functions are not as appropriate or accurate as, for example, Legendre polynomials, and these cases have rarely, if ever, been used). We therefore discuss only the $(-\infty, \infty)$ range case in which case the basis becomes the infinite set of sinc functions.

In the infinite-range case the kinetic energy operator becomes, of course, translationally invariant, depending only on the distance between the grid points in question. This DVR corresponds to an infinite basis of sinc functions

$$s_n(x) = \frac{\sin\left(\pi(x - x_n)/\Delta x\right)}{\pi(x - x_n)} \tag{2.20}$$

where $x_n = n\,\Delta x$, and $n = 0, \pm 1, \pm 2, \ldots$

On the surface, these DVR functions and the associated quadrature would seem ideal since the kinetic energy matrix elements are given analytically, on the DVR points; the basis can be orthonormal, both analytically and on the uniform quadrature; the discrete orthonormality holds [Eq. (2.10)]; and the coordinate operator, x, is given exactly in this localized basis by $X_{ij}^{DVR} = x_i\,\delta_{ij}$. The problem, however, is that in application one cannot use an infinite basis. For a finite set of points, the kinetic energy is *not* given exactly by the truncated representation *and* as with all DVRs the potential energy is also not given exactly. The sinc DVR is usually truncated by limiting it to regions in which the potential energy is less than some maximum value. It has, however, the advantage of great simplicity (equally spaced points and analytic kinetic energy matrices) and has been quite widely used.

F. Generalized DVRs?

As noted above and discussed in Section III, DVRs are useful primarily for multidimensional problems. If direct product bases are used for the multidimensional system, the basis for each dimension may be separately transformed to the DVR. In this representation, the multidimensional potential energy is diagonal, and the kinetic energy terms may be evaluated simply from the one-dimensional representations of each momentum or second derivative operator. In the case of orthogonal coordinates, this leads to very sparse Hamiltonian matrices in the multidimensional DVR.

However, there exist coordinate systems, spherical polar coordinates being the most familiar, where eigenfunctions of the kinetic energy operators do not form direct product bases, such as the spherical harmonics. For multidimensional coupled bases such as the two-dimensional spherical harmonics, an accurate quadrature and DVR–FBR transformation cannot be generated by diagonalization of a coordinate. Thus no simple 1 : 1 mapping of an arbitrary number of basis functions to DVR-type functions by unitary or orthogonal transformations exists.

A few attempts to overcome this difficulty have been made without great success. In particular the collocation method generates a non-orthogonal transformations between the basis and a set of points, but leads to a generalized eigenvalue problem with attendant problems of some complex eigenvalues. Light et al. [25] generated an orthogonal transformation using the metrics of the collocation matrix. Although this is briefly described below, the accuracy is greatly dependent on the choice of points and this avenue has not been used for real problems.

If the N basis functions $\{\phi_n\}$ are evaluated at N "independent" points, $\{q_\alpha\}$ a square collocation matrix may be formed:

$$\mathbf{R}_{n\alpha} = \phi_n(q_\alpha) \tag{2.21}$$

If one takes the collocation matrix to define the DVR–FBR transformation, then the metrics of the two representations are related by

$$\mathbf{S} = \mathbf{R}(\Delta)^{-1}\mathbf{R}^t \tag{2.22}$$

where $\mathbf{S}$ and Δ are the metrics in the FBR and DVR, respectively.

Thus, if the functions are orthonormal and we want to maintain the orthonormality over the quadrature implied in the DVR, then $\mathbf{S}$ is the unit matrix, and Δ must be chosen accordingly:

$$\Delta = \mathbf{R}^t\mathbf{R} \tag{2.23}$$

If Δ is a diagonal and positive definite, then a good quadrature is defined using the diagonal values of δ^{-1}. For general points and basis sets this will not be true. If Δ^{-1} exists, however, an orthogonal transformation can still be defined by

$$\mathbf{T} = \mathbf{R}\Delta^{-1/2} \tag{2.24}$$

However, in multidimensional systems not only is Δ not diagonal, but it often has very small eigenvalues indicating that on the set of points chosen, the set of functions is not independent (or vice versa). In this case the attempt to represent the Hamiltonian accurately in the DVR fails disastrously.

Szalay [22] has investigated this problem further and has shown that with careful choice of points a mixed representation of the Hamiltonian can be used that yields some eigenvalues with quite acceptable accuracy. The final GDVR method proposed is in essence a collocation method and impressive accuracy was obtained for one-dimensional (1D) problems. (The accuracy was much higher than for the GDVR described above.) The costs, however, are similar to those encountered in collocation methods that include dealing with the generalized eigenvalue problem. In the two-dimensional example given, complex eigenvalues were encountered before all bound-state eigenvalues were determined.

It has become apparent from a number of studies [31–35] that such nondirect product multidimensional bases are best treated by direct product quadratures, with enough points in each dimension to assure accurate evaluation of the matrix elements. (The basis representation is used.) The use of direct product quadratures (i.e., the points and weights in each dimension are independent of the other dimensions) greatly simplifies the transformations between the basis and grid representations. In addition, one can choose the quadratures sufficiently large that all matrix elements are evaluated accurately. These can, of course, be combined with DVRs in other coordinates in a direct product fashion. This is discussed more fully in the section on applications below.

III.   MULTIDIMENSIONAL DVRs AND APPLICATIONS

A.   Introduction

In the previous section we defined and discussed various DVRs. In this section we consider how the favorable properties of DVRs facilitate solving the Schrödinger equation. Many problems of interest in chemical dynamics and spectroscopy involve large-amplitude motions of polyatomic systems. For such problems numerical representations are required and standard harmonic normal-mode approximations are not adequate. If one is to solve the Schrödinger equation accurately for such systems, it is important to have

1. Accurate potential-energy surface(s)
2. An appropriate coordinate system for the Hamiltonian operator
3. An adequate and efficient basis in which to represent the Hamiltonian
4. Appropriate means of extracting the desired information

To extract the desired information, the first thing one must do is to write down equations for the matrix elements of the Hamiltonian. This is addressed in subsections III. B and III. C. In Section III. D we present procedures for exploiting symmetry. In Sections III. E and III. F, we discuss ideas for coping with the huge number of basis functions required to obtain converged solutions for polyatomic large-amplitude motion problems. Even for total angular momentum zero there are $3n - 6$ degrees of freedom for an n atom system and it is therefore imperative that one either devise schemes to minimise the number of basis functions or employ iterative methods that depend only on matrix–vector products. Explicit sequential diagonalization truncation schemes for obtaining a compact, high-quality basis from a multidimensional direct product DVR are treated in Section III. E. The use of iterative methods with DVRs and basis sets obtained from DVRs is considered in Section III. F.

B. Orthogonal Coordinates

Orthogonal coordinates are coordinates in terms of which the kinetic energy operator has no mixed second derivatives (no cross terms). Because the kinetic energy operator is simpler in orthogonal coordinates they are often preferable (however, it is possible that basis functions that are functions of coordinates that simplify the kinetic energy will be poorer basis functions and that the tradeoff to simplify the kinetic energy operator is an increase in the size of the Hamiltonian matrix).

DVRs are useful primarily for multidimensional problems. If a direct product basis is used for the multidimensional system, the basis for each dimension may be separately transformed to the DVR. In this DVR the multidimensional potential energy is diagonal. In the case of orthogonal coordinates, the kinetic energy terms may be evaluated simply from the one-dimensional representations of each momentum or second derivative operator. This leads to very sparse Hamiltonian matrices in the multi-dimensional DVR.

The great simplification of Hamiltonians using direct product DVRs are most easily demonstrated for a three-dimensional Cartesian system. In a three-dimensional Cartesian DVR where the DVR points for x, y, and z are labeled by α, β, and γ respectively, the Hamiltonian matrix elements would be

$$\mathbf{H}_{\alpha,\alpha',\beta,\beta',\gamma,\gamma'} = \mathbf{K}_{x\alpha,\alpha'}\delta_{\beta,\beta'}\delta_{\gamma,\gamma'} + \mathbf{K}_{y\beta,\beta'}\delta_{\alpha,\alpha'}\delta_{\gamma,\gamma'} + \mathbf{K}_{z\gamma,\gamma'}\delta_{\alpha,\alpha'}\delta_{\beta,\beta'} \tag{3.1}$$

$$+ V(x_\alpha, y_\beta, z_\gamma)\delta_{\alpha,\alpha'}\delta_{\beta,\beta'}\delta_{\gamma\gamma'} \tag{3.2}$$

where $\mathbf{K}_i$ is kinetic energy operator for the i coordinate.

For triatomic molecules, orothogonal coordinate systems such as Jacobi coordinates, Radau coordinates, and hyperspherical coordinates all permit one to avoid mixed second derivatives. If kinetic energy singularities are unimportant ("unimportant" singularities are singularities for which all wavefunctions are extremely small where the kinetic energy operator is infinite), they can always be dealt with by simply not putting quadrature points close to the singularity. It is straightforward to construct a direct product DVR for the triatomic Jacobi, and Radau Hamiltonians.

For coordinates with more complex kinetic energy operators, direct product DVRs are still highly advantageous for iterative methods in some cases. For four-atom molecules it is not possible to choose coordinates to eliminate all terms with mixed second derivatives from the kinetic energy operator and one is forced (if the calculation is to be exact) to deal with complicated kinetic energy operators.

C. Complicated Kinetic Energy Operators

A *quadrature FBR* is defined above as the representation obtained by evaluating potential matrix elements by Gauss quadrature and taking exact, analytic kinetic energy matrix elements. One transforms from the FBR to the DVR using a transformation matrix that may be obtained by diagonalizing a matrix representation of a coordinate function. If exact (FBR) kinetic energy matrix elements are known, it is easy to determine the DVR of a kinetic energy operator. However, obtaining a Hermitian FBR (or DVR) of a complicated kinetic energy operator whose terms involve both (noncommuting) derivatives and functions of coordinates is somewhat tricky because analytic kinetic energy matrix elements are generally not available. We denote a kinetic energy operator as "complicated" if either (1) it is factorizable and matrix elments of a factor that involves derivatives must be calculated numerically, (2) it is not factorizable and matrix elments of a term must be evaluated numerically, or (3) matrix representations of derivatives are not anti-Hermitian.

One way to obtain a Hermitian FBR for a complicated kinetic energy operator is to use Gauss quadrature to compute kinetic energy matrix elements. A DVR matrix can then be obtained by transforming the FBR matrix whose elements are computed by quadrature. As originally defined, for simple kinetic energy operators, one can compute matrix elements of the DVR Hamiltonian matrix without calculating integrals by quadrature. DVR matrices for coordinate functions are obtained dirrectly from values of the functions at the DVR points and not by first computing an FBR matrix that is then transformed to the DVR. Not having to compute integrals makes the DVR simple and convenient. It would be advantageous not to have to evaluate quadratures even for complicated kinetic energy operators.

Quadrature may be avoided only by using some sort of product approximation. We refer to replaing the matrix representation of a product of operators with a product of representations as a *product approximation* (see Section II. C). Unfortunately, a simple product approximation Hamiltonian matrix may be non-hermitian. In this section we show how the product approximation can be used to construct a Hermitian DVR of a general kinetic energy operator without evaluating quadratures.

Consider a one-dimensional (1D) Hamiltonian (in atomic units) with a complicated kinetic energy operators:

$$H = K + V, \qquad K = -\left(\frac{1}{2}\right)\frac{d}{dx}G(x)\frac{d}{dx}, \qquad x \in (x_1, x_2) \tag{3.3}$$

with a unit weight factor so that orthogonal basis functions are normalized as

$$\int_{x_1}^{x_2} dx\,\theta_m(x)^*\theta_n(x) = \delta_{mn} \tag{3.4}$$

where G and V are real and G is positive and bounded. If G satisfies the boundary condition

$$G(x_j) = 0 \tag{3.5}$$

then even if $\theta_n(x_j)$ and $(d/dx)\,\theta_n(x_j)$ are nonzero at the boundary point x_j, the $\theta_n(x)$ matrix representation of the Hamiltonian is Hermitian. The matrix representing the derivative with respect to x is not anti-Hermitian if the "surface term" $\theta_m(x)^*\theta_n(x)|_{x_1}^{x_2}$, obtained by integrating by parts, does not vanish. Although the derivative matrix is not anti-Hermitian, the kinetic energy matrix *is* Hermitian, *if matrix elements are evaluated exactly*. We now show that, with care, the product approximation can be used to construct a Hermitian DVR of the kinetic energy operator.

To define a prescription for building DVR matrices for complicated kinetic energy operators without evaluating quadratures, it would be natural to use a product approximation and to replace representations of products with products of representations. Applying the product approximation in this way yields an "ordinary" FBR for the 1D kinetic energy operators above:

$$K_{mn}^{\text{OFBR}} = -\left(\frac{1}{2}\right)\sum_{j,k=0}^{N-1} D_{mj}\langle j|G|k\rangle D_{kn}, \tag{3.6}$$

where

$$D_{mn} = \int_{x_1}^{x_2} dx\, \theta_m(x)^* \frac{d}{dx} \theta_n(x) \tag{3.7}$$

and $\langle j|G|k \rangle$ is evaluated by quadrature

$$\langle j|G|k \rangle = G_{jk}^{\text{FBR}} = \sum_{\alpha=0}^{N-1} \theta_j(x_\alpha)^* G(x_\alpha) \theta_k(x_\alpha) \frac{\omega_\alpha}{\omega(x_\alpha)}$$

$$= (\mathbf{T}^\dagger \mathbf{G}^{\text{DVR}} \mathbf{T})_{jk} \tag{3.8}$$

with

$$T_{\alpha j} = \sqrt{\frac{\omega_\alpha}{\omega(x_\alpha)}}\, \theta_j(x_\alpha)^*, \qquad G_{\alpha\beta}^{\text{DVR}} = G(x_\alpha)\delta_{\alpha\beta} \tag{3.9}$$

and ω_α represents a quadrature weight.

With a unit weight factor the derivative matrix $\mathbf{D}$ is anti-Hermitian if the basis functions $\theta_n(x)$ are zero at x_1 nd x_2, the largest and smallest allowed values of the coordinate. If $\mathbf{D}$ is anti-Hermitian, the product FBR Hamiltonian matrix $\mathbf{K}^{\text{0FBR}}$ is Hermitian. However, if $\mathbf{D}$ is not anti-Hermitian the product FBR Hamiltonian matrix is not Hermitian. Although an exact variational basis representation (VBR) matrix (with no product and no quadrature approximation) is Hermitian, $\mathbf{H}^{\text{0FBR}}$ is not.

To obtain a Hermitian FBR kinetic energy matrix by invoking the product approximation, one must write the kinetic energy operators in the explicitly Hermitian form [36]:

$$\breve{K} = \left(\frac{1}{2}\right)\left(\frac{d}{dx}\right)^\dagger G(x) \frac{d}{dx} \tag{3.10}$$

with

$$\breve{K}_{mn} = \langle m|\breve{K}|n \rangle = \left(\frac{1}{2}\right) \int_{x_1}^{x_2} dx\, \theta_m(x)^* \left(\frac{d}{dx}\right)^\dagger G(x) \frac{d}{dx} \theta_n(x) \tag{3.11}$$

where $(d/dx)^\dagger = \overleftarrow{d}/dx$ and the arrow denotes differentiation to the left. If integrals are evaluated exactly, then

$$K_{mn} = \breve{K}_{mn} \tag{3.12}$$

because at x_1 and x_2 either $G(x_j) = 0$ or $\theta_n(x_j) = 0$ or $(d/dx)\theta_n(x_j) = 0$ and the surface term obtained when one integrates by parts is zero. It is clear, therefore, that one may replace K with $\check{K}$. We define the FBR of $\check{K}$ as

$$\check{\mathbf{K}}^{\text{FBR}} = \tfrac{1}{2}\mathbf{D}^\dagger \mathbf{G}^{\text{FBR}}\mathbf{D} \tag{3.13}$$

where $\mathbf{G}^{\text{FBR}}$ is the matrix of G with elements evaluated by quadrature. Both the exact VBR matrix $\mathbf{K}$ and the FBR matrix $\check{\mathbf{K}}^{\text{FBR}}$ obtained via the product approximation are Hermitian.

The DVR matrix is obtained from $\mathbf{K}$ by premultiplying by $\mathbf{T}$ and postmultiplying by $\mathbf{T}^\dagger$

$$\begin{aligned}
\mathbf{K}^{\text{DVR}} &= \mathbf{T}\check{\mathbf{K}}^{\text{FBR}}\mathbf{T}^\dagger = \tfrac{1}{2}\mathbf{T}\mathbf{D}^\dagger\mathbf{G}^{\text{FBR}}\mathbf{D}\mathbf{T}^\dagger \\
&= \tfrac{1}{2}\mathbf{T}\mathbf{D}^\dagger\mathbf{T}^\dagger\mathbf{G}^{\text{DVR}}\mathbf{T}\mathbf{D}\mathbf{T}^\dagger = \tfrac{1}{2}(\mathbf{D}^{\text{DVR}})^\dagger\mathbf{G}^{\text{DVR}}\mathbf{D}^{\text{DVR}}
\end{aligned}$$

where

$$\mathbf{D}^{\text{DVR}} = \mathbf{T}\mathbf{D}\mathbf{T}^\dagger \tag{3.15}$$

Here $\mathbf{K}^{\text{DVR}}$ is an explicitly Hermitian operator. To calculate the DVR of $\check{K}$ it is not necessary to first calculate $\check{\mathbf{K}}^{\text{FBR}}$. $\mathbf{K}^{\text{DVR}}$ is computed without calculating $\mathbf{G}^{\text{FBR}}$, without quadratures, without differentiating G matrix elements, from the values of $G(x)$ at the DVR points and $\mathbf{D}^{\text{DVR}}$. Note that although no quadratures are evaluated to obtain the DVR, the quadrature approximation is implicit in our assumption that $\mathbf{G}^{\text{DVR}}$ is diagonal. For a multidimensional application of this DVR for complicated kinetic energy operators, see Ref. 36.

D. Symmetry-Adapted DVRs

It is almost always advantageous to block-diagonalize the Hamiltonian matrix by exploiting symmetry. (Note that for floppy molecules the permutation inversion groups are more appropriate than space groups, although isomorphisms exist.) If basis functions are chosen so that they transform like irreducible representations of the group of the molecule being studied, matrix elements in off-diagonal blocks, connecting basis functions of different symmetries, are zero. Block-diagonalizing the Hamiltonian matrix reduces the size of the matrices one works with and aids assignment of energy levels. In this section we shall discuss methods of constructing DVRs that take advantage of such symmetry.

It is simplest to construct DVR basis functions adapted to symmetry operations that affect just one coordinate. Examples of such symmetry operations are (1) the permutation of homonuclear diatom nuclear labels in the triatomic Jacobi Hamiltonian for an X_2Y molecule (affecting only θ), (2) space-fixed inversion of a four-atom molecule (affecting only the out-of-plane coordinate), and (3) permutation of the identical nuclei of an X_2Y molecule (affecting only the antisymmetric Radau coordinate). For such symmetry operations a method of devising a symmetry adapted DVR was suggested by Whitnell and Light [15]. They chose VBR functions that have the desired symmetry properties and defined separate sets of DVR functions for each symmetry by diagonalizing the VBR of a symmetric operator. For example, in option 1 above, one designs a symmetry adapted DVR for the operation having the effect $\theta \rightarrow \pi - \theta$ by separating symmetric $(j - |m|$ even) and antisymmetric $(j - |m|$odd) associated Legendre functions and diagonalizing $\cos^2(\theta)$ separately in the symmetric and antisymmetric basis sets to obtain separate sets of points (in the half-range $(0, \pi/2)$ for the symmetric and antisymmetric blocks. For example 2 one diagonalizes $\cos\phi$ in a basis of $\cos(m\phi)$ functions and basis of $\sin(m\phi)$ to generate points [in the half-range $(0, \pi)$] for each symmetry block.

Another (perhaps better) method to devise symmetry-adapted DVRs was first used by McNichols and Carrington [37] and Wei and Carrington [36]. Rather than constructing symmetry adapted DVR functions as linear combinations of symmetry-adapted VBR function, they suggested taking symmetry combinations of DVR functions that do not have the desired symmetry properties. For example 1, one would construct symmetry-adapted DVR functions by taking symmetric and antisymmetric combinations of the DVR functions obtained by diagonalizing $\cos(\theta)$ in a Legendre basis. For example 2, one takes symmetric and antisymmetric combinations of the DVR functions obtained by diagonalizing $\sin(\phi/2)$ in the $[\cos(m\phi), \sin(m\phi)]$ basis. This methods of devising symmetry-adapted DVRs has the advantage that the symmetry-adapted DVR functions for both symmetry blocks are localized about the same values of the coordinate. This simplifies calculating integrals of antisymmetric operators between symmetric and antisymmetric functions. The second method can also be used to to construct DVR basis functions adapted to symmetry operations that affect more than one coordinate. One simply takes linear combinations of 1D DVR functions (without the symmetry properties) to obtain new DVR functions with the symmetry properties [36].

E.   Sequential Diagonalization–Truncation or Adiabatic Reduction

The DVR is a powerful tool because it facilitates multidimensional calculations [13,14,38]. The simplest multidimensional DVR basis is a basis each of whose functions is a product of 1D DVR functions for the separate degrees of freedom. This is a direct product DVR. Direct product DVRs are useful for several reasons: (1) the potential and functions of coordinates in the kinetic energy operators are diagonal in the DVR and hence very easy to construct (no integrals are computed) and (e.g., with iterative methods); and (2) for many molecules the direct product DVR basis may be used to efficiently construct more compact basis functions using the sequential diagonalization/truncation method.

The structured sparseness of the DVR Hamiltonian matrix can be exploited to generate in a sequential fashion good contracted basis sets in an increasing number of dimensions. The final result is a Hamiltonian matrix expressed in an optimized contracted basis for all dimensions. This approach, known as the *sequential diagonalization/truncation method* or *sequential adiabatic reduction* (SAR) *method*, was applied to large amplitude vibrational motions of polyatomic molecules by Bacic et al. [11–14] and has since been used in numerous studies. It has the advantage that a relatively large number of accurate eigenvalues and eigenfunctions can be evaluated, even for "floppy" molecules in which a basis of harmonic oscillator functions, for example, is inadequate. It has the disadvantage that at each stage of reduction the Hamiltonians for the reduced-dimension problems must be diagonalized.

The approach is most easily understood in terms of adiabatic decompositions of the degrees of freedom, although the SAR method does not actually use adiabatic approximations. If we consider two orthogonal degrees of freedom, and the motions in one coordinate, say, y, are much lower frequency than those in x, then one is tempted to use an approximate adiabatic decomposition of the Hamiltonian and the solutions.

The approximate (not the SAR) adiabatic procedure is as follows. Let the exact Hamiltonian be

$$H(x,y) = K_x + K_y + V(x,y) \tag{3.16}$$

and find the solutions for x for fixed values of y:

$$h(x;y) = K_x + V(x,y) \tag{3.17}$$

$$h(x;y)\phi_n(x;y) = \epsilon_n(y)\phi_n(x;y) \tag{3.18}$$

Then the adiabatic approximation for the solutions is obtained by solving for the y motion in the potentials defined by the $\epsilon(y)$ values

$$(K_y + \epsilon_n(y))\theta_m^n(y) = \lambda_m^n \theta_m^n(y) \tag{3.19}$$

and the full wavefunction for an n, m state is *approximated* by

$$\Theta_{n,m}(x,y) = \theta_m^n(y)\phi_n(x;y) \tag{3.20}$$

with energy λ_m^n. The error in this procedure, of course, is that the kinetic energy operator in y is not allowed to operate on the $\phi_n(x:y)$. However, if we use a DVR for y, these terms may be included exactly (within the basis), and simply.

If we are using a DVR for y and an arbitrary basis for x, then at each y_α the solutions for x of Eq. (3.18) are obtained as a matrix of eigenvectors (a transformation matrix), $\mathbf{T}^\alpha$. The full hamiltonian matrix can now be transformed *exactly* to this basis:

$$H^{2D}_{\alpha,n;\alpha',m} = (\mathbf{T}^{\alpha T}\mathbf{T}^{\alpha'})_{n,m}(K_y)_{\alpha,\alpha'} + \epsilon^\alpha_n \delta_{n,m}\delta_{\alpha,\alpha'} \qquad (3.21)$$

The adiabatic *reduction* is easily accomplished by *eliminating* from this basis some of the eigenvectors of $h(x; y_\alpha)$ at some (or all) of the y_α. This makes the transformation matrices, T^α *rectangular* of dimension $n_x \times n^\alpha_x$, where n_x is the number of basis functions in x, and n^α_x is the number of eigenvectors *retained* at y_α. A minimum number of eigenvectors of $h(x; y_\alpha)$ are retained at each y_α (those with lowest energy). Some of the remaining eigenvectors of $h(x; y_\alpha)$ are usually eliminated on the basis of energy; with an energy cutoff of $E_{\max,1D}$ those vectors with $\epsilon^\alpha_n > E_{\max,1D}$ are discarded. It has been found empirically and repeatedly [12,13,39] that convergence of the final eigenvalues of multidimensional problems is greatly enhanced by keeping *all* the localized DVR basis functions available to the final states. If the energy criterion alone is used to eliminate eigenvectors [of $h(x, y_\alpha)$ in this case], then for some y_α all the eigenvalues of $h(x, y_\alpha)$ may lie above the cutoff, and the DVR function corresponding to y_α would be eliminated from the basis. This has unfortunate consequences for accuracy. A better basis fo the 2D Hamiltonian is obtained by keeping 3–10 eigenvectors at each y, and using a lower cutoff energy to limit the size of the resulting Hamiltonian matrix.

Obviously this process may be extended to higher dimension, and is used frequently for three-dimensional problems. In this case H^{2D} would be evaluated as above for each value of the third coordinate, say, z_γ, the two dimensional basis vectors would be truncated by an $E_{\max,2D}$ cutoff again keeping a minimum number at each z point. The form of H^{3D} is identical to that of H^{2D} above, with obvious substitution of the K_z for K_y and the 2D eigenvectors and eigenvalues for those of the 1D Hamiltonian in Eq. (3.21).

Although the CPU requirements and scaling depend on the actual reductions used, one can get a general picture of the scaling with n, the number of basis functions per dimension (assumed equal for all dimensions), and the truncation fraction, $q = (1 - f), 0 < f < 1$ ($f = 1$ means no truncation). In d dimensions the basis size would be n^d, and the diagonalization time would scale as n^{3d}. In the SAR method, there are n^{d-1} diagonalizations of the first coordinate, which itself scales as n^3. Thus the time for this scales as n^{d+2}. After truncation, there are n^{d-2} diagonalizations of average dimension $n^2 f$. Thus the time for this second set of diagonalizations scales as $n^{d+4}f^3$, and so on. For 3D, the CPU time for the third (and last) diagonalization would scale as n^9f^6. The transformations [see Eq. (3.21)] also require amounts of computer time comparable to the diagonalizations themselves. A substantial savings in time can also be obtained if diagonalizations are carried out only at every third point in some dimensions, and that basis, suitably truncated, is used for neighboring points. The residual potential is, of course, retained and transformed to the reduced basis [39].

It should be noted that the general approach above is not restricted to DVR bases. First, the representation for the initial diagonalizations (of coordinate x above) can be in any basis, one- or higher-dimensional. In a number of problems coupled angular bases must be used for some degrees of freedom, and, after diagonalization of this basis at each DVR point, the adiabatic reduction may be carried out as above. The initial SAR implementations [11,12] were with a distributed Gaussian basis coupled with a DVR. In the 6D rigid water dimer problem, as an extreme example, a 5D coupled angular basis is used together with a 1D DVR in the water–water distance [40–42] to calculate the bound states accurately.

Bowman and Gazdy [43] noted that direct product bases of any sort may lend themselves to a reduction procedure as above; merely diagonalize the block-diagonal components of the Hamiltonian corresponding to one or more fixed basis functions [$\phi_i(y)$, say], truncate the eigenvector bases for each, and transform the Hailtonian to this truncated basis. The reduced Hamiltonian may then be diagonalized. In this approach (without a DVR), however, the full potential matrix must be evaluated.

F. Iterative Methods

Methods for solving the time-independent Schrödinger equation equation (whether designed to calculate S matrix elements or spectra) are often impeded by the number of basis functions required for convergence. For example, using a conventional time-independent method, one calculates a spectrum by diagonalizing a matrix representation of the Hamiltonian; energy levels converge from above (in the VBR) as the basis set is enlarged. To calculate highly excited states of triatomic molecules or more than a few levels for four-atom molecules, one has to diagonalize large matrices. Large matrices are difficult to deal with for two reasons: (1) conventional diagonalization algorithms explicitly modify the matrix as it is being diagonalized and therefore require that it be stored in the core memory of the computer (we presently know of no conventional diagonalization calculations with matrices larger than approximately $10,000 \times 10,000$ and (2) the cost of conventional diagonalization algorithms scales as N^3, where N is the size of the matrix [44].

To alleviate these problems one may either (1) devise good basis functions that represent wavefunctions compactly to minimize the number of basis functions (and the size of the matrix) required to obtain converged energy levels, (2) use simple product basis functions and an iterative method that exploits the simplicity of the basis, or (3) use good (and therefore not simple) basis functions and an iterative method. In the previous sections we discussed in detail sequential diagonalization/truncation methods for devising good basis functions that enable one to use conventional algorithms for solving the time-independent Schrödinger equation equation. In this section we shall discuss using DVRs with iterative methods.

To use an iterative method to compute eigenvalues and eigenvectors of a matrix representation of the Hamiltonian, $\mathbf{H}$, one does not need to store $\mathbf{H}$; one merely needs to store vectors and to evaluate the product of $\mathbf{H}$ with a vector at each iteration of the recursion. If the basis is a product basis (option 2 above), each matrix–vector product may be evaluated with a cost scaling as n^{f+1}, where n is a representative number of basis functions for a single degree of freedom and f is the number of degrees of freedom. Most of the best methods for solving the time-dependent Schrödinger equation also obviate the need to store $\mathbf{H}$ and allow one to propagate by computing matrix-vector products. [45]

1. Efficient Matrix–Vector Products

If the basis is a product basis, each of whose functions is a product of functions of a single coordinate, and the Hamiltonian is factorizable, that is, if it can be written in the form

$$\hat{H} = \sum_{l=1}^{g} \prod_{k=1}^{f} \hat{h}^{(k,l)}(q_k) \qquad (3.22)$$

a sum of g terms each with f factors (for a single term no more than two of the factors contain derivatives), then it is always possible to evaluate matrix–vector products so that their cost scales as n^{f+1}. Almost all kinetic energy operators have this factorizable form, and most have fewer than f factors in each term. Owing to the factorizability of $\hat{H}$ and the product structure of the basis, a matrix–vector product $\mathbf{Hu} = \mathbf{u}'$ is evaluated most efficiently by doing sums sequentially [46–48]

$$\sum_{l=1}^{g} \sum_{i_1} h_{i_1',i_1}^{(1,l)} \sum_{i_2} h_{i_2'i_2}^{(2,l)} \cdots \sum_{i_f} h_{i_f',i_f}^{(f,l)} u_{i_1,i_2,\dots,i_f}^{(f,l)} = u'_{i_1',i_2',\dots,i_f'} \qquad (3.23)$$

where $h_{i_k'i_k}^{(k,l)}$ is an element of the $n \times n$ matrix representation of the factor $\hat{h}^{(k,l)}(q_k)$ (involving a single coordinate) and for simplicity we have assumed that the basis is a direct product basis:

$$\Phi_{i_1,i_2,\ldots,i_f} = \phi_{i_1}(q_1)\phi_{i_2}(q_2)\cdots\phi_{i_f}(q_f) \tag{3.24}$$

Note that for notational simplicity we have assumed that there are n basis functions for each degree of freedom but that this is not necessary. If the lth term has c_l nonidentity $\mathbf{h}^{(k,l)}$ matrices the number of multiplications required to evaluate the matrix–vector product for that term is $c_l n^{f+1}$.

The favorable n^{f+1} scaling relation is not due the sparsity of the Hamiltonian matrix or to the sparsity of the $\mathbf{h}^{(k,l)}$ matrices, it is not due to the absence of mixed second-order differential operators, and it is not due to choosing special single-coordinate functions from which to build the product basis functions. The n^{f+1} scaling relation is a result of using a product basis and having a factorizable operator [46]. Note that if one evaluates matrix–vector products by doing sums sequentially, one never actually calculates (or stores) matrix elements of the Hamiltonian (not even on the fly): one uses matrix elements of the factors of the terms of the Hamiltonian to build up the Hamiltonian matrix–vector product. One stores only (small) one-dimensional representations of the factors.

Although matrix–vector product can be evaluated so that their cost scales as n^{f+1} regardless of whether one chooses FBR or DVR single-coordinate functions, it is often advantageous to choose DVR functions for three reasons:

1. If one uses DVR functions, some of terms in the Hamiltonian can be combined before computing matrix–vector products so that the effective number of terms for which matrix–vector products must be computed is reduced. This is most obvious for the potential. If matrix–vector products were evaluated for each term in the potential separately, the total cost of an iterative calculation would depend on the number of potential terms, but in the DVR the entire potential can be treated as one term; thus the number of terms is irrelevant and the cost of the potential matrix–vector product scales as only n^f. In general, if two terms share the same nondiagonal $\mathbf{h}^k$ matrices (i.e., the same differential operators), they can be added together prior to the matrix–vector multiplication.

2. If one uses the DVR, multiplicative operators need not be factorizable to attain n^{f+1} scaling.

3. Since functions of coordinates in the kinetic energy operator are also diagonal in the DVR, then for each term there are at most two non-diagonal $\mathbf{h}^{(k,l)}$ matrices. Neglecting n^f compared to n^{f+1}, evaluating a DVR matrix–vector product for product for a kinetic energy term costs

either n^{f+1} or $2n^{f+1}$ depending on whether the term has derivatives with respect to one or with respect to two coordinates.

2. Efficient FBR Matrix–Vector Products

Sometimes it is preferable (see the section on nondirect product FBRs below) to use an FBR rather than a DVR basis. In the previous section we explained that using DVR basis enables one to combine all the terms of the potential before evaluating matrix vector products. It would be very costly to compute matrix–vector products for each potential term separately, but even in the FBR this is not necessary. The best way to evaluate matrix–vector products in the FBR is to use quadratures and to evaluate sums sequentially. This technique is sometimes referred to as the *psuedospectral method* [49]. Using this method, one can evaluate matrix–vector products for terms which are not factorizable at a cost which scales as n^{f+1}.

Consider the matrix–vector product for the 2-degree-of-freedom (unfactorizable) term

$$h = \frac{1}{a^2 + q_1^2 + q_2^2} \qquad (3.25)$$

This operator cannot be written as a product (or a simple sum of products) of functions of a single coordinate, and it might therefore appear that the cost of the corresponding matrix–vector product

$$\sum_{i_1} \sum_{i_2} h_{i_1' i_2', i_1 i_2} u_{i_1, i_2} = u'_{i_1', i_2'} \qquad (3.26)$$

should necessarily scale as n^{f+2} (in this case n^4). However, if one introduces sums over the quadrature points that would be required to compute the $h_{i_1' i_2', i_1 i_2}$ integral and then performs all of the sums sequentially, one can achieve n^{f+1} scaling.

The integral

$$h_{i_1' i_2', i_1 i_2} = \int dq_1 \, dq_2 \, \phi_{i_1'}(q_1) \phi_{i_2'}(q_2) \frac{1}{a^2 + q_1^2 + q_2^2} \phi_{i_1}(q_1) \phi_{i_2}(q_2) \qquad (3.27)$$

$$\approx \sum_\alpha \sum_\beta \sqrt{\frac{\omega_\alpha}{\omega(q_1)_\alpha}} \sqrt{\frac{\omega_\beta}{\omega(q_2)_\beta}} \phi_{i_1'}(q_1)_\alpha \phi_{i_2'}(q_2)_\beta$$

$$\times \frac{1}{a^2 + (q_1)_\alpha^2 + (q_2)_\beta^2} \qquad (3.28)$$

$$\times \phi_{i_1}(q_1)_\alpha \phi_{i_2}(q_2)_\beta \sqrt{\frac{\omega_\alpha}{\omega(q_1)_\alpha}} \sqrt{\frac{\omega_\beta}{\omega(q_2)_\beta}} \qquad (3.29)$$

where $(q_1)_\alpha$ and $(q_2)_\beta$ are quadrature points for the q_1 and q_2 coordinates, ω_α and ω_β are the corresponding quadrature weights, and $\omega(q_1)$ and $\omega(q_2)$ are the corresponding weight functions (which together with the coordinate ranges determine the quadrature points and weights).

In terms of the **T** matrices, we obtain

$$h_{i'_1 i'_2, i_1 i_2} \approx \sum_\alpha \sum_\beta (T^\dagger)_{i'_1,\alpha} (T^\dagger)_{i'_2,\beta} \frac{1}{a^2 + (q_1)_\alpha^2 + (q_2)_\beta^2} (T)_{\alpha,i_1} (T)_{\beta,i_2} \qquad (3.30)$$

The matrix–vector product can now be written

$$\sum_{i_1} \sum_{i_2} \sum_\alpha \sum_\beta (T^\dagger)_{i'_1,\alpha} (T^\dagger)_{i'_2,\beta} \frac{1}{a^2 + (q_1)_\alpha^2 + (q_2)_\beta^2} (T)_{\alpha,i_1} (T)_{\beta,i_2} u_{i_1,i_2} = u'_{i'_1,i'_2}$$

$$(3.31)$$

and doing the sums sequentially

$$\sum_\alpha (T^\dagger)_{i'_1,\alpha} \sum_\beta (T^\dagger)_{i'_2,\beta} \frac{1}{a^2 + (q_1)_\alpha^2 + (q_2)_\beta^2} \sum_{i_1} (T)_{\alpha,i_1} \sum_{i_2} (T)_{\beta,i_2} u_{i_1,i_2} = u'_{i'_1,i'_2}$$

$$(3.32)$$

the matrix vector–product is evaluated at a cost of $4n^3$, which scales as n^{f+1}. Note that the $h_{i'_1 i'_2, i_1 i_2}$ integrals are never actually computed.

Provided the basis is a product basis, all matrix–vector products can therefore be evaluated at a cost that scales as n^{f+1} (regardless of whether the Hamiltonian or its terms are factorizable) using the pseudospectral method or a combination of the pseudospectral method and the DVR.

3. Effecient Matrix–Vector Products for Nondirect Product and Nonproduct Representations

For some problems a direct product basis

$$\Phi_{i_1,i_2,\ldots,i_f} = \phi_{i_1}(q_1)\phi_{i_2}(q_2)\cdots\phi_{i_f}(q_f) \qquad (3.33)$$

for which matrix–vector products were discussed above, is adequate. Often, however, it will be advantageous to use either a product basis that is not a direct product basis or to use a basis each of whose functions is not a product of functions of a single coordinate. In this section we discuss evaluating matrix–vector products for such basis sets.

Before we explain how the matrix–vector products are done efficiently, we briefly address the issue of why it is useful to choose nondirect product or nonproduct basis sets. The Lanczos algorithm is the most useful iterative method for calculating a large number of energy levels. The number of Lanczos iterations required to converge energy levels of interest is strongly correlated to the spacing of the close energy levels relative to the spacings of the most widely separated eigenvalues of the Hamiltonian matrix; widely spaced energy levels are converged first, whereas bunches of closely spaced eigenvalues are converged slowly. The Lanczos algorithm also tends to preferentially converge eigenvalues at the top and at the bottom of the spectrum of the matrix. For both these reasons the Lanczos algorithm converges enthusiastically for systems with very large, widely spaced energy levels. Convergence is also poorer if the spectral range (the difference between the largest and the smallest eigenvalues) of the matrix is large [50]. (The number of Chebyshev polynomials and hence the number of matrix–vector products required to accurately propagate a wavepacket using a Chebyshev series also directly depends on the spectral range of the Hamiltonian matrix [51].)

If the Lanczos algorithm is used, as suggested by Cullum and Willougby [50], without reorthogonalization of Lanczos vectors, to calculate energy levels, the highest (and usually widely spaced) eigenvalues of the Hamiltonian matrix converge first, and as one iterates further, they are wastefully reproduced. Unfortunately, the eigenvalues computed most easily by the Lanczos algorithm (the largest) are often those one cares least about since they may not be converged with respect to the basis size and may not have physical significance. For many molecules this problem is not really debilitating but it is exacerbated by any important singularities in the kinetic energy operator. An important singularity is one for which the wave-functions one wishes to calculate have significant amplitude at points in configuration space at which the kinetic energy operator is infinite. Because they magnify the spectral range, singularities also make application of most iterative propagation methods more difficult.

4. Nondirect Product Representations

To deal with important singularities, one chooses basis functions to reduce the size of the kinetic energy matrix elements and thereby reduce the spectral range of the Hamiltonian matrix. Optimal basis functions may be nondirect product functions that are eigenfunctions of part of the Hamiltonian that includes the kinetic energy term(s) with the singularity. We use the term *nondirect product function* to mean a function that is a product of functions of different coordinates but for which at least one of the single-coordinate functions is labeled not only by the index for its coordinate but also by the index for another coordinate, meaning that there is a shared index. The most familiar example of a nondirect product basis function is a spherical harmonic: $Y_{lm} = \Theta_l^m(\theta)\Phi_m(\phi)$, where m is the shared index. Singularities occur whenever one coordinate takes a limiting value and another becomes undefined. If there is an important singularity good basis functions are always nondirect product functions that are products of functions (one with a shared index) of the coordinate that becomes undefined and the coordinate that takes a limiting value.

To evaluate the matrix–vector product efficiently for basis functions that have the form $f_n^m(q_1)g_m(q_2)$ (m is the shared index and the singularity occurs when q_1 takes a limiting value and q_2 becomes undefined), one uses the pseudospectral sequential summation method described above. Two issues require some thought: (1) how one should order the factors in the nondirect product equivalent of Eq. (3.31) and (2) whether one should use different sets of q_1 quadrature points for different values of the shared index m or whether it is better to use the same set of q_1 quadrature points for all values of m. If one chooses to use different points for different values of m, it is clear that one should use the appropriate (m-dependent) Gauss quadrature points for the polynomial associated with $f_n^m(q_1)$. If, on the other hand, one chooses to use the same q_1 quadrature points for all values of m, what points should one choose?

The nondirect product basis functions will usually be eigenfunctions of either part of the kinetic energy operator or the sum of part of the kinetic energy operator and part of the potential, and therefore the only nontrivial matrix–vector product will be for the potential or for part of the potential. One may always attain n^{f+1} scaling by choosing the same set of q_1 quadrature points for all values of m.

If one chooses m-independent points, the quadrature approximation for the integral

$$V_{n'm',nm} = \int dq_1\, dq_2 f_{n'}^{m'}(q_1)g_{m'}(q_2)V(q_1,q_2)f_n^m(q_1)g_m(q_2) \tag{3.34}$$

$$= \sum_\alpha \sum_\beta \sqrt{\frac{\omega_\alpha}{w((q_1)_\alpha)}}\sqrt{\frac{\omega_\beta}{w((q_2)_\beta)}} f_{n'}^{m'}((q_1)_\alpha)g_{m'}((q_2)_\beta) \tag{3.35}$$

$$\times V((q_1)_\alpha,(q_2)_\beta)f_n^m((q_1)_\alpha)g_m((q_2)_\beta)\sqrt{\frac{\omega_\alpha}{w((q_1)_\alpha)}}\sqrt{\frac{\omega_\beta}{w((q_2)_\beta)}} \tag{3.36}$$

where $(q_1)_\alpha$ and $(q_2)_\beta$ are quadrature points for the q_1 and q_2 coordinates, ω_α and ω_β are the corresponding quadrature weights, and $w(q_1)$ and $w(q_2)$

are the corresponding weight functions. In terms of the **T** matrix elements

$$({}^{[q_2]}T)_{\beta,m} = \sqrt{\frac{\omega_\beta}{w((q_2)_\beta)}}\, g_m((q_2)_\beta) \tag{3.37}$$

$$({}^{[q_1]}T^m)_{\alpha,n} = \sqrt{\frac{\omega_\alpha}{w((q_1)_\alpha)}}\, f_n^m((q_1)_\alpha) \tag{3.38}$$

the potential integral is

$$V_{n'm',nm} \approx \sum_\alpha \sum_\beta (({}^{[q_1]}T^{m'})^\dagger)_{n',\alpha}\, ({}^{[q_2]}T^\dagger)_{m',\beta}\, V((q_1)_\alpha, (q_2)_\beta)\, ({}^{[q_1]}T^m)_{\alpha,n}\, ({}^{[q_2]}T)_{\beta,m}$$

$$\tag{3.39}$$

where for clarity we have indicated at the left of each T whether it is for q_1 or q_2. As the notation implies there is a different ${}^{[q_1]}T^m$ for each value of m (because there is a different set of q_1 function for each m). The matrix–vector product can now be written

$$\sum_\alpha (({}^{[q_1]}T^{m'})^\dagger_{n',\alpha}\, \sum_\beta ({}^{[q_2]}T^\dagger)_{m',\beta}\, V((q_1)_\alpha, (q_2)_\beta) \sum_m ({}^{[q_2]}T)_{\beta,m}$$

$$\times \sum_m ({}^{[q_1]}T^m)_{\alpha,n}\, u_{nm} = u'_{n'm'} \tag{3.40}$$

Note that because u_{nm} on the lefthand side is indexed by m, the fact that the matrix elements $({}^{[q_1]}T^m)_{\alpha,n}$ are indexed not only by α and n but also by m does not mean that the result of the summation $\sum_n ({}^{[q_1]}T^m)_{\alpha,n} u_{nm}$ is labeled by more indices than would be necessary if the T matrix elements were independent of m. The sum over n yields a vector labeled by α, and m (and costs n^3). If the ${}^{[q_2]}T$ matrix were placed to the right of the ${}^{[q_1]}T^m$ matrix, the matrix–vector product would be more costly because one could not sum over the elements of one T matrix, store the result, and then sum over the elements of the second T matrix (because both are labelled by m). To attain n^{f+1} scaling, the T matrices whose elements are labeled by three indices should therefore be on the outside.

We must still consider the choice of the m-independent q_1 quadrature points. For each value of m, the f_n^m functions are products of polynomials and the square root of a weight function and the polynomials are associated with Gauss quadrature points. What (fixed) value of m should one choose to determine the m-independent points? Clearly the value of m should be chosen so that the associated quadrature accurately approximates the potential and overlap integrals. Regardless of what m is chosen, it will almost always be true that some potential integrals will not be exact, but it is important to choose the fixed value of m for which q_1 Gauss quadrature points are determined so that all overlap integrals (for all values of m) are exact. If overlap integrals were not exact, one would have to take a non-unit-overlap matrix into account.

Examples of important $f_n^m(q_1)$ functions are Jacobi basis functions and spherical oscillator functions. Each function is proportional to the product of a polynomial and the square root of a weight function

$$f_n^m(q_1) \propto \sqrt{w^m(q_1)}p_{n,m}(q_1) \tag{3.41}$$

where $w^m(q_1)$ is the weight function with respect to which the polynomials $p_{n,m}(q_1)$ are orthogonal. In the Jacobi case with $a = b = m, w^m(q_1) = (1 - (q_1)^2)^m$ and $p_{n,m}(q_1) = P_{n-m}^{(m,m)}$, a Jacobi polynomial. In general, the overlap integral is

$$\int dq_1 \, dq_2 f_n^m(q_1) g_m(q_2) f_{n'}^{m'}(q_1) g_{m'}(q_2) \tag{3.42}$$

$$= \int dq_1 f_n^m(q_1) f_{n'}^m(q_1) \delta_{m,m'} \tag{3.43}$$

which is proportional to

$$\int dq_1 \, w^m(q_1) p_{n,m}(q_1) p_{n',m}(q_1) \delta_{m,m'} \tag{3.44}$$

It is simple to rewrite this integral as

$$\int dq_1 \, w^{m_{\text{fix}}}(q_1) \left[\frac{w^m(q_1)}{w^{m_{\text{fix}}}(q_1)} \right] p_{n,m}(q_1) p_{n',m}(q_1) \delta_{m,m'} \tag{3.45}$$

where m_{fix} is a chosen, fixed value of m. This integral can be evaluated exactly by the Gauss quadrature associated with $p_{n,m_{\text{fix}}}(q_1)$ if $\{[w^m(q_1)/w^{m_{\text{fix}}}(q_1)]\}p_{n,m}(q_1)p_{n',m}(q_1)$ is a polynomial of finite degree. Therefore, if the weight function ratio $R = \{[w^m(q_1)/w^{m_{\text{fix}}}(q_1)]\}$ can be written as a polynomial of finite degree, that is, if m_{fix} is the smallest possible value of m, the overlap ingetrals can be evaluated exactly with the q_1 quadrature appropriate for $m = m_{\text{fix}}$. If polynomials are chosen so that the degree of the weight function ratio, R, increases by the same amount that the degree of $p_{n,m}(q_1 p_{n',m}(q_1)$ decreases as m is increased [so that the degree of the integrand of Eq. (3.45), does not depend on the value of m], then the quadrature for $m_{fix} = m_{smallest}$ will do all the overlap integrals correctly. Corey and Lemoine [31] were the first to point out the importance of choosing the points for $m_{smallest}$ if m independent points are used.

Note that although it is possible to choose one set of quadrature points to evaluate all the overlap integrals exactly, it may not, in general, be possible to use one set of quadrature points to evaluate overlap integrals and all the potential matrix elments accurately. For example, if one uses spherical harmonic basis functions up to j_{max}, all the 2D overlap integrals are evaluated exactly using $(2j_{max} + 1)$ Fourier points in ϕ and $j_{max} + 1$ Gauss Legendre $(m = 0)$ points in θ [Note that $(2j_{max} + 1)(j_{max} + 1)$ points are required to compute the overlap integrals for $(j_{max} + 1)^2$ basis functions.] However, if one wishes to calculate a potential integral such as $z = \cos(\theta)$

$$\int dz \, d\phi \, \Theta_{lm}(\theta) \cos{(m\phi)} \cos{(\phi)} \cos{(m'\phi)}\Theta_{l'm'}(\theta) \qquad (3.46)$$

it is necessary to calculate

$$\int dz \, \Theta_{lm}(\theta)\Theta_{l'm+1}(\theta) \qquad (3.47)$$

which is difficult to evaluate accurately with a Gauss Legendre quadrature because

$$\int dz \, \Theta_{lm}(z)\Theta_{l'm+1}(z) \propto \int dz(1 - z^2)^{(1/2)}(1 - z^2)^m P_{j-m}^{(m,m)} P_{j-m-1}^{(m+1,m+1)} \qquad (3.48)$$

and $(1 - z^2)^{(1/2)}$ is not a polynomial in z. To evaluate such an integral (off-diagonal in m), one would need a Gauss Jacobi (with $a = b = \frac{1}{2}$) quadrature. With spherical harmonic basis functions an $a = b = \frac{1}{2}$ Gauss Jacobi quadrature will be required to evaluate integrals of $\cos(n\phi)\sin(n'\theta)$, where $n + n'$ is odd.

Consider now whether it is necessary to choose m-independent q_1 points to attain n^{f+1} scaling. To calculate $V_{n'm',nm}$, one might, for example, choose either different q_1 quadrature points for each (m, m') pair or different q_1 quadrature points for each m'. In either case at least one q_1 **T** matrix would depend on both m and m' and to calculate the matrix–vector product one would be obliged to evaluate sums such as

$$\sum_n {(}^{[q_1]}T^{m,m'}{)}_{\alpha,n} u_{nm} = w'_{n'm'm} \qquad (3.49)$$

the cost of which scales as n^4 (and in general as n^{f+2}). If, however, $V_{n'm',nm} \propto \delta_{mm'}$, one can use different q_1 quadrature points for each value of m without jeopardizing the n^{f+1} scaling relation. This is the case, for example, for the bending basis functions for a triatomic molecule. The bending basis functions depend on the index for the body-fixed component of the angular momentum, but the potential does not depend on the Euler angles, and one may therefore use different sets of q_1 quadrature points with impunity.

The use of the underlying direct product angular DVR [based on Legendre $(m = 0)$ polynomials and a Fourier grid in ϕ] instead of spherical harmonics was examined by Dai and Light [52]. In the direct product DVR with the set of points given above, the j^2 operator was diagonalized. As expected from the preceding analysis, the lowest eigenvalues for even values of m are given exactly since the exact associated Legendre eigenfunctions (m even) can be represented exactly by Legendre polynomials ($m = 0$). For odd m values, however, the eigenvalues converge to the exact values slowly (where $m = 1$ is the slowest to converge). The exact eigenfunctions for m odd cannot be given exactly in a finite basis of Legendre polynomials. For angular scattering problems, however, this direct product angular DVR was shown to be quite accurate.

5. Using Lanczos with Sequential Diagonalization–Truncation

We have discussed in detail evaluating matrix–vector products because the efficacy of several of the best methods for solving the time-dependent (e.g., the Chebyshev and Lanczos propagation schemes) and the time-independent (e.g., Lanczos' and Davidson's algorithm [53]) methods for solving the Schrödinger equation is contingent on the efficiency of matrix–vector products. If one uses a product basis (it may be either a nondirect or a direct product basis), matrix–vector products can always be evaluated at a cost that scales as n^{f+1}. The n^{f+1} scaling relation is attained by exploiting the special structure of the product basis and evaluating summations (over basis set or quadrature indices) sequentially. If one were not able to exploit the structure of the basis and had to do the sums sequentially, the cost of the matrix–vector product (for an $n^f \times n^f$ matrix) would scale as n^{2f}. Although n^{f+1} is certainly favorable compared to n^{2f}, it is clear that as f (or n) increases, the cost of the matrix–vector product calculation will become prohibitive. In addition, although iterative methods obviate the need to store a Hamiltonian matrix, one does need to store vectors, with as many components as there are basis functions, and eventually, as f (or n) is increased, one finds that the core memory of the computer is not large enough to do so. To mitigate both these problems, it is natural to consider using iterative methods with basis functions that represent wavefunctions or wavepackets more compactly (so that one will require fewer basis functions than would be needed if product basis functions were used). If fewer basis functions are required, one will surely have less to store, and it might appear that the cost of each matrix–vector product should also be reduced. Basis functions obtained from the sequential diagonalization/truncation method of Light and co-workers are excellent and might fruitfully be combined with iterative methods. In this section we discuss using DVRs to evaluate matrix–vector products for nonproduct basis functions.

It is clear that using a more compact basis will reduce storage requirements. Although one might expect that using a compact basis should somehow reduce n and hence the cost of each matrix–vector product, compact basis matrix–vector products are actually more and not less expensive than product basis matrix–vector products. They are more expensive because the n^{f+1} scaling relation is due to the simplicity of the product basis; better basis functions are necessarily more complicated (the structure of the basis is less simple), and the corresponding matrix–vector products are more costly. The memory advantage of better basis functions is manifest, but if one considers only the cost of a single matrix–vector product, it appears that using better basis functions may increase (and not decrease) the computer time required. However, the cost of an iterative calculation depends not only on the cost of a matrix–vector product for a single term but also on the number of times the Hamiltonian must be applied to a vector (to converge either a series representation of the evolution operator or the desired energy levels) and the effective number of terms for which matrix–vector products must be evaluated for each application of the Hamiltonian. For energy-level calculations, sequential diagonalization/ truncation basis functions will, in general, not only reduce memory requirements but also make Lanczos calculations more efficient. This was first exploited by Wu and Hayes [54,55]. For many molecules a sequential diagonalization/truncation basis Lanczos calculation is more efficient than its product basis counterpart because (1) using a better basis reduces the spectral range of the Hamiltonian matrix and uniformizes the gaps between neighboring eigenvalues and thus decreases the number of iterations required to converge the energy levels of interest and (2) sequential diagonalization/truncation basis functions can be chosen so that only one term in the Hamiltonian is not diagonal and therefore so that only one nontrivial matrix vector product must be evaluated.

We define the primary representation as the representation in which the iteration is performed (i.e, the representation in which the wavepacket is propagated or the representation in which wavefunctions are computed). To efficiently evaluate matrix–vector products for a sequential diagonalization/ truncation primary representation, one uses matrices (much like the quadrature T matrices to transform between the sequential diagonalization/ truncation representation and a representation in which matrix elements of Hamiltonian terms can be computed easily and inexpensively. For this (secondary) representation, it is usually best to choose a DVR. The number of labels on the transformation matrices and the number of values each label assumes determine the cost of the sequential diagonalization/truncation matrix–vector products. The more refined the sequential diagonalization/ truncation basis is, the greater the number of labels on the transformation matrices.

Consider first the matrix–vector product for a triatomic sequential diagonalization/truncation basis of the type discussed earlier. We label functions of q_1 and q_2 obtained by diagonalizing a two-dimensional Hamiltonian for each DVR point γ for coordinate q_3 by the index j. If the two-dimensional Hamiltonian is diagonalized in a direct product $q_1 q_2$ basis and α and β are DVR labels for q_1 and q_2 DVR basis functions, then the matrix of eigenvectors is the transformation matrix, $C^{\gamma}_{\alpha\beta, j}$. (Instead of diagonalizing the two-dimensional Hamiltonian in a direct product DVR basis, one might use products of optimized 1D functions for q_1 and DVR functions for q_2. In a basis of functions labeled as j and γ the Hamiltonian (written in Radau, symmetrized Radau, or Jacobi coordinates) matrix elements are

$$\langle j'\gamma'|\hat{H}|j\gamma\rangle = E^{\gamma}_j \delta_{j'j}\delta_{\gamma\gamma'} + \sum_{\alpha\beta} C^{\gamma'}_{\alpha\beta, j'}\mu(r_{1\alpha}r_{2\beta})l_{\gamma'\gamma}C^{\gamma}_{\alpha\beta, j} \qquad (3.50)$$

where μ is an inverse moment of inertia function and $l_{\gamma'\gamma}$ is a DVR matrix element of an operator proportional to $(\partial^2/\partial\theta^2) + \cot\theta\,(\partial/\partial\theta)$.

In this representation applying the Hamiltonian to a vector requires only one nontrivial matrix–vector product. When sums are evaluated sequentially the cost of the matrix–vector product

$$\sum_{\alpha\beta} C^{\gamma'}_{\alpha\beta, j'}\mu(r_{1\alpha}r_{2\beta}) \sum_{\gamma} l_{\gamma'\gamma} \sum_{j} C^{\gamma}_{\alpha\beta, j}v_{\gamma j} \qquad (3.51)$$

scales as $2(n_j n_\alpha n_\beta n_\gamma) + n^2_\gamma n_\alpha n_\beta + n_\alpha n_\beta n_\gamma$. If $n_\alpha = n_\beta = n_\gamma = n$ and it were necessary to retain all n^2 eigenvectors of the two-dimensional Hamiltonians, then the cost would scale as n^5 (and in general as n^{2f-1}). This should be compared to the cost of the product basis matrix–vector product, n^{f+1}. Clearly, if at each stage of the sequential diagonalization procedure it were necessary to retain all the eigenvectors of the reduced-dimensional Hamiltonians, the sequential diagonalization and the product matrices would have the same eigenvalues (same spectral range, same gap structure), and the same number of iterations would be required for both representations and it would be much more expensive to use the sequential diagonalization basis. However, the utility of the sequential diagonalization basis is derived from the truncation, and if it is truncated, one reduces *both* the cost of each individual matrix–vector product *and* the number of iterations required. For the three-dimensional example, if n_j is reduced from n^2 to n (by retaining only n of the n^2 eigenvectors of the two-dimensional Hamiltonians), the cost of each matrix–vector product is reduced from n^5 to n^4. The cost of product basis matrix–vector product also scales as n^4, but because the spectral range and/or gap structure of the sequential diagonalization basis are much more favorable, the number of required iterations is reduced and the calculation is less costly. If the number of required iterations is large and truncation enables one to reduce n_j significantly, it might be advantageous to multiply the **C 1 C** matrices prior to iterating. Although it is not possible to know a priori how effective the sequential diagonalization procedure will be (how much one will be able to truncate), it seems clear that for Lanczos calculations of energy levels, the sequential diagonalization Lanczos method will, for many, but not all, triatomic molecules, be more efficient than the product basis Lanczos method [56,57].

For molecules with four or more atoms, the advantages of the sequential diagonalization basis are even more convincing; as the number of degrees of freedom increases, it becomes increasingly important to reduce the size of the product basis and to exclude poor (unimportant) functions from the basis.

The best contracted basis functions are obtained from a contraction scheme that involves diagonalizing reduced-dimensional Hamiltonians whose dimension is almost as large as the original Hamiltonian. Unfortunately, the larger the dimension of the reduced-dimensional Hamiltonian, the more costly the matrix–vector product for the associated contracted basis. This has motivated Antikainen et al. [58] to develop a sequential diagonalization/truncation Lanczos method using less than optimal basis functions but basis functions that allow one to evaluate matrix–vector products efficiently. Each of the contracted basis functions of Antikainen et al. is obtained by diagonalizing a one-dimensional Hamiltonian for fixed values of indices of other basis functions. Because their basis functions are eigenfunctions of one- (and not multi-) dimensional reduced Hamiltonians, the cost of each matrix–vector product is reduced. Such a scheme seems very promising for molecules with more than three atoms.

The cost of the sequential diagonalization matrix–vector product can also be reduced (at the price of decreasing the quality of the basis functions and hence increasing the number of basis functions and the number of iterations required to converge), while using multidimensional basis functions, if eigenfunctions of the reduced-dimension Hamiltonian are not recomputed for all the values of the DVR index for the lowest-frequency degree of freedom [39,57]. For example, for a triatomic molecule, rather than choosing a new set of (r_1, r_2) functions for every θ DVR function, one could use one set of (r_1, r_2) functions computed for (say) the equilibrium value of θ. This reduces the number of integrals to calculate in the contracted basis, but more importantly, it reduces the number of labels on the C matrix and hence allows one to construct (explicitly) a matrix representation of μ labelled by only two indices so that the cost of the matrix–vector product is reduced. Alternatively, for every third value of θ one might compute the contracted basis in (r_1, r_2) and use this basis also for the two neighboring values of θ. (Please see Refs. 39 and 57 for details and numeric tests.)

IV. CAVEATS

In any approach to the numeric solution of complex problems, many opportunities for subtle errors exist, and DVR methods are no exception. Since DVRs are basis representations, the common problems of convergence with basis size, and so on all occur in DVRs. There are additional somewhat subtle problems, however, associated with DVRs that are seldom encountered with normal basis representations and that must be considered. They are associated with quadrature error versus convergence (variational) error: criteria for basis-set reduction (particularly in the sequential adiabatic reduction approach), and questions of boundary conditions and ranges that may not be so obvious in DVRs.

A. Boundary Conditions and Ranges, and Symmetries

In general, all DVRs are defined in terms of some basis set that spans some range and satisfies certain boundary conditions. The underlying basis for the DVR should be appropriate to the problem and to the operators being used. As discussed below, the use of DVRs corresponding to bases satisfying the wrong boundary conditions can lead to substantial and persistent quadrature errors.

Since DVRs are useful primarily for multidimensional problems in which solutions spanning a rather large region of space are desired, assuring that the DVRs span the appropriate range with an appropriate density of points is a major concern. In particular, if the range of interest in one coordinate varies substantially with the value of another coordinate, then the range of the DVR basis must span the entire range. One way of assuring this is to find the minimum of the potential for one coordinate (say, x) as the others (say, y and z) are varied. This defines a one-dimensional "global minimum" potential for x V_g (x). The same can be done for the other coordinates. These "global minimum" potentials define the ranges required for each coordinate in order to represent the system up to given energies, and DVRs can be chosen in each coordinate to satisfy this. These "global minimum" potentials are also the potentials of choice in defining potential optimized DVRs (PODVRs) since the PODVR points will then span the appropriate range.

When convergence of DVR bases is checked, both the density of DVR points (which affects the quality of the quadratures) and the range of the DVR points (if it can vary) must be considered separately.

B. Quadrature Error and Variational Error

Since DVRs are closely related to numeric quadratures and contain inherent quadrature approximations, their accuracy depends on the size of the DVR basis (for a given range) in two ways: (1) the accuracy of the quadrature increases (the quadrature error decreases) as the basis is increased and (2) the variational error due to the finite basis size also decreases. The variational error is always positive; that is, the convergence to eigenvalues of the Hamiltonian is from above. However, the quadrature error can be of either sign, depending on the potential and the type and size of the DVR, but should decrease in magnitude as the basis size is increased.

We have often observed that the quadrature error causes *negative* errors in the calculated energies. (The sign may be due to the common existence of regions of large repulsive nonpolynomial potentials.) If this is the case, then one often observes first increases in the energy levels with basis size as the quadrature error decreases more rapidly than the variational error, then for some range of basis size, increases in the basis size cause changes of comparable magnitude in the quadrature error and the variational error, leading to very stable, but not quite accurate, results, and finally the quadratures become very accurate and the reduction of variational error causes convergence from above for very large basis size.

Another tempting, but questionable, practice is to replace infinite integrals with finite quadrature results. Kinetic energy and effective potential terms often contain repulsive singularities; the repulsive effective potentials from angular momentum are the most common. Since these singularities exclude the wavefunctions in any case, one is tempted to use a DVR for which the quadrature points miss the singularity and thus ignore the problem. This may lead to *very* slow convergence and not necessarily to the correct answer. The best way to treat such situations is to use basis functions that have the correct boundary conditions and in which the singular operators are treated exactly.

The most common example is the singularity of the angular momentum operator in polar coordinates at $\theta = 0, \pi$. The matrix elements of the angular momentum operator are finite in a basis of spherical harmonics, containing associated Legendre polynomials. If only Legendre polynomial ($m = 0$) basis functions are used, the exact matrix elements of the angular momentum operator are infinite. However, because a Legendre polynomial DVR has no points at $0, \pi$, the quadrature approximations to the integrals are always finite, that is, the quadrature error is infinite! If the wavefunction amplitude is very small near the singularities, then the error in energy may not be large, but it will be persistent; For example, $m = 1$, the eigenvalues of the j^2 operator converge very slowly as the Legendre DVR basis size is increased. Thus it may be better to forego the pleasures of a DVR in these cases, and use (usually nondirect product) variational bases for these degrees of freedom. The use of both the DVR and nondirect product bases were discussed above.

C. Truncation of Primitive DVRs

Another, perhaps more subtle, problem has to do with the truncation of primitive DVRs in multidimensional problems. In using direct product DVRs in several dimensions, there are always DVR points located in regions of space where the wavefunctions of interest will be very small; regions of very large repulsive potentials are the most common. It is tempting to eliminate such points from the DVR basis in order to reduce both the size of the basis and the spectral range of the Hamiltonian operator. However, DVR basis functions are not truly localized, and kinetic energy operators in DVR bases give long-range coupling. Thus, as indicated above, it is best *not* to eliminate DVR points themselves, but to permit all DVR points to mix in the eigenfunctions (at a given level) and then to eliminate high-energy functions if desired to reduce the basis size. This means that in sequential reduction procedures, an energy criterion alone should not be used to eliminate functions at intermediate levels, but a minimum number of functions should be kept at each DVR point being considered explicitly in the basis. In the case of the sinc function DVRs however, explicit truncation must be made since the basis is always infinite.

V. SUMMARY AND CONCLUSIONS

The DVR has had an enormous impact on the way we calculate vibrational spectra, rate constants, state-to-state transition probabilities, and other properties that characterize the motion of atoms in or between molecules. To calculate spectra, rate constants, and other parameters, one almost always proceeds by choosing basis functions, calculating integrals, and solving a linear algebra problem. The most important and the most striking advantage of the DVR is that it drastically simplifies the calculation of integrals—in the DVR, matrix elements of functions of coordinates do not need to be calculated at all because matrix representations of functions are diagonal and the diagonal elements are simply values of the function. Because the DVR obviates the need to calculate integrals, it significantly simplifies most dynamical calculations.

It is easy to construct a DVR Hamiltonian matrix for a one-dimensional problem. There is no need to master the details of Gauss quadrature or finite-difference methods to use the DVR. To construct a DVR Hamiltonian matrix, one must merely (1) choose a VBR basis set, $\theta_k(z)$ [e.g., harmonic oscillator functions with $z = x - x_e$ or Legendre functions with $z = \cos(\theta)$]; (2) diagonalize z in the VBR basis set; (3) form the potential matrix by building a diagonal matrix whose diagonal elements are $V(z_\alpha)$, where the z_α are the eigenvalues of the VBR z matrix; and (4) form the kinetic energy matrix by transforming the VBR kinetic energy matrix with the eigenvectors of the z matrix. It is easy! The most straightforward way to handle multidimensional problems is with direct product DVRs.

To use the DVR is one thing; to understand why it works is another. The DVR Hamiltonian matrix is *not* unitarily equivalent to the VBR Hamiltonian matrix. VBR and DVR eigenvalues are not equal. Instead, the DVR Hamiltonian matrix is unitarily equivalent to the FBR Hamiltonian matrix. FBR and DVR eigenvalues are equal. Since one would like to have the VBR eigenvalues, but the advantages of the DVR enables one to compute DVR eigenvalues easily, one wants to know why the DVR and VBR eigenvalues are different and the extent of this difference. DVR and VBR eigenvalues are different simply because the FBR and the VBR are not identical. The FBR may be thought of in two ways. If the VBR basis functions are classical polynomial functions (e.g., simple harmonic oscillator functions), the FBR may be regarded as the representation in which potential matrix elements are computed with a Gauss quadrature with as many quadrature points as basis functions. The FBR may also be considered as the representation obtained by replacing matrix representations of products with products of matrices. The difference between DVR (or FBR) and VBR eigenvalues becomes smaller and smaller as the size of the basis is increased. In general, near the top of the spectrum a few of the DVR eigenvalues will be significantly in error, but most eigenvalues are essentially exact.

The DVR is simply an alternative representation. A matrix representation is obtained from a set of basis functions. Like any representation, the DVR is associated with a set of basis functions. The DVR basis functions are linear combinations of the VBR basis functions, chosen so that the DVR functions diagonalize the coordinate matrix. The basis functions are localized about the eigenvalues of the coordinate matrix, the "DVR points."

The most obvious and striking advantage of the DVR is that it eliminates the calculation of integrals. It has other advantages:

1. The use of the DVR facilitates the construction of good contracted basis functions. For example, for a triatomic molecule, one can devise an excellent basis using the sequential diagonalization truncation method. Although it is possible to use other (non-DVR) functions to build a contraction schemes, it is often best to form different functions of $q_2, q_3, \ldots$ for each DVR function localized about a q_1 point.

2. Because DVR functions are localized, it is sometimes possible to discard DVR functions localized about points in configuration space for which the potential is high (see, however, the discussion in Ref. 46).

3. The DVR facilitates the use of iterative methods (e.g., Chebyshev expansions of the time evolution operator or the Lanczos method to calculate spectra) because it reduces the number of non-zero terms in the Hamiltonian matrix and increases the fraction of diagonal terms.

One potential disadvantage of the DVR is that it works best if the VBR basis functions are classical polynomial functions, but it might not always be efficient to choose such functions. The accuracy of the DVR eigenvalues is equivalent to the accuracy of FBR eigenvalues; this accuracy is, in turn, determined by the accuracy of the product approximation used to construct the FBR potential matrix. If one uses N classical polynomial functions as a basis, then FBR matrix elements $\langle \theta_k(z)|z^d|\theta_{k'}(z)\rangle$ with $k = 0, 1, \ldots, N$ are exact if $k + k' + d \le 2N + 1$. If one uses VBR functions that are not classical polynomial functions, one will increase the number of FBR matrix elements that are not exact. Nonetheless, it is often worth accepting this disadvantage in order to decrease the dimension of the matrices. If classical polynomial basis functions are poor in the sense that many of them are required to represent the wavefunctions one wishes to calculate, it is better to use basis functions adapted to the potential. For examploe, rather than using Legendre functions and the associated DVR for a bending angle, it is often better to use eigenfunctions of the one-dimensional bending problem and a PODVR.

As is the case with any representation, it is useful to have basis functions that transform as irreducible representations of the appropriate symmetry group. Two ways of obtaining symmetrized DVR functions have been described in this chapter; one can obtain symmetrized DVR functions by taking either (1) linear combinations of symmetrized VBR functions or (2) symmetrized linear combinations of DVR functions obtained from VBR functions that do not transform like irreducible representations.

Despite the benefits of the DVR, it is sometimes better to use the FBR. This is true, for example, if (multidimensional) nondirect product VBR functions are ideal basis functions. In this case it is difficult to devise a good DVR, and one is better off using direct product or product quadratures.

The DVR will continue to dominate the way theorists calculate spectra, photodissociation cross sections, rate constants, and so on. Earlier theoretical methods were mostly based on models for which matrix elements could be computed analytically. Increasingly at least for small molecules, it is possible to use realistic but complicated potential functions or interpolations for which matrix elements cannot be computed analytically. The DVR enables one to use such potentials without needing to compute difficult integrals. New methods that will be developed to couple exact quantum calculations and approximate classical or semiclassical approaches to handle more complex problems will also surely exploit the advantages of the DVR.

ACKNOWLEDGMENTS

JCL acknowledges partial support from the NSF (CHE-9877086) and the DOE (DEFG02-87ER13679). TC is grateful for the support of NSERC.

References

1. C. Lanczos, *Applied Analysis*, Prentice-Hall, Engelwood Cliffs, NJ, 1956.

2. L. Fox, in *Methods of Numerical Approximation*, D. Handscomb, ed., Pergamon Press, New York, 1996, p. 27.

3. L. Fox and I. B. Parker, *Chebyshev Polynomials in Numerical Analysis*, Oxford Univ. Press, New York, 1968

4. D. O. Harris, G. G. Engerholm, and W. D. Gwinn, *J. Chem. Phys.* **43**, 151 (1965).

5. A. S. Dickinson and P. R. Certain, *J. Chem. Phys.* **49**, 4209 (1968).

6. J. V. Lill, G. A. Parker, and J. C. Light, *Chem. Phys. Lett.* **89**, 483 (1982).

7. J. V. Lill, G. A. Parker, and J. C. Light, *J. Chem. Phys.* **85**, 900 (1986).

8. R. W. Heather and J. C. Light, *J. Chem. Phys.* **79**, 147 (1983).

9. B. Shizgal and R. Blackmore, *J. Comp. Phys.* **55**, 313 (1984).

10. R. Blackmore and B. Shizgal, *Phys. Rev. A* **31**, 1855 (1985).

11. Z. Bacic and J. C. Light, *J. Chem. Phys.* **85**, 4594 (1986).

12. Z. Bacic and J. C. Light, *J. Chem. Phys.* **86**, 3065 (1987).

13. J. C. Light, and Z. Bacic, *J. Chem. Phys.* **87**, 4008 (1987).

14. Z. Bacic, R. M. Whitnell, D. Brown, and J. C. Light, *Comp. Phys. Commun.* **51**, 35 (1988).

15. R. M. Whitnell and J. C. Light, *J. Chem. Phys.* **89**, 3674 (1988).

16. J. C. Light, R. M. Whitnell, T. J. Park, and S. E. Choi, in *Supercomputer Algorithms for Reactivity, Dynamics and Kinetics of Small Molecules*, A. Lagana, ed., Kluwer Dordrecht, 1989, Vol. 277, pp. 187–214, NATO ASI Series C.

17. D. E. Manolopoulos and R. E. Wyatt, *Chem. Phys. Lett.* **152**, 23 (1988).
18. J. Echave and D. C. Clary, *Chem. Phys. Lett.* **190**, 225 (1992).
19. H. Wei and T. Carrington Jr., *J. Chem. Phys.* **97**, 3029 (1992).
20. J. T. Muckerman, *Chem. Phys. Lett.* **173**, 200 (1990).
21. D. T. Colbert and W. H. Miller, *J. Chem. Phys.* **96**, 1982 (1992).
22. V. Szalay, *J. Chem. Phys.* **105**, 6940 (1996).
23. Z. Bacic and J. C. Light, *Annu. Rev. Phys. Chem.* **40**, 469 (1989).
24. J. C. Light, in *Time Dependent Quantum Molecular Dynamics: Experiments and Theory*, J. Broeckhove and L. Lathouwers, eds., Plenum press, New York, 1992, pp. 185–199, NATO ARW 019/92.
25. J. C. Light, I. P. Hamilton, and J. V. Lill, *J. Chem. Phys.* **82**, 1400 (1985).
26. P. Dennery and A. Krzywicki, *Mathematics for Physicists*, Harper and Row, New York, 1967.
27. M. Abramowitz and I. Stegun, *Handbook of Mathematical Functions*, Vol. NBS Applied Mathematics Series, 55 of *NBS Applied Mathematics Series*, U.S. Government Printing Office, Washington, DC, 1964.
28. J. Stoer and R. Bulirsch, *Introduction to Numerical Analysis*, Springer-Verlag, New York, 1980.
29. V. Szalay, *J. Chem. Phys.* **99**, 1978 (1993).
30. H. Wei, *J. Chem. Phys.* **106**, 6885 (1997).
31. G. C. Corey and D. Lemoine, *J. Chem. Phys.* **97**, 4115 (1992).
32. C. Leforestier, *J. Chem. Phys.* **101**, 7357 (1994).
33. G. C. Corey and J. W. Tromp, *J. Chem. Phys.* **103**, 1812 (1995).
34. O. A. Sharafeddin and J. C. Light, *J. Chem. Phys.* **102**, 3622 (1995).
35. M. J. Bramley, W. Tromp, T. Carrington, Jr., and G. C. Corey, *J. Chem. Phys.* **100**, 6175 (1994).
36. H. Wei and T. Carrington, Jr., *J. Chem. Phys.* **101**, 1343 (1994).
37. A. McNichols and T. Carrington, Jr., *Chem. Phys. Lett.* **202**, 464 (1993).
38. J. Tennyson and J. R. Henderson, *J. Chem. Phys.* **91**, 3815 (1989).
39. S. E. Choi and J. C. Light, *J. Chem. Phys.* **97**, 7031 (1992).
40. S. Althorpe and D. C. Clary, *J. Chem. Phys.* **101**, 3603 (1994).
41. S. C. Althorpe and D. C. Clary, *J. Chem. Phys.* **102**, 4390 (1995).
42. H. Chen, S. Liu, and J. C. Light, *J. Chem. Phys.* **110**, 168 (1999).
43. J. M. Bowman and B. Gazdy, *J. Chem. Phys.* **94**, 454 (1991).
44. B. N. Parlett, *The Symmetric Eigenvalue Problem*, Prentice-Hall, Englewood Cliffs, NJ, 1980.
45. R. Kosloff, *J. Phys. Chem.* **92**, 2087 (1988).
46. M. J. Bramley and T. Carrington, Jr., *J. Chem. Phys.* **99**, 8519 (1993).
47. R. A. Friesner, R. E. Wyatt, C. Hempel, and B. Criner, *J. Comp. Phys.* **64**, 220 (1986).
48. U. Manthe and H. Koppel, *J. Chem. Phys.* **93**, 345 (1990).
49. D. Gotlieb and S. Orszag, *Analysis of Spectral Methods: Theory and Applications*, SIAM, Philadelphia, 1977.

50. J. K. Cullum and R. A. Willoughby, *Lanczos Algorithms for Large Symmetric Eigenvalue Computations,* Birkhaeuser, Boston, 1985.

51. R. Kosloff, in *Dynamics of Molecules and Chemical Reactions,* R. E. Wyatt and J. Z. H. Zhang, eds., Marcel Dekker, New York, 1996, Chapter 5, p. 185.

52. J. Dai and J. C. Light, *J. Chem. Phys.* **107**, 8432 (1997).

53. E. R. Davidson, *J. Comp. Phys.* **17**, 87 (1975).

54. X. T. Wu and E. F. Hayes, *J. Chem. Phys.* **107**, 2705 (1997).

55. P. P. Korambath, X. T. Wu, and E. F. Hayes, J. Phys. Chem. **100**, 6116 (1996).

56. R. A. Friesner, J. A. Bentley, M. Menou, and C. Leforestier, *J. Chem. Phys.* **99**, 324 (1993).

57. M. J. Bramley and T. Carrington, Jr., *J. Chem. Phys.* **101**, 8494 (1994).

58. J. Antikainen, R. A. Friesner, and C. Leforestier, *J. Chem. Phys.* **102**, 1270 (1995).

ABOVE AND BELOW THE WANNIER THRESHOLD

ARLENE M. LOUGHAN

*Department of Applied Mathematics and Theoretical Physics,
The Queen's University of Belfast, Belfast, Northern Ireland*

CONTENTS

Advances in Chemical Physics, Volume 114, Edited by I. Prigogine and Stuart A. Rice.
ISBN 0-471-39267-7   © 2000 John Wiley & Sons, Inc.

I. WANNIER THRESHOLD IONIZATION

A. Introduction

The problem of three particles linked by inverse square central forces is known mainly from celestial mechanics as the "problem of three bodies." One of the most basic atomic processes that occurs in many natural and laboratory plasmas is the three-body ionization problem

$$e^- + A^{(q)+} \rightarrow e^- + e^- + A^{(q+1)+}$$

of an atom $(q = 0)$ or ion $(q > 0)$ by electron impact where two of the three residual bodies are identical. The minimum energy required by the incident electron for this process to occur is referred to as the *threshold energy*. The existence of such a cutoff point is indicative of the need for a quantum rather than classical mechanical description of this process. The theoretical approach adopted when considering such a problem is dictated mainly by the energy of the incident electron relative to the threshold energy, whether it is high, intermediate, or, as is the case of the current study, low. This "low" energy region is referred to as the *near-threshold region*, which can be extended beyond energy impact values just above threshold to those just below threshold in which case a capture process ensues according to

$$e^- + A^{(q)+} \rightarrow (A^{(q-1)+})^*$$

At high energies, the single ionization of an electron from an atom is treated theoretically as a perturbative process, and the Born or Coulomb–Born approximations produce solutions to the nonseparable Schrödinger equation that agree well with experiment. At lower energies these approximations become invalid as they fail to model the delicate situation where all the interactions must be taken into account, peculiar to threshold processes. The low energy of the reaction products, coupled with the low residual charge of the ion, results in significant electron–electron long-range Coulombic interaction. Therefore a proper description of the correlated motion of the electrons is fundamental to our understanding of the threshold three-body problem. This correlated electron motion is pertinent for electron impact energies both above and below the ionization threshold. In this work we consider both these cases, namely threshold ionization and threshold capture processes.

The three-body ionization process has been the subject of extensive theoretical and experimental research for many years, and yet a complete solution still evades us. Of particular interest is the dependence of the cross section on the energy in excess of the threshold energy which is known as the *threshold law*. The most widely accepted threshold law is the famous $E^{1.127}$ Wannier law [2]. The Wannier theory, which amounts to a classical analysis of the equations of motion of the two escaping electrons for the simplest case of zero total angular momentum $L = 0$, is thoroughly discussed in Section I.B. There have been a number of theories developed over the years based on the fundamental Wannier principles [5,6,9], which have reproduced semiclassically and quantum-mechanically the Wannier law. However, not all of these approaches produced an absolute law, which is of the utmost importance for comparison with absolute experimental results. This requires a final-state wavefunction that obeys the boundary conditions for threshold ionization remaining finite in the limit as $E \rightarrow 0$ for a specific particle configuration. This wavefunction must also be a proper two-electron wavefunction despite the nonseparability of the Schrödinger equation and cannot simply be a product of two single electron Coulomb wavefunctions as in Bates et al. [10]. Such a well-behaved, nonsingular two-electron wavefunction can be used to determine the magnitude of the scattering amplitude and hence the absolute cross section. Another aspect of interest, regarding threshold escape, is a quantitative description of the angular and radial correlations of the escaping particles. This information is given by calculating the *triple-differential cross section* (TDCS), which gives the probability of each electron emerging at a particular angle with a particular energy. The theoretical description of the threshold ionization process is further complicated if angular momentum terms are retained in the original Schrödinger equation. The contribution made from P, D, and $L \neq 0$ states requires careful consideration of the symmetry properties of the wavefunction.

The Wannier theory deals specifically with two-electron breakup just above threshold and has become the primary focus for all discussions of processes involving two electrons near the threshold for double escape from the field of an ion. This includes the study of doubly excited resonant states formed as a result of electron capture where the energy of the incident electron lies just below the threshold energy. The connection between the above-threshold theory and below-threshold doubly excited resonant states was suggested many years ago [9,11–15], culminating in the potential ridge resonance theory proposed by Fano [3,4]. The theoretical and experimental arguments establishing such resonances as *Wannier ridge resonances* are given in Section II.B. The main objective in studying such states is to provide accurate resonance energy position values and lifetimes or, in summary, complex eigenenergies that will verify the accuracy of the theoretical methods applied to these threshold problems. A successful below-threshold extension of an above-threshold theory will reinforce the reliability of a theoretical approach. We begin our investigations with a look at the Wannier theory, which provides the fundamental physical principles on which this work is based.

B. Wannier Theory

The problem of the threshold law for electron impact single ionization of atoms was first considered by Wannier [2]. His work was inspired by the two-body threshold law of Wigner [1], who considered single-particle threshold escape. Wigner highlighted the difficulty in "accounting theoretically for detailed observations on reactions or transmutations where the reactants are passing through an intermediate state in which all constituents are tightly coupled." He proposed that it was a relatively easier problem to derive the threshold law. "In such a derivation the final escape products may be separated from the reaction proper which is confined to a small reaction zone. It is then shown that the threshold law arises from a feature of the escape process where this feature is amenable to calculation even when the reaction proper is not." In other words, both Wannier and Wigner realised that in order to determine a threshold power law, one need not solve the two-electron Schrödinger equation for the region of space in which the reactants are close together and interact strongly, namely, the reaction zone. This would be difficult mathematically, but fortunately it is not necessary to determine a threshold power law. We must consider only the events that occur when the reactants are far apart (i.e., the Coulomb zone), since it is believed that the threshold behavior of the particles is independent of reaction zone events. Wannier extended the two-body threshold law of Wigner to the more complicated three-body problem. He argued that in the Coulomb zone, where the kinetic energies of all particles are comparable with their Coulomb interactions, the de Broglie wavelengths of the escaping electrons are small compared with the variation of the Coulomb interactions, and hence the asymptotic motion for large separations of the ion and electrons is essentially classical. The basic Wigner proposal is, however, also applicable to semiclassical and quantum mechanical approaches.

1. The Coordinate System

We assume throughout this work, as did Wannier, that the residual ion is sufficiently heavy so that it may be considered at rest during the reaction. It will have a charge Z and be located at the origin of a space-fixed frame of reference. We require six independent coordinates to locate the two electrons in space at $\mathbf{r}_1$ and $\mathbf{r}_2$, where $\mathbf{r}_i = \mathbf{r}_i(r_i, \theta_i, \phi_i)(i = 1, 2)$ are the usual spherical coordinates. The Schrödinger equation for the two escaping electrons for $Z = 1$ is given by

$$\left(-\frac{1}{2}\nabla^2_{\mathbf{r}_1} - \frac{1}{2}\nabla^2_{\mathbf{r}_2} - \frac{1}{r_1} - \frac{1}{r_2} + \frac{1}{r_{12}} - E \right)\Psi(\mathbf{r}_1, \mathbf{r}_2) = 0 \qquad (1.1)$$

where

$$\nabla^2_{\mathbf{r}_i} = \frac{1}{r_i^2}\frac{\partial}{\partial r_i}\left(r_i^2 \frac{\partial}{\partial r_i} \right) - \frac{l^2(\hat{\mathbf{r}}_i)}{r_i^2} \qquad (1.2)$$

and the individual angular momentum terms are given by

$$l^2(\hat{\mathbf{r}}_i) = -\frac{1}{\sin\theta_i}\frac{\partial}{\partial\theta_i}\left(\sin\theta_i\frac{\partial}{\partial\theta_i} \right) - \frac{1}{\sin^2\theta_i}\frac{\partial^2}{\partial\phi_i^2} \qquad (1.3)$$

where clearly we have used atomic units. We come now to the crux of the near-threshold three-body problem. On consideration of (1.1) it becomes clear that the interelectronic Coulomb potential term, $1/r_{12}$, renders this equation nonseparable. For high-energy collision theory this does not pose a great problem since the electron's kinetic energy will dominate their Coulomb repulsion. However, for low-energy collisions where the incident electron has energy just above the threshold required to ionize the target (low excess energy), the kinetic energy of the escaping electrons will be small compared with their Coulomb interactions. Thus the $1/r_{12}$ term plays a significant role, which is further emphasised by the fact that for the single ionization of a neutral atom the residual charge is low. Wannier recognized the importance of the electron–electron interaction, and to effectively describe the electron correlated motion, he introduced the hyperspherical coordinates $(\rho, \alpha, \theta_{12})$ given by

$$\rho^2 = r_1^2 + r_2^2, \qquad \alpha = \tan^{-1}\left(\frac{r^2}{r^1}\right), \qquad \theta_{12} = \cos^{-1}(\hat{\mathbf{r}}_1 \cdot \hat{\mathbf{r}}_2) \qquad (1.4)$$

These coordinates define the shape of the three-body triangle converting the motion of the two electrons to that of a single particle in six-dimensional space. In this system ρ is the only coordinate with the dimension of length and it characterizes the root-mean-square size of the system; the other angles are dimensionless. The hyperspherical angle α is associated with the radial correlated motion of the two electrons, while the mutual polar angle θ_{12} obviously reflects the angular correlation between the motion of the electrons. In this new coordinates system Eq. (1.1) becomes

$$\left(h_0 - \frac{l^2(\hat{\mathbf{r}}_1)}{\rho^2\cos^2\alpha} - \frac{l^2(\hat{\mathbf{r}}_2)}{\rho^2\sin^2\alpha} + X^2 + \frac{2\zeta(\alpha,\theta_{12})}{\rho}\right)\Psi(\mathbf{r}_1,\mathbf{r}_2) = 0 \qquad (1.5)$$

where

$$h_0 = \frac{1}{\rho^5}\frac{\partial}{\partial\rho}\left(\rho^5\frac{\partial}{\partial\rho}\right) + \frac{1}{\rho^2\sin^2\alpha\cos^2\alpha}\frac{\partial}{\partial\alpha}\left(\sin^2\alpha\cos^2\alpha\frac{\partial}{\partial\alpha}\right) \qquad (1.6)$$

The potential energy of the system is $V = \zeta(\alpha,\theta_{12})/\rho$, where

$$\zeta(\alpha,\theta_{12}) = \frac{Z}{\cos\alpha} + \frac{Z}{\sin\alpha} - \frac{1}{\left(1 - \cos\theta_{12}\sin 2\alpha\right)^{1/2}} \qquad (1.7)$$

with $X^2 = 2E$ and for certain reactions such as $e^- + $ He and $\gamma + H^-$ and so on $Z = 1$. Since the hyperspherical radius is simply a scale factor, the potential energy can be represented by $\zeta(\alpha,\theta_{12})$ [16] (sometimes known as the effective charge), which is a function purely of the hyperspherical angles θ_{12} and α. The energy term E used here, and throughout, is the energy in excess of the threshold energy. A further three coordinates are required to give the orientation of the ion–electron–electron triangle in space. These are known as the *Euler angles*, and the wavefunction is given as a sum of products of functions of the Euler angles, and functions of the three internal $(\rho,\alpha,\theta_{12})$ coordinates. In effect, this is a separation of the angular momentum part of the wavefunction for a system in which the interaction is taken into account and the individual particles have no specified angular momentum. For the case of zero total angular momentum $L = 0$ the motion becomes confined to a fixed plane and the Euler angles become redundant. In Section IV, we consider the electron capture for $L = 0, 1$ and 2 states and employ the Euler angles adopted by Selles et al. [17]. Many choices of Euler angle are possible, but the original treatment of separating the wavefunction into products of functions was given by Breit [18]. We give here a brief account of the Breit–Euler angles, sufficient for future reference; a more complete discussion is given in Morse and Feshbach [19].

The Breit–Euler angles are given schematically in Fig. 1, from which it can be seen that the body-fixed axis Z_B is chosen along $\mathbf{r}_1$ so that the first two Euler angles giving the triangle orientation are the first-electron spherical angles, θ_1 and ϕ_1 in the laboratory frame; thus $\mathbf{r}_1$ is the new polar axis for $\mathbf{r}_2$. The third Euler angle ψ_B is an azimuthal angle around Z_B, and more precisely the angle between the $(Z_B,\mathbf{r}_2)$ and (Z_B,Z_L) planes. The

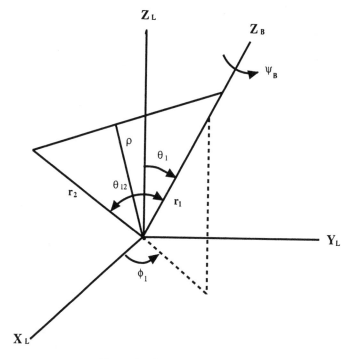

Figure 1. The Breit–Euler angles.

projection of the total angular momentum L onto the Z_L and Z_B axis corresponds to the quantum numbers M and K, respectively.

From the preceding discussion we seek a wavefunction of the form

$$\Psi_{LM} = \sum_K D^L_{LM}(\phi_1, \theta_1, \psi_B) f^{LM}_K(\rho, \alpha, \theta_{12}) \qquad (1.8)$$

where the Euler-angle factor in the wavefunction D^L_{MK}, which has eigenvalue B, is given by the equation

$$\left[(1 + \cot^2 \theta_1) \frac{\partial}{\partial \phi_1} + \frac{\partial^2}{\partial \theta_1^2} + \frac{1}{\sin^2 \theta_1} \frac{\partial^2}{\partial \psi_B^2} - \frac{2 \cot \theta_1}{\sin \theta_1} \frac{\partial^2}{\partial \phi_1 \partial \psi_B} + \cot \theta_1 \frac{\partial}{\partial \theta_1} + B \right]$$
$$\times D^L_{MK} = 0 \qquad (1.9)$$

[18]. The angles ϕ_1 and ψ_B are cyclic; therefore the factors involving these angles will be trigonometric functions, so that we can set

$$D^L_{MK} = \exp(iM\phi_1 + iK\psi_B) H^L_{MK}(\theta_1) \qquad (1.10)$$

where the remaining equation for $H_{MK}^L(\theta_1)$ is transformed into that for the hypergeometric equation by setting

$$H_{MK}^L(\theta_1) = \sin^d\left(\frac{1}{2}\theta_1\right)\cos^s\left(\frac{1}{2}\theta_1\right)F(\theta_1) \qquad (1.11)$$

where $d = |M - K|$ and $s = |M + K|$. Then, if we set $z = \frac{1}{2}(1 - \cos\theta_1)$, the equation for F becomes

$$z(1-z)\frac{d^2F}{dz^2} + [c - (a + b + 1)z]\frac{dF}{dz} - abF = 0 \qquad (1.12)$$

where

$$c = 1 + d; \qquad a + b = 1 + d + s; \qquad ab = (d+s)(d+s+2) - B \qquad (1.13)$$

The solution at the origin is finite and also at $\theta_1 = \pi$ if either a or b are taken as negative integers, $-v$. This yields the following finite solutions

$$D_{MK}^L = \exp(iM\phi_1 + iK\psi_B)\sin^d\left(\frac{1}{2}\theta - 1\right)\cos^s\left(\frac{1}{2}\theta_1\right)_2$$

$$\times F_1\left(-v, 1 + d + s + v | 1 + d | \sin^2\frac{1}{2}\theta_1\right) \qquad (1.14)$$

with the corresponding eigenvalue given by $B = L(L + 1)$. There will be $2L + 1$ different eigenfunctions for the given values of M and of L, for K can go from $-L$ to L. Thus there will be nine possible factors of D_{MK}^L when $L = 1$ since there are three possible relative orientations of the individual angular momenta which produce $L = 1$ and three orientations of L_Z. The properties of these eigenfunctions are discussed more fully in Morse and Feshback [19], but for the purposes of the present study we are concerned with the $M = 0P^0$ wave case only which corresponds to one particle having angular momentum either one greater or one less than the other with the projection on the Z_L axis being zero. The aim is to construct a wavefunction with the proper symmetry and a definite parity which will have the general form of Eq. (1.8), and in order to do so we need to deduce the radial equation for the function f_K^{LM}. On inversion $\theta_1 \to \pi - \theta_1$, $\phi_1 \to \pi + \phi_1$ and $\psi_B \to \pi - \psi_B$, so that the linear combination of D_{0K}^1 factors with $M = 0$ odd parity results in wavefunctions of the form

$$\Psi = f_0\cos\theta_1 + f_1\sin\theta_1\cos\psi_B \qquad (1.15)$$

We wish to remove ψ_B from the Euler factors to allow proper symmetrization of this wavefunction, which will inevitably result in the introduction of θ_{12} functions. The Pauli principle then leads to the relation

$$f_0 = \pm(\tilde{f}_0 \cos\theta_{12} - \tilde{f}_1 \sin\theta_{12}) \tag{1.16}$$

where the upper and lower signs refer to the singlet and triplet states, respectively (this convention will be utilised throughout this work) and where the interchanged function is $\tilde{f}(\rho, \alpha, \theta_{12}) = f(\rho, \pi/2 - \alpha, \theta_{12})$. The wavefunction may be made more symmetric by the substitution

$$f_0 = F_1 + \cos\theta_{12} F_2$$
$$f_1 = -\sin\theta_{12} F_2$$

which leads to the symmetrical form

$$\Psi = F_1 \cos\theta_1 + F_2 \cos\theta_2 \tag{1.17}$$

with $F_2 = \pm\tilde{F}_1$. The general form of the system Hamiltonian with Breit coordinates is given by

$$H = H_0 + A_1 + A_2 \tag{1.18}$$

where H_0 is the S-states Hamiltonian, given by

$$H_0 = h_0 + \frac{1}{\rho^2 \sin^2\alpha \cos^2\alpha} \frac{1}{\sin\theta_{12}} \frac{\partial}{\partial\theta_{12}} \left(\sin\theta_{12} \frac{\partial}{\partial\theta_{12}} \right) \tag{1.19}$$

and where A_1 and A_2 are differential operators in Breit–Euler angles and θ_{12}, given in Breit [18] with h_0 given by (1.6). This then leads, for P^0 states, letting $f = F_1$ with the interchanged function $\tilde{f}(r_1, r_2, \theta_{12}) = f(r_2, r_1, \theta_{12})$, to the equation

$$\left(H_0 + X^2 + \frac{2\zeta}{\rho} \right)f + \frac{2}{\rho^2 \cos^2\alpha} \left(\cot\theta_{12} \frac{\partial f}{\partial\theta_{12}} - f \right) \mp \frac{2}{\rho^2 \sin^2\alpha} \frac{1}{\sin\theta_{12}} \frac{\partial\tilde{f}}{\partial\theta_{12}} = 0 \tag{1.20}$$

In this system the Euler angles are not symmetrical with respect to the two electrons. As noted by Bhatia and Temkin [20], this is a theoretical disadvantage when compared to systems that are completely symmetric [20–22]. However, the system does allow a straightforward connection with experiments where θ_1 and ϕ_1 locate the first detector in the laboratory frame with respect to the known ion position and antisymmetrisation of the total two-electron wavefunction then introduces the second detector position θ_2 and ϕ_2. The unsymmetric nature of these Euler angles starts to cause significant problems in the construction of symmetrized wavefunctions for $L > 1$. We will discuss these problems in Section III, when we will consider the form of the $M = 0, D^e$ wavefunctions as well as P^0 in conjunction with an examination of the theoretical method of Selles et al. [17].

We return now to the early theory of Wannier [2] and complete our discussion of the mathematical description of the position of the particles in space by considering the specific stages of an escape process at threshold for the case of ionization. In view of the argument of Wigner [1], the six-dimensional configuration space is divided into three specific regions described below, and as shown in Fig. 2.

Reaction Zone. The first region is taken as the reaction zone with a boundary at $\rho = b$, where b may be assumed to be of the order of the magnitude of the Bohr radius a_0. It is within this zone that the finer details of the reaction occur, where the reactants are close and interacting strongly, and whose limit, according to Wannier, marks the transition from quantal to classical behavior.

Coulomb Zone. This second zone marks the first of the two asymptotic regions. In the Coulomb zone $k^2 \ll \zeta/\rho$, where the upper bound of this region is marked by the point at which the magnitude of the Coulomb potential energy of the system starts to become less than the combined kinetic energies. In the threshold limit of $E \to 0$ this upper limit tends

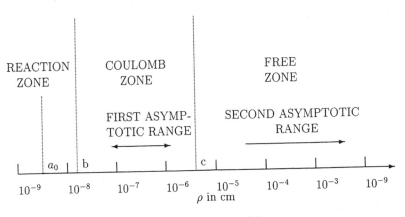

Figure 2. Representation of the zones.

to infinity, and this zone therefore dominates near threshold so that the threshold law is a feature of the wavefunctions in this zone.

Free Zone. This is the second asymptotic zone, in which the electrons move essentially independently of each other.

2. The Classical Wannier Arguments

The derivation of Wannier's threshold law involves a purely classical analysis of the three-body Hamiltonian. It is argued that the basic requirement for ionization, that the kinetic energy of either electron be larger than the negative of its potential energy, results in appreciable wavepacket formation. This argument is not significantly disturbed by the effect of the electron–electron interaction, and classical mechanics is employed. The equations of motion of the Hamiltonian (1.1) for zero angular momentum are given by

$$m\ddot{\rho} = m\rho\dot{\alpha}^2 + m\rho\sin^2\alpha\cos^2\alpha\dot{\theta}_{12}^2 - \left(\frac{Ze^2}{\rho^2}\right)\zeta(\alpha,\theta_{12})$$

$$(1.21)$$

$$\frac{d}{dt}\left(\frac{1}{2}m\rho^2\dot{\alpha}\right) = \frac{1}{4}m\rho^2\sin2\alpha\cos2\alpha\dot{\theta}_{12}^2 + \frac{Ze^2}{2\rho}\frac{\partial\zeta}{\partial\alpha}$$

$$(1.22)$$

$$\frac{d}{dt}(m\rho^2\sin^2\alpha\cos^2\alpha\dot{\theta}_{12}) = \frac{Ze^2}{\rho}\frac{\partial\zeta}{\partial\theta_{12}}$$

$$(1.23)$$

where $\dot{}$ signifies differentiation with respect to time and where the charge and mass of an electron are given by their standard notation. In this section Z is no longer taken as unity, so that the potential (1.7), on dividing through by Z, now reads

$$\zeta(\alpha,\theta_{12}) = \frac{1}{\cos\alpha} + \frac{1}{\sin\alpha} - \frac{1/Z}{(1-\cos\theta_{12}\sin2\alpha)^{1/2}}$$

$$(1.24)$$

The equations of motion are transformed into trajectory equations by the elimination of time and the substitution $\rho = b\exp(q)$ in (1.21)–(1.23). What follows is an interpretative study of the classical two-electron orbits where the first theorem to be proved is that almost all orbits end up asymptotically at $\alpha = 0$ or $\alpha = \pi/2$, which correspond to minimum values of the potential energy (1.24). Such configurations give rise to the "valley states," in which one electron is close to the nucleus and the other is far away so that all these orbits lead to single escape only. This can be seen by temporarily "switching off" the interaction between the two electrons. However, as discussed

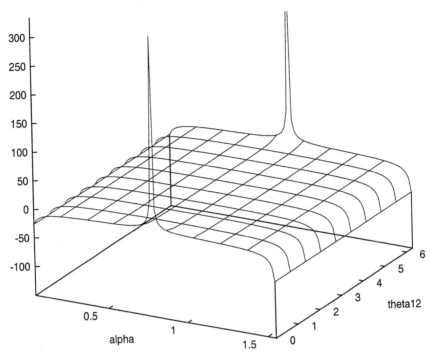

Figure 3. The Wannier ridge.

previously, the electron–electron interaction plays a significant role in threshold processes, although its inclusion makes the establishment of any theorem or law more difficult.

We come now to the central point of the Wannier theory, which was deduced by careful consideration of the potential surface $\zeta(\alpha, \theta_{12})$, which is given schematically in Fig. 3. In looking for orbits leading to double escape at zero energy, the point $\alpha = \pi/4, \theta_{12} = \pi$ results in the equations for the orbit vanishing. Thus, Wannier proposed that this point led to the symmetric escape of the two electrons in opposite directions, retaining this character for all time and resulting in double escape. This point coincides with the saddle point of the potential of the combined electron–nucleus and electron–electron interactions defined by $\zeta(\alpha, \theta_{12})$, in Eq. (1.24) and can be seen from (1.4) to correspond to the region $\mathbf{r}_1 = -\mathbf{r}_2$ in six-dimensional space. This saddle point is stable in θ_{12} and unstable in α. Connected with this saddle point is an unstable ridge given by $\alpha = \pi/4 \forall \theta_{12}$, known as the *Wannier ridge*, where double escape ensues for those orbits that approach the point $\alpha = \pi/4, \theta_{12} = \pi$ asymptotically.

The physical arguments leading to this hypothesis are now presented. They are based on the significant correlated electron motion, discussed previously, resulting directly from the fact that for a threshold process with low residual charge, the motion of the electrons is relatively slow and the three body Coulomb interactions are maximised. It can be seen from (1.7) that the point $\theta_{12} = \pi$ minimizes the interelectronic potential term and that mutual repulsion of the electron favors the π value for the mutual polar angle. The reason for the $\pi/4$ value of the hyperspherical angle is a little less obvious. As earlier, stated the α angle reflects the radial correlation between the two escaping electrons that share a total residual energy E. As the electrons try to escape from the attractive pull of the positive ion, with each electron moving in a field screened by the other, the total energy will be partitioned between the two. The main feature of this problem is that the mutual screening between the two electrons determines and is determined by the partition of the available energy between them. The way in which the available energy is partitioned into the kinetic energies of the electrons will determine their velocities and hence their distance from the nucleus. The configuration at any instant will, in turn, determine the subsequent energy exchange between the two electrons and hence to a new configuration. Thus there will be a sequence of such configurations in time, representing the escape process. This phenomenon of continual repartitioning of energy, known as *dynamical screening*, is particularly relevant to threshold ionization. If energies are not equal, $|\mathbf{r}_1| \neq |\mathbf{r}_2|$, one electron will be farther from the ion than the other and will therefore be screened more effectively from the nuclear field. This results in the outer electron gaining more energy at the expense of the inner electron and ultimately results in single escape. Therefore, it is imperative that the energy be equally shared between the two electrons, specifically, $|\mathbf{r}_1| = |\mathbf{r}_2|$, which implies $\alpha = \pi/4$ so that double escape occurs with the electrons emerging in opposite directions. This assumes classical behavior where $\mathbf{r}_i = \mathbf{v}_i t$ with $\mathbf{v}_i$ fixed so that (1.4) becomes

$$\tan\alpha = \left(\frac{E_2}{E_1}\right)^{1/2} \tag{1.25}$$

It is clear from the preceding remarks that the motion along the Wannier ridge is unstable with respect to the asymmetric stretch mode, namely, deviations of α from $\pi/4$. That the equal distribution of energy is more likely than an asymmetric distribution is exactly the opposite the usual behavior at energies high above threshold.

Regarding the physical significance of the point $\alpha = \pi/4, \theta_{12} = \pi$ the potential $\zeta(\alpha, \theta_{12})$ is expanded as a two-dimensional Taylor series according to

$$
\zeta(\alpha, \theta_{12}) = 2\sqrt{2} - \frac{1}{2}\sqrt{2}\frac{1}{Z} + \frac{1}{2}\left(6\sqrt{2} - \frac{1}{2}\sqrt{2}\frac{1}{Z}\right)\left(\alpha - \frac{\pi}{4}\right)^2
$$
$$
- \frac{1}{2}\frac{1}{8}\sqrt{2}\frac{1}{Z}(\theta_{12} - \pi)^2 \tag{1.26}
$$

It can be seen from Fig. 3 that the Wannier ridge is relatively flat over a considerable range of α and θ_{12}, making this expansion valid over a relatively wide range. Employing the linearized potential (1.26) to the transformed equations of motion [Eqs. (17) and (18) or Wannier [2] where the independent variable is given by $r = b\exp(q)$], we get the following second-order differential equations:

$$
\alpha'' + \frac{1}{2}\alpha' - \frac{12Z - 1}{8Z - 2}\left(\alpha - \frac{\pi}{4}\right) = 0 \tag{1.27}
$$
$$
\theta_{12}'' + \frac{1}{2}\theta_{12}' + \frac{1}{8Z - 2}(\theta_{12} - \pi) = 0 \tag{1.28}
$$

where the prime sign ($'$) denotes differentiation with respect to q. The asymptotic motion in α is separated from the one in θ_{12} with the general solutions to (1.27) and (1.28) given respectively by

$$
\alpha - \frac{\pi}{4} = \exp(-\frac{1}{4}q)\left(C_1\exp\left(-\frac{1}{2}\mu_1 q\right) + C_2\exp\left(\frac{1}{2}\mu_1 a\right)\right) \tag{1.29}
$$
$$
\theta_{12} - \pi = C_3\exp(-\frac{1}{4}q)\cos\left(\frac{1}{2}\mu_2 q + C_4\right) \tag{1.30}
$$

Here we have set for brevity

$$
\mu_1 = \frac{1}{2}\left[\frac{(100Z - 9)}{(4Z - 1)}\right]^{1/2} \tag{1.31}
$$
$$
\mu_2 = \frac{1}{2}\left[\frac{(9 - 4Z)}{(4Z - 1)}\right]^{1/2} \tag{1.32}
$$

and C_i ($i = 1$–4) are constants. All trajectories converge in θ_{12} to $\theta_{12} = \pi$, whereas in α, there are both converging and diverging trajectories with respect to $\alpha = \pi/4$. The former always leads to ionization because they reach $\alpha = \pi/4$ as $\rho \to \infty$, i.e., $r_1 = r_2 = \infty$. It is clear, from (1.29), that the solution will diverge from the ridge $\alpha = \pi/4$ unless $C_2 = 0$, so that the total set of orbits leading to double escape, for zero energy, is given by this one constraint. However, another feature in the Wannier theory has the consequence that at any finite energy, the diverging orbits can also lead to double escape. This is because the orbits given above only extend up to the limits of the Coulomb zone beyond which the electrons are essentially free, so that any orbit reaching this boundary in the Wannier direction will lead to double escape. Therefore the range of C_2 is now extended beyond zero and the way in which this widening takes place is central to the Wannier law and is, of course, energy-dependent. It is important to note, for future reference, that this concept of trajectory bundles diverging smoothly from the ridge, or converging to it, implies a Jeffreys–Wentzel–Kramers–Brillouin (JWKB)-like assumption that all relevant parameters vary smoothly along the ridge. The limitation imposed on C_2 is

$$(C_2)_{max} \propto E^{(1/2)\mu_1 - (1/2)} \tag{1.33}$$

where the form of the exponent should be noted. The main difficulty encountered in the development of a threshold law from this classical analysis arises when one attempts to apply Wigner's two-body method without modification to the three-body problem. The two-body wavefunction can be fully specified but the three-body wavefunction is restricted by continuity requirements at the reaction/Coulomb zone boundary, which demands a closer knowledge of the ionization process. This is somewhat contradictory to the original assumption that detailed knowledge of the reaction zone processes was not necessary. In order to overcome this problem, Wannier made an ergodic assumption for all redundant degrees of freedom; that is, he assumed that in the neighborhood of the reaction zone the density of representative points of the system in phase space is constant in the mean. We seek the energy dependence of the probability of ionization, which is clearly given by the ratio of the number of representative points leaving the reaction zone per unit time for double escape to the same number for all orbits at a fixed density. Wannier computes only the numerator of the ratio defined above, which is found to be limited in the same way as C_2 (1.33) and thus the Wannier law obtains

$$\sigma \propto E^{(1/2)\mu_1 - (1/2)} \tag{1.34}$$

where, from (1.31) for $Z = 1$, the Wannier exponent is 1.127 and for $Z = 2$ it is reduced to 1.056. In other words, the incident and ejected electron can simultaneously escape the attractive nuclear field only by moving in their joint phase space along classical trajectories within a certain narrow bundle. The width of this bundle shrinks as E decreases causing the ionization cross section to become proportional to E^n where n is Z-dependent. The reduction in size of the exponent with increasing Z is expected. If we return to our earlier discussion on the phenomenon of dynamic screening it is clear that the escape process is a continuous competition between the nuclear attraction and the interelectronic repulsion. The former is Z-dependent but the latter is not; hence the decreasing effect of the dynamic screening as Z increases accounts for the decreasing Wannier exponent. Therefore the reduced significance of the electron–electron interaction results in a tendency toward a linear law. One expects such a linear threshold law to hold for uncorrelated electrons. The ergodic assumption of uniform distribution, which Wannier argued to be reasonable beyond any doubt, has since been proven to within a 15% deviation [23]; however, as we shall see in Section II, it is an unnecessary axiom for the semiclassical derivation of this threshold law. A point to note at this stage is the failure of this classical analysis of the equations of motion to produce an absolute power law. Thus the need arose to find a method of solution of the two-electron Schrödinger equation producing a well-behaved, nonsingular two-electron final-state wavefunction from which absolute total and differential cross sections could be calculated. The semiclassical approach of Peterkop [5] and its uniformly amended version, given by Crothers [6], is just one of many approaches taken to reach this goal. The Peterkop–Crothers method forms the basis of this work and is discussed in detail in Section II. There are many other threshold laws to be found in the literature, but the most widely accepted, by both theorists and experimentalists, is the Wannier law. In the next section (I.C) we take a brief look at some of these alternative laws and the experimental evidence supporting the Wannier argument.

C. Experimental Evidence

Since the publication of Wannier's paper in 1953 a considerable amount of research has been conducted in this 'low energy' region of three-body atomic processes. Tables I and II summarize a few alternative laws presented over years [2,5,6,9,16,23–27]. The Wannier threshold law (1.34) has been verified classically [2], quantum-mechanically [9], semiclassically [5,6], and experimentally [23]. Other detailed aspects of the Wannier theory have also been considered with a shift in focus from the threshold law toward the angular and energy distribution functions, $P_\theta(\theta_{12})$ and $P_E(E_1)$, respectively.

It can be seen clearly from Table I that two threshold "extremes" exist, namely, the linear law and the $E^{3/2}$ law with the Wannier law nestled between the two. These results are indicative of the models used by the authors. In the close-coupling calculation of Burke and Taylor [26], the effective charges are taken to be Z and $Z - 1$, which amounts to a model in which one electron is fully screened by the other from the Coulomb

TABLE I

A Few of the Available Theoretical Results for the Near-Threshold Energy Dependence of Electron Impact Ionization Cross Sections[a]

Theory	Prediction
Wannier [2]	$\sigma \propto (E)^{1.127}$
Rau [9]	$\sigma \propto (E)^{1.127}$
Peterkop [5]	$\sigma \propto (E)^{1.127}$
Crothers [6]	$\sigma \propto (E)^{1.127}$
Rudge and Seaton [16]	$\sigma \propto E$
Temkin [24]	$\sigma \propto E(\ln E)^{-2}[1 - C\sin(\alpha \ln E + \mu)]$
Geltman [25]	$\sigma \propto E$
Burke and Taylor [26]	$\sigma \propto (E)^{3/2}$

[a] C and α are quantities related to the dipole moment created by the ion and the ejected electron; μ is constant of integration [24].

attraction of the nucleus. This is similar to the Born and Born–Oppenheimer approximations, and it is therefore not surprising that they both produce the "full screening" $E^{3/2}$ cross-sectional dependency. At the other end of the scale is the linear law produced by the assumption that no screening exists. The linear law derived by Rudge and Seaton [16] is based on the premiss that only a negative total potential energy in the asymptotic region contributes to the threshold cross section. This requires that the effective charges $z_i (i = 1, 2)$ satisfy

$$\frac{z_1}{k_1} + \frac{z_2}{k_2} = \frac{1}{k_1} + \frac{1}{k_2} - \frac{1}{|\mathbf{k}_1 - \mathbf{k}_2|} \qquad (1.35)$$

where $k_i (i = 1, 2)$ represents momentum which implies that z_1 and z_2 are both positive. A JWKB ansatz for the final-state wavefunction is adopted, and the values of the classical action S and the classical density P are taken as real. As we shall see later, this is not the most appropriate method for a system with such a small potential in the limit of double escape. The linear threshold law is subsequently deduced from an integral expression for the

TABLE II

A Few of the Available Experimental Results for the Near Threshold Energy Dependence of the Electron Impact Ionization Cross Sections[a]

Experiment	Prediction
Cvejanović and Read [23]	$\sigma \propto E^{1.131 \pm 0.019}$
Marchand and Cardinal [27]	$\sigma \propto E^{1.16 \pm 0.03}$

[a] The range of validity for these values varies.

scattering amplitude. It is therefore reasonable, from the arguments discussed above, to assume that when the correlated electron motion is described accurately, the threshold law exponent will lie between these two extreme values of 1 and $\frac{3}{2}$. We know from Section I.B.2 that this is the case with the effect of the correlation phenomenon of dynamic screening producing the 1.127 Wannier exponent.

The theory that poses the greatest opposition to Wannier's arguments is that of Temkin [24], which produced the modulated linear law given in Table I. The theory is based on the hypothesis that the most important configuration for double-electron escape at threshold is given by the region in which the two electrons have very different energies, namely, that $E_1 > 4E_2$; this is referred to as the *Coulomb dipole region* and is in total contradiction to the physical arguments presented previously. In the Coulomb dipole region one electron is believed to occupy a position very close to the nucleus experiencing its full Coulomb attraction while the other electron resides much farther away, influenced by the Coulomb dipole potential of the inner "core." As such the potential is taken as

$$V = -\frac{2}{r_1} + \frac{2}{r_1 + r_2}$$

$$\approx -\frac{2r_2}{r_1^2} \tag{1.36}$$

There is little evidence to support the Temkin dipole law. In fact, the unequal energy sharing test carried out by Selles et al. [28] shows no contribution from Coulomb dipole states. Absolute fits to the Wannier law and the Temkin law of measured total cross sections for the double photoionization of H$^-$ at threshold [29] were found to be equally good, with fewer parameters required for the Wannier fit. The experimental evidence of Fournier-Lagarde et al. [30] and the unequal energy sharing results of Selles et al. [17] supports the central assumption of the Wannier theory, where the region of space near the Wannier point ($\theta_{12} = \pi$) is supposed to be dominant at threshold. This is in contradiction with Temkin's theory, which leads only to a weak angular correlation of the electrons showing no evidence for a contribution from Coulomb dipole states. Also Temkin was not specific about $P_\theta(\theta_{12})$ or $P_E(E_1)$.

Experimentally, the first detailed information on the threshold exponent was reported by the group based at Manchester [23] using a coincidence time-of-flight experiment. The most detailed information about the electron impact ionization of atoms is available from $(e, 2e)$ electron coincidence spectroscopy, in which both the final-state electrons, following an ionization event, are detected in coincidence. In such experiments the flight times are given by $1/v_i (i = 1, 2)$, where the difference in time of arrival at the exit holes is $t = 1/v_1 - 1/v_2$. The energy and momentum of both final-state electrons, namely, E_1, E_2 and k_1, k_2, respectively, can be immediately deduced from this so that all the kinematics of each ionizing event is fully determined. On measuring the yield of electrons with a specific energy that is proportional to the differential cross section, it was found that, on average

$$\sigma \propto E^{1.131 \pm 0.019} \tag{1.37}$$

where the theoretical prediction of Wannier lies well within the limits of error. Equation (1.37) was found to be valid for the energy range $0.2 < E(\text{eV}) < 1.7$. The complete findings of this group are totally consistent with all the Wannier predictions. The probability distribution $P_\theta(\theta_{12})$ of angles θ_{12} has a maximum at $\theta_{12} = \pi$, with the width (FWHM) of the angular probability function varying as $E^{1/4}$ [31] and the distribution $P_E(E_1)$ of electron energies is found to be constant in line with the ergodic assumption. This work is supported by the results of Selles et al. [28], who obtain a departure from the double differential cross section (DDCS) $E^{0.127}$ threshold law and the $E^{1.127}$ integral cross-sectional behavior for energies $E \geq 2$ (where E is measured in electronvolts and also observe maximum points of calculated TDCS to be located on the $\theta_{12} = \pi$ line.

II. THEORY

A. The Semiclassical Method

In this section, we present a detailed summary of the semiclassical JWKB (Jeffreys–Wentzel–Kramers–Brillouin) theory, which forms the basis of this current research. We begin with a summary [5] of the JWKB treatment of the near-threshold electron impact ionization problem with an emphasis on the analytic techniques used and the early failings of this approach. We will then consider how this standard JWKB method has been adopted by Crothers [6] to produce a uniform semiclassical approximation which gave the first absolute theoretical total, partial, and differential cross sections for the three-body ionization problem.

1. Peterkop's JWKB method

The work of Peterkop is restricted to $L = 0$ for the final-state two electrons. The Schrödinger equation in hyperspherical coordinates is given by

$$\left(H_0 + X^2 + \frac{2\zeta(\alpha, \theta_{12})}{\rho} \right) \Psi(\mathbf{r}_1, \mathbf{r}_2) = 0 \tag{2.1}$$

where H_0 is given by (1.19) and where as usual the potential surface on which the particles are moving is given by

$$\zeta(\alpha, \theta_{12}) = \frac{1}{\cos\alpha} + \frac{1}{\sin\alpha} - \frac{1}{(1 - \cos\theta_{12}\sin 2\alpha)^{1/2}} \tag{2.2}$$

for nuclear charge $Z = 1$. The JWKB method is valid for a system that has a large slowly varying potential. It can be seen clearly that for $\rho \to \infty$ the potential for this three-body system certainly is not large. Nevertheless, Peterkop proceeds by taking the three-dimensional JWKB ansatz, where the asymptotic form of the wavefunction is given by

$$\Psi = P^{1/2}\exp\left(\frac{iS}{\hbar}\right) \tag{2.3}$$

The action S and density P are given respectively by the Hamilton–Jacobi equation

$$\left(\frac{\partial S}{\partial\rho}\right)^2 + \frac{1}{\rho^2}\left(\frac{\partial S}{\partial\alpha}\right)^2 + \frac{4}{\rho^2\sin^2 2\alpha}\left(\frac{\partial S}{\partial\theta_{12}}\right)^2 = 2E + \frac{2\zeta(\alpha,\theta_{12})}{\rho} \tag{2.4}$$

and the continuity equation

$$D_0\left[P\frac{\partial S}{\partial\rho}\right] + \frac{1}{\rho^2}\left[D_1\left(P\frac{\partial S}{\partial\alpha}\right) + D_2\left(P\frac{\partial S}{\partial\theta_{12}}\right)\right] = 0 \tag{2.5}$$

with

$$\begin{aligned}
D_0 f &= \frac{1}{\rho^5}\frac{\partial}{\partial\rho}(\rho^5 f) \\
D_1 f &= \frac{1}{\sin^2 2\alpha}\frac{\partial}{\partial\alpha}(f\sin^2 2\alpha) \\
D_2 f &= \frac{4}{\sin^2 2\alpha\sin\theta_{12}}\frac{\partial}{\partial\theta_{12}}(f\sin\theta_{12}).
\end{aligned} \tag{2.6}$$

Following Wannier, Peterkop expands S and P about the most important region for the threshold escape of two electrons. specifically, at $\alpha = \pi/4, \theta_{12} = \pi$, corresponding to the Wannier line $\mathbf{r}_1 = -\mathbf{r}_2$ such that

$$S = S_0(\rho) + \tfrac{1}{2}S_1(\rho)(\Delta\alpha)^2 + \tfrac{1}{8}S_2(\rho)(\Delta\theta_{12})^2 \tag{2.7}$$

$$P = P_0(\rho)\cdots \tag{2.8}$$

The deviation of the hyperspherical angle α from its Wannier ridge value of $\pi/4$ is denoted by $\Delta\alpha = \alpha - \pi/4$, and the deviation of the mutual polar angle θ_{12} from its Wannier ridge value of π is denoted by $\Delta\theta_{12} = \pi - \theta_{12}$. Similarly, the Taylor expansion of $\zeta(\alpha, \theta_{12})$ to second order is given by

$$\zeta(\alpha, \theta_{12}) = Z_0 + \tfrac{1}{2}Z_1(\Delta\alpha)^2 + \tfrac{1}{8}Z_2(\Delta\theta_{12})^2 \tag{2.9}$$

where

$$Z_0 = \frac{3}{\sqrt{2}} \tag{2.10}$$

which is the charge $\zeta(\alpha, \theta_{12})$ evaluated on the Wannier ridge and

$$Z_1 = \frac{11}{\sqrt{2}}, \qquad Z_2 = -\frac{1}{\sqrt{2}} \tag{2.11}$$

are the second-order derivatives of the charge with respect to the hyperspherical and mutual polar angles, respectively, evaluated on the ridge. Substituting (2.7) and (2.9) into (2.4), we have

$$\left[S_0' + \frac{1}{2}S_1'(\Delta\alpha)^2 + \frac{1}{8}S_2'(\Delta\theta_{12})^2 \right]^2 + \frac{1}{\rho^2}[S_1(\Delta\alpha)]^2$$
$$+ \frac{4}{\rho^2 \sin^2 2\alpha}\left[\frac{1}{4}S_2(\Delta\theta_{12}) \right]^2 = 2E + \frac{2Z_0}{\rho} + \frac{Z_1(\Delta\alpha)^2}{\rho}$$
$$+ \frac{Z_2(\Delta\theta_{12})^2}{4\rho} \tag{2.12}$$

where the prime (') denotes differentiation with respect ρ. Equating prowers of $\Delta\alpha$ and $\Delta\theta_{12}$ yields

$$\frac{dS_0}{d\rho} = \omega(\rho) \tag{2.13}$$

and

$$\omega\frac{dS_i}{d\rho} + \frac{S_i^2}{\rho^2} = \frac{Z_i}{\rho}, \qquad i = 1, 2 \tag{2.14}$$

where

$$\omega = \left(X^2 + \frac{2Z_0}{\rho} \right)^{1/2} \tag{2.15}$$

Equation (2.14) is a Riccati equation, which has the general form

$$y' = P + Qy + Ry^2 \tag{2.16}$$

where P, Q, and R are functions of the differential variable. Riccati equations are solved by removing the nonlinear term with the substitution

$$y = -\frac{u'}{Ru} \quad y' = -\frac{u''}{Ru} + \frac{u'^2}{Ru^2} + \frac{R'u'}{R^2u} \tag{2.17}$$

[32]. Thus Peterkop proceeds by taking

$$S_i = \rho^2 \omega \frac{1}{u_i} \frac{du_i}{d\rho}, \qquad i = 1, 2 \tag{2.18}$$

transforming (2.14) into

$$(\rho^2 X^2 + 2Z_{0\rho}) \frac{d^2 u_i}{d\rho^2} + (2\rho X^2 + 3Z_0) \frac{du_i}{d\rho} - \frac{Z_i u_i}{\rho} = 0 \tag{2.19}$$

thus removing the nonlinear S_i^2 term. It is worth noting at this stage that equation (2.19) has three regular singular points at $\rho = 0, \rho = \infty$, and $\rho = -Z_0/E$. This is to be compared with the result obtained in Section III.B, when the case of $L \neq 0$ is considered. The following substitution is then taken:

$$u_i = \rho^{m_i} F\left(-\frac{E_\rho}{Z_0}\right) \tag{2.20}$$

where m_i is some parameter and F is a function of $-E_\rho/Z_0$, yielding

$$-2E\left[-\left(\frac{E_\rho}{Z_0}\right)^2 + \left(-\frac{E_\rho}{Z_0}\right)\right]F'' - 2E\left[\left(2m_i + \frac{3}{2}\right) - (2m_i + 2)\left(-\frac{E_\rho}{Z_0}\right)\right]F$$
$$+ 2E\left[m_i(m_i + 1) + \frac{m_i}{2E_\rho}\left(2Z_0 m_i + Z_0 - \frac{Z_i}{m_i}\right)\right]F = 0$$

The $E_\rho \rightarrow 0$ limit, applied to the last term of (2.21), requires that the parameter m_i satisfy

$$2Z_0 m_i^2 + Z_0 m_i - Z_i = 0 \tag{2.22}$$

thus

$$m_i = -\frac{1}{4} \pm \frac{1}{4}\left(1 + \frac{8Z_i}{Z_0}\right)^{1/2} \tag{2.23}$$

and (2.21) can now be written as

$$\left(-\frac{E_\rho}{Z_0}\right)\left(1 - \left(-\frac{E_\rho}{Z_0}\right)\right)F'' + \left[\left(2m_i + \frac{3}{2}\right) - (2m_i + 2)\left(-\frac{E_\rho}{Z_0}\right)\right]F'$$
$$- m_i(m_i + 1)F = 0 \tag{2.24}$$

It can be seen from (2.23) that for $i = 1, 2$ there are two possible values of m_i. They are labeled m_{i1} and m_{i2}. Therefore, for each value of i, the full solution is a linear combination of the two solutions giving

$$u_i = C_{i1}u_{i1} + C_{i2}u_{i2}, \qquad i = 1, 2 \tag{2.25}$$

where $C_{ij}\, j = 1, 2$ are obviously arbitrary constants. It is clear from (2.24) that the function F is of the form of the Gauss hypergeometric function so that

$$u_{ij} = \rho_2^{m_{ij}} F_1\left(m_{ij}, m_{ij} + 1; 2m_{ij} + \frac{3}{2}; \frac{-E\rho}{Z_0}\right) \tag{2.26}$$

where

$$m_{i1} = -\tfrac{1}{4} - \tfrac{1}{2}\mu_i, \qquad m_{i2} = -\tfrac{1}{4} + \tfrac{1}{2}\mu_i \tag{2.27}$$

with

$$\mu_i = \frac{1}{2}\left(1 + \frac{8Z_i}{Z_0}\right)^{1/2} \tag{2.28}$$

It should be noted here that from (2.11) $Z_1 = 11/\sqrt{2}$ and $Z_2 = -1/\sqrt{2}$, which subsequently results in μ_1 being real and μ_2 being complex. Their respective values are given by

$$\mu_1 = \frac{1}{2}\sqrt{\frac{91}{3}}; \qquad \mu_2 = \frac{i}{2}\sqrt{\frac{5}{3}} \tag{2.29}$$

so that

$$m_{11} = -\frac{1}{4} - \frac{1}{4}\left(1 + \frac{8Z_1}{Z_0}\right)^{1/2} = -1.627$$

$$m_{12} = -\frac{1}{4} + \frac{1}{2}\left(\frac{1}{2}\sqrt{\frac{91}{3}}\right) = 1.127 \qquad (2.30)$$

$$m_{21} = -\frac{1}{4} - i\frac{1}{4}\sqrt{\frac{5}{3}}$$

where $m_{22} = m_{21}^*$. Clearly the Wannier exponent is given by $m_{12} = 1.127$. Since m_{2j} is complex $S_2(\rho)$ is a complex-valued function. However, for residual charge Z

$$\mu_1 = \frac{1}{2}\sqrt{\frac{100Z - 9}{4Z - 1}}, \qquad \mu_2 = \frac{1}{2}\sqrt{\frac{4Z - 9}{4Z - 1}} \qquad (2.31)$$

so that for $Z > 9/4$, μ_2 and subsequently S_2 becomes real. Thus Peterkop's analysis becomes invalid because of the effect of increasing residual charge. This abrupt cutoff point can also be seen clearly in the original Wannier analysis from equations (1.29)–(1.32), which surprisingly persists even if the theory is extended from second to fourth order [33]. The functions S_1 and S_2 are now fully specified. Equating zeroth-order powers of $(\Delta\alpha)$ and $(\Delta\theta_{12})$ in (2.12) yields

$$\frac{dS_0}{d\rho} = \omega(\rho) \qquad (2.32)$$

This is solved for S_0 by integrating by parts so that

$$S_0 = \rho\omega + Z_0\int\frac{1}{\rho\omega(\rho)}d\rho$$

which on a change of dependent variable gives

$$S_0 = \rho\omega - \int\rho\,d\omega \qquad (2.33)$$

Using (2.15), we can write (2.33) as

$$S_0 = \rho\omega - 2Z_0 \int \frac{1}{\omega^2 - X^2} d\omega$$

$$= \rho\omega + \frac{Z_0}{X} \int \left(\frac{1}{\omega + X} - \frac{1}{\omega - X} \right) d\omega$$

$$= \rho\omega + \frac{Z_0}{X} \ln \left(\frac{(\omega + X)^2}{\omega^2 - X^2} \right) + C$$

$$= \rho\omega + \frac{Z_0}{X} \ln \left(\frac{\rho(\omega + X)^2}{2Z_0} \right) \tag{2.34}$$

were the constant of integration is chosen such that $S_0 = (8Z_0\rho)^{1/2)}$ for $E = 0$. In a similar way, the continuity equation (2.5) for the density function P is solved to leading order. In other words, the density function P is expanded as

$$P = P_0 + \text{terms involving } ((\Delta\alpha)^2, (\Delta\theta_{12})) \tag{2.35}$$

so that Eq. (2.5) becomes

$$\frac{1}{\rho^5} \frac{\partial}{\partial\rho} \left(\rho^5 P_0 \frac{\partial S_0}{\partial\rho} \right) = 0$$

$$\frac{1}{2\rho^2} \frac{1}{\sin^2 2\alpha} \frac{\partial}{\partial\alpha} (\sin^2 2\alpha P_0 S_1) \frac{\partial}{\partial\alpha} (\Delta\alpha)^2 = 0 \tag{2.36}$$

$$\frac{1}{8\rho^2} \frac{4}{\sin^2 2\alpha} \frac{1}{\sin\theta_{12}} \frac{\partial}{\partial\theta_{12}} \left(\sin\theta_{12} P_0 S_2 \frac{\partial}{\partial\theta_{12}} (\Delta\theta_{12})^2 \right) = 0$$

Solving (2.36) for values of α and θ_{12} close to their Wannier ridge values gives

$$P_0 = \frac{C}{\rho^5 \omega(\Delta\alpha)(\Delta\theta_{12})^2} \tag{2.37}$$

It is noted in Peterkop [5] that the set of orbits described by $S(\rho)$ is given by

$$\frac{d\mathbf{r}_i}{dt} = \nabla_i S \tag{2.38}$$

which in hyperspherical coordinates reads

$$\frac{d\rho}{dt} = \frac{\partial\alpha}{\partial\rho}, \qquad \frac{d\alpha}{dt} = \frac{1}{\rho^2}\frac{\partial S}{\partial\alpha}, \qquad \frac{d\theta}{dt} = \frac{4}{\rho^2\sin^2 2\alpha}\frac{\partial S}{\partial\theta_{12}} \qquad (2.39)$$

By considering only the leading terms in the $S(\rho)$ expansion and eliminating time

$$\frac{d\alpha}{d\rho} = \frac{\Delta\alpha}{u_1}\frac{du_1}{d\rho}, \qquad \frac{d\theta_{12}}{d\rho} = \frac{\Delta\theta_{12}}{u_2}\frac{du_2}{d\rho} \qquad (2.40)$$

so that close to the ridge

$$\Delta\alpha = u_1(\rho), \qquad \Delta\theta_{12} = u_2(\rho) \qquad (2.41)$$

Therefore (2.37) now becomes

$$P_0 = \frac{C}{\rho^5 \omega u_1 u_2^2} \qquad (2.42)$$

Thus, Peterkop's semiclassical JWKB wavefunction is now fully described. Peterkop was able to extract the Wannier threshold law by considering the asymptotic behaviour of his semiclassical wavefunction. This requires the use of the analytic continuation formula of the hypergeometric function [34] in Eq. (2.26), which, in the limit as $E\rho \to \infty$, becomes

$$u_{ij} \simeq \gamma_{ij}\left(1 - \frac{Z_i}{2E\rho}\ln\frac{E\rho}{Z_0}\right) \qquad (2.43)$$

where

$$\gamma_{ij} = \left(\frac{Z_0}{E}\right)^{m_{ij}}\frac{\Gamma(2m_{ij} + \frac{3}{2})}{\Gamma(m_{ij} + \frac{3}{2})\Gamma(m_{ij} + 1)} \qquad (2.44)$$

where $\Gamma(x)$ denotes the gamma function [35]. Substituting (2.43) into (2.25) yields a constant u_i provided $C_{i1}\gamma_{i1} + C_{i2}\gamma_{i2} \neq 0$. Hence, the asymptotic form of $S(\rho)$ is

$$S(\rho) \simeq X\rho + \frac{1}{X}\left(Z_0 + \frac{1}{2}Z_1(\Delta\alpha)^2 + \frac{1}{8}Z_2(\Delta\theta_{12})^2\right)\ln\rho\cdots \qquad (2.45)$$

Having a fully specified wavefunction, analytically continued asymptotically, Peterkop introduces the normalization condition

$$P_0 \to \frac{1}{\rho^5} \quad \text{as} \quad \rho \to \infty \qquad (2.46)$$

which gives, in (2.37)

$$C = X(C_{11}\gamma_{11} + C_{12}\gamma_{12})(C_{21}\gamma_{21} + C_{22}\gamma_{22})^2 \qquad (2.47)$$

where solutions corresponding to spherically outgoing waves, specifically, the finite values of condensation points R_1 and R_2 for which $u_i(R_i) = 0$ in (2.41), are considered. It can be seen from (2.47), (2.44), and (2.30) that in the threshold limit of $E \to 0$ there is one dominant term:

$$C \simeq C_{12}E^{1-m_{12}} \qquad (2.48)$$

The threshold law can be determined by the use of a matching procedure, where the exact wavefunction is matched with an approximate one of known energy dependence. Peterkop applies the JWKB wavefunction at $\rho \geq r_0$, assuming it to be valid there, and introduces a matching coefficient f. Thus, in this region the wavefunction takes the form

$$\Psi = f P_0^{1/2} \exp\left(\frac{i}{\hbar}S\right) \qquad (2.49)$$

which, on employing (2.37) and (2.45), gives, in the limit as $\rho \to \infty$

$$\Psi \simeq \frac{f}{\rho^{5/2}} \exp\left(\frac{i}{\hbar}X\rho + \frac{iZ(\alpha, \theta_{12})}{\hbar X}\ln\rho\right) \qquad (2.50)$$

The matching procedure takes the wavefunction to be finite at a finite distance r_0 as $E \to 0$:

$$\Psi(r_0) \to \text{const} \quad \text{as} \quad E \to 0 \qquad (2.51)$$

which yields

$$|f|^2 \simeq \frac{\text{const}}{|P_0(\rho)|} \quad \text{as} \quad E \to 0 \qquad (2.52)$$

The differential cross section is defined as

$$\sigma_d = \frac{X}{k_0}|f|^2 \qquad (2.53)$$

so that the energy dependence of the differential cross section evaluated on the Wannier line is given by

$$\sigma_d \simeq E^{m_{12}-(1/2)} \qquad (2.54)$$

Integrating for the total cross section

$$\sigma_{\text{tot}} = 2\pi^2 \int \sigma_d(\alpha, \theta_{12}) \sin^2 2\alpha \sin \theta_{12} d\alpha d\theta_{12} \qquad (2.55)$$

and approximating σ_d with the model

$$\begin{aligned}
\sigma_d(\alpha, \theta_{12}) &= \sigma_d(\pi/4, \pi) \quad \text{for} \quad |\Delta\theta_{12}| \leq |\Delta\theta_{12}|_{\text{max}} \\
&= 0 \quad \text{for} \quad |\Delta\theta_{12}| > |\Delta\theta_{12}|_{\text{max}}
\end{aligned} \qquad (2.56)$$

where with the argument of Rau [9] that $|\Delta\theta_{12}|_{\text{max}} = \text{const} E^{1/4}$, gives

$$\sigma_{\text{tot}} \simeq E^{1/2}\sigma_d(\pi/4, \pi) \simeq E^{1.127} \qquad (2.57)$$

confirming the Wannier threshold law.

Peterkop succeeds in deriving the correct Wannier exponent but fails to derive a constant of proportionality for this threshold law. On closer consideration of his arguments, the reason for this failing, despite the fact that a fully specified final-state two electron wavefunction exists, becomes apparent. The problem lies in the form of P_0, which from (2.42) is proportional to the inverse of $u_2(\rho)$. The function $u_2(\rho)$ tends to zero on the Wannier ridge in the ionization limit of $E \to 0, \rho \to \infty$ irrespective of the order in which the double limit is imposed. Thus, $\Psi \to \infty$ on the Wannier ridge in the limits of threshold ionization. Consequently, the matching procedure was conducted at a finite hyperradius r_0 as opposed to infinity, where any matching procedure should be conducted. Therefore it is not surprising that absolute results were not obtained from this method.

Another wavefunction formulation was presented, in the same year, by Rau [9] in a quantum-mechanical realisation of Wannier's classical results. Again, solutions of the two-electron Schrödinger equation for $L = 0$ were sought with the final-state wavefunction taking the form

$$\Psi(\rho, \alpha, \theta_{12}) = \rho^{-5/2} \csc 2\alpha \phi(\rho, \alpha, \theta_{12}) \qquad (2.58)$$

Terms were expanded about the Wannier direction as in (2.9) and retained up to the first nontrivial order in $\Delta\alpha$ and $\Delta\theta_{12}$. The function $\phi(\rho, \alpha, \theta_{12})$ is written as

$$\phi(\rho, \alpha, \theta_{12}) = \exp\left(ic\rho^{1/2} + \frac{1}{2}iac(\Delta\alpha)^2\rho^{1/2} - bc\rho^{1/2}(\Delta\theta_{12})^2 \right)\chi(\rho)$$

$$(2.59)$$

which is valid in the Coulomb zone on the Wannier ridge. The angular dependence of χ has been dropped since the derivatives acting on this function are of the order of $1/\rho^2$, and in this treatment these higher-order terms are neglected. This approach leads to a wavefunction containing two arbitrary constants that can be determined only if a complete solution of reaction zone events is obtained. Therefore the determination of an absolute power law is not possible. Rau does, however, produce wavefunctions that are accurate in the Coulomb zone, and on following Wannier's suggestion that the energy dependence of the amplitude of the diverging solution as opposed to the converging solution is required, confirms the Wannier threshold law. Subsequent studies by Rau [36] have also produced angular and energy distribution functions, $P_\theta(\theta_{12})$ and $P_E(E_1)$, respectively in agreement with Wannier and experiment. The Wannier theory and its quantal formulation [9,36] have undergone further developments consisting in the generalization to other $(LS\pi)$ states [14,15,37,38]. Such derivations, extended beyond the $L = 0$ case, show that (2.57) is equally applicable to the escape of two electrons with $L => 0$ [37]. Employing the Breit–Euler angles discussed in Section I.B.1, Roth argues that the angular momentum terms having a factor $1/\rho^2$, as one would expect for a centrifugal potential, tend to zero more rapidly as $\rho \to \infty$ than do the $L = 0$ terms and will affect the wavefunction less. This analysis becomes increasingly more complicated as L increases. The theoretical approaches that we have so far discussed have come to be known collectively as the *Wannier–Peterkop–Rau* (WPR) theory.

2. Crothers' Uniform Semiclassical Approximation

In this section we describe in detail the derivation of Crothers' uniform semiclassical JWKB wavefunction for near threshold electron impact ionization. This derivation, although in many respects is similar to that of

Peterkop's, avoids any ergodic hypothesis, as seen in Section I.B.2, and ill-defined matching procedures as seen in Section III.A.1. Most importantly, Crothers' uniform semiclassical wavefunction does not break down in the region of configuration space and in the physical limits most pertinent for threshold ionization. The details of this method form the basis of our future analysis, and for convenience we refer to Crothers' [6] method as "method I" throughout this work. We begin, as in the previous section, with the $L = 0$ two-electron Schrödinger equation in hyperspherical coordinates,

$$
\left[\frac{1}{\rho^5} \frac{\partial}{\partial \rho} \left(\rho^5 \frac{\partial}{\partial \rho} \right) + \frac{1}{\rho^2 \sin^2\alpha \cos^2\alpha} \frac{\partial}{\partial \alpha} \left(\sin^2\alpha \cos^2\alpha \frac{\partial}{\partial \alpha} \right) \right.
$$
$$
\left. + \frac{1}{\rho^2 \sin^2\alpha \cos^2\alpha} \frac{1}{\sin\theta_{12}} \frac{\partial}{\partial \theta_{12}} \sin\theta_{12} \frac{\partial}{\partial \theta_{12}} + X^2 + \frac{2\zeta(\alpha, \theta_{12})}{\rho} \right] \Psi = 0
$$

$$(2.60)$$

This Schrödinger equation, as we shall see later, applies equally well to bound states as to the continuum; the only difference is the form of the energy. On realising that the potential of this two-electron problem is certainly not large for large values of ρ the JWKB ansatz given in (2.3) is not taken immediately. Instead, Crothers realises that a change of dependent variable is required in order to provide a uniform solution avoiding semiclassical breakdown, on the Wannier ridge, in the limit of double ionization. To use the Sturm–Liouville form for the final-state wavefunction

$$
\Psi_f^{-*} = \frac{\Phi}{\rho^{5/2} \sin\alpha \cos\alpha (\sin\theta_{12})^{1/2}}
$$

$$(2.61)$$

seems the natural solution. We have adopted the notation Ψ_f^{-*} since we seek a final-state ingoing wavefunction. The equation for Φ is then given by

$$
\left[\frac{\partial^2}{\partial \rho^2} + \frac{1}{\rho^2} \frac{\partial^2}{\partial \alpha^2} + \frac{1}{\rho^2 \sin^2\alpha \cos^2\alpha} \frac{\partial^2}{\partial \theta_{12}^2} + X^2 \right.
$$
$$
\left. + \frac{2\zeta}{\rho} + 4 + \frac{\frac{1}{4} + \csc^2(\Delta\theta_{12})/4}{\rho^2 \sin^2\alpha \cos^2\alpha} \right] \Phi = 0
$$

$$(2.62)$$

The pseudopotential $\csc^2(\Delta\theta_{12})$ term in (2.62) is certainly large for $\theta_{12} = 0$ or π increasing the validity of the JWKB ansatz at this stage. However, if we proceed as in the previous section by adopting the usual expanded form of S, given by (2.7), and solve the continuity equation for P_0, we find that

$$P_0 \propto \frac{1}{\Delta\alpha}$$
$$= \frac{1}{u_1}, \quad \text{classically}$$
$$\propto E^{m_{12}} \tag{2.63}$$

This will require the constant of proportionality in the original equation for P_0 to be given by

$$C \propto E^{-m_{12}} \tag{2.64}$$

producing a singularity in the wavefunction as $E \to 0$ for a given finite ρ value. Thus we must look elsewhere for a solution. The answer was to consider the single-particle system, in which the two electrons are represented in six dimensional hyperspherical coordinates, as having a precessing cusplike behavior in the region of the Wannier ridge. This cusplike motion is to be compared to that of a precessing top [39], where we take $\alpha = \pi/4$ as a natural barrier. Hence, we introduce a change of dependent variable, namely

$$\Psi_f^{-*} = \frac{x|\sin(\alpha - \pi/4)|^{1/2}}{\rho^{5/2}\sin\alpha\cos\alpha(\sin\theta_{12})^{1/2}} \tag{2.65}$$

resulting in x being given by

$$\left[\frac{\partial^2}{\partial\rho^2} + \frac{1}{\rho^2\sin|\Delta\alpha|}\frac{\partial}{\partial\alpha}\left(\sin|\Delta\alpha|\frac{\partial}{\partial\alpha}\right) + \frac{1}{\rho^2\sin^2\alpha\cos^2\alpha}\frac{\partial^2}{\partial\theta_{12}^2} + X^2 + \frac{2\zeta}{\rho}\right.$$
$$\left. + \frac{\frac{1}{4} + \csc^2(\Delta\theta_{12})/4}{\rho^2\sin^2\alpha\cos^2\alpha} - \frac{\csc^2|\Delta\alpha|}{4\rho^2}\right]x = 0 \tag{2.66}$$

We now have two pseudopotentials that are both large near the region most pertinent to threshold double escape, that is, the saddle of the potential for which $\Delta\theta_{12} = \Delta\alpha = 0$. This makes (2.66) more suited to a JWKB-type solution. It is interesting to note that the potential term in $\Delta\theta_{12}$ is attractive reflecting the stable nature of the θ_{12} coordinate while the potential in $\Delta\alpha$ is negative reflecting the metastable nature of the α coordinate. Cancellation of the two inverse-square potentials is limited to a region of zero measure in the $(\Delta\alpha, \Delta\theta_{12})$ plane. The $1/4\rho^2$ term is considered to be small compared to the

$\csc^2(\Delta\theta_{12})/4\rho^2$ term and so can be neglected. We are now in a much better position to apply the JWKB ansatz where, similarly to Peterkop, we take

$$x = P^{1/2} \exp\left(\frac{iS}{\hbar}\right) \tag{2.67}$$

On adopting atomic units, in which $\hbar = 1$, the Hamilton–Jacobi and continuity equations for the action S and density P are now given by

$$\left(\frac{\partial S}{\partial\rho}\right)^2 + \frac{1}{\rho^2}\left(\frac{\partial S}{\partial\alpha}\right)^2 + \frac{4}{\rho^2 \sin^2 2\alpha}\left(\frac{\partial S}{\partial\theta_{12}}\right)^2$$
$$= 2E + \frac{2\zeta(\alpha, \theta_{12})}{\rho} + \frac{\csc^2\theta_{12}}{4\rho^2 \sin^2\alpha \cos^2\alpha} - \frac{\csc^2(\Delta\alpha)}{4\rho^2} \tag{2.68}$$

and

$$D_0\left[P\frac{\partial S}{\partial\rho}\right] + \frac{1}{\rho^2}\left[D_1\left(P\frac{\partial S}{\partial\alpha}\right) + D_2\left(P\frac{\partial S}{\partial\theta_{12}}\right)\right] = 0 \tag{2.69}$$

with

$$D_0 = \frac{\partial}{\partial\rho}$$
$$D_1 = \frac{1}{\sin(\Delta\alpha)}\frac{\partial}{\partial\alpha}\sin(\Delta\alpha) \tag{2.70}$$
$$D_2 = \frac{4}{\sin^2 2\alpha}\frac{\partial}{\partial\theta_{12}}.$$

The action S is now generalized to

$$S = s_0 \ln|\Delta\alpha| + s_1 \ln(\Delta\theta_{12}) + S_0(\rho) + \frac{1}{2}S_1(\rho)(\Delta\alpha)^2 + \frac{1}{8}S_2(\rho)(\Delta\theta_{12})^2 \tag{2.71}$$

where the extra logarithmic terms are necessary to take into account the long-range nature of the centripetal and centrifugal potentials and hence account for explicit correlation. The role of this generalized action becomes clearer when we come to consider the resulting Hamilton–Jacobi equation

$$\omega^2(\rho) + \frac{Z_1}{\rho}(\Delta\alpha)^2 + \frac{Z_2}{4\rho}(\Delta\theta_{12})^2 + \frac{1}{\rho^2(\Delta\theta_{12})^2} - \frac{1}{4\rho^2(\Delta\alpha)^2}$$

$$= \left[S_0' + \frac{1}{2}S_1'(\Delta\alpha)^2 + \frac{1}{8}S_2'(\Delta\theta_{12})^2 \right]^2 + \frac{1}{\rho^2}\left(\frac{s_0}{\Delta\alpha} + S_1(\Delta\alpha) \right)^2 \quad (2.72)$$

$$+ \frac{4}{\rho^2}\left(\frac{s_1}{\Delta\theta_{12}} + \frac{1}{4}S_2(\Delta\theta_{12}) \right)^2$$

where $'$ denotes differentiation with respect to ρ. Equation (2.72) is solved, in the usual manner, by equating the coefficients of powers of $\Delta\alpha$ and $\Delta\theta_{12}$ to give

$$s_0^2 = -\frac{1}{4}, \qquad s_1^2 = \frac{1}{4} \quad (2.73)$$

$$\omega^2(\rho) = (S_0')^2 + \frac{2s_0 S_1}{\rho^2} + \frac{2s_1 S_2}{\rho^2} \quad (2.74)$$

$$\frac{Z_1}{\rho} = S_0' S_1' + \frac{S_1^2}{\rho^2} \quad (2.75)$$

$$\frac{Z_2}{\rho} = S_0' S_2' + \frac{S_2^2}{\rho^2} \quad (2.76)$$

Obviously the values of s_0 and s_1 can be taken as either the positive or negative square root which introduces a certain degree of freedom. Thoughtful consideration of the desired form of the wavefunction, in particular the symmetry properties of the wavefunction, led to selection of the values of s_0 and s_1 as

$$s_0 = -\frac{1}{2}i, \qquad s_1 = \frac{1}{2} \quad (2.77)$$

so that (2.74) becomes

$$(S_0')^2 = \tilde{\omega}^2 = X^2 + \frac{2Z_0}{\rho} + \frac{iS_1}{\rho^2} - \frac{S_2}{\rho^2} \quad (2.78)$$

with (2.75) and (2.76) left unaltered. Thus the equation for S_0 now involves S_1 and S_2, and so (2.78) requires a perturbative solution. This involves setting $S_1 = S_2 = 0$ in (2.74) to obtain S_0, which is used to solve (2.75) and (2.76), where both equations take the form of a Riccati equation and are

solved in direct analogy with Peterkop, Eqs. (2.16) and (2.17). These
solutions are then iterated back into (2.78) to give

$$\tilde{\omega}^2 \simeq \omega^2 - \omega \frac{d}{d\rho}(\ln u_2) + i\omega \frac{d}{d\rho}(\ln u_1) \tag{2.79}$$

S_0 is then given by

$$S_0 \simeq \int_0^\rho d\tilde{\rho}\, \tilde{\omega}(\tilde{\rho}) \tag{2.80}$$

which can be approximated, for the ionization limit of large ρ, by

$$S_0 \simeq \int_0^\rho d\tilde{\rho} \left(\omega - \frac{1}{2}\frac{d}{d\tilde{\rho}}(\ln u_2) + \frac{i}{2}\frac{d}{d\tilde{\rho}}(\ln u_1) \right)$$

$$\simeq \rho\omega + \frac{Z_0}{X}\ln\left(\frac{\rho(\omega + X)^2}{2Z_0}\right) - \frac{1}{2}\ln u_2 + \frac{1}{2}i\ln u_1 \tag{2.81}$$

where a binomial expansion of $\tilde{\omega}$ is taken. The continuity equation is solved
by the method of Peterkop, where the generalized action of Eq. (2.71) now
gives the density to leading order as

$$P_0 \propto \frac{1}{\tilde{\omega}} |\Delta\alpha|^0 (\Delta\theta_{12}) \tag{2.82}$$

Thus we can see from (2.65) that the wavefunction takes the form

$$\Psi_f^{-*} = \frac{C^{1/2} |\sin(\Delta\alpha)|^{1/2}}{\rho^{5/2}\tilde{\omega}^{1/2}\sin\alpha\cos\alpha(\sin\theta_{12})^{1/2}} (\Delta\theta_{12})^{1/2} \exp\left[-\frac{1}{2}\ln|\Delta\alpha| \right.$$

$$\left. -\frac{1}{2}i\ln(\Delta\theta_{12}) - i\left(S_0 + \frac{1}{2}S_1(\Delta\alpha)^2 + \frac{1}{8}S_2(\Delta\theta_{12})^2 \right) \right]. \tag{2.83}$$

It is clear from (2.83) that the given choice of s_0 and s_1 will result in a
cancellation of the $\Delta\alpha$ terms. This is an important point since we seek a
singlet wavefunction and therefore we expect that on particle exchange our
final-state wavefunction remains unchanged. In this coordinate system
particle exchange $(\mathbf{r}_1 \leftrightarrow \mathbf{r}_2)$ results in $\Delta\theta_{12} \to \Delta\theta_{12}$ and $\Delta\alpha \to -\Delta\alpha$,
making it imperative that the $\Delta\alpha$ terms cancel. The asymptotic form of this

wavefunction may be expressed as

$$\Psi_f^{-*} \simeq \frac{C^{1/2}u_1^{1/2}}{X^{1/2}\rho^{5/2}\sin\alpha\cos\alpha}\exp\left[-iX\rho - \frac{i}{X}\left(Z_0 + \frac{1}{2}Z_1(\Delta\alpha)^2\right.\right.$$
$$\left.\left. + \frac{1}{8}Z_2(\Delta\theta_{12})^2\right)\ln\rho + \frac{4i}{(8Z_0\rho)^{1/2}(\Delta\theta_{12})^2}\right] \tag{2.84}$$

where (2.45) has been employed and where in the classical interpretation of $\rho \gg 1$

$$\ln(\Delta\theta_{12}) = \ln u_2(\rho) - \frac{4}{(2Z_0\rho)^{1/2}(\Delta\theta_{12})^2} \tag{2.85}$$

Equation (2.84) gives a $L = 0$ two-electron semiclassical singlet wavefunction that remains finite in the limits of threshold double escape. Because it is a JWKB wavefunction, it becomes invalid near the classical turning point, which is given by $\rho = 0$. The constant C is found by the current normalization at $\rho = +\infty$ as opposed to the method of Peterkop, where the matching procedure was carried out at a finite ρ value. The current probability density (CPD) is matched to that of Rudge and Seaton [16], which requires that

$$\frac{1}{2}\lim_{\rho\to+\infty}\Im\int\left[(\Psi_f^{-*})^*\frac{\partial}{\partial\rho}\Psi_f^{-*} - \Psi_f^{-*}\frac{\partial}{\partial\rho}(\Psi_f^{-*})^*\right]dS = 4\pi^2 \tag{2.86}$$

The wavefunction may be written as

$$\Psi_f^{-*} = \frac{C^{1/2}u_1^{1/2}\exp(-iS)}{\tilde{\omega}^{1/2}\rho^{5/2}\sin\alpha\cos\alpha} \tag{2.87}$$

with S given by

$$S = S_0 + \frac{1}{2}S_1(\Delta\alpha)^2 + \frac{1}{8}S_2(\Delta\theta_{12})^2 \tag{2.88}$$

so that (2.86) becomes

$$\lim_{\rho\to+\infty}\int\frac{Cu_1\exp(2\Im S)}{\tilde{\omega}}\frac{\partial}{\partial\rho}(\mathrm{Re}S)d\alpha\,d\hat{\mathbf{r}}_1\,d\hat{\mathbf{r}}_2 = 4\pi^2 \tag{2.89}$$

Here we have taken the element of surface area in hyperspherical coordinates as

$$dS = \rho^5 \sin^2\alpha \cos^2\alpha \, d\hat{\mathbf{r}}_1 d\hat{\mathbf{r}}_2. \tag{2.90}$$

It is clearly seen from (2.89) that the complex nature of S_2 is very important when it comes to the normalization of this JWKB wavefunction. However, as discussed earlier, S_2 is not always complex. From (2.31) it was seen that for residual integer charge $Z > 2\mu_2$ become real and so subsequently did S_2. This highlights the detrimental effect that increasing the residual charge has on the validity of the preceding analysis. Since S_0 and S_1 are both real, $\Im S$ is simply $\Im S_2$. Equation (2.89) is solved to give

$$C = \frac{8\pi^3}{\Upsilon} cE^{m_{12}} \tag{2.91}$$

where

$$\Upsilon = \int \exp(-2\eta^2)(\Delta\theta_{12})^2) d\alpha \, d\hat{\mathbf{r}}_1 \, d\hat{\mathbf{r}}_2 \tag{2.92}$$

and

$$\eta^2 = -\frac{1}{8} \Im (S_2) \tag{2.93}$$

with

$$c = \frac{1}{2\pi} \frac{\Gamma(m_{12} + \frac{3}{2})\Gamma(m_{12} + 1)}{Z_0^{m_{12}} \Gamma(2m_{12} + \frac{3}{2})} \tag{2.94}$$

From Section II.A.1 we know that for $E \neq 0$ we may choose $C_{11} = C_{22} = 0$ and $C_{12} = C_{21} = 1$ so that from (2.25) $u_2 = u_{21}$. It is essential that the correct asymptotic form of u_{21} is determined if (2.92) is to be evaluated correctly with regard to the threshold double-escape limits $E \to 0, \rho \to \infty$. The asymptotic form of $u_{ij}(i,j = 1,2)$ given in (2.43) is incomplete; extra complex terms will have been neglected that convert (2.43) to

$$u_{21} \underset{\rho\to\infty}{\sim} \gamma_{12}\left[1 - \frac{Z_2}{2E\rho}\ln\left(\frac{E\rho}{Z_0} + h_{21}\right)\right] \tag{2.95}$$

for $i = 2$, $j = 1$, with γ_{21} given by (2.44). The extra h_{21} term consists of a sum of digamma functions $\psi(x)$ [35] of either integer or complex argument given by

$$h_{21} = \psi(1) + \psi(2) - \psi(m_{12} + \tfrac{1}{2}) - \psi(m_{21} + 1) \qquad (2.96)$$

[33,40]. The definition of S_2, (2.19) gives

$$S_2 \sim \frac{Z_2}{(2E)^{1/2}} \left(\ln\left(\frac{E\rho}{Z_0}\right) - 1 + \psi(1) + \psi(2) - \psi\left(m_{12} + \frac{1}{2}\right) - \psi(m_{21} + 1) \right)$$
$$(2.97)$$

where the asymptotic form of $\omega(\rho) \underset{\rho \to \infty}{\sim} (2E)^{1/2}$ has been used. Therefore

$$\Im(S_2) \underset{\rho \to \infty}{\sim} \frac{Z_2}{(2E)^{1/2}} \Im\left[-\psi(m_{12} + \frac{1}{2}) - \psi(m_{21} + 1) \right] \qquad (2.98)$$

$$= -\frac{\pi Z_2 \tanh(2\pi \Im m_{12})}{(2E)^{1/2}} \qquad (2.99)$$

where we have used equations (6.3.8) and (6.3.12) of Abramowitz and Stegun [35]. Hence from (2.93) we have

$$\eta^2 = \frac{\pi Z_2 \tanh(2\pi \Im m_{12})}{8(2E)^{1/2}} \qquad (2.100)$$

where clearly $\eta^2 \to \infty$ as $E \to 0$. Therefore, for small E, the most significant contribution to (2.92) will be given by $(\Delta\theta_{12}) = 0$. The integral in (2.92) is solved by referring $\hat{\mathbf{r}}_2$ to $\hat{\mathbf{r}}_1$ so that given

$$d\hat{\mathbf{r}}_1 d\hat{\mathbf{r}}_2 = \sin\theta_1 d\theta_1 d\phi_1 \sin(\Delta\theta_{12}) d(\Delta\theta_{12}) d\psi_B \qquad (2.101)$$

we have, in the limit as $E \to 0$

$$\Upsilon = \frac{\pi^3}{\eta^2} \qquad (2.102)$$

where ψ_B is described in Section I.B.1. This gives the asymptotic form of the normalisation constant as

$$C = \frac{\pi Z_2 \tanh(2\pi \Im m_{21})}{(2E)^{1/2}} c E^{m_{12}} \qquad (2.103)$$

In this threshold ionization process the two escaping electrons will have specific asymptotic directions that are represented, in method I, by a factor

$$\delta(\hat{\mathbf{k}}_1 - \hat{\mathbf{r}}_1)\delta(\hat{\mathbf{k}}_2 - \hat{\mathbf{r}}_2) \qquad (2.104)$$

This factor will project out the required outgoing scattering amplitude and amounts to the inclusion of

$$Y_{l_1}^{m_1}(\hat{\mathbf{r}}_1)Y_{l_2}^{m_2}(\hat{\mathbf{k}}_2) \qquad (2.105)$$

for arbitrary angular momentum states for each of the two electrons. At this stage the natural question to ask is how one can include this angular momentum term when the original solution was based on a $L = 0$ Schrödinger equation. It is argued in I that the preceding approximation is valid for low L values since the cross-terms that arise in the exact Hamiltonian [19] do not affect the Hamilton–Jacobi equation for S, and hence the threshold law. They do, however, manifest themselves in the continuity equation for P. Nevertheless, a subtle combination of a change of dependent variable and solution of the continuity equation ensures a regular JWKB description, as in the case for $L = 0$. The final-state two-electron wavefunction has the asymptotic form

$$\Psi_f^{-*} \underset{\rho\to\infty}{\sim} \frac{c^{1/2}E^{(1/2)m_{12}}u_1^{1/2}}{X^{1/2}\rho^{5/2}\sin\alpha\cos\alpha}\exp\left\{-iX\rho - \frac{i}{X}\left[Z_0 + \frac{1}{2}Z_1(\Delta\alpha)^2\right.\right.$$
$$\left.\left. + \frac{1}{8}Z_2(\Delta\theta_{12})^2\right]\ln\rho + 4i(8Z_0\rho)^{-1/2}(\Delta\theta_{12})^{-2}\right\}\delta(\hat{\mathbf{k}}_1 - \hat{\mathbf{r}}_1)\delta(\hat{\mathbf{k}}_2 - \hat{\mathbf{r}}_2)$$
$$\times\left(\frac{\pi Z_2\tanh(2\pi\Im m_{12})}{(2E)^{1/2}}\right)^{1/2}\exp\left\{\left(\frac{-iZ_2}{8X}\right)(\Delta\theta_{12})^2\right.$$
$$\times\left.\left[\ln\left(\frac{E}{Z_0}\right) + \psi(2) + \psi(1) - 1 - \psi(m_{12} + 1) - \psi\left(m_{21} + \frac{1}{2}\right)\right]\right\}$$

$$(2.106)$$

which is employed, in method I, in the Kohn variational principle. For the first time the singlet ionization amplitude was found in method I for the electron impact ionization of helium at threshold, and hence the absolute values of the total, doubly differential and triply differential singlet cross sections. The variational principle is applied perturbatively. We begin by defining the functional L as

$$L = \Psi^*\left(-\frac{1}{2}\nabla_{r_1}^2 - \frac{1}{2}\nabla_{r_2}^2 - \frac{1}{r_1} - \frac{1}{r_2} + \frac{1}{r_{12}} - E\right)\Psi \qquad (2.107)$$

where we have employed the two-electron time-independent Schrödinger equation. The principle requires that $\delta I = 0$, where

$$I = \int \int L \, d\mathbf{r}_1 d\mathbf{r}_2 \tag{2.108}$$

Therefore, in effect, we replace the two-electron Schrödinger equation by the six-dimensional Kohn second-order variational principle, $\delta I = 0$, where

$$I = \int \int \Psi^* (H - E) \Psi \, d\mathbf{r}_1 d\mathbf{r}_2 \tag{2.109}$$

Applying the principle to our problem, we take $\Psi = \Psi_i^+$ to represent the physical solution of the scattering problem and $\Psi^* = \Psi_f^{-*}$, which is our semiclassical solution to the Schrödinger equation. The exact total wavefunction, including the direct ionization amplitude f is given by

$$\Psi_i^+ \underset{\rho \to \infty}{\sim} \exp(i\mathbf{k}_0 \cdot \mathbf{r}_1)\varphi_i(\mathbf{r}_2) + f(\hat{\mathbf{r}}_1, \hat{\mathbf{r}}_2)\psi_i^+ \tag{2.110}$$

where

$$\psi_i^+ =$$

$$\frac{\exp\left\{iX\rho + i/X[Z_0 + \tfrac{1}{2}Z_1(\Delta\alpha)^2 + \tfrac{1}{8}Z_2(\Delta\theta_{12})^2]\ln\rho - 4i(8Z_0\rho)^{-1/2}(\Delta\theta_{12})^{-2}\right\}}{X^{1/2}\rho^{5/2}\sin\alpha\cos\alpha c^{1/2}E^{(1/2)m_{12}}u_i^{1/2}}$$

$$\times \left(\frac{\pi Z_2 \tanh(2\pi \Im m_{12})}{(2E)^{1/2}}\right)^{1/2} \exp\left\{(-iZ_2/8X)(\Delta\theta_{12})^2\left[\ln\left(\frac{E}{Z_0}\right)\right.\right.$$

$$\left.\left. + \psi(2) + \psi(1) - 1 - \psi(m_{12}^* + 1) - \psi\left(m_{21}^* + \frac{1}{2}\right)\right]\right\} \tag{2.111}$$

Since Ψ_f^{-*} (2.106) is an asymptotic solution of the two-electron Schrödinger equation (2.109) may be rewritten as

$$I = \int \int \left[\Psi_f^{-*}(H - E)\Psi_i^+ - \Psi_i^+(H - E)\Psi_f^{-*}\right] d\mathbf{r}_1 d\mathbf{r}_2$$

$$= \frac{1}{2}\lim_{\rho \to \infty} \int \int \left(\Psi_f^{-*}\frac{\partial}{\partial\rho}\Psi_i^+ - \Psi_i^+\frac{\partial}{\partial\rho}\Psi_f^{-*}\right)\rho^5 \sin^2\alpha \cos^2\alpha \, d\hat{\mathbf{r}}_1 d\hat{\mathbf{r}}_2 \tag{2.112}$$

where we have used Green's theorem and taken the surface of the hemisphere to be at infinity. This gives, in the limit as $E \to 0$

$$I = \frac{1}{2}\pi i f(\hat{\mathbf{k}}_1, \hat{\mathbf{k}}_2). \tag{2.113}$$

We then apply first-order perturbation theory to the asymptotic from of Ψ_i^+ so that with (2.112), we have

$$\delta I = \int \int \Psi_f^{-*}(H - E)\delta f(\hat{\mathbf{r}}_1, \hat{\mathbf{r}}_2)\psi_i^+ d\mathbf{r}_1 d\mathbf{r}_2 \tag{2.114}$$

which when equated to zero gives

$$f(\hat{\mathbf{k}}_1, \hat{\mathbf{k}}_2) \simeq \frac{2i}{\pi} \int \int \Psi_f^{-*}(H - E)\exp(i\mathbf{k}_0 \cdot \mathbf{r}_1)\varphi_i(\mathbf{r}_2)d\mathbf{r}_1 d\mathbf{r}_2 \tag{2.115}$$

for the direct ionization scattering amplitude. As can be seen from (2.110), φ_i represents the target wavefunction, and the incoming electron is represented by the plane-wave expansion

$$\exp(i\mathbf{k}_0 \cdot \mathbf{r}_1) = \sum_{L=0}^{\infty} i^L (2L + 1)j_L(\mathbf{k}_0 \cdot \mathbf{r}_1)P_L(\cos\theta_1). \tag{2.116}$$

For the further development of this theory we take φ_i to be the ground-state hydrogenic wavefunction and then later adopt the form of φ_i for the case of helium. In method I, the $E\rho \to 0$ limit is taken since it is argued that (2.115) is dominated by small ρ values due to the exponential decay of φ_i. Thus we set

$$u_1 = \rho^{m_{12}} \tag{2.117}$$

on which it becomes quite clear that u_1 is the Wannier "message carrier". The $E\rho \to 0$ limit of S_0 and $S_i, i = 1, 2$ are also taken where from (2.34) and (2.18) we have

$$S_0 \sim (8Z_0\rho)^{1/2} \tag{2.118}$$

$$S_i \sim (2Z_0\rho)^{1/2}m_{ij}, \qquad i = 1, 2, \quad j = 2, 1 \tag{2.119}$$

The basic Wannier model requires that in the limit of $\rho \to \infty$ the mutual polar angle behavior will be $\delta(\Delta\theta_{12})$, ensuring that the electrons escape in opposite directions. This behavior is obtained by including the factor

$$\rho^{1/4}(2Z_0)^{1/4}(-\Im m_{12})^{1/2} \tag{2.120}$$

which is an effective renormalization of $\exp[-\frac{1}{8}i(\Delta\theta_{12})^2 m_{21}(2Z_0\rho)^{1/2}]$. The precise form of the incoming final-state wavefunction is given by

$$\Psi_f^{-*} = \frac{c^{1/2}E^{m_{12}/2}\rho^{(m_{12}/2)+(1/2)}(2Z_0)^{(1/4)}(-\Im m_{12})^{(1/2)}}{(2Z_0)^{1/4}\rho^{5/2}\sin\alpha\cos\alpha}\delta(\hat{\mathbf{k}}_1 - \hat{\mathbf{r}}_1)\delta(\hat{\mathbf{k}}_2 - \hat{\mathbf{r}}_2)$$

$$\times\exp\left(\frac{4i}{(8Z_0\rho)^{1/2}}(\Delta\theta_{12})^{-2}\right)\left\{\exp\left[-i(8Z_0\rho)^{1/2} - i\frac{1}{2}(\Delta\alpha)^2(2Z_0\rho)^{1/2}m_{12}\right.\right.$$

$$\left.\left. - i\frac{1}{8}(\Delta\theta_{12})^2(2Z_0\rho)^{1/2}m_{21} - i\frac{\pi}{4}\right] - \text{conjugate}\right\} \qquad (2.121)$$

where we have include both incoming and outgoing JWKB waves and where the $\pi/4$ in the exponential term is a results of the Gans–Jeffreys connection formula applied at the classical turning point $\rho = 0$. We now use the outgoing current probability density (CPD) of Ψ_i^+ as given in (2.110) to find the cross sections. The CPD for the first electron is given by

$$2\pi\int_0^{\pi/2} d\alpha \int\int d\hat{\mathbf{r}}_1 d\hat{\mathbf{r}}_2 \frac{\pi Z_2\tanh(2\pi\Im m_{21})}{(2E)^{1/2}}$$

$$\times\exp\left(-\frac{Z_2}{4(2E)^{1/2}}(\Delta\theta_{12})^2\pi\tanh(2\pi\Im m_{21})\right)|f(\hat{\mathbf{r}}_1,\hat{\mathbf{r}}_2)|^2 \qquad (2.122)$$

so that the CPD for the second electron is obtained by replacing $\mathbf{r}_1$ with $\mathbf{r}_2$. Hence the total cross section for distinguishable particles is given by

$$\sigma = \frac{\pi^2 a_0^2}{k_0}\int\int d\hat{\mathbf{k}}_1 d\hat{\mathbf{k}}_2 \frac{\pi Z_2\tanh(2\pi\Im m_{21})}{(2E)^{1/2}}$$

$$\times\exp\left(-\frac{Z_2}{4(2E)^{1/2}}(\Delta\theta_{12})^2\pi\tanh(2\pi\Im m_{21})\right)|f(\hat{\mathbf{k}}_1,\hat{\mathbf{k}}_2)|^2 \qquad (2.123)$$

The triple-differential cross sections are given by

$$\frac{d^3\sigma}{d\hat{\mathbf{k}}_1 d\hat{\mathbf{k}}_2 d(\frac{1}{2}k_i^2)} = \frac{2\pi^2 a_0^2}{k_0}\frac{d}{dE}\frac{\pi Z_2\tanh(2\pi\Im m_{21})}{(2E)^{1/2}}$$

$$\times\exp\left(-\frac{Z_2}{4(2E)^{1/2}}(\Delta\theta_{12})^2\pi\tanh(2\pi\Im m_{21})\right)|f(\hat{\mathbf{k}}_1,\hat{\mathbf{k}}_2)|^2$$

$$(2.124)$$

where it is worth noting that we do not integrate with respect to α since we are specifying E_1 and therefore E_2. We therefore obtain vital information about the probability of a given electron emerging at a specific angle with a particular energy.

The theory, which has so far been developed for the hydrogen atom, is now extended to the ionozation of helium

$$e^- + He(1s^2\,{}^1S) \rightarrow e^- + e^- + He^+(1s\,{}^2S)$$

When dealing with the electron impact ionization of helium in its S^e ground state and limiting the angular momentum of the incident electron to $l \leq 3$, it follows from the conservation of parity, total angular momentum and its projection that eight states for the two outgoing electrons can contribute, namely, the ${}^{1,3}S^e$, ${}^{1,3}P^o$, ${}^{1,3}D^e$, and ${}^{1,3}F^o$ states. In order to account for indistinguishable particles (particles with different spin components are distinguishable), we require with f given by (2.115) that the flux $|f(\hat{\mathbf{k}}_1, \hat{\mathbf{k}}_2)|^2$ be replaced by

$$\frac{1}{4}|f(\hat{\mathbf{k}}_1, \hat{\mathbf{k}}_2) + f(\hat{\mathbf{k}}_2, \hat{\mathbf{k}}_1)|^2 + \frac{3}{4}|f(\hat{\mathbf{k}}_1, \hat{\mathbf{k}}_2) - f(\hat{\mathbf{k}}_2, \hat{\mathbf{k}}_1)|^2 \qquad (2.125)$$

[16,40]. Clearly the $1:3$ ratio of the symmetric and antisymmetric sum of the direct and exchange amplitude reflects the *para/ortho* nature of helium. As such, we redefine $f(\hat{\mathbf{k}}_1, \hat{\mathbf{k}}_2)$ as

$$f(\hat{\mathbf{k}}_1, \hat{\mathbf{k}}_2) \sim \frac{2i}{\pi} \int \Psi_f^{-*}(\mathbf{r}_1, \mathbf{r}_2)\varphi(\mathbf{r}_3, 2)(H - \xi)\exp(i\mathbf{k}_0 \cdot \mathbf{r}_1)\varphi_i(\mathbf{r}_2, \mathbf{r}_3)d\mathbf{r}_1 d\mathbf{r}_2 d\mathbf{r}_3$$

$$(2.126)$$

and similarly for $f(\hat{\mathbf{k}}_2, \hat{\mathbf{k}}_1)$ with $(\mathbf{r}_1, \mathbf{r}_2, \mathbf{r}_3) \rightarrow (\mathbf{r}_2, \mathbf{r}_3, \mathbf{r}_1)$. In (2.126) $\varphi_i(\mathbf{r}_2, \mathbf{r}_3)$ denotes the target wavefunction that now represents bound-state helium, for which we use a simple open-shell independent-electron wavefunction, namely

$$\varphi_i(\mathbf{r}_2, \mathbf{r}_3) = \frac{\varphi(\mathbf{r}_2, z_0)\varphi(\mathbf{r}_3, \beta) + \varphi(\mathbf{r}_3, z_0)\varphi(\mathbf{r}_2, \beta)}{[2(1 + S)]^{1/2}} \qquad (2.127)$$

where $\beta = 2$ and $z_0 = (1.8072)^{1/2}$, and

$$\varphi(\mathbf{r}, z_0) = z_0^{3/2}\pi^{-1/2}\exp(-z_0 r) \qquad (2.128)$$

with

$$S = \int \varphi(\mathbf{r}, z_0)\varphi(\mathbf{r}, \beta)\,d\mathbf{r} = \left(\frac{4z_0\beta}{(z_0 + \beta)^2}\right)^3 \qquad (2.129)$$

The physical parameter z_0 is given by the square root of twice the single-ionization potential of the ground-state helium. In this simple model we assume that the outer electron experiences a reduced charge given by z_0 while the inner electron experiences the full nuclear Coulomb attraction with $\beta = 2$. It is clear from (2.126) that we have ignored classical exchange and from (2.127) that the target wavefunction does not contain any explicit electron correlation. In $f(\hat{\mathbf{k}}_1, \hat{\mathbf{k}}_2)$, the incoming electron e_1 introduces the following perturbation term:

$$-\frac{2}{r_1} + \frac{1}{r_{12}} + \frac{1}{r_{13}} \qquad (2.130)$$

which, when taken as $(H - \xi)$ in Eq. (2.126), yields

$$f = \int_0^\infty d\rho^{3/2 + m_{12}/2 + 1/4} \sum_{L=0}^{L_{\max}} i^L (2L + 1) j_L\left(\frac{\rho z_0}{\sqrt{2}}\right) P_L(\cos\Theta_1)$$

$$\times \exp\left[\frac{1}{8}\Im m_{21}(\Theta_{12} - \pi)^2 (2Z_0\rho)^{1/2}\right]$$

$$\times \cos\left[(8Z_0\rho)^{1/2} + \frac{1}{8}\mathrm{Re}\, m_{21}(\Theta_{12} - \pi)^2 (2Z_0\rho)^{1/2}\right] r(\rho, \Theta_{12}) \qquad (2.131)$$

for the ionization amplitude f and similarly for the interchanged function g (with Θ_1 replaced by Θ_2). The purpose of the delta functions (2.104) becomes clearer now, where it can be seen that specific asymptotic values of the spherical polar angles of $\mathbf{r}_i$, namely, θ_1, θ_2, and of θ_{12}, have been projected out. This amounts to the exchange of the lowercase $\mathbf{r}_i$ characters by the uppercase $\mathbf{k}_i$ characters, namely, Θ_1, Θ_2, and Θ_{12}. The integration over $\mathbf{r}_3$ introduced the function $r(\rho, \Theta_{12})$ given by

$$(1 + S)^{1/2} r(\rho, \Theta_{12}) = \exp\left(-\frac{\rho z_0}{\sqrt{2}}\right)\left(2^{1/2} z_0 - \frac{1}{(1 - \cos\Theta_{12})^{1/2}}\right) + \exp(-\sqrt{2}\rho)$$

$$\times \left[\frac{64\sqrt{2}(z_0 - 1)}{(z_0 + 2)^3} + \frac{32}{(z_0 + 2)^3}[2^{3/2} + (z_0 + 2)\rho]\right.$$

$$\left. \times \exp\left(-\frac{\rho(z_0 + 2)}{\sqrt{2}}\right)\right] \qquad (2.132)$$

wherein correlation is included implicitly and the two terms on the righthand side are directly associated with the simple two-electron model $\varphi(\mathbf{r}_2, z_0)\varphi$ $(\mathbf{r}_3, \beta)$ and its exchanged counterpart $\varphi(\mathbf{r}_3, z_0)\varphi(\mathbf{r}_2, \beta)$. It is interesting to note that a more recent calculation employing an explicit correlated wavefunction [41] has shown a slight improvement in results over the elementary helium bound-state wavefunction employed here. The integration over α was carried out by the method of steepest descent, which was a natural progression of the JWKB method, in which we have expanded functions in terms of small parameters as measured from the saddle point. The resultant TDCS expression is given by

$$
\frac{d^3\sigma}{d\hat{\mathbf{k}}_1 d\hat{\mathbf{k}}_2 d(\frac{1}{2}k_1^2)} = \frac{2 \times 280 \times cz_0^2(2Z_0)^{1/2}(-\Im m_{21})}{27.21 Z_0 m_{12}} \frac{\pi Z_2 \tanh(2\pi\Im m_{21})}{(2)^{1/2}}
$$

$$
\times \frac{d}{dE} E^{m_{12}-\frac{1}{2}} \exp\left(-\frac{Z_2(\Delta\Theta_{12})^2 \pi \tanh(2\pi\Im m_{21})}{4(2E)^{1/2}}\right)
$$

$$
\times \left(\frac{1}{4}|f+g|^2 + \frac{3}{4}|f-g|^2\right)(10^{-19}\,\mathrm{cm}^2\,\mathrm{sr}^{-2}\,\mathrm{eV}^{-1}) \quad (2.133)
$$

which can be seen to have the required $E^{-0.373}$ energy dependence. The angular distribution around $\pi - \theta_{12}$ is clearly in the form of a Gaussian whose FWHM [which is given by $2(\ln 2)^2$ divided by the square root of the coefficient of $-(\pi - \Theta_{12})^2$ in the exponential] is

$$
(\pi - \Theta_{12})_{1/2} = \left(\frac{32\ln 2}{\pi \tanh(1/2\pi\sqrt{5/3})}\right)^{1/2} E^{1/4} \quad (2.134)
$$

which displays the quarter-law dependence predicted by Wannier. In method I, the triplet contribution to (2.133) given by $\frac{3}{4}|f-g|^2$ is set to zero and only the singlet contribution to the TDCSs is calculated. This was inherently sensible since any attempt to apply it to a triplet calculation gave results that were much too large. Assuming coplanar geometry ($\Theta_{12} = \Theta_1 + \Theta_2$, $\Phi_2 - \Phi_1 = \pi$) singlet TDCSs are presented for $L_{max} = 0, 1, 2, 3$ where for $L_{max} = 3$, which includes $^1S^e$, $^1P^o$, $^1D^e$ and $^1F^o$ contributions, an absolute singlet maximum of $0.85 \times 10^{-19}\mathrm{cm}^2\,\mathrm{sr}^{-2}\,\mathrm{eV}^{-1}$ is obtained for excess of threshold energy $E = 1$ and $0.65 \times 10^{-19}\,\mathrm{cm}^2\,\mathrm{sr}^{-2}\,\mathrm{eV}^{-1}$ for $E = 2$.

Good qualitative agreement in the shape of the polar plotted TDCSs with the relative experimental results of Fournier-Lagarde et al. [30] was achieved emphasizing the fact that even at threshold the shape of the experimental TDCSs cannot possibly be accounted for by $L = 0$ states alone [14,30]. However, experimentalists have recorded features of the TDCSs that have not been reproduced theoretically. It is believed that such features may be accounted for by the inclusion of triplet states with the distorted-wave Born approximation (DWBA) results of Pan and Starace [42] suggesting that $^3F^o$ may be important at $\Theta_1 = \pi/6$ and $5\pi/6$, where "peaking" has been observed [17] for supplementary values of Θ_2. We shall reconsider the determination of total and differential ionization cross sections in Section IV, specifically addressing the inclusion of triplet states. The total singlet cross section is given by

$$
\sigma = \frac{32\pi z_0^2 \Gamma\left(m_{12} + \tfrac{3}{2}\right) \Gamma\left(m_{12} + 1 E^{m_{12}} a_0^2\right)}{Z_0^{m_{12}+1} m_{12} \Gamma\left(2m_{12} + \tfrac{3}{2}\right)}
$$

$$
\times \int_0^\infty d\rho \rho^{3/2 + m_{12}/2} \cos(8Z_0\rho)^{1/2} r(\rho, \pi)
$$

$$
\times \int_0^\infty dR R^{3/2 + m_{12}/2} \cos(8Z_0 R)^{1/2} r(R, \pi)
$$

$$
\times 2\left[j_0\left(\frac{z_0\rho}{\sqrt{2}}\right) j_0\left(\frac{z_0 R}{\sqrt{2}}\right) + 5 j_2\left(\frac{z_0\rho}{\sqrt{2}}\right) j_2\left(\frac{z_0 R}{\sqrt{2}}\right) \right] \qquad (2.135)
$$

which for $L_{max} = 3$ confirms the Wannier law according to

$$
\sigma = 2.37 E^{m_{12}} a_0^2 \qquad (2.136)
$$

where the constant of proportionality has been claculated for the first time. Equation (2.136) remains in good agreement with the experimental results of Pichou et al. [43] for values of $E \le 3.6\,\mathrm{eV}$.

In this section we have presented the mathematical details of the uniform semiclassical approximation and seen how it has been used to produce absolute results. Since its first publication in 1986, this method has been successfully applied and adopted to a variety of atomic threshold processes for instance the analogous application to photodouble detachment $(\gamma, 2e)$ [44–46] with the threshold ionization of helium extended to $^3P, ^3D,$ and ^3F states [47]. As outlined in the introduction, the main recent development of this theory is its extension below threshold to address the problem of doubly excited states. The viability of this extension is discussed in the next section. In Section IV we return to the above-threshold ionization problem of helium and present preliminary theoretical investigations into the $L \ne 0$ contribution to the TDCSs and the subsequent results.

B. Potential Ridge Resonances

In this section we consider the theoretical background that has inspired our extension of Crothers' semiclassical above-threshold JWKB theory to the below-threshold study of doubly excited states (DES). The relevance of Wannier's study to a broader classes of problems was suggested many years ago [48,49]. It is now well established that the essence of the WPR analysis transcends its application from ionization threshold laws to become the main focus for all discussions of processes involving two electrons near the threshold for double escape from the field of an ion. If we consider the simple independent-electron configuration of DES, $X^{q+}nln'l'$, where for $q = 1$ we have a negative ion and for $q \geq 2$ we have a neutral atom, it is clear that the two highly excited electrons $(nln'l')$, contained within the continuum of the ionic core, occupy a position below threshold analogous to the two escaping electrons above threshold. The structure and dynamics of DES are therefore dominated by the angular and radial electron correlations, and we should expect an intimate connection between the ionization problem and the DES of the same two-electron atom that lie on either side of the threshold [9]. Experimental studies have shown that long series of resonating states exist in negative ions that can be supported only by a long-range Coulomb field. It has been suggested [3,4] that such a field can be generated only if the two highly excited electrons assumed the Wannier ridge position. In the next section we will consider the theoretical evidence of Fano [3,4] that these doubly excited states fashion themselves as "Wannier ridge resonances," a proposal confirmed by the experimental evidence of Buckman et al. [7,8].

1.   Fano's Theory

In 1980 Fano proposed that the form of the Wannier two-electron potential ridge might give rise to a series of quasi-standing-wave patterns formed when the wavepacket representing the two-electron system propagates along the ridge and becomes reflected at the classical turning points of the system. These standing waves could manifest themselves as resonances in electron–atom scattering at energies below the ionization energy. They are considered analogs of the Landau levels of free electrons in magnetic fields, which have been studied by Edmonds [50] and Starace [51]. Fano believed that motion along a potential ridge may provide the main mechanism responsible for the formation of eigenmodes of motion that remain quasistationary while propagating along the ridge, accounting for notable resonance phenomena observed experimentally (initially observed by Garton and Temkin [52] in the absorption spectra of atoms in strong magnetic fields. He referred to the occurrence of these resonances as due to the "Wannier phenomenon". His aim was to study the Wannier phenomenon by adopting Peterkop's treatment in order to provide a mathematical model for the description of resonances in the absorption spectra of atoms in strong magnetic fields and the numerous analogs of this setup, including two-electron excitations.

We consider here the theory developed by Fano [3] for the propagation of waves on a potential ridge in order to obtain a clear insight into the formation of standing-wave resonant features. As a prototype for an extended Wannier analysis, we consider the motion of a single Rydberg (high-energy) electron in a magnetic field. The sum of the Coulomb and diamagnetic potentials given, in polar coordinates, by

$$V(r, \theta) = -\frac{1}{r} + \frac{1}{2}\gamma^2 r^3 - \frac{1}{2}\gamma^2 r^2 \cos^2\theta \qquad (2.137)$$

has a ridge along $\theta = \pi/2$ that separates two valleys centred at $\theta = 0$ and $\theta = \pi$. The frequency of Larmor rotation for a magnetic field B is given by $\gamma [\gamma/B = 2.13 \times 10^{-10}$ au/G (atomic units per gauss)]. This potential rises to infinity as r increases along the ridge in contrast to the Wannier ridge (Fig. 3), where the potential surface flattens out. The quasi-Landau resonances are believed to arise from formation of standing waves along this ridge with the electron's motion extending about the ridge under conditions of unstable equilibrium leading to eventual escape through one of the valleys [53]. Let us consider the Schrödinger equation

$$(\nabla^2 + k^2)\psi = 0 \qquad (2.138)$$

where the form of the ridge suggests representing the wavenumber by the expansion

$$k^2(r.\xi) = k_0^2(r) + \frac{1}{2}k_1^2(r)\xi^2 + \cdots \qquad (2.139)$$

$$k_0^2 = 2E + \frac{2}{r} - \gamma^2 r^2, \qquad k_1^2 = 2\gamma^2 r^2 \qquad (2.140)$$

similar to (1.26). Here we have set $\xi = 0 - \pi/2$, and in this present context the r axis corresponds in Wannier's problem to the line $r_1 = r_2$, which is the axis of the bundle of trajectories, which leads to double escape. The Wannier law stated that the incident and ejected electron can simultaneously escape the field only by moving in their joint phase space along classical trajectories within a narrow bundle (Section I.B.2). These bundles of trajectories constitute eigenmodes of motion of the particles in the region about a potential ridge. The trajectories of the three-body problem are given by (1.29), where two eigenbundles diverge from (converge to) the ridge, at distances proportional to $r^{\beta\pm}$. The exponent $\beta_{\pm}$ is given by the positive (negative) square root of the second-degree equation [Eg. (1.27)], which is $-\frac{1}{4} \pm \frac{1}{2}\mu_1$. The essential point of the Fano treatment consists of redefining

ARLENE M. LOUGHAN

the Wannier parameter $\beta_{\pm}$ for each value of r in terms of the expansion coefficients k_0 and k_1, regardless of their analytic form. The desired solution of (2.138) oscillates along the ridge $\xi = 0$ according to

$$\psi(r, 0) \propto \sin \int^r k_0(r')dr' \tag{2.141}$$

[51], where this phase integral form and the fact that the problem deals with large quantum numbers, results in a JWKB approach, mirroring the treatment of Peterkop [5]. The phase function, denoted as usual by S, is expanded in powers of ξ, and the Wannier parameter is redefined as

$$\beta(r) = \frac{S_1(r)}{k_0(r)r} \tag{2.142}$$

so that

$$\beta_{\pm}(r) = -\frac{d\ln[k_0(r)r]^{1/2}}{d\ln r} \pm \left[\left(\frac{d\ln[k_0(r)r]^{1/2}}{d\ln r} \right)^2 + \frac{k_1^2}{2k_0^2} \right]^{1/2} \tag{2.143}$$

where the trajectories of the two Wannier eigenbundles ($\pm$) are represented by

$$\xi_{T_{\pm}}(r) = T \int^r \beta_{\pm}(r')d\ln r' + O(\xi^3) \tag{2.144}$$

where T is a scale parameter that labels a trajectory within its bundle [4]. Each eigenbundle ξ_{T_1} includes the trajectory $\xi_0(r) = 0$ that lies exactly on the ridge and a resonance is expected at energies $E = E_T$. The turning points occur where $k_0(r)$ vanishes not just for $\xi = 0$ but all along a turning line defined by

$$k^2(r_T, \xi) = 0 \tag{2.145}$$

It is clear from (2.143) that as one approaches a turning point, the logarithmic derivatives will diverge. However, this divergence cancels out for β_- while β_+ diverges as r approaches the turning point. The divergence of β_+ results in a sharp deflection of the trajectories (2.144). In contrast to this, propagation along the converging eigenbundle $\xi_{T_-}(r)$ results in reflection at the turning line because this bundle hits the line orthogonally. Similar circumstances occur at the lower turning point $r = 0$, where the logarithmic derivative term in (2.143) now has a positive sign and the situation is reversed. This two eigenmodes propagate along most of the ridge length traveling outward on the divergent eigenbundle $\xi_{T+}(r)$ and backward on the convergent bundle $\xi_{T-}(r)$ and the excited electron appears to bounce back and forth between $r = 0$ and the turning line resulting in standing-wave formation. There has thus emerged a plausible picture of the formation of standing waves in the propagation along a potential ridge using the initial basis of the Wannier theory and a Peterkop JWKB approach. If a centrifugal term replaces the magnetic term in (2.140) the lower turning point moves away from the origin and a similar process is predicted.

Lin [12,13] studied the position probability density (squared wavefunction) for DES, for a number of wavefunctions determined by various methods, to find that a dominant group of states concentrates in the saddle region. This strongly supports the Fano theory with evidence that the localized motion along the two-electron potential ridge persists and even becomes sharper toward the limit of breakup [14,15]. This has been supported by experimental findings, namely, by the work of Buckman et al. [7] and Buckman [8], who provided the first quantitative experimental information on highly excited two-electron states (preliminary indications were contained in Heddle et al. [54] and Brunt et al. [55]). As argued by Rau [56], this provided direct confirmatory evidence for this picture of energy levels quantized into the Wannier saddle point. A more detailed discussion of experimental methods and findings is given in Section III.B.2. We are therefore confident that the doubly excited resonant states of He$^-$ manifest themselves as "Wannier ridge resonances" and may be studied as a below-threshold extension of Crothers above-threshold theory. Details of this study are given in Section III.

III. DOUBLY EXCITED STATES

A. Introduction

In this section we present the theoretical details and subsequent results of the application of the Crothers above-threshold electron impact ionization theory to the below-threshold study of doubly excited states (DES). Our inaugural investigation considers the near-threshold capture excitation process

$$e^- + \mathrm{He}(1s^2\,{}^1S) \rightarrow \mathrm{He}^-(1s^2S(n_1\,sn_2\,s))^2S \tag{3.1}$$

where the total angular momentum L of the final state is zero. This is extended to include states for $L = 1$ and $L = 2$:

$$e^- + \mathrm{He}(1s^2\,{}^1S) \rightarrow \mathrm{He}^-(1s^2S(n_1\,sn_2\,p^3P^o))^2P \tag{3.2}$$

and

$$e^- + He(1s^2\,{}^1S) \rightarrow He^-(1s^2S(n_1\,sn_2\,d^1\,D^e))^2D \tag{3.3}$$

We further extend our below-threshold analysis beyond the negative-ion case to consider the electron capture by the singly charged helium cation:

$$e^- + He^+(1s^2S) \rightarrow He(n_1\,sn_2\,s)^1S \tag{3.4}$$

In each process the energy of the incident electron lies below the first ionization potential (IP) of the target atom/ion but above the first IP of the resulting ionic/atomic product. The outcome is the capture of the incident electron so that a doubly excited state ensues that is highly unstable. Doubly excited configurations are degenerate with the continuum built on lower single ionization thresholds; in particular, there is almost always present the single-electron continuum of the ground state of the neutral atom or positive ionic core. Thus, DES are not stable bound states but decay by autoionization to the lower continua with the ejection of an electron. They have the character of discrete states embedded in a continuum [57]. The $(He^-)^{**}$ is perceived as consisting of a positively charged He^+ ($1s$) core (core charge $Z = 1$), with two excited electrons in a Rydberg state. This arrangement is referred to as the "grandparent model" [58], which provides an appropriate framework within which to view the Wannier ridge resonances. An alternative classification at this stage is given by the "parent model", which views the resonating species it terms of a single electron bound to the "parent" core. This is not conducive to the theoretical modeling of correlated electron motion. In direct analogy with the He^- case, we consider the electron capture by the "parent" $He^+(1s)$ to form the DES of atomic He with the grandparent core He^{2+}, obviously has a charge $Z = 2$. As we have seen in Section II.B, these DES are viewed as *Wannier ridge resonances* because they are regarded as a feature of localized motion along the two-electron potential ridge. The term *resonance* is used throughout this work as meaning no more than an excited stated of a negative ion for the $(He^-)^{**}$ case and an excited state of an atom for the He^{**} case.

The main objective is the determination of the complex eigenvalues E_N of the resonant features given by

$$E_N = E_R^N - \frac{i}{2}\Gamma_N \tag{3.5}$$

in which E_R^N is the resonance position and Γ_N is the lifetime of the metastable state. This eigenvalue is used to describe loosely bound autoionising states so that $E_R^N < 0$.

In Section III.B we present the details of our study of He$^-$ $L = 0$ DES [59]. In Section III.C we present the details of the extension of this work beyond the $L = 0$ case to He$^-$ $L = 1$ and $L = 2$ DES [60]. Finally, in Section III.D we consider high-energy resonances in He**; details of this study are to be published in a review of this work [61].

B. Wannier Below-Threshold $L = 0$ Doubly Excited He$^-$ States

1. Theory

We now consider the analytic continuation of Crothers' above-threshold ionization theory to below-threshold in order to calculate the complex eigenenergies E_N of the high-lying doubly excited Rydberg states of He$^-$ for $L = 0$. The process that we wish to describe is given by Eq. (3.1) We begin with the $L = 0$ two-electron Schrödinger equation in hyperspherical coordinates (2.60)

$$\left[\frac{1}{\rho^5} \left(\frac{\partial}{\partial \rho} \right) \left(\rho^5 \frac{\partial}{\partial \rho} \right) + \frac{1}{\rho^2 \sin^2\alpha \cos^2\alpha} \frac{\partial}{\partial \alpha} \left(\sin^2\alpha \cos^2\alpha \frac{\partial}{\partial \alpha} \right) \right.$$
$$\left. + \frac{1}{\rho^2 \sin^2\alpha \cos^2\alpha} \frac{1}{\sin\theta_{12}} \frac{\partial}{\partial \theta_{12}} \sin\theta_{12} \frac{\partial}{\partial \theta_{12}} + X^2 + \frac{2\xi(\alpha, \theta_{12})}{\rho} \right] \Psi = 0.$$

$$(3.6)$$

which, we have seen in Section II. A.2 has been solved using a JWKB approximation to produce a wavefunction of the form

$$\Psi_f^{-*} = \frac{C^{1/2} \exp\{ -\frac{1}{2} i \ln \Delta\theta_{12} - i[S_0 + \frac{1}{2} S_1 (\Delta\alpha)^2 + \frac{1}{8} S_2 (\Delta\theta_{12})^2] \}}{\tilde{w}^{1/2} \rho^{5/2} \sin\alpha \cos\alpha} \quad (3.7)$$

The asymptotic form of this wavefunction, given by (2.84), describes the final-state two electron threshold escape from the field of an ion. By simply changing the form of the energy and avoiding any threshold escape limits such as $\rho \rightarrow \infty$, Eq. (3.7) can be amended to describe two highly excited electrons with energy $E_R^N < 0$ just below the threshold for double escape. The high-Rydberg electrons that have positions $\mathbf{r}_1$ and $\mathbf{r}_2$ are described by the following below - threshold JWKB wavefunction:

$$\Psi_f^{-*} = \frac{C_N^{1/2} Y_{LM}(\hat{\mathbf{r}}_1, \hat{\mathbf{r}}_2)}{\tilde{w}^{1/2} \rho^{5/2} \sin\alpha \cos\alpha} \left(\exp\left\{ -\frac{1}{2} i \ln \Delta\theta_{12} - i \left[S_0 + \frac{1}{2} S_1 (\Delta_\alpha)^2 \right. \right. \right.$$
$$\left. \left. \left. + \frac{1}{8} S_2 (\Delta\theta_{12})^2 - i \frac{\pi}{4} \right] \right\} - \text{conjugate} \right) \quad (3.8)$$

This is an analytically continued version of (3.7) to negative energy with the inclusion of both ingoing and outgoing waves connected by the Gans–Jeffreys [62,63] connection formula applied at the classical turning point $\rho = 0$. The wavefunction includes, for $L \in [|l_1 - l_2|, l_1 + l_2]$ and $M = m_1 + m_2$, the spherical harmonic factor $Y_{LM}(\hat{\mathbf{r}}_1, \hat{\mathbf{r}}_2)$, representing (2.104). We wish to represent He$^-$ in terms of the grandparent model and therefore require a suitable description of the core electron as well as the two excited electrons. The core electron, with position vector $\mathbf{r}_3$, is represented by the ground-state one-electron atom eigenfunction given by

$$u_{100}(r, \theta, \phi) = \frac{1}{\sqrt{\pi}} 2^{3/2} \exp(-2r_3) \tag{3.9}$$

Therefore the final-state $L = 0$ $M = 0$ wavefunction representing He$^-$ is given by

$$\Psi_f^{-*} = \frac{C_N^{1/2} Y_{LM}(\hat{\mathbf{r}}_1, \hat{\mathbf{r}}_2) 2^{3/2} e^{-2r_3 - (i/2)\ln \Delta\theta_{12}} / \sqrt{\pi}}{\rho^{5/2} \sin\alpha \cos\alpha}$$

$$\times \frac{\sin\left[\int_0^\rho d\tilde{\rho}(\omega^2 - \omega\{\ln u_2 - i\ln u_1\}')^{1/2} \right.}{\left[(\omega^2 - \omega\{\ln u_2 - i\ln u_1\}']^{1/4}}$$

$$\left. + \frac{\rho^2}{2}\omega(\ln u_1)'(\Delta\alpha)^2 + \frac{\rho^2}{8}\omega(\ln u_2)'(\Delta\theta_{12})^2 + \frac{\pi}{4} \right] \tag{3.10}$$

which is a product of (3.8) and (3.9), where the sin function appears as a result of the inclusion of both ingoing and outgoing JWKB solutions ($d/d\tilde{\rho}$ or $d/d\rho$ is indicated by $'$). The complex eigenenergy given in (3.1) is now employed so that (2.15) becomes

$$\omega^2 = 2E_N + \frac{2Z_0}{\rho} \tag{3.11}$$

and the Wannier functions (2.26) take the form

$$u_1 = \rho_2^{m_{12}} F_1\left(m_{12}, m_{12} + 1; 2m_{12} + \frac{3}{2}; -\frac{E_{N\rho}}{Z_0}\right) \tag{3.12}$$

$$u_2 = \rho_2^{m_{22}} F_1\left(m_{22}, m_{22} + 1; 2m_{22} + \frac{3}{2}; -\frac{E_{N\rho}}{Z_0}\right) \tag{3.13}$$

where the Frobenius indices m_{12} and m_{22} are given by (2.27) and (2.28). In (3.10) we have included the exact form of S_0, which was determined perturbatively in Section II.A.2, namely

$$S_0 \simeq \int_0^\rho d\tilde{\rho}\tilde{\omega}(\tilde{\rho}) \qquad (3.14)$$

where

$$\tilde{\omega}^2 \simeq \omega^2 - \omega\frac{d}{d\rho}(\ln u_2) + i\omega\frac{d}{d\rho}(\ln u_1) \qquad (3.15)$$

The approximation given by (2.81) is applicable when the ionisation limit of large ρ is taken but is to be avoided in this case since the two electrons do not "escape" to infinity but remain "bounded" in a metastable state.

As discussed in Section I.C, a crucial feature of the threshold region is the $E^{1/4}$ dependence of the angular distribution; such tight angular correlation $\theta_{12} = \pi$ can be viewed, even for $L = 0$, as the ability of the individual orbital angular momenta of the two electrons to vary over a large range [9]. In this case $L = 0$ by virtue of $m_1 = -m_2$, and the only restrictions on the magnitudes of the l values of the individual electrons is that they should be equal and less than the individual principal quantum numbers.

Viewing the DES of He$^-$ as Wannier ridge resonances, where standing waves are produced as a result of wavepacket reflection at the classical turning points, exponential decay of the wavefunction is required beyond the two transition points of this resonating system. The transition points, given by $\omega = 0$, are $\rho = 0$ and $\rho = -Z_0/E_N$ so that the regions, beyond these values, are classically forbidden and the physical solution must decrease in this region. We clearly have one complex and one real transition point for this $L = 0$ problem, where, in the regions beyond the two turning points, the solution is required to be the decreasing JWKB solution, in order to satisfy the boundary conditions at 0 and ∞. Continuity across these points requires that

$$\int_0^{-(z_0/E_N)} d\rho\sqrt{\omega^2 - \omega\left(\frac{d}{d\rho}\ln u_2 - i\frac{d}{d\rho}\ln u_1\right)} = N\pi + \frac{\pi}{2} \qquad (3.16)$$

This takes the form of the Bohr–Sommerfeld quantization rule from old quantum mechanics.

It is easily seen that N, a hyperspherical radial quantum number for the two excited electrons confined to the ridge, is the number of nodes of the JWKB wavefunction between the two turning points. In order to simplify the solution of (3.16) the dummy variable was changed according to

$$\rho = \frac{-Z_0 x}{E_N} \qquad (3.17)$$

with the result that (3.16) becomes

$$c \int_0^1 dx \left(\frac{1}{x} - 1\right)^{1/4} \sqrt{c^2 \left(\frac{1}{x} - 1\right)^{1/2} - \frac{d}{dx} \ln \frac{u_2}{u_1^i}} = N\pi + \frac{\pi}{2} \qquad (3.18)$$

where

$$c = \left(\frac{-2Z_0^2}{E_N}\right)^{1/4} \qquad (3.19)$$

Clearly this introduces a removable singularity at the lower endpoint for the integrand in (3.18), which is easily removed by the change of dummy variable $x = X^6$, but what is a little less obvious is the fact that for the upper limit of this integral the Gauss hypergeometric functions in (3.12) and (3.13) diverge when the form

$$_2F_1(a, b; c; z) = \sum_{n=0}^{\infty} \frac{(a)_n (b)_n}{(c)_n n!} z^n \qquad |z| < 1 \qquad (3.20)$$

is employed; $(\cdot)_n$ is the Pochhammer symbol [35]. The behavior of this series on its circle of convergence, that is, at $z = 1$, is such that absolute convergence is obtained only when

$$\Re (c - a - b) > 0 \qquad (3.21)$$

If we consider the logarithmic derivatives that take the form

$$\frac{d}{dx} \ln u_1 = \frac{m_{12} u_1}{x} + \left(\frac{-Z_0 x}{E_N}\right)^{m_{12}} \frac{m_{12}(m_{12} + 1)}{2m_{12} + \frac{3}{2}} {}_2F_1$$

$$\times \left(m_{12} + 1, m_{12} + 2; 2m_{12} + \frac{5}{2}; x\right) \qquad (3.22)$$

(and similarly for u_2) it is clear that the condition given by (3.21) is violated at the upper limit, and we therefore require an alternative form for $_2F_1$ that will be valid in this region of the complex plane. The appropriate analytically continued form of (3.20), convergent at $x = 1$, is

$$_2F_1(a, b; d; z) = \frac{\Gamma(d)\Gamma(d - a - b)}{\Gamma(d - a)\Gamma(d - b)} {}_2F_1(a, b; a + b - d + 1; 1 - z)$$

$$+ (z - 1)^{d - a - b} \frac{\Gamma(d)\Gamma(a + b - d)}{\Gamma(a)\Gamma(b)} {}_2F_1$$

$$\times (d - a, d - b; d - a - b + 1; 1 - z)$$

$$(|arg(1 - z)| < \pi) \qquad (3.23)$$

see Eq. 15.3.6 in Ref. 35. The gamma function $\Gamma(z)$ is defined by

$$\Gamma(z) = \int_0^\infty dt \exp(-t) t^{z-1} \tag{3.24}$$

Taking further account of any other removable singularities that arise at this upper limit, which can be trivially removed by the substitution $x = 1 - X^6$ (or one similar of higher order), we find that (3.18) becomes

$$c \int_0^{(1/2)^{1/6}} dx\, 6x^{1/2}(1 - x^6)^{1/4} \sqrt{c^2 x^3 (1 - x^6)^{1/2} - \frac{x}{6}\frac{d}{dx}\ln\left(\frac{u_2}{u_1^i}\right)} +$$

$$c \int_0^{(1/2)^{1/6}} dx\, 6x^5 \left(\frac{x^6}{1 - x^6}\right)^{1/4} \sqrt{c^2 \left(\frac{x^6}{1 - x^6}\right)^{1/2} + \frac{1}{6x^5}\frac{d}{dx}\ln\left(\frac{u_2}{u_1^i}\right)}$$

$$= \left(N + \frac{1}{2}\right)\pi \tag{3.25}$$

which we have solved for c using the complex Newton–Raphson method. The Newton–Raphson method is a well-known iterative procedure used for determining the roots of an equation. Here, we set the function $f(c)$ equal to the sum of the two integrals in equation (3.25) minus the righthand-side term and look for the solutions of $f(c) = 0$. We employ the Newton–Raphson formula

$$c_{j+1} = c_j - \frac{f(c_j)}{f'(c_j)} \tag{3.26}$$

for a suitably chosen initial value of c_0 and proceed until a convergence limit is reached. The derivative of $f(c)$ with respect to c, $f'(c)$ is given by

$$f'(c) = \frac{f(c_j) + (N + \frac{1}{2})\pi}{c} + c^2 \int_0^{(1/2)^{1/6}} dx \frac{6x^{7/2}(1 - x^6)^{3/4}}{\sqrt{c^2 x^3 (1 - x^6)^{1/2} - \frac{x}{6}\frac{d}{dx}\ln\left(\frac{u_2}{u_1^i}\right)}}$$

$$+ c^2 \int_0^{(1/2)^{1/6}} dx \frac{6x^7 \left(\frac{x^6}{1 - x^6}\right)^{3/4}}{\sqrt{c^2 x^4 \left(\frac{x^6}{1 - x^6}\right)^{1/2} - \frac{1}{6x^5}\frac{d}{dx}\ln\left(\frac{u_2}{u_1^i}\right)}} \tag{3.27}$$

Equation (3.18) is solved for c for a large number of positive integer values of N, where the complex eigenenergies are extracted using (3.19). As stated previously, the real part of the complex eigenenergies E_R^N gives the resonance position for a given N. The results obtained from this calculation are given in Table III, where they are compared with the experimental results of Buckman et al. [7] (Theoretical and experimental values of resonance position energies and relative intensities are listed in Tables III–XI [7,8,56,64–70].) In Table IV, we compare our results (theory a) with those given on p. 576 of Buckman and Clark [71]. These results are presented and discussed in Section III.B.3.

The experimentalists Buckman and Newman [8] not only determined the energy position of high-lying He$^-$ resonances but also recorded the intensity values of these resonating states given experimentally as the ratio of the mean peak–dip height to the mean background count. We now tentatively approach the issue of the intensity values of these states in terms of our theoretical description. By definition, the intensity is given as the average rate of energy flow normal to the direction of propagation of radiation per unit cross-sectional area. Therefore it is clear that the probability of a transition from say state a to b will be dictated by the intensity of the energy either emitted or absorbed in that transition. In our case, of the metastable DES, we have capture and subsequent excitation so that two electrons reside in discrete states degenerate with the continuum. The reverse of this capture process, by detailed balance, is the radiationless decay of one of these

TABLE III

Resonance Position Energies for $L = 0, n_1 = n, N_2 = n_1 \Rightarrow N = n_1 + n_2 - 1 = 2n - 1$, and $L = 0', n_1 = n, n_2 = n_1 + 1 \Rightarrow N = n_1 + n_2 - 1 = 2n$. Since $N = n_1 + n_2 - |l_1 - l_2| - 1$

Where $L = |l_1 - l_2|^a$

	Present Results Resonance Position Energies (eV)				Theory	
					a	b
n	N	$L = 0$	N	$L = 0'$	$L = 0$	$L = 0'$
2	3	19.492	—	—	19.367(5)	—
3	5	22.330	6	22.963	22.451(10)	22.881(05)
4	7	23.368	8	23.641	23.435(10)	23.667(05)
5	9	23.833	10	23.973	23.850(10)	23.983(10)
6	11	24.077	12	24.158	24.080(10)	24.176(10)
7	13	24.220	14	24.271	24.217(10)	24.288(10)
8	15	24.311	—	—	24.307(15)	—
9	17	24.372	—	—	24.387(15)	—

^{a}The experimental values are from Buckman et al. [7] (theory a) and Buckman and Newman [8] (theory b) using their notation. (Numbers in parentheses indicated errors in second and third decimal places).

TABLE IV
Further Comparison Resonance Position Energies[a]

			Theory					
n	N	Experiment	a	b	c	d	e	f
3	5	22.451(10)	22.330	—	—	22.432	22.774	22.439
4	7	23.435(10)	23.368	—	—	23.408	23.578	23.434
5	9	23.850(10)	23.833	23.857	23.865	23.843	23.879	—
6	11	24.080(10)	24.077	24.087	24.095	24.077	24.090	—
7	13	24.217(10)	24.220	24.223	24.230	24.213	24.219	—
8	15	24.307(15)	24.311	24.310	24.316	24.301	24.304	—
9	17	24.387(15)	24.372	24.369		24.361	24.362	—

[a] The experimental values (numbers in parentheses indicate errors in second and third decimal places) are from Buckman et al. [7] and Buckman and Newman [8]. Theories = (a) current results (this chapter); (b) Rau [56]; (c) Lin and Watanabe [64]; (d) Komninos et al. [65]; (e) Rost and Briggs [66]; (f) Fon et al. [67].

excited electrons with the simultaneous ejection of the other. This process is known as *radiationless autoionization*:

$$e^- + He(1s^2 \, {}^1S) \leftrightarrow He^-(1sn_1 \, sn_2 s \, {}^2S) \qquad (3.28)$$

This equation illustrates the familiar description of a resonance as a discrete state embedded in the continuum. The preceding discussion leads us directly to the autoionization theory of Fano [72], which considers the transition probability rate of autoionizing states. Let us consider the interaction of a

TABLE V
Relative Intensity Values for $L = 0$ Resonances[a]

	Relative Intensity		
N	Experiment	$\frac{1}{2}\Gamma_N * |E_R^n - \frac{i}{2}\Gamma_N|$	Ratio
5	0.53(1)	0.0050	0.0094
7	0.048(1)	0.00099	0.021
9	0.0105(2)	0.00029	0.028
11	0.0021(3)	0.00011	0.052
13	0.0011(1)	0.000049	0.045
15	0.00054(12)	0.000024	0.044
17	0.00030(12)	0.000013	0.043

[a] The experimental values numbers in parentheses indicate errors in last and second-last decimal places) are from Buckman and Newman [8].

TABLE VI
Relative Intensity Values For $L = 0$ Resonances[a]

N	Experiment	$\frac{1}{2}\Gamma_N * \left|E_R^N - \frac{i}{2}\Gamma_N\right| \times k$
5	0.53(1)	0.12
7	0.048(1)	0.023
9	0.0105(2)	0.0067
11	0.0021(3)	0.0025
13	0.0011(1)	0.0011
15	0.00054(12)	0.00055
17	0.00030(12)	0.00030

[a] The experimental values (numbers in parentheses indicate errors in last and second-last decimal places) are from Buckman and Newman [8]. The normalization factor is given as k, which is the ratio of the theoretical and experimental $N = 17$ intensity value.

TABLE VII
Resonance Position Energies for $\left(1s^2 S(n_1 s n_2 p^3 P^0)\right)^2 P, L = 1, M = 0, n_1 = n, n_2 = n_1$
$$\Rightarrow N = 2n - \left(\frac{17^{1/2}-1}{2}\right) - 1^a$$

	Resonance Energies (eV)	
n	Present Results	Experiment
3	22.639	22.600(10)
4	23.518	23.518(1)
5	23.915	23.907(10)

[a] The experimental values are from Buckman et al. [7]. (Numbers in parentheses indicate errors in second and third decimal places.)

TABLE VIII
Resonance Position Energies for $\left(1s^2 S(n_1 s n_2 d^1 D^e)\right)^2 D, L = 2,$
$M = 0, n_1 = n, n_2 = n_1 \rightarrow N = 2n - 4^a$

	Resonance Energies (eV)	
n	Present Results	Experiment
3	22.715	22.660(10)
4	23.544	23.579(10)
5	23.927	23.952(10)
6	24.133	24.144(15)
7	24.256	24.261(15)

[a] The experimental values are from Buckman et al. [7]. (Numbers in parentheses indicate errors in second and third decimal places.)

TABLE IX

Widths and Associated Energy Positions for $(1s^2S(n_1sn_2d^1D^e))^2D, L = 2, M = 0,$
$n_1 = n, n_2 = n_1 \Rightarrow N = 2n - 4(in\,eV)$

n	Γ^a	$E_{\Gamma}{}^a$	Experiment
4	0.3401	23.578	23.579
5	0.2313	23.952	23.952
6	0.1252	24.144	24.144
7	0.0680	24.261	24.261

TABLE X

Resonance Position Energies for He $L = 0, n_1 = n, n_2 = n_1 \Rightarrow N = 2n - 1$

ns^2	$(K, T)^A$	N	Resonance Energies (au)	
			Present Results	Theory a^a
$7s^2$	$(6,0)^+$	13	0.073668	0.066716
$8s^2$	$(7,0)^+$	15	0.055528	0.051207
$9s^2$	$(8,0)^+$	17	0.043310	0.040538
$10s^2$	$(9,0)^+$	19	0.034702	0.032887
$11s^2$	$(10,0)^+$	21	0.028415	0.027365
$12s^2$	$(11,0)^+$	23	0.023688	0.023205
$13s^2$	$(12,0)^+$	25	0.020044	0.01964
$14s^2$	$(13,0)^+$	27	0.017179	0.01695
$15s^2$	$(14,0)^+$	29	0.014885	0.01478

^{a}Theory a: Rost and Briggs [66,68].

TABLE XI

Resonance Position Energies for He $L = 0, n_1 = n, n_2 = n_1 + 1 \Rightarrow N = 2n$

n_1sn_2s	$(K,T)^A$	N	Resonance Energies (au)	
			Present Results	Other Theories
$5s6s$	$(4,0)^+$	10	0.122978	0.10964^a
$6s7s$	$(5,0)^+$	12	0.086213	0.07865^b
$7s8s$	$(6,0)^+$	14	0.063650	0.0599^b
$8s9s$	$(7,0)^+$	16	0.048856	—
$9s10s$	$(8,0)^+$	18	0.038652	—
$10s11s$	$(9,0)^+$	20	0.031325	—
$11s12s$	$(10,0)^+$	22	0.025891	—
$12s13s$	$(11,0)^+$	24	0.021753	—
$13s14s$	$(12,0)^+$	26	0.018529	—
$14s15s$	$(13,0)^+$	28	0.015970	—
$15s16s$	$(14,0)^+$	30	0.013906	—

^{a}Theory due to Ho [69].
^{b}Theory due to H. Fukuda, N. Koyama, and M. Matsuzawa, *J. Phys. B* **20**, 2959 (1987).

discrete state with the continuum where the configuration interaction matrix elements are given as

$$\langle \phi | H | \psi_\epsilon \rangle = V_\epsilon \tag{3.29}$$

The unperturbed discrete state is described by the wavefunction ϕ, and a mutually orthogonal set of continuum wavefunctions is given by ψ_ϵ for an energy ϵ. The wavefunction Ψ^ϵ of the Hamiltonian H with eigenvalue ϵ

$$H\Psi^\epsilon = \epsilon \Psi^\epsilon \tag{3.30}$$

must be expandable in the complete set $\{\phi, \psi_\epsilon\}$ so that

$$\Psi^\epsilon = a^\epsilon \phi + \int b^\epsilon_{\epsilon'} \psi_{\epsilon'} d\epsilon' \tag{3.31}$$

Fano has shown that the substitution of (3.31) into (3.30), with some manipulation, yields

$$(a^\epsilon)^2 = \frac{V_\epsilon^2}{(\epsilon - \epsilon_\phi - F(\epsilon))^2 + \pi^2 V_\epsilon^4} \tag{3.32}$$

which gives the fractional contribution of the pure discrete state ϕ to the wavefunction Ψ^ϵ where the function $F(\epsilon)$ is the perturbation experienced by the discrete level due to the presence of the continuum. The configuration interaction CI has spread the discrete state out into a resonance lineshape whose center is at $\epsilon_\phi + F(\epsilon_\phi)$ and that has a half-width at half-maximum (HWHM) (in Rydbergs) of

$$\Gamma_a = \pi V_\epsilon^2 \tag{3.33}$$

The half-life τ_a with which an atom prepared initially in the discrete state autoionizes into the continuum state can be inferred from the uncertainty principle $(\Delta E)(\Delta t) = \hbar$. Thus the half-life and the corresponding auto-ionization transition probability rate are given by

$$A^a = \frac{1}{\tau_a} = \frac{2\Gamma_a}{\hbar} = \frac{2\pi V_\epsilon^2}{\hbar} \tag{3.34}$$

where the unit of A_a is obviously in reciprocal seconds (s^{-1}). Thus, for autoionizing states the strength of the interaction of these states with the continuum determines the lifetime of the discrete states. If we now return to

(3.29), it is clear from (3.34) that the autoionization transition probability rate A^a is proportional to the square of the CI matrix elements, $\langle\phi|H|\psi_\epsilon\rangle$. Applying this to the forward direction problem in (3.28) ϕ corresponds to our final-state wavefunction Ψ_f^{-*}, and ψ_ϵ corresponds to the target $He(1s^2\,{}^1S)$ wavefunction ϕ_i. Therefore, in order to determine a relationship between the intensity of these states and the calculated eigenenergies, we seek the form of $|\langle\Psi_f^{-*}|H|\phi_i\rangle|^2$. First, we must determine the energy behavior of $\langle\Psi_f^{-*}|\Psi_f^{-*}\rangle$, which, if we consider the form of C_N given in (2.91), can be simplified to

$$\langle\Psi_f^{-*}|\Psi_f^{-*}\rangle \sim |E_N|^{m_{12}} \int_0^{-\frac{z_0}{E_N}} d_\rho\,\rho^{m_{12}}\sin^2\left[\int_0^\rho d\tilde{\rho}(\omega^2 - \omega\{\ln u_2 - i\ln u_1\}')^{1/2}\right]$$

$$\sim \frac{1}{E_N} \tag{3.35}$$

where we have averaged over the rapidly varying $\sin^2$ term and where the integration over the mutual polar angle and the hyperspherical angle have been carried out by the method of steepest descent. Therefore, the transition probability rate A^a is found to be proportional to $|E_N|$, and so we take the intensity as being proportional to the product of this probability rate and the lifetime, $\frac{1}{2}\Gamma_N$:

$$I_N \propto |E_N|\frac{1}{2}\Gamma_N \tag{3.36}$$

The results of this calculation are given in Section III.B:3.

2. Experimental Work and Classification Schemes

In this section we consider some of the experimental methods used in the study of resonating states with particular emphasis on the work of Buckman et al. [7], whose measurements of resonance energy positions and intensity are compared with the corresponding theoretical results in the next section. In conjunction with the experimental observation of these resonances, many attempts were made, by both experimentalists and theorists, to classify them. The more widely accepted classification schemes are also briefly discussed.

Since the pioneering observation of the $He^-1s(2s^2)\,{}^2S$ resonance by Schulz [72], resonances in the helium atom from 19.3 eV to the ionization threshold, have been studied in a variety of electron-scattering experiments. Electron impact spectroscopy has provided the majority of the recent information on atomic negative-ion resonances that are studied by detecting reaction products in one or more of the decay channels energetically available to the compound state. Some of the many possible decay routes are given by

$$e^- + A \rightarrow A^{-**} \begin{cases} \nearrow e_s^- + A \\ \rightarrow e_s^- + A^* \\ \searrow e_s^- + A^+ + e_e^- \end{cases} \tag{3.37}$$

The first illustrates the decay of the excited negative-ion to produce an elastically scattered electon and stable atom where, in this process, the detection product is the elastically scattered electron. In the second decay route, both the inelastically scattered electron and metastable atom are detected as is the photon emitted in the decay of the metastable atom; in the third process, ionization results in a scattered and ejected electron with the positive-ion core, all of which may be detected experimentally. Such studies involve the measurement of either the total or differential scattering cross section (or a quantity proportional to these) for the reaction products as a function of the incident electron energy. A number of intermediate and final states involved in the decay process have been studied extensively over the years in a wide variety of crossed-beam studies by the detection of meta-stable atoms [7,55,73], decay photons [57,74], and in the inelastic electron channel [43,75], to name only a few. An alternative to these crossed-beam studies involves studying the passage of an electron beam through the target gas and detecting those electrons that are not scattered; this is the so-called transmission technique. In the appropriate geometry, the transmitted current I_t is related to the incident current I_0 by the relation $I_t = I_0 \exp(-N\sigma L)$, where N is the gas density, σ is the total cross section, and L is the length of the collision chamber. This technique was used extensively by Schulz and his co-workers to provide the bulk of the early information on atomic and molecular negative-ion resonances. More recently this method has been used mainly in the study of molecular systems. A related technique, which in many aspects is very similar to the transmission technique but actually detects the scattered electrons is the trapped-electron method, first used by Schulz [76]. The conventional trapped-electron technique detects only those electrons that have been inelastically scattered. Trans-mission experiments are arguably the most sensitive means of detecting the presence of narrow negative-ion resonances, but all information about the energy and the scattering angle of those that have undergone collisions is lost. As a result, transmission experiments cannot provide conclusive evidence leading to the unambiguous classification of negative-ion resonan-ces or their decay modes. Such information can be obtained only from measurements involving particles that arise directly from the decay of the resonant state.

In the study of temporary negative-ions (resonances), important features of the experiment are energy resolution, sensitivity, and precision of the incident energy calibration. Experimental techniques have been refined such that low-energy electron beams with energy spreads < 25 meV (FWHM) can now be readily obtained as opposed to the very early days when an energy resolution of ~100 meV was the best experimentalists could obtain [77]. In the late 1970s this improvement in energy resolution provided a wealth of detail at the higher end of the He$^-$ resonance spectrum. More recent experiments are now producing resolutions below 10 meV [78,79]. The first detailed studies of the higher-lying reaches of the He$^-$ resonance spectrum were made by Heddle and co-workers [54,80], who measured optical excitation functions, namely, photon detection, a method particularly suitable for detecting resonances at higher energies, with tentative classifications. However, the most extensive study of these features was made by Buckman et al. [7] and subsequently Buckman and Newman [8], who measured the yield of metastable atoms resulting from electron impact excitation of helium, in the range of incident electron energy 19.8–24.6 eV. Briefly, the apparatus consisted of an electron energy selector that uses a 180° hemispherical electrostatic deflector and a series of lenses to produce an electron beam at the target. This electron beam has an energy that may be adjusted over the range 5–60 eV and a typical energy spread of 20 meV. The electron beam is crossed at right angles with a beam of helium atoms emerging from a hypodermic needle. The use of a gas beam, as opposed to a gas cell, is preferred in order to reduce Doppler broadening. The detector for metastable atoms consists of a channel electron multiplier (CEM) which is located at a 14° angle with respect to the initial gas beam direction to allow for the recoil of gas atoms due to electron impact. Metastable excitation functions are due almost exclusively to negative-ion resonances that decay either directly into the metastable levels or into higher excited states that subsequently decay into metastable levels, that is, directly or via cascading. The transit time for gas atoms between the interaction region and the metastables detector is typically 10^{-4} s, and the time spread in arrival of simultaneously excited metastables at the detector is of the same order. For a particular energy resolution it is therefore necessary to ensure that the range of incident electron energies swept within this time is small compared with the required maximum energy spread. Buckman and co-workers present the yield of metastable atoms as a function of the incident electron energy schematically from which distinctive groups in the resonant structure can be clearly seen and are labeled by the appropriate quantum number. Many high-lying partially overlapping sharp resonance structures spanning ≈22.4–24 eV can be seen in the spectrum, but the experimental difficulty of energy resolution means that the widths are often limited by the energy width of the incident beam. The narrow widths indicating relatively long lifetimes are surprising since, as we have seen, there are many channels into which such states can decay.

Buckman et al. [7] traced the occurrence of resonances from $n = 3$ to $n = 7$, where n is the lower of the two principal quantum numbers n_1 and n_2 of the two excited electrons. The observed resonances were classified into four prominent symmetry classes: $^2S, {}^2P^o, {}^2D$, and $\bar{s}\,^2S$ [7,55,75,81]. Theoretical calculations [70,81] indicate that these resonances can be further classified into two distinct groups. The first of these, proposed by Nesbet [81] describe narrow features that are associated with the threshold of a neutral excited state. They are formed by the attachment of an electron in the polarization potential associated with the neutral excited state (parent) and are referred to as "non valence" or *intershell resonances*. Resonances in the second class are broader and, in general, are not closely associated with any threshold, but are mixtures of configurations. These states are referred to as "grandparent" or *intrashell resonances*. The lowest three (in energy), of the four symmetry classifications, are intrashell resonances [55,75,81], with $n_1 = n_2$ while the final resonance is of the intershell type, with $n_1 \neq n_2$. Buckman and Newman [8] traced all four of these features to $n = 6$ and the $^2S, {}^2P$, and $\bar{s}\,^2S$ features to $n = 8$ with an additional tentative observation of the lowest ^2S feature at $n = 9$. In order to reveal the weak structures in this high-energy region ($E > 24.28\,\text{eV}$), several hundred hours of data accumulation were required.

Buckman et al. use the standard independent electron configurations to describe the observed resonances, which, in the following section, we also use for the sake of brevity and comparison with experiment. However, following the early observation of the absorption spectra of DES of helium [82], it was immediately recognized that a complete understanding of these states requires a fundamental departure from the conventional independent electron approach. One of the goals of studying DES is to find a new way of characterizing electron correlations or more precisely, to find a new set of quantum numbers that characterize the correlations between two excited electrons. A number of semiempirical techniques have been developed as an aid in the classification of negative-ion resonances, including the use of a modified Rydberg formula [83] which we will discuss further in the next section, but the most widely accepted two-electron classification scheme is the so-called (K, T, A) classification proposed by Lin (see Ref. 13 and references cited therein). In this scheme, a given state of a two-electron atom is designated by the notation

$$_\alpha(K, T)_\beta^{A\ 2S+1}L^\pi \tag{3.38}$$

where L and S are the total angular momentum and spin respectively, and π is the parity. The quantum number β, usually denoted by N, is the principal quantum number of the inner electron and α, usually denoted by n, is the principle quantum number of the outer electron so that $n \geq N$. The quantum numbers K and T have been determined by the use of a group theoretical method and are obtained by diagonalizing the square of $|\mathbf{A}_1 - \mathbf{A}_2|$, where $\mathbf{A}_1$ and $\mathbf{A}_2$ are the Runge–Lentz vectors. The physical meanings for K and T can be described briefly as follows: K is related to $- < \cos\theta_{12} >$ so that the larger the positive K, the closer $- < \cos\theta_{12} >$ is to unity and the electrons assume positions on opposite sides of the nucleus. The quantum number T describes the orientations between the orbitals of the two electrons, specifically, the projection of L onto the axis $\mathbf{r}_1 - \mathbf{r}_2$. For example, a state with $T = 0$ implies that the two electrons are moving in the same plane. The possible values of K and T for given β, L, and π are determined by

$$T = 0, 1, 2, \ldots, \min(L, \beta - 1)$$
$$K = \beta - 1 - T, \beta - 3 - T, \ldots, -(\beta - 1 - T) \tag{3.39}$$

The quantum number A provides a direct indication of the radial correlation in the two-electron wavefunction and can take on the values $0, \pm 1$ and is defined by

$$A = \begin{cases} \pi(-1)^{S+T} & \text{for} \quad K > L - \beta \\ 0 & \text{otherwise} \end{cases} \tag{3.40}$$

In terms of these correlation quantum numbers, K, T and A, all doubly excited states of two-electron atoms can be uniquely designated. In the next section, we mainly use the conventional notation of Buckman et al., which is admittedly less accurate than the two-electron quantum numbers discussed here. However, recognizing the wide acceptance of the (K, T, A) classification, we have also expressed the theoretically determined eigenenergies in terms of this scheme.

3. Results and Discussion

In this section we present and discuss the results of the calculation outlined in Section III.A.1 concerning the $L = 0$ DES of He$^-$. In the conventional independent electron notation, we present resonance energy positions for He$^-$ $(1s, {}^2S(n_1 s n_2 s, {}^1S))^2 S$, which in terms of the (K, T, A) classification are given by

$$_\alpha(K, T)^{A\ 2S+1}_\beta L^\pi = n_2(n_1 - 1, 0)^0_{n_1}\ {}^1 S^e \tag{3.41}$$

In Table (III) we present our resonance energy positions for both intrashell and intershall ^2S states that are compared with experiment. Our calculated eigenvalues are found to correspond to those of Buckman and colleagues [7,8] for

$$N = n_1 + n_2 - L - 1 \qquad (3.42)$$

The lower of the ^2S features are given by N values where $n_1 = n_2 = n$, whereas those of the higher ^2S features correspond to values of N for $n_1 = n$, $n_2 = n + 1$, where N is the hyperspherical radial quantum number introduced in Section III. A.1 Thus our classification is consistent with those of Nesbet [81], and an equal sharing of energy, $n_1 = n_2$, or almost equal sharing, $n_2 = n_1 + 1$, is obtained. This reflects the energy repartitioning to increase stability expected in Wannier ridge features. Essentially any partitioning of E, into (E_1, E_2), where $E = E_1 + E_2$, is possible; equally well any partitioning of $N + L + 1$, into (n_1, n_2) couples where $N + L + 1 = n_1 + n_2$, is possible. As n increases motion becomes increasingly localised near $\theta_{12} = \pi$ as the electrons move to minimize their mutual repulsive energy and strong mixing of higher l values is expected. However, we are concerned with the partitioning relevant to the central resonance peak while recognizing that other (n_1, n_2) couples are involved in the broadening of the resonance.

In Table IV, we compare our results (theory a) with those of Rau [56] theory and Lin and Watanabe [64] (theory c), both of which are semi-empirical, based on generalized Rydberg quantum-defect formulas; with those of Komninos et al. [65] (theory d), which is a multiconfiguration Hartree–Fock theory; with those of Rost and Briggs [66] (theory e), which is a diabatic molecular treatment; and finally with those due to Fon et al. [67] (theory (f), who conducted an R-matrix calculation. For $n \in [6, 9]$, our results lie within experimental error. For $n \in [3,5]$, our results are a little on the low side by 0.111, 0.057, and 0.007 respectively. For $n \in [3, 4]$ and R-matrix theory (theory e [67]) gives the best agreement. By the very semiclassical nature of our near-threshold analysis, the accuracy of our results at the higher n values would be expected to exceed that for the lower n values.

In Table (V) we compare our relative intensities, given by $|E_N|\frac{1}{2}\Gamma_N$ (3.36), with those of Buckman and Newman [8]. For $N = 13, 15, 17$—that is, for the highest Rydberg states and for $n = 7, 8, 9$—we see that the ratio of our Wannier results to the experimental results has reached a stable converged value, namely, $4.4(\pm 0.1) \times 10^{-2}$ for $N \in [13, 17]$, which embraces an error tolerance of $< 30\%$. Further evidence of the suitability of (3.36), especially for higher N, can be seen in Table VI, which presents the calculated intensities normalized with respect to the highest experimental intensity value.

We have also investigated the variation in the intensities of these resonances as a function of N. By running our code up to $N \approx 300$, we determined that for $N \to \infty$, $E_R^N \sim N^{-2}$ and $\Gamma_N \sim N^{-3}$ with the net effect that $I_N \sim N^{-5}$. This N^{-2} and N^{-3} behavior is to be expected, for instance, by taking the small x behavior of u_1 and u_2 in (3.12) and (3.13), evaluating as a sum of two $_3F_2$ hypergeometric functions, and analytically continuing via Barnes complex contour integrals. However, for lower N values the result is quite different: $N^{-4.89}$ for $N \in [21, 29]$ and $N^{-4.88 \pm 0.01}$ for $N \in [13, 17]$. This can be compared to the results of Feagin and Macek [84], namely $N^{-5.254} = N^{-3-2m_{12}}$, Rau [] (1984), namely $N^{-6.254}$, Heim and Rau [] (1995), namely $N^{-6.5}$ and Buckman and Newman [8], namely for $N \in [13, 17] N^{-5.2 \pm 0.7}$.

Essentially we have solved the Bohr–Sommerfeld quantization formula for a modified Coulomb potential. This can be seen if we reconsider Eq. (3.18) and remove the last term under the square-root sign to yield

$$c^2 \int_0^1 dx \left(\frac{1}{x} - 1 \right)^{\frac{1}{2}} = \left(N + \frac{1}{2} \right) \pi \tag{3.43}$$

which, if we return to our original variables ρ and E_N using (3.17) and (3.19), becomes

$$\int_0^{-\frac{Z_0}{E_N}} d\rho \left(2E_N + \frac{2Z_0}{\rho} \right)^{\frac{1}{2}} = \left(N + \frac{1}{2} \right) \pi. \tag{3.44}$$

This is clearly the Bohr–Sommerfeld quantisation rule for a purely Coulomb potential. In other words, if, at the outset, we had neglected the $(\Delta\alpha)^2$ and $(\Delta\theta_{12})^2$ expansion terms in (2.9) by assuming strict localization on the Wannier ridge, the sequence of eigenvalues would be as described by (3.44). However, the potential has a rather flat saddle surface, particularly in the θ_{12} variable, so that deviation in α and θ_{12} away from the saddle is not completely negligible. The inclusion of the $(\Delta\alpha)^2$ and $(\Delta\theta_{12})^2$ terms in the expansion of the potential result in a modified Coulomb potential quantization rule where the modification is in the form of logarithmic derivatives of the Wannier functions (3.12) and (3.13). This is not surprising since, as we have seen in Section II.A.2, u_1 is the Wannier message carrier. It is well known that the Coulomb potential Z_0/ρ has discrete eigenvalues given by the generalized Bohr energy expression

$$E_n = I - \frac{Z_0^2}{2(n + \frac{5}{2})^2}, \qquad n = 0, 1, 2, \ldots \tag{3.45}$$

where $I = 24.588\,\text{eV}$ for helium. This is altered to

$$E_n = I - \frac{1}{2}\left[\frac{Z_0(Z - \sigma)}{(n + \frac{5}{2} - \delta)}\right]^2 \qquad (3.46)$$

for a modified Coulomb potential that is used to parametrize the energies of two-electron configurations and has proved beneficial as an aid in the classification of negative-ion resonances. The situation is such that the two electrons mutually screen each other from the core charge Z so that they each experience an effective charge $(Z - \sigma)$, where σ is a screening parameter. The effect of core penetration by either of the two Rydberg electrons is accounted for by the quantum defect δ and the function Z_0 is given by $Z_0 = 2\sqrt{2}(Z - \frac{1}{4})$ [56]. Using fixed values for both the screening constant and the quantum defect, Rau was able to fit, within experimental error, the energies as determined by Buckman et al. for the lowest member of each multiplet for values of n from 3 to 7. Feagin and Macek [84] and Macek and Feagin [85] have also shown the existence of a Rydberg series of resonances converging on the first ionization threshold by extrapolating Wannier continuum wavefunctions, which describe two-electron escape, to the energy region just below the ionization threshold. However, their formula required unrealistic values of the screening parameter in order to fit the experimental data over the whole range of $n = 2$–7. They also obtain an equation for the partial widths of these resonance states that is based on an argument that the observed intensity, $\langle I_n \rangle$, for these resonances can be related to the Wannier threshold law at high n:

$$\langle I_n \rangle = \frac{\Gamma_n}{E_{n+1} - E_n} \underset{n \to \infty}{\approx} E_n^{1.127} \qquad (3.47)$$

where Γ_n is the partial width. From this they show that the partial widths of these states should vary as $n^{5.254}$ for large n. This disagrees with the result of Rau [56] who obtains an $n^{-6.254}$ variation based on the normalization of a Coulomb function in six dimensions.

C. Analytic Continuation to Below-Threshold $L = 1$ and $L = 2$ States of He$^-$

1. Introduction

We now consider the analytic continuation of Crothers' near-threshold ionization theory to below threshold for $L = 1$ and $L = 2$ states of He$^-$ [60]. For the $L = 0$ case the singlet wavefunction developed in method I, as described in Section II.B was extended to negative energy. This was considered to be suffice for the determination of $L = 0$ He$^-$ eigenvalues since, in view of the Pauli exclusion principle, the $^3S^e$ symmetry requiring $l_1 = l_2$ and $s_1 = s_2$ is suppressed and we need be concerned only with the ^1S states. The partial cross section associated with the $^3S^e$ state, which is purely antisymmetric, has been proposed by Greene and Rau [14] to follow a E^n law where n is greater than 1.127 (viz., $n = 3.881$), and is therefore suppressed at threshold. It is therefore essential that, at the outset, we clarify which symmetries we expect to be of importance, in terms of the Wannier ridge configuration, for $L = 1$ and $L = 2$ states.

According to the $LS\pi$ conservation scheme, where the conservation of partity eliminates the states with $\pi(-1)^L = -1$, the allowed symmetries for the given values of total angular momentum are $^{1,3}P^o$ and $^{1,3}D^e$. However, it is argued [14] that the $^1P^o$ and $^3D^e$ states are suppressed on the Wannier ridge. This can be seen if we consider the behaviour of the $^1P^o$ wavefunction under inversion, $\mathbf{r}_i \rightarrow -\mathbf{r}_i$ and exchange, $\mathbf{r}_1 \leftrightarrow \mathbf{r}_2$:

$$\Psi(\mathbf{r}_1, \mathbf{r}_2) = -\Psi(-\mathbf{r}_1, -\mathbf{r}_2)$$
$$\Psi(\mathbf{r}_1, \mathbf{r}_2) = \Psi(\mathbf{r}_2, \mathbf{r}_1) \tag{3.48}$$

which on restriction of the region to the Wannier ridge $\mathbf{r}_1 = -\mathbf{r}_2$ becomes

$$\Psi(\mathbf{r}_1, -\mathbf{r}_1) = \Psi(-\mathbf{r}_1, \mathbf{r}_1)$$
$$\Psi(\mathbf{r}_1 - \mathbf{r}_1) = \Psi(-\mathbf{r}_1, \mathbf{r}_1) \tag{3.49}$$

Clearly the contradiction shown in (3.49) concludes that the symmetry properties of the $^1P^o$ state are not consistent with the Wannier ridge configuration. This result will also hold for $^3D^e$. Therefore, in this section, we consider the $^3P^o$ and $^1D^e$ symmetries: states in which ($LS\pi$) are all odd or all even with the aim of producing solutions of the two-electron Schrödinger equation with the retention of angular momentum terms that are of the order of ρ^{-2}. What is significant in this calculation, since we are to apply a Bohr–Sommerfeld-type quantization (3.16), is the overall contribution made from terms of the order of ρ^{-2}. The final-state wavefunction describing the Wannier ridge motion of two electrons with negative energy must also avoid any asymptotic approximations, since the electrons are essentially bound.

As discussed in Section I.B.1, the angular momentum contribution to the wavefunction may be considered in terms of the Euler angles $(\theta_1, \phi_1, \psi_B)$ that will describe the orientation of the three-body triangle (electron–electron–core) in space. The shape of this triangle is, as usual, defined by the hyperspherical coordinates $(\rho, \alpha, \theta_{12})$. As we have seen in Section I.B.1, the appropriate symmetrised wavefunction for $^3P^o$ states is given by

$$\Psi = f\cos\theta_1 - \tilde{f}\cos\theta_2$$
$$= f P_1(\cos\theta_1) - \tilde{f}P_1(\cos\theta_2) \tag{3.50}$$

where the tilde marks the interchanged function $\tilde{f}(\mathbf{r}_1, \mathbf{r}_2) = f(\mathbf{r}_2, \mathbf{r}_1)$ and particle exchange results in $\Delta\theta_{12} \to \Delta\theta_{12}$ and $\Delta\alpha \to -\Delta\alpha$. The Legendre polynomials are, as usual, denoted by P_L. The functions f and $\tilde{f}$ satisfy the radial equation

$$\left[H_0 + X^2 + \frac{2\zeta(\alpha, \theta_{12})}{\rho} \right] f$$
$$+ \frac{2}{\rho^2} \left[\sec^2\alpha(\cot\theta_{12}\frac{\partial}{\partial\theta_{12}} - 1)f + \csc^2\alpha\csc\theta_{12}\frac{\partial\tilde{f}}{\partial\theta_{12}} \right] = 0 \tag{3.51}$$

For the $L = 2$ case the derivation of the wavefunction using the Breit–Euler angles and working backward for the reduced radial equations is a difficult and arduous task. This has been done by Selles et al. [17], who find that the $L = 2$, $M = 0$ wavefunction may be written as

$$\Psi = f_0(3\cos^2\theta_1 - 1) + f_1(3\cos^2\theta_2 - 1) + f_2(3\cos\theta_1\cos\theta_2 - \cos\theta_{12})$$
$$= f_0 P_2(\cos\theta_1) + f_1 P_2(\cos\theta_2) + f_2(3P_1(\cos\theta_1)P_1(\cos\theta_2) - \cos\theta_{12})$$
$$\tag{3.52}$$

where for the singlet case

$$f_2 = \tilde{f}_1, \qquad f_3 = \tilde{f}_3 \tag{3.53}$$

The reduced radial equations therefore involve only f_1 and f_3 and are given as

$$\left[H_0 + X^2 + \frac{2\zeta(\alpha, \theta_{12})}{\rho} \right] f_1 + \frac{1}{\rho^2\cos^2\alpha} \left[4\frac{\partial f_1}{\partial\theta_{12}}\cot\theta_{12} - 6f_1 \right]$$
$$+ \frac{1}{\rho^2\sin^2\alpha} \left[-\frac{2}{\sin\theta_{12}}\frac{\partial f_3}{\partial\theta_{12}} \right] = 0$$
$$\tag{3.54}$$

$$\left[H_0 + X^2 + \frac{2\zeta(\alpha, \theta_{12})}{\rho} \right] f_3 + \frac{1}{\rho^2\cos^2\alpha} \left[2\cot\theta_{12}\frac{\partial f_3}{\partial\theta_{12}} - 2f_3 - \frac{4}{\sin\theta_{12}}\frac{\partial f_3}{\partial\theta_{12}} \right]$$
$$+ \frac{1}{\rho^2\sin^2\alpha} \left[-2f_3 + 2\cot\theta_{12}\frac{\partial f_3}{\partial\theta_{12}} - \frac{4}{\sin\theta_{12}}\frac{\partial f_3}{\partial\theta_{12}} \right] = 0$$
$$\tag{3.55}$$

This is in agreement with Roth [37], who predicts an increasing number of coupled equations as L increases above the P^o states. In contrast the coordinates used by Feagin and Macek [84] give uncoupled equations for all values of L. Selles et al. proceed to determine the Wannier wavefunctions as $E = 0$. For the P^o case they follow Green and Rau [15] splitting the wavefunction into symmetric (*gerade*) and antisymmetric (*ungerade*) parts and solve in direct analogy with Rau [9].

Following Selles et al. [17], we employ the Breit–Euler angles for $M = 0$, $L = 1, 2$. We proceed with the ansatz

$$\Psi = f(\rho, \alpha, \theta_{12}) P_L(\cos\theta_1) \tag{3.56}$$

which on symmetrizing becomes

$$\Psi = f(\rho, \alpha, \theta_{12}) P_L(\cos\theta_1) \pm \tilde{f}(\rho, \alpha, \theta_{12}) P_L(\cos\theta_2) \tag{3.57}$$

where the upper (lower) sign is for the singlet–symmetric (triplet–antisymmetric) case. For the $^3P^o$ case, Eq. (3.57) is consistent with (3.50). However, in this treatment we have simplified the $L = 2$ case by neglecting the coupling term given by the coefficient of f_3 in (3.52). If we consider this in terms of a spherical harmonic representation (2.104), and (2.105), it is clear we have taken components comprising of products of Y_{L0} and Y_{00} where

$$Y_{L0}(\hat{\mathbf{r}}_1) = \left(\frac{2L+1}{4\pi}\right)^{\frac{1}{2}} P_L(\cos\theta_1) \tag{3.58}$$

and, for the $L = 2$ case, neglected the cross product terms such as $Y_{1-1}Y_{11}$. This simplifies the problem in that we avoid coupled equations of the type (3.54) and (3.55), as we shall see in the next section.

2. Theory

Having established the form of the wavefunction as (3.57), we now seek the radial equation for the function $f(\rho, \alpha, \theta_{12})$ by employing the Schrödinger equation

$$\left[\frac{1}{\rho^5}\left(\rho^5\frac{\partial}{\partial\rho}\right) + \frac{1}{\rho^2\sin^2\alpha\cos^2\alpha}\frac{\partial}{\partial\alpha}\left(\sin^2\alpha\cos^2\alpha\frac{\partial}{\partial\alpha}\right)\right.$$
$$\left.- \frac{L^2(\hat{\mathbf{r}}_1)}{\rho^2\cos^2\alpha} - \frac{L^2(\hat{\mathbf{r}}_2)}{\rho^2\sin^2\alpha} + X^2 + \frac{2\zeta(\alpha, \theta_{12})}{\rho}\right]\Psi = 0 \tag{3.59}$$

where

$$L^2(\hat{\mathbf{r}}_i) = -\left[\frac{1}{\sin\theta_i}\frac{\partial}{\partial\theta_i}\left(\sin\theta_i\frac{\partial}{\partial\theta_i}\right) + \frac{1}{\sin^2\theta_i}\frac{\partial^2}{\partial\phi_i^2}\right], \qquad i = 1,2 \quad (3.60)$$

[19]. The substitution of (3.56) into (3.59) results in the following equation for the radial wavefunction f

$$\left[\frac{1}{\rho^5}\frac{\partial}{\partial\rho}\rho^5\frac{\partial}{\partial\rho} + \frac{1}{\rho^2\sin^2 2\alpha}\frac{\partial}{\partial\alpha}\sin^2 2\alpha\frac{\partial}{\partial\alpha} + \frac{4}{\rho^2\sin\theta_{12}}\frac{\partial}{\partial\theta_{12}}\sin\theta_{12}\frac{\partial}{\partial\theta_{12}}\right.$$
$$\left. + 2E + \frac{2\zeta(\alpha,\theta_{12})}{\rho} - \frac{2L(L+1)}{\rho^2}\right]f = 0$$

$$(3.61)$$

where the factor of 2 in the last term comes from $\sec^2\alpha$ evaluated at $\alpha = \pi/4$. This substitution also results in the cross-term

$$\frac{4\sin\theta_1(-\sin\theta_1\cos\theta_2 + \cos\theta_1\sin\theta_2\cos(\phi_1 - \phi_2))}{\rho^2\sin\theta_{12}}\frac{P_L'(\cos\theta_1)}{P_L(\cos\theta_1)}\frac{\partial f}{\partial\theta_{12}} \quad (3.62)$$

where the prime indicates $d/[d(\cos\theta_1)]$. This term can be shown to vanish on the Wannier line ($\alpha = \pi/4, \theta_{12} = \pi$) where the singularity due to $\sin\theta_{12}$ in the denominator requires the use of l' Hôpital's rule with the numerator expressed in terms of the following identity:

$$\cos\theta_{12} = \cos\theta_1\cos\theta_2 + \sin\theta_1\sin\theta_2\cos(\phi_1 - \phi_2) \quad (3.63)$$

This term, which would otherwise inconsistently contain Euler angles, is therefore neglected, and it follows that on the Wannier ridge $f = \tilde{f}$.

Equation (3.61) is equivalent to (3.51) or $L = 1$ with $\partial/\partial\theta_{12}$ terms vanishing on the Wannier ridge. Equation (3.51) has appeared frequently in the literature, such as in Eq. (8) of Selles et al. [17] Eq. (10) of McCann and Crothers [44], and Eq. (3) of Roth [37] with the cross-term

$$\frac{2}{\rho^2}\left[2\cot\theta_{12}\frac{\partial f}{\partial\theta_{12}} + \frac{2}{\sin\theta_{12}}\frac{\partial\tilde{f}}{\partial\theta_{12}}\right] \quad (3.64)$$

equivalent to (3.62) vanishing on the Wannier ridge for $f = \tilde{f}$ according to

$$\frac{2(1 + \cos\theta_{12})}{\sin\theta_{12}}\frac{\partial f}{\partial\theta_{12}}$$
$$\rightarrow -\frac{2\sin\theta_{12}}{\cos\theta_{12}}\frac{\partial f}{\partial\theta_{12}}$$
$$= 0 \quad \text{for} \quad \theta_{12} = \pi \quad (3.65)$$

A consistency is also obtained between Eq. (3.61) and the radial equations of Selles et al. for the case of $L = 2$ with the first of the two coupled equations (3.55) containing the same ρ^{-2} factor, namely, $2L(L+1)/\rho^2 = 12/\rho^2$.

We are now confident that the correct radial equation has been obtained and can be solved in direct analogy with Eq. (2.60) by method I, as outlined in Section II.A.2, if the ω term is taken to be L-dependent according to

$$\omega^2 = 2E_N + \frac{2Z_0}{\rho} - \frac{2L(L+1)}{\rho^2} \tag{3.66}$$

Therefore the analysis from Section II.A.2 is applicable with $\rho = 0$ becoming even more classically forbidden. However, the inclusion of the L term in ω does require a careful look at the intricate details of Crothers' theory. First, consider the Riccati equation given in (2.14), which is reduced to a linear second-order equation by removing the nonlinear S_i^2 term with the substitution given by Eq. (2.18). With employment of (3.66) it becomes

$$\rho^2 w^2 \frac{d^2 u_i}{d\rho^2} + \left(2\rho w^2 - Z_0 + \frac{2L(L+1)}{\rho} \right) \frac{du_i}{d\rho} = \frac{Z_i u_i}{\rho}, \qquad i = 1, 2 \tag{3.67}$$

which is transformed to

$$[2E_N \rho^2 + 2Z_0 \rho - 2L(L+1)]F_i''$$

$$+ \left[4(m_i + 1)E_n \rho + (4m_i + 3)Z_0 - \frac{2(2m_i + 1)L(L+1)}{\rho} \right] F_i'$$

$$t; \quad + \left[2m_i(m_i + 1)E_N - \frac{2m_i^2 L(L+1)}{\rho^2} \right] F_i = 0 \tag{3.68}$$

with the following substitution:

$$u_i = \rho^{m_i} F_i(\rho), \qquad i = 1, 2 \tag{3.69}$$

As before, m_i is given by the roots

$$m_{i1} = -\frac{1}{4} - \frac{1}{2}\mu_i, \qquad m_{i2} = -\frac{1}{4} + \frac{1}{2}\mu_i \tag{3.70}$$

with

$$\mu_i = \frac{1}{2}\left(1 + \frac{8Z_i}{Z_0} \right)^{1/2} \tag{3.71}$$

Eqs. (2.27) and (2.28). This is linear differential equation of the Heun type [86], which has four regular singular points at $\rho = 0, \infty$ and the two solutions of $\omega^2 = 0$, namely, $\rho\pm$, where

$$\rho\pm = \frac{-Z_0 \pm \sqrt{Z_0^2 + 4E_N L(L+1)}}{2E_N} \tag{3.72}$$

Detailed knowledge of the Heun equations is limited so an approximate solution to (3.68) is required. We simplify matters by assuming that

$$\rho \sim \frac{1}{|E_N|} \tag{3.73}$$

which is certainly reasonable, so that all the nonazimuthal terms in the coefficients of (3.68) are of the order of $1/|E_N|$. The result is, that for the threshold limit of $E_N \to 0$, we may neglect the L-dependent terms. This $E_N \to 0$ threshold limit is tended to from negative-energy values and is perfectly applicable since we are concerned with the high-lying DES. We can therefore approximate the solutions of (3.68) with those obtained for the $L = 0$ case. For $L = 0$ with $F = F(-E_\rho/Z0)$, a three regular singular point differential equation obtains, given by (2.24), which takes the form of the Gauss hypergeometric differential equation resulting in the Wannier–Peterkop functions given in (3.12) and (3.13). We have now essentially derived the required wavefunction with the simple amendment to the $L = 0$ below-threshold case of an L-dependent ω term. The asymptotic form of the final-state wavefunction for $L \neq 0$ He$^-$ is given by

$$\Psi_f^{-*} = \frac{C_N^{1/2} Y_{LM}(\hat{\mathbf{r}}_1, \hat{\mathbf{r}}_2) 2^{(3/2)} e^{-2r_3 - (i/2)\ln \Delta_0 12}/\sqrt{\pi}}{\rho^{5/2} \sin\alpha\cos\alpha}$$

$$\times \frac{\sin\left[\int_{\rho_+}^{\rho} d\tilde{\rho}(\omega^2 - \omega\{\ln u_2 - i \ln u_1\}')^{1/2}\right.}{[\omega^2 - \omega\{\ln u_2 - i \ln u_1\}']^{1/4}} \tag{3.74}$$

$$\frac{\left. + \frac{\rho^2}{2}\omega(\ln u_1)'(\Delta\alpha)^2 + \frac{\rho^2}{8}\omega(\ln u_2)'(\Delta\theta_{12})^2 + \frac{\pi}{4}\right]}{[\omega^2 - \omega\{\ln u_2 - i \ln u_1\}']^{1/4}}$$

which is completely analogous to (3.10) with two significant changes. The first is the obvious change in the integration limits from $0 \to \rho$ to $\rho_+ \to \rho$ since the classical turning point moves away from the origin for L-dependent states. The second is the change in the form of ω according to (3.66). Of course, the inclusion of angular momentum terms dictates the need to retain all terms of order ρ^{-2}. The substitution of Eq. (2.65) into (3.59) results in such terms, namely, $-15/4\rho^2$ and $4/\rho^2$, previously neglected in the above-threshold analysis. However, these terms do not play a significant role since they are canceled by the $-1/4\rho^2$ Langer modification correction [87] The L-dependent form of ω further complicates the solution of the Bohr–Sommerfeld quantisation expression

$$\int_{\rho+}^{\rho-} d\rho \sqrt{ w^2 - w \left(\frac{d}{d\rho} \ln u_2 - i \frac{d}{d\rho} \ln u_1 \right) } = N\pi + \frac{\pi}{2} \qquad (3.75)$$

which is completely analogous to (3.16) with two complex transition points. Following the method for the $L = 0$ case, we simplify (3.75) with a change of dummy variable according to

$$\rho = \frac{-(Z_0 + Zx)}{2E_N} \qquad (3.76)$$

where

$$Z = \sqrt{Z_0^2 - 2c^2 L(L+1)} \qquad (3.77)$$

and

$$c = \sqrt{-2E_N}. \qquad (3.78)$$

Equation (3.75) then becomes

$$\mathscr{L} \int_{-1}^{1} dx \sqrt{ \varpi^2 - \frac{c}{Z} \varpi \left(\frac{d}{dx} \ln \frac{u_2}{u_1^i} \right) } = c \left(N + \frac{1}{2} \right) \pi \qquad (3.79)$$

where

$$\varpi^2 = -1 + \frac{2Z_0}{Z_0 + \mathscr{L}x} - \frac{2c^2 L(L+1)}{(Z_0 + \mathscr{L}x)^2} \qquad (3.80)$$

which has been solved, numerically, for c by the complex Newton–Raphson method as described in (3.26).

3.   Results and Discussion

The results obtained from this calculation for the $L = 1$ and $L = 2$ resonant energy positions are given in Tables VII and VIII, respectively. They have been compared with the experimental results of Buckman et al. [7] using their notation. In the now-standard notation [13], these states are represented by

$$_\alpha(K,T)^A\,_\beta^{2S+1}L^\pi =_{n_2} (n_1 - 1, 0)^0_{n_1}\,^3P^0, ^1D^e \tag{3.81}$$

It has been found that the calculated eigenvalues for both $L = 1$ and $L = 2$ resonant states correspond to those of experiment for

$$N = 2n - \mathscr{L} - 1 \tag{3.82}$$

where for $L = 1$, $\mathscr{L} = (-1 + \sqrt{17})/2$ and for $L = 2$, $\mathscr{L} = 3$ and where $2n = n_1 + n_2$ in line with the intrashell classification as discussed in Section III.B.2. Thus, for $L = 0$ and $L = 2$, we have an integer hyperspherical azimuthal quantum number but for $L = 1$, the mapping of $2L(L + 1)$ to $\mathscr{L}(\mathscr{L} + 1)$ in the Bohr–Sommerfeld quantization rule results in an irrational hyperspherical azimuthal quantum number.

Many reports have been published by theoretical physicists on the discovery of noninteger quantizations; in particular, numerous references to fractional quantum numbers can be found in the literature [88,89]. An interesting example is the fractional quantum Hall effect (FQHE) where electrons in solids in strong magnetic fields at cryogenic temperatures form a liquid with fractional quantum numbers. This discovery has resulted in a growing interest in fractional quantum numbers among solid-state physicists, whose attention is turning toward quantum mechanical many-body descriptions for electrons in solids. Many solid-state calculations neglected to take account of the interactions between electrons until a new line of thinking was introduced in the early 1980s emphasising the need to include electron interaction with the introduction of a many-body quantum-mechanical wavefunction by Laughlin [90]. Subsequent calculations [89] have successfully determined the odd-integer fractional quantisations observed in the FQHE using the Laughlin many-body wavefunction and the Wannier–Hubbard model. Therefore, despite memories of our early quantum-mechanics teaching, we feel that in view of the preceding discussion and the low-energy high-interaction nature of this calculation, the appearance of a noninteger quantum number is not surprising. Infact, in Rau's quantum-mechanical treatment of threshold ionization, discussed in Section II.A.2, irrational orbital quantum numbers occur when the $L = 0$ above-threshold theory is extended to $L = 1$ [37]. For an orbital angular momentum term of the form $-l(l + 1)/\rho^2$, Eq. (8) of Roth [37] the orbital quantum number is $l = \frac{3}{2}$ for the $L = 1$, $M = 0$ odd-parity symmetric and antisymmetric states and $l = \frac{2}{3}\sqrt{3} - \frac{1}{2}$ for the even-parity $L = 1$, $M = 0$ state.

The P and D resonances are close together and overlap significantly so that the width for a given E_R^N is smeared by the neighboring resonances [75] As such, in our search for the resonance energy position, we took the narrowest (central and highest) peak, corresponding to the complex eigen-energy E_N closest to the real axis. On allowing Γ_N to tend to zero, a convergence limit was reached yielding the energy positions tabulated in Tables VII and VIII. Agreement with experiment is good, improving for higher values of N. In particular, for $L = 1$ and $n = 4$ or 5 and for $L = 2$ and $n = 6$ or 7, experiment and theory agree within the experimental error. This is to be expected because of the semiclassical nature of the calculation and our perturbative solution of the Heun differential equation.

In order to determine Γ_N for these "smeared" overlapping states, a sum over the unobserved final states [84] was taken in order to obtain an average width Γ^a. Since our intensity is taken as being proportional to $\frac{1}{2}\Gamma_N|E_N|$, [Eq. (3.36)], this sum takes the form

$$\Gamma^a = \frac{\frac{1}{n}\sum_{i=1}^{n} \Gamma_N^i |E_N^i|}{\frac{1}{n}\sum_{i=1}^{n} |E_N^i|} \tag{3.83}$$

In Table IX we present the Γ^a values for the ^2D case. The lower limit of the sum is determined by the convergence limit for a given n; the upper limit is determined by the position of the neighboring observed resonance. Admittedly, the choice of the upper limit is somewhat arbitrary in that the extent of the overlapping is not definite. Nevertheless, with an appropriate choice of upper limit, the energy positions E_{Γ^a} corresponding to width Γ^a show excellent agreement with experiment. Convergence difficulties have restricted the application of (3.83) to higher n values of the ^2D case only.

D. Doubly Excited States of He

1. Introduction

In this section we apply the $L = 0$ below-threshold theory developed in Section III.B to the doubly excited states of atomic helium. In direct analogy with the He$^-$ case, we consider the electron capture by the parent He$^+(1s)$ to form the DES of atomic He with the grandparent core He^{2+}. The reaction under consideration is given by

$$e^- + \text{He}^+(1s\,^2S) \rightarrow \text{He}(n_1s n_2 s\,^1S) \tag{3.84}$$

where the energy of the incident electron lies below the first ionization potential of He and above the first ionization potential of He$^+$.

The continuum between the first and second ionization thresholds of the helium atom has been investigated theoretically in a number of diverse studies (Fukuda) [66,70,91–94]. This work was initiated by Madden and Codling [82], who, in a photoabsorption experiment, were the first to observe doubly excited autoionizing states of atomic helium. Subsequent photoabsorption experiments [95,96] have yielded further information about states of ^1P symmetry with electron impact experiments [97,98] revealing the full spectrum of S, P, D, F, G states. An infinite number of Rydberg series of autoionizing states exist within the continuum. However, these studies have concentrated predominantly on the lower Rydberg series that converge to the He^+ $n = 2 - 6$ thresholds, with the more recent diabatic molecular approach of Rost and Briggs [68] producing intrashell S resonance positions for $n \leq 15$.

We present intrashell and intershell S resonance energy positions for the higher quantum numbers $5 \leq n \leq 15$, with the semiclassical limitations of our theory producing poor results for lower n values. Minor adjustments are made to the theory presented in Section III.B these are outlined in the next section.

2. Theoretical Amendments

Clearly the core charge in this case is $Z = 2$, which reflects the increasing Coulomb attraction of the He^{2+} core over the $Z = 1$ He^+ core. The three-body potential (1.7) for a core charge Z is given as

$$\xi(\alpha, \theta_{12}) = \frac{Z}{\cos \alpha} + \frac{Z}{\sin \alpha} - \frac{1}{(1 - \cos \theta_{12} \sin 2\alpha)^{1/2}} \quad (3.85)$$

which yields the zeroth- and second-order expansion coefficients, when expanded according to (2.9), as

$$Z_0 = \frac{4Z - 1}{\sqrt{2}} \qquad Z_1 = \frac{12Z - 1}{\sqrt{2}} \qquad Z_2 = -\frac{1}{\sqrt{2}}. \quad (3.86)$$

Therefore, for $Z = 2$ we must replace previous $Z_i (i = 0 - 2)$ values, given by Eqs. (2.10) and (2.11), with

$$Z_0 = \frac{7}{\sqrt{2}}, \qquad Z_1 = \frac{23}{\sqrt{2}}, \qquad Z_2 = -\frac{1}{\sqrt{2}}. \quad (3.87)$$

This subsequently affects the magnitude fo the Wannier indices according to Eqs. (2.27) and (2.28):

$$m_{i1} = -\frac{1}{4} - \frac{1}{2}\mu_i, \qquad m_{i2} = -\frac{1}{4} + \frac{1}{2}\mu_i \quad (3.88)$$

where

$$\mu_1 = \frac{\sqrt{100Z - 9}}{4Z - 1}, \qquad \mu_2 = \frac{\sqrt{4Z - 9}}{4Z - 1}. \qquad (3.89)$$

Having established the minor changes required as a result of the increased core charge, we now proceed to determine the complex eigenenergies E_N in exactly the same manner as that outlined in Section III.B.

3. Results and Discussion

The results from the present calculation, given in Tables X and XI, show poor agreement with the available experimental and theoretical results for values of $n \leq 7$, for both the intrashell $(n_1 = n_2)$ and intershell $(n_1 \neq n_2)$ states. Other theories, such as the complex coordinate rotation theory [69, 92] close-coupling method [94], and hyperspherical approach (Fukuda) have produced accurate energy positions for lower n but require greater numeric effort for higher values of the principal quantum number. Rost and Briggs [66] have produced intrashell energy positions for $n \leq 15$ using an adiabatic molecular potential. For $n \leq 7$ these results are in excellent agreement with the highly accurate results of Ho [69, 92] We have compared our resonance energy positions for He** $^1S^e$ intrashell states for $7 \leq n \leq 15$ with those of Rost and Briggs [66]. We find the results for $7 \leq n \leq 10$ to be in agreement within 10% and for $n \geq 11$ within 4%. Thus agreement with other theory is seen to improve for increasing n. Rost and Briggs have employed a simple hydrogenic wavefunction, of the form $e^{-\alpha(r_1 + r_2)}$, which is clearly symmetric in r_1 and r_2. Thus, this approach does not allow for the determination of resonance positions where the energy of the two electrons is not evenly distributed; that is, only energy positions for intrashell states could be calculated.

IV. NEAR-THRESHOLD IONIZATION

A. Introduction

In Section II.A.2 we saw how the Kohn variational principle was applied perturbatively in method I with the uniform semiclassical wavefunctions of Crothers, to find an expression for the total and differential cross section for the threshold ionization of helium. This inaugural calculation was restricted to singlet states and $L \leq 3$. As a first-order perturbation theory was used, approximations were made on the final-state angular dependence, which, as we have seen, leads to the elimination of terms such as the $(3\cos\theta_1 \cdot \cos\theta_2 - \cos\theta_{12})$ component of the $^1D^e$ state. The results of this calculation are in good agreement with experimental data for $\Theta_1 = 60°$, $90°$, and $120°$ with some discrepancies occurring for the remaining angles, namely, $30°$

and 140°. In Section III we have seen, in the context of a below-threshold study of doubly excited states for $L \neq 0$, a return to first principles producing a significant angular momentum term with the symmetrization of states taken as $Y_{LM}(\hat{\mathbf{r}}_1, \hat{\mathbf{r}}_2) \pm Y_{LM}(\hat{\mathbf{r}}_2, \hat{\mathbf{r}}_1)$ in accordance with Selles et al. [17]. This new development of Crothers' theory has produced impressive resonance energy positions for the below-threshold case of DES for $L = 1\,{}^3P$ and $L = 2\,{}^1D$ symmetries. Therefore, our aim in this section is to reconsider the above-threshold ionization theory and to calculate both singlet and triplet contributions to the triple-differential cross sections. The determination of the relative contribution made from triplet and singlet states provides a sound testing ground for any ab initio calculation.

As we have seen from method I, the $L = 0$ Wannier theory is insufficient to account for the high angular complexity in electron impact threshold ionization, where even at threshold an incident electron with sufficient energy to ionize an atom has an appreciable number of partial waves higher than zero. This has been recognized for some time now [14,17,28,30]; however, the relative contribution of certain symmetries to threshold processes has not yet been conclusively established and is the subject of much conjecture in the literature. The triplet contribution is believed to be much weaker than the singlet, and some authors [99] conclude that all triplet amplitudes are essentially zero. Of course, if terms of the order of ρ^{-2} are discarded, as in the case of Altick [99], because they are asymptotically small, then all the angular information is lost. This leads to an $L = 0$ reduced radial equation with the erroneous conclusion that the triplet states do not contribute. As in Section III.C, the symmetries that we expect to contribute on the Wannier ridge are ${}^1S^e, {}^3P^o, {}^1D^e, {}^3F^o$, and so on, with the relative contribution decreasing as L increases. In this work the ${}^3P^o$ and ${}^3F^o$ contributions are considered significant [42], especially for $\Theta_1 = 30°$ and $\Theta_1 = 140°$, where experimentalists have recorded "peaking" of the TDCS at the supplementary value of Θ_2. This peaking has not yet been reproduced theoretically and is the source of the main discrepancy between the results of method I and experiment [17,28,30,100].

In the next section we present the theoretical amendments to method I required to calculate both the singlet and triplet contributions. The proper electron orientation term, which, as we have seen from Section III.C, takes the form $-2L(L + 1)/\rho^2$, is retained as are terms of the order of ρ^{-2}. The results of this calculation are presented and discussed in Section IV.C.

B. Theory

In Sectios III.C.1 and III.C.2 we adopted the ansatz of Selles et al. [17], given by (3.56) as

$$\Psi = f(\rho, \alpha, \theta_{12}) P_L(\cos\theta_1) \tag{4.1}$$

in order to obtain the correct symmetrized wavefunctions for $^3P^o$ and $^1D^e$ doubly excited states. In this section we wish to amend the theory of method I to allow for the determination of triplet-state contributions to TDCS for the electron impact threshold ionization of helium. Therefore, we begin as in Section III.C.1 with the preceding ansatz which, on symmetrizing, becomes

$$\Psi = f(\rho, \alpha, \theta_{12})P_L(\cos\theta_1) \pm \tilde{f}(\rho, \alpha, \theta_{12})P_L(\cos\theta_2) \qquad (4.2)$$

(3.57), where the upper (lower) sign is for the singlet-symmetric (triplet-antisymmetric) case. The substitution of (4.1) into the two-electron Schrödinger equation given by (3.59) produces the following radial equation for the function f:

$$\left[\frac{1}{\rho^5}\frac{\partial}{\partial\rho}\rho^5\frac{\partial}{\partial\rho} + \frac{1}{\rho^2\sin^2 2\alpha}\frac{\partial}{\partial\alpha}\sin^2 2\alpha\frac{\partial}{\partial\alpha} + \frac{4}{\rho^2\sin\theta_{12}}\frac{\partial}{\partial\theta_{12}}\sin\theta_{12}\frac{\partial}{\partial\theta_{12}} \right.$$
$$\left. + 2E + \frac{2\zeta(\alpha,\theta_{12})}{\rho} - \frac{2L(L+1)}{\rho^2} \right]f = 0 \qquad (4.3)$$

The solution to this equation has been considered in Section III.C.2, which can, to a good approximation, be given by the original solution of method I with the ω term (2.15) replaced by the following L-dependent version:

$$\omega^2 = 2E_N + \frac{2Z_0}{\rho} - \frac{2L(L+1)}{\rho^2} \qquad (4.4)$$

We must now consider the effect of this amended ω term on the determination of an expression for the TDCS with the retention of terms of the order ρ^{-2}.

We begin by considering the original expression for the TDCS derived in Section II.A.2 and given by Eq. (2.124) according to

$$\frac{d^3\sigma}{d\hat{\mathbf{k}}_1 d\hat{\mathbf{k}}_2 d(\frac{1}{2}k_1^2)} = \frac{2\pi^2 a_0^2}{k_0}\frac{d}{dE}\frac{\pi Z_2\tanh(2\pi\Im m_{21})}{(2E)^{1/2}}$$
$$\times \exp\left(-\frac{Z_2}{4(2E)^{1/2}}(\Delta\Theta_{12})^2\pi\tanh(2\pi\Im m_{21}) \right)|f(\hat{\mathbf{k}}_1,\hat{\mathbf{k}}_2)|^2$$
$$(4.5)$$

For the case of exchange with helium, which has two indistinguishable electrons, the existence of triplet $(S = 1)$ and singlet $(S = 0)$ states requires that $|f(\hat{\mathbf{k}}_1,\hat{\mathbf{k}}_2)|^2$ in (4.5) be replaced by

$$\frac{1}{4}|f(\hat{\mathbf{k}}_1,\hat{\mathbf{k}}_2) + f(\hat{\mathbf{k}}_2,\hat{\mathbf{k}}_1)|^2 + \frac{3}{4}|f(\hat{\mathbf{k}}_1,\hat{\mathbf{k}}_2) - f(\hat{\mathbf{k}}_2,\hat{\mathbf{k}}_1)|^2 \qquad (4.6)$$

(2.125). In the subsequent calculation we will retain the complete form of this combination of scattering and exchange amplitudes, thereby determining both singlet and triplet contributions. Previous calculations have omitted the second term in (4.6) [6,41]. The scattering amplitude $f(\hat{\mathbf{k}}_1, \hat{\mathbf{k}}_2)$, and by permutation $f(\hat{\mathbf{k}}_2, \hat{\mathbf{k}}_1)$ is calculated using

$$
f(\hat{\mathbf{k}}_1, \hat{\mathbf{k}}_2) \sim \frac{2i}{\pi} \int \Psi_f^{-*}(\mathbf{r}_1, \mathbf{r}_2)\varphi(\mathbf{r}_3, 2)(H - \xi)
$$
$$
\times \exp(i\mathbf{k}_0 \cdot \mathbf{r}_1)\varphi_i(\mathbf{r}_2, \mathbf{r}_3)d\mathbf{r}_1 d\mathbf{r}_2 d\mathbf{r}_3, \qquad (4.7)
$$

where $\varphi_i(\mathbf{r}_2, \mathbf{r}_3)$ and $\varphi(\mathbf{r}_3, 2)$ are given by (2.127) and (2.128), respectively. As usual, the energy operator $(H - \xi)$ is given by

$$
\left(\frac{1}{r_{12}} + \frac{1}{r_{13}} - \frac{2}{r_1} \right) \qquad (4.8)
$$

We now address the matter of the wavefunction Ψ_f^{-*}, which must be adopted to allow for the inclusion of terms of the order ρ^{-2}. We will begin by considering the final-state wavefunction of method I given by (2.121):

$$
\Psi_f^{-*} = \frac{c^{1/2}E^{m_{12}/2}\rho^{\frac{m_{12}}{2}+\frac{1}{2}}(2Z_0)^{\frac{1}{4}}(-\operatorname{Im} m_{12})^{1/2}}{(2Z_0)^{1/4}\rho^{5/2}\sin\alpha\cos\alpha}\delta(\hat{\mathbf{k}}_1 - \hat{\mathbf{r}}_1)\delta(\hat{\mathbf{k}}_2 - \hat{\mathbf{r}}_2)
$$
$$
\times \exp\left(\frac{4i}{(8Z_0\rho)^{1/2}}(\Delta\theta_{12})^{-2} \right)
$$
$$
\times \left\{ \exp\left[-i(8Z_0\rho)^{1/2} - i\frac{1}{2}(\Delta\alpha)^2(2Z_0\rho)^{1/2}m_{12} \right. \right.
$$
$$
\left. \left. -i\frac{1}{8}(\Delta\theta_{12})^2(2Z_0\rho)^{1/2}m_{21} - i\frac{\pi}{4} - \text{conjugate} \right] \right\} \qquad (4.9)
$$

where the threshold ionization limit of $E\rho \to 0$ has been taken. In the 1986 treatment of method I, the $E\rho \to 0$ limit of u_{ij} in Ψ_f^{-*} was taken since in (2.115), it is assumed that the integral is dominated by small values of ρ, due to the exponential decay of φ_i. This limiting procedure reduces S_0 and $S_i, i = 1, 2$ to the following forms:

$$
S_0 \sim (8Z_0\rho)^{1/2} \qquad (4.10)
$$
$$
S_i \sim (2Z_0\rho)^{1/2}m_{ij}, \qquad i = 1, 2, \quad j = 2, 1 \qquad (4.11)
$$

In this treatment we avoid such limits, ensuring that essential terms of the order ρ^{-2} are not discarded and we use the fully extended version of u_{12} [101],

$$u_{12} = \rho_2^{m_{12}} {}_2F_1\left(m_{12}, m_{12} + 1; 2m_{12} + \frac{3}{2}; -\frac{E\rho}{Z_0}\right) \qquad (4.12)$$

As such, it is reasonable to expect that if an improvement in results is to be achieved, it will be seen primarily for the higher values of E. The final-state wavefuntion then takes the form

$$
\begin{aligned}
\Psi_f^{-*} = {}& \frac{c^{1/2}E^{m_{12}/2}u_1^{1/2}S_2^{1/2}}{\tilde{\omega}^{1/2}\rho^{5/2}\sin\alpha\cos\alpha}\delta(\hat{\mathbf{k}}_1 - \hat{\mathbf{r}}_1)\delta(\hat{\mathbf{k}}_2 - \hat{\mathbf{r}}_2) \\
& \times \exp\left(\frac{4i}{(8Z_0\rho)^{1/2}}(\Delta\theta_{12})^{-2}\right) \\
& \times \exp\left[-i\left(S_0 + \frac{1}{2}S_1(\Delta\alpha)^2 + \frac{1}{8}S_2(\Delta\theta_{12})^2 + \frac{\pi}{4}\right)\right] - \text{conjugate}
\end{aligned}
$$

$$(4.13)$$

where S_1 and S_2 are defined by (2.18) with ω taking the L-dependent form (4.4).

As we have seen in Section III, the inclusion of the angular momentum term moves the classical turning point away from the origin so that, in this description, S_0 is no longer given by (2.80) but is redefined to be

$$S_0 \simeq \int_{\rho_+}^{\rho} d\tilde{\rho}\,\tilde{\omega}(\tilde{\rho}) \qquad (4.14)$$

where

$$\rho_+ = \frac{-Z_0 + \sqrt{Z_0^2 + 4E_N L(L+1)}}{2E_N} \qquad (4.15)$$

Applying Eq. (3.3.33) of Abramowitz and Stegun [35] to this integral results in the replacement of (2.81) with

$$
\begin{aligned}
S_0 = {}& \rho\omega + \frac{Z_0}{X}\ln\left[\frac{X\rho\omega + X^2\rho + Z_0}{X^2\rho^+ + Z_0}\right] \\
& - \sqrt{2L(L+1)}\cos^{-1}\left[\frac{2L(L+1) - Z_0\rho}{\rho(Z_0^2 + 4EL(L+1))^{1/2}}\right]
\end{aligned}
$$

$$(4.16)$$

The integral over dr_3 reduces (4.7) to

$$f = \frac{2i}{\pi} \int \Psi_f^{-*} \exp(i\mathbf{k}_0 \cdot \mathbf{r}_1) r(\rho, \theta_{12}) d\mathbf{r}_1 d\mathbf{r}_2 \qquad (4.17)$$

where the function $r(\rho, \theta_{12})$ is defined by (2.132). Using

$$d\mathbf{r}_1 d\mathbf{r}_2 = \rho^5 \, d\rho \sin^2\alpha \cos^2\alpha \, d\alpha \, d\hat{\mathbf{r}}_1 d\hat{\mathbf{r}}_2 \qquad (4.18)$$

and applying (4.13) gives the scattering amplitude as

$$f = \frac{2i}{\pi} c^{1/2} E^{m_{12}/2} \int \frac{u_1^{1/2} S_2^{1/2}}{\tilde{\omega}^{1/2}} \rho^{5/2} \sin\alpha \cos\alpha$$

$$\times r(\rho, \Theta_{12}) \sum_{L=0}^{\infty} i^L (2l+1) j_L(\mathbf{k}_0 \cdot \mathbf{r}_1) P_L(\cos\Theta_{\frac{1}{2}})$$

$$\times (\exp(-iS_f) - \exp(iS_f^*)) \, d\alpha \, d\rho \qquad (4.19)$$

where

$$S_f = S_0 + \frac{1}{2} S_1 (\Delta\alpha)^2 + \frac{1}{8} S_2 (\Delta\theta_{12})^2 + \frac{\pi}{4} \qquad (4.20)$$

and where the delta functions $\delta(\hat{\mathbf{k}}_1 - \hat{\mathbf{r}}_1)\delta(\hat{\mathbf{k}}_2 - \hat{\mathbf{r}}_2)$ in Ψ_f^{-*} have resulted in the conversion from θ_1, θ_2, and θ_{12} (corresponding to $\hat{\mathbf{r}}_i$) to Θ_1, Θ_2, and Θ_{12} (corresponding to $\hat{\mathbf{k}}_i$), respectively. The integration over α has been conducted using the method of steepest descent, where the point of stationary phase is given by the saddle point $\alpha = \pi/4$, in line with the Wannier theory. This results in a factor

$$\frac{1}{2} \sqrt{\frac{2\pi}{S_1}} \exp\left(\frac{i\pi}{4}\right) \qquad (4.21)$$

so that the difference between the two exponential terms in the integrand of (4.19) becomes

$$(\exp(-iS_f) - \exp(iS_f^*)) = \exp\left(\frac{1}{8} \Im S_2 (\Delta\theta_{12})^2\right)$$

$$\times -2i\cos\left(S_0 + \frac{1}{8} \Re S_2 (\Delta\theta_{12})^2\right) \qquad (4.22)$$

Therefore the expression for the scattering amplitude is given by

$$
f = \int_{\rho+}^{\rho} \frac{\rho^{3/2+1/2} u_1 \left(\dfrac{d}{d\rho} \ln u_2 \right)^{1/2}}{\rho^{1/2} \tilde{\omega}^{1/2} \left(\dfrac{du_1}{d\rho} \right)^{1/2}}
$$

$$
\times r(\rho, \Theta_{12}) \sum_{L=0}^{\infty} i^L (2L+1) j_L(\mathbf{k}_0 \cdot \mathbf{r}_1) P_L(\cos\Theta_1)
$$

$$
\times \exp\left(\frac{1}{8} \Im S_2 (\Delta\theta_{12})^2 \right) \cos\left(S_0 + \frac{1}{8} S_2 (\Delta\theta_{12})^2 \right) d\rho \qquad (4.23)
$$

and the corresponding TDCS takes the form

$$
\frac{d^3\sigma}{d\hat{\mathbf{k}}_1 d\hat{\mathbf{k}}_2 d\left(\frac{1}{2} k_1^2\right)} = \frac{4 \times 280 \times c z_0^2 \pi Z_2 \tanh(2\pi \Im m_{21})}{2^{1/2} k_0 \cdot 27.21}
$$

$$
\times E^{m_{12}-3/2} \left(m_{12} - \frac{1}{2} + \frac{Z_2}{8(2E)^{1/2}} (\Delta\Theta_{12})^2 \pi \tanh(2\pi \Im m_{21}) \right)
$$

$$
\times \exp\left(-\frac{Z_2}{4(2E)^{1/2}} (\Delta\Theta_{12})^2 \pi \tanh(2\pi \Im m_{21}) \right)
$$

$$
\times \left(\frac{1}{4} |f+g|^2 + \frac{3}{4} |f-g|^2 \right) \quad (10^{-19} \mathrm{cm}^2 \mathrm{sr}^{-2} \mathrm{eV}^{-1})
$$

$$
(4.24)
$$

Clearly we have used the trivial differential result

$$
\frac{d}{dE} E^{m_{12}-1/2} \exp\left(-\frac{Z_2}{4(2E)^{1/2}} (\Delta\Theta_{12})^2 \pi \tanh(2\pi \Im m_{21}) \right)
$$

$$
= E^{m_{12}-3/2} \left(m_{12} - \frac{1}{2} + \frac{Z_2}{8(2E)^{1/2}} (\Delta\Theta_{12})^2 \pi \tanh(2\pi \Im m_{21}) \right)
$$

$$
\times \exp\left(-\frac{Z_2}{4(2E)^{1/2}} (\Delta\Theta_{12})^2 \pi \tanh(2\pi \Im m_{21}) \right) \qquad (4.25)
$$

in (4.24). Taking $\tilde{\omega} = \omega$ in (4.23) results in a removable singularity at the lower limit, which is readily dealt with by the substitution $\rho - \rho_+ = \xi^2$.

This substitution produces the final form for the scattering amplitude according to

$$
f = \int_0^\infty d\xi \frac{2\xi^{1/2}(\xi^2 + \rho_+)}{(2E)^{1/4}(\xi^2 + \rho_+ - \rho_-)^{1/4}} \frac{u_1 \left(\Im \dfrac{d}{d\xi} \ln u_2 \right)^{1/2}}{\left(\dfrac{du_1}{d\xi} \right)^{1/2}}
$$

$$
\times \cos\left(S_0 + \frac{1}{8} \Re S_2 (\Delta\theta_{12})^2 \exp(i\mathbf{k}_0 \cdot \mathbf{r}_1) r(\xi^2 + \rho_+, \Theta_{12}) \right) \qquad (4.26)
$$

It is easily seen that equations (4.24) and (4.26) revert back to (2.133) and (2.131) for $L = 0$ with the limit $E\rho \to 0$. We have therefore successfully maintained the significant ρ^{-2} terms with the correct angular momentum form, namely, $-2L(L+1)/\rho^2$ required for the determination of $L \neq 0$ contributions. The results of this calculation are presented and discussed in the next section.

C.   Results and Discussion

We begin this section with a brief discussion of the techniques used by some of the major contributors to the experimental study of electron impact threshold ionization of helium [17,30,100]. The *relative* results obtained by Selles et al. [17] and the *absolute* results of Rösel et al. [100] will also be discussd and compared with the present theoretical calculation.

The choice of geometry is crucial in such experiments. In the case of coplanar geometry, $\mathbf{k}_0, \mathbf{k}_1$, and $\mathbf{k}_2$ are in the same plane, specifically, $\phi_1 - \phi_2 = 180°$, only the $^1S^e, ^3P^o$, and $^1D^e$ states contribute while symmetric kinematics $(\theta_1 = \theta_2 = 90°)$ are governed entirely by singlet scattering. The coplanar geometry minimizes the interaction of the incoming electron with the target. An alternative to this is the case when the momentum vector of one of the escaping electrons is perpendicular to the plane of the incident and other escaping electron, which, for obvious reasons, is referred to as the *perpendicular geometry*. The coplanar and perpendicular geometries are, of course, two extremes with many out-of-plane geometries existing between the two. These may be considered in order to examine the effect of projectile core interaction. As can be seen from Eq. (4.24), the TDCS is differential in the momenta $\mathbf{k}_1, \mathbf{k}_2$ of the two continuum electrons in the final state. All our results focus on the geometry $\mathbf{k}_1 = -\mathbf{k}_2$, in which the electrons depart in opposite directions. Thus $\hat{\mathbf{k}}_1 \cdot \hat{\mathbf{k}}_2 = \cos\theta_{12} = -1$, from which arises the terminology $\theta_{12} = \pi$ geometry with respect to the incident electron direction (taken as the z axis).

The experimental work of the group based in Paris was initially conducted by Fournier–Lagarde et al. [30] and later improved by Selles and co-workers [17,28]. In both cases a coincidence technique in coplanar geometry was employed, to determine the TDCS for the electron impact threshold ionization of helium, with the latter group working in an energy range of 0.5–2 eV above threshold. The aim of any experimentalist involved in a near-threshold reaction study is to investigate the physical phenomenon as close as possible to the threshold. However, as the threshold is approached, the experiments become increasingly more difficult because of the decreasing magnitude of the cross section and the many problems inherent in detecting low-energy electrons. Low-energy electrons require highly efficient optical systems for detection, which in this case was typically 1 eV. The experimental setup consists essentially of a single electrostatic selector and two detection systems, each with two electrostatic analyzers in series. The whole system is maintained at a high temperature (typically 200°C) to keep all surfaces clean, ensuring stability. The incident electron beam is crossed perpendicularly with a gas beam emanating from a fine needle, and the two outgoing electrons are detected in coincidence and analyzed in energy and angle with a time resolution of <1 ns. Selles et al. present their relative triple-differential cross sections (TDCS) in the form of polar plots for a range of total energies $E = 0.5, 1, 2$ eV, where $E_1 = E_2$ and for unequal energy sharing values of $E_1 = 1.75$ eV and $E_2 = 0.25$ eV. For each value of E, five values of Θ_1 are considered: $\Theta_1 = 30°, 60°, 90°, 120°$, and $140°$ with $180° \geq \Theta_2 \geq 0°$. Their results have a statistical accuracy of 10% when random coincidences and normalization procedures are taken into account.

Many aspects of the Wannier theory have been confirmed by their findings, including a global localization around $\theta_{12} = 180°$, where for $\Theta_1 = 90°$ it is clearly visible that the TDCS become more and more concentrated around its maximum near $\Theta_{12} = 180°$ as the energy above threshold decreases. This supports the very central assumption of the Wannier theory, as does the symmetric distribution obtained at $\Theta_1 = 90°$ for unequal energy sharing. Their results also show a decrease in the width of the lobes as E decreases, which is consistent with the energy behavior of the Gaussian distribution. A deviation from these trends is observed for $E = 2$ eV, and as such Selles et al. take the range of validity of the Wannier configuration to be approximately 1 eV above threshold. Indeed, at the much higher energy of 8 eV, the symmetry with respect of $\Theta_1 = 90°$ disappears, and the TDCS becomes much stronger when the higher-energy electron is detected in the forward direction compared with the backward direction, as expected by continuity with high-energy results.

A true test for any theory is a comparison with absolute experimental results. Rösel et al. [100] have studied the electron impact ionization of helium at 2 eV above threshold, producing absolute experimental TDCS. The nature of their apparatus allowed the study of many different geometries, including the coplanar geometry described above. It consisted mainly of an electron gun, a gas inlet above which is a movable ion collector, and two spectrometers, one of which is mounted on a track that rises in a semicircular arc, thus allowing out-of-plane measurements. Measurement of absolute threshold TDCS is notoriously difficult, not only because the energies involved are low but also because it requires knowing simultaneously the absolute values of the coincidence energy resolution (ΔE), the incident beam intensity (N_e) (or electron beam current), the acceptance angles $(\Delta\Omega)$ efficiencies (ε) of the spectrometer, the effective scattering length (l), and the effective target gas density (n) of the interaction region. The first two quantities are measurable, and the product $N_e \cdot l \cdot n$ may be found by measuring the count rate of ions produced, using known absolute total ionization cross sections at a particular energy (well above threshold), and the product $\Delta\Omega \cdot \varepsilon$ may be found similarly by comparing the count rate of scattered electrons with known absolute double differential cross sections at a total energy of 100 eV. Their results, obtained at an excess-of-threshold energy $E = 2$, have an absolute error of $< 22\%$.

The results of the current calculation for two cases of electron impact threshold ionization of helium, namely, $E_1 = E_2 = 0.5\,\text{eV}, E = 1$ eV, and $E_1 = E_2 = 1\,\text{eV}, E = 2$ eV for a range of Θ_1 values, are presented in Figs. 4–9 and 10–15, respectively. We have assumed equal energy sharing, and the geometry is coplanar, which is the natural choice for the semiclassical approximation since a plane wave, $\exp(i\mathbf{k}_0 \cdot \mathbf{r}_1)$, is used to represent the incident electron. The results obtained are compared with the relative experimental results of Selles et al. [17,28] for excess-of threshold energy $E = 1$ eV and with the absolute experimental data of Rösel et al. [100] for $E = 2\,\text{eV}$. We have also plotted the calculated results of method I and Copeland and Crothers [41] against the experimental results for both values of E for comparison with the current calculation. In this calculation, angular momentum states of $L \leq 5$ have been included for six values of Θ_1: 30°, 45°, 60°, 90°, 120°, 140°/150°, where Θ_2 is varied from 0° to 180° in 5° increments. The results are presented in the form of polar plots with an absence of experimental data at $\Theta_1 = 45°$.

For $E = 1$ eV, the relative experimental results of Sellus et al. have been normalized with respect to the maximum observed in the present calculation for $\Theta_1 = 90°$, which occurs at $\Theta_2 = 90°$ and takes the value of $1.18 \times 10^{-19}\,\text{cm}^2\,\text{sr}^{-2}\,\text{eV}^{-1}$. This is in poor agreement with Crothers' original result of $0.85 \times 10^{-19}\,\text{cm}^2\,\text{sr}^{-2}\,\text{eV}^{-1}$. However, this calculation has produced the notable peaking at $\Theta_1 = 30°, 140°, \Theta_{12} = 180°$, which has been recorded by experiment and as yet has not been reproduced by the

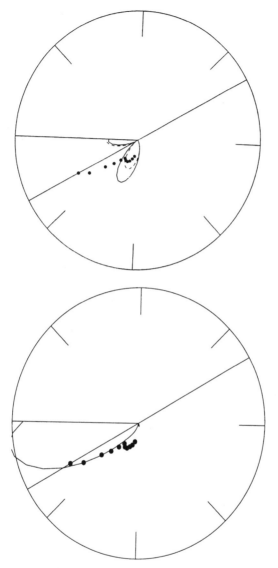

Figure 4. Coplanar geometry, $\Theta_1 = 30°, E_1 = E_2 = 0.5$ eV and $E = 1$ eV. The upper plot shows Crothers [6] (dotted line) and Copeland and Crothers [41] (full curve) calculated TDCS in polar coordinates, with polar coordinate Θ_2, in comparison with the relative experimental results of Selles et al. [17,28] (●). The lower plot shows the present calculated TDCS (full curve) in comparison with the relative experimental results of Selles et al. [17,28] (●). The radius of the upper circle is 1.0×10^{-19} cm^2 sr^{-2} eV^{-1}; the radius of the lower circle is slightly higher, 1.18×10^{-19} cm^2 sr^{-2} eV^{-1}.

ARLENE M. LOUGHAN

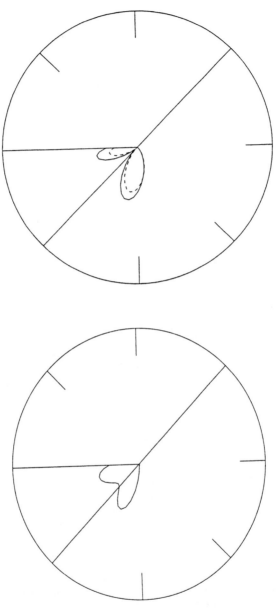

Figure 5. As for Fig. 4, but with $\Theta_1 = 45°$. Experimental results are not available for this value of Θ_1.

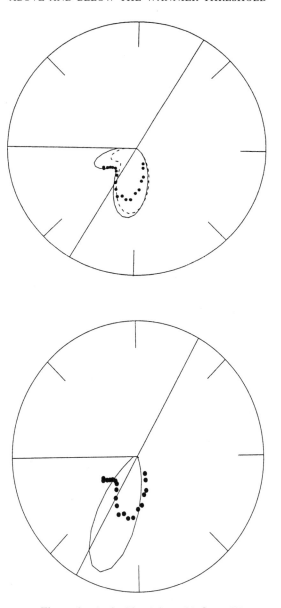

Figure 6. As for Fig. 4, but with $\Theta_1 = 60°$.

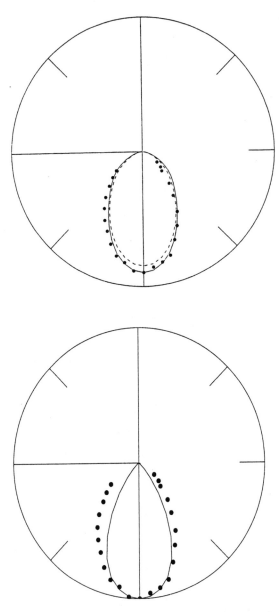

Figure 7. As for Fig. 4, but with $\Theta_1 = 90°$.

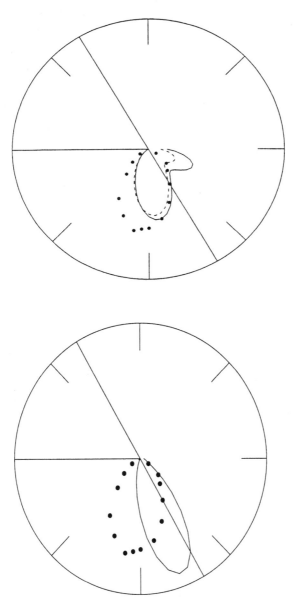

Figure 8. As for Fig. 4, but with $\Theta_1 = 120°$.

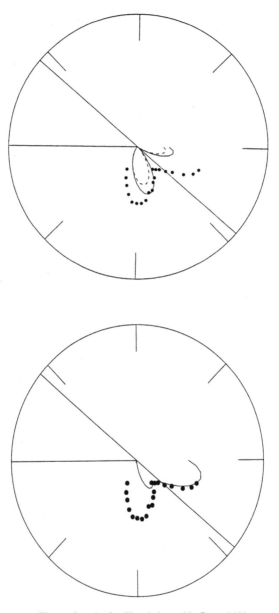

Figure 9.   As for Fig. 4, but with $\Theta_1 = 140°$.

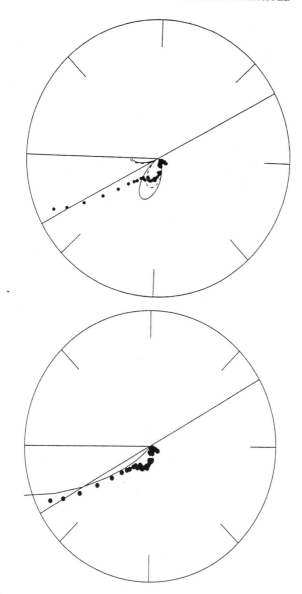

Figure 10. Coplanar geometry, $\Theta_1 = 30°, E_1 = E_2 = 1.0$ eV and $E = 2$ eV. The upper plot shows Crothers [6] (dotted line) and Copeland and Crothers [41] (full curve) calculated TDCS in polar coordinates, with polar coordinate Θ_2, in comparison with the absolute experimental results of Rösel et al. [100] (●). The lower plot shows the present calculated TDCS (full curve) in comparison with the absolute experimental results of Rösel et al. [100] (●). The radius of each circle 1.0×10^{-19} cm^2 sr^{-2} eV^{-1}.

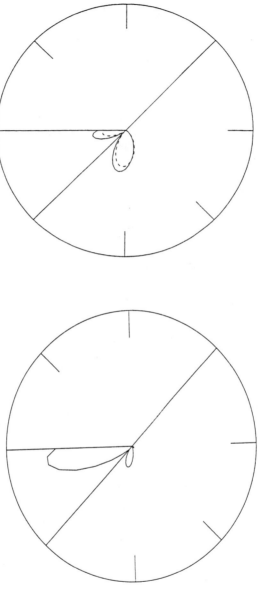

Figure 11. As for Fig. 10, but with $\Theta_1 = 45°$. Experimental results are not available for this value of Θ_1.

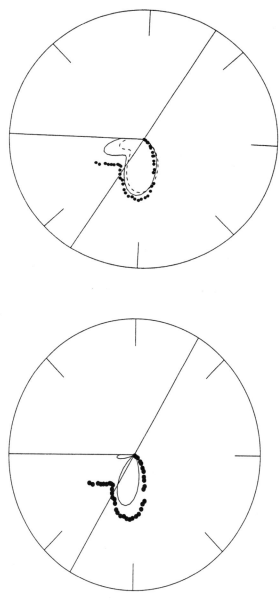

Figure 12. As for Fig. 10, but with $\Theta_1 = 60°$.

ARLENE M. LOUGHAN

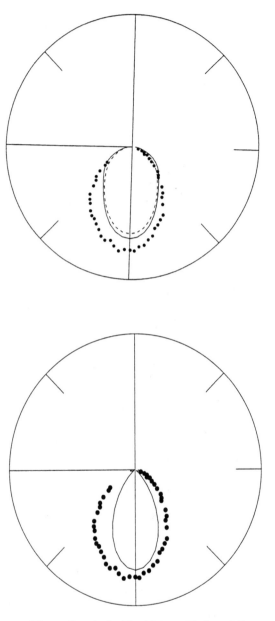

Figure 13.   As for Fig. 10, but with $\Theta_1 = 90°$.

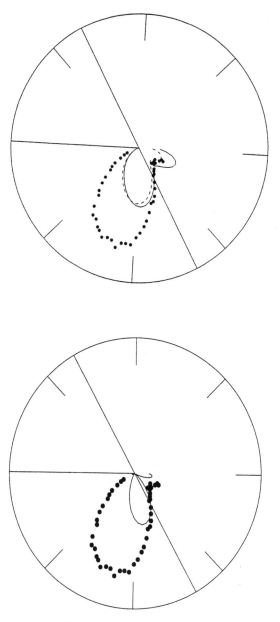

Figure 14. As for Fig. 10, but with $\Theta_1 = 120°$.

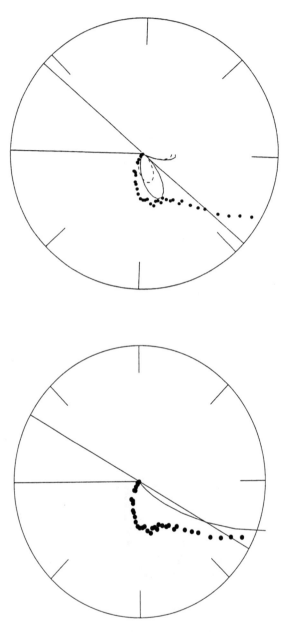

Figure 15.   As for Fig. 10, but with $\Theta_1 = 150°$.

semiclassical approximation. There is no doubt, however, that for $\Theta_1 = 60°$ and $\Theta_1 = 120°$ our agreement with experiment, relative to the previous calculation of Copeland and Crothers [41], is poor. The main lobe for both angles, and especially for $\Theta_1 = 60°$, is notably larger than experiment, which is reminiscent of the calculation of method I, in which the inclusion of triplet states produced TDCS that were much too large. The general form of the TDCS for $\Theta_1 = 90°$ has indeed been maintained, but discrepancies clearly exist. The fact that we are now seeing some kind of peaking at the angles mentioned above, as opposed to previous calculations that observed a sharp dip, reinforces our contention that we are including the correct angular momentum terms, significant at these angles. In general, the poor agreement for this value of E may be explained by the fact that we did not apply the $E \to 0$ limit, taken in the previous calculations [6,41]. As such, an improvement should be seen in going from the $E = 1$ to the $E = 2$ results. If we consider Figs. 10–15, this can clearly be seen to be the case. For $E = 2$, the maximum calculated cross section is found to be 0.73×10^{-19} cm^2 sr^{-2} eV^{-1}, which is well within the error bars of corresponding absolute experimental value of 0.78×10^{-19} cm^2 sr^{-2} eV^{-1} and also compares well with 0.65×10^{-19} cm^2 sr^{-2} eV^{-1} for Crothers' 1986 paper. The $E = 2$ value is expected to be smaller than the $E = 1$ value because of the $E^{m_{12}-3/2}$ TDCS dependence. The $E = 2$ results show a marked improvement over the $E = 1$ values with the significant "peaking" recorded experimentally at $\Theta_1 = 30°, 150°, \Theta_{12} = 180°$ being reproduced, within the error bars of the absolute experimental results. However, yet again, agreement at $\Theta_1 = 60°$ and $\Theta_1 = 120°$ is not as good as the previous calculation. The DWBA calculations of Pan and Starace [42] and Jones et al. [102] have, to date, only been significantly better than the semiclassical approximation results in the geometry where the large peaks have been found by Rösel et al. They successfully reproduced the peaks by including electron–electron interaction in the final state via the use of effective charges in the Schrödinger equation. Thus their models differ from the standard DWBA, which treats a large part of the electron–atom interaction in the initial state, and the electron–ion interaction in the final state to all orders of perturbation but treats the electron–electron interaction in the final state only to first order. Previous attempts to extrapolate the DWBA to a few tens of electronvolts above threshold have therefore been unsuccessful [103]. The more recent calculations have also taken exchange into account in the calculation of the continuum wavefunctions and restrict themselves to the case of Θ_{12} geometry.

Clearly, much work can be done to improve this preliminary investigation. However, despite the failures, the success of generating the peaks at the lowest and highest values of Θ_1, for each value of E studied, is highly encouraging. Future work to improve these results may be focused on many areas; details of this and further concluding remarks are given in Section V.

V. CONCLUSIONS

We begin this final Section with a brief summary of the work presented in Sections III and IV, on which we shall draw conclusions from our results to finally finish with a few ideas for future study. This work has been divided into two main areas both centred on near-threshold processes: (1) a below-threshold study of electron capture into an excited state and (2) an above-threshold electron impact ionization investigation. In both cases the target is atomic or ionic helium.

In Section III we presented the above-threshold Wannier quantal ionization theory of method I analytically continued to below threshold in order to study Wannier quantal doubly excited states. Our first concern was the calculation of $L = 0\,^2S$ and $\bar{s}\,^2S$ doubly excited states of He$^-$, formed as a result of electron capture. We presented results for the principal series, $n_1 = n_2 = n$ and $N = 2n - 1, L = 0$, while for the subsidiary series, $n_2 = n_1 + 1 = n + 1$ and $N = 2n, L = 0'$, in the notation of Buckman et al. [7]. We extended this theory to $L = 1$ and $L = 2\,^2P^o$ and ^2D states, where with the inclusion of angular momentum terms evaluated on the Wannier ridge an irrational principal quantum number was obtained for $L = 1$, namely, $N = 2n - [(17^{1/2} - 1)/2] - 1$, whereas for $L = 2, N = 2n - 4$. The width and intensity of these states were also considered using the imaginary parts of the calculated complex eigenenergies that directly give the lifetime of these Wannier doubly excited Rydberg states. This is unique in that other theories address the problem of resonance position only. We also applied the $L = 0$ below-threshold analysis to the DES of He, where a simple change in the magnitude of the core charge produced reasonable results for high values of principal quantum number, n. As far as we know, the intershell energies are the first to be presented for these higher quantum numbers; other theories are restricted by computational demands at the higher end of the spectrum.

There is no doubt that this below-threshold theory is an exciting development in the Wannier description of doubly excited states (DES) where we can, conversely, view the theory of method I for above-threshold ionization as the analytic continuation of our new below-threshold theory. This reinforces the above-thresholds theory of I as a fully fledged quantal treatment of ionization. The fact that we use semiclassical asymptotic methods to derive the form of Ψ_f^{-*} on the hyperspherical surface at infinity, at which the detectors of the two electrons must be placed, does not obviate our contention that we have formed a uniform semiclassical exact quantal description of the Wannier ridge states. As we have seen, the calculated resonance positions were found to be in good agreement with experiment, and the agreement improved for higher n. This is to be expected because of the semiclassical nature of the calculation and strongly supports the notion that these features are examples of highly correlated, ridge-riding states. However, there is, of course, room for many improvements in the initial total wavefunction Ψ_i^+ and ψ_ϵ or ϕ_i, which is only a plane wave times a target wavefunction. For instance, and of relevance especially to lower principal quantum numbers, exchange could be included.

There are many theoretical approaches to the study of doubly excited states with several semiempirical forms of the modified Rydberg formula, correlated wavefunction expansions, and R-matrix techniques, to name only a few. It is not surprising that the R-matrix theory, which is among the most widely used methods for the computation of atomic collision phenomena, produces the most accurate results for the lower values of n. However, such calculations are highly computational relative to the current ab initio method, which provides impressive quantitative accuracy with the minimum of computational effort even with $n \leq 9$. The modelistic Wannier approach adopted in this study can, provide not only accurate results but also useful qualitative insights and play a key role in the classification scheme of two electron resonances. There are numerous possibilities for the future use and improvement of this below-threshold theory. We need not restrict ourselves to electron capture by helium in the ground state but may alter our description of ϕ_i to allow the calculation of eigenvalues of other symmetries. This approach may also be extended to fourth order and has opened the way for future investigations into the determination of the intensity of a resonance from the complex part of its eigenvalue.

The most significant advancement made in Section III was in our description of $L \neq 0$ states, which, as we have seen in Section IV, has been utilized in our above-threshold study of electron impact threshold ionization. We have included in our calculation of threshold TDCS suitably symme- terized states taken as $Y_{LM}(\hat{\mathbf{r}}_1, \hat{\mathbf{r}}_2) \pm Y_{LM}(\hat{\mathbf{r}}_2, \hat{\mathbf{r}}_1)$, with $M = 0$. In principle, all angular momentum terms are accessible where, of course, the delta functions can be resolved into an infinite set of spherical harmonics and the plane-wave initial-state projects out the $P_L(\cos\theta_1)$. By including allowed singlet and triplet states for $L \leq 5$, the peaking observed at $\Theta_1 = 30°, 140°,$ $150°$ for $\Theta_{12} = 180°$ has been reproduced for $E = 1$ and $E = 2$. This confirms our belief that all possible states for $L \leq 5$ make a significant contribution to threshold TDCS, including both singlet and triplet states. We also feel that the improved description of the $^1D^e$ symmetry with the basic angular form

$$P_2(\cos\theta_1) + P_2(\cos\theta_2) \tag{5.1}$$

has enhanced our accuracy. This seems reasonable; after all, there is no doubt that the contribution from such a term will be a maximum when $\Theta_2 = 180° - \Theta_1$, and as Θ_1 and Θ_2 increase, so as to be zero when $\Theta_1 = \Theta_2 = 90°$, thus not affecting the local maximum at $\Theta_1 = \Theta_2 = 90°$. This is in line with the secondary peak being largest when $\Theta_1 = 30°$, $\Theta_2 = 150°$ and decreasing through $\Theta_1 = 60°, \Theta_2 = 120°$ to zero at $\Theta_1 = \Theta_2 = 90°$. The improvement in our results for $E = 2$ is unfortunately not mirrored in the results for $E = 1$. Peaking is observed at $\Theta_1 = 30°$, $140°, \Theta_{12} = 180°$ in line with experiment; however, the TDCS for the remaining angles, namely, $\Theta_1 = 60°, 90°$ and $120°$, are much too large.

Previous attempts to amend the discrepancies between the theoretical TDCS results of method I and experiment [41,47] produced small but definitive improvements; however, they failed to resolve the significant differences at $\Theta_1 = 30°, 140°, 150°, \Theta_{12} = 180°$. To this end, Copeland and Crothers [47] employed an explicitly correlated bound-state wavefunction for the helium target developed by Le Sech [104]; that is a wavefunction containing explicitly the inter-electronic distance r_{23}. This allowed for the inclusion of small but significant physical processes such as "shakeoff" in which the incident electron collides with one of the target electrons or nucleus and the untouched electron is ejected. The small but notable improvements made, highlighted the need to consider the contribution from electron correlation in the initial state. A previous attempt to include the $^3P^o$ contribution [47] also produced a slight improvement; however, the radial equation employed in that calculation contained, as we have now discovered, the wrong angular momentum dependence, namely, $-L(L+1)/\rho^2$, as opposed to the correct form, $-2L(L+1)/\rho^2$. They obtained a complete final-state wavefunction of the form

$$\Psi_f = \frac{N(\eta_f)u_1^{1/2}(\cos\theta_1 - \cos\theta_2)}{\omega^{1/2}\rho^{5/2}\sin\alpha\cos\alpha}\left[\exp(-i\tilde{S}_f) - \exp(i\tilde{S}_f^*)\right] \qquad (5.2)$$

where

$$\tilde{S} \simeq (8Z_0\rho)^{1/2}\left(1 + \frac{1}{4}m_{12}(\Delta\alpha)^2 + \frac{1}{16}m_{21}(\Delta\theta_{12})^2\right) - \left(l + \frac{1}{2}\right)\pi + \frac{\pi}{4}$$

$$(5.3)$$

by following the method of McCann and Crothers [44] and by considering the total angular momentum dependence of the connection formula.

It is possible that future work may improve our theoretical description for the lower value of E; the preliminary results, do indeed, show the necessary peaking but suffer from the original problem of method I in that the triplet contribution is proving too large for the other angles. Of course, notable improvements are to be obtained if an explicitly correlated bound-state wavefunction is employed, as discussed above. The logical next step to the calculation of TDCS is the determination of total differential cross sections from which the asymmetry parameter A, which is given by

$$A = \frac{\sigma_S - \sigma_T}{\sigma_S + 3\sigma_T} \tag{5.4}$$

can be investigated. Clearly, σ_S and σ_T are the singlet and triplet total cross section, respectively. In the conventional Wannier theory, the cross section asymmetry parameter would be constant throughout the energy region for which the threshold law is valid. However, the experimental results of Friedman et al. [105] appear to contradict this Wannier prediction, even at only 0.5 eV above threshold. This matter certainly must be addressed in order to validate the Wannier theory.

This present theoretical treatment has firmly placed the uniform semi-classical approximation as a fully fledged quantal treatment of Wannier threshold processes both above and below the first ionization threshold. Our study of high-lying doubly excited states confirms the dominant geometry as that of the Wannier ridge configuration. The emphasis throughout this study has been on the final-state electron–electron interaction with the physical process of dynamic screening dictating the Wannier configuration. Although some results have been impressive, others have merely acted as indicators to what may be achieved in the future. This work has answered some questions concerning near-threshold processes, a large and interesting area of study, and we firmly believe that further development of this powerful theory will answer many more.

ACKNOWLEDGMENTS

The author would like to thank Professor DSF Crothers many helpful discussions and EPSRC for financial assistance.

References

1. E. P. Wigner, *Phys. Rev.* **73**, 1002 (1948).
2. G. H. Wannier, *Phys. Rev.* **90**, 817 (1953).
3. U. Fano, *J. Phys. B* **13**, L519 (1980).
4. U. Fano, *Phys. Rev. A* **22**, 2660 (1980).
5. R. K. Peterkop, *J. Phys. B* **4**, 513 (1971).
6. D. S. F. Crothers, *J. Phys. B* **19**, 463 (1986).
7. S. J. Buckman, P. Hammond, F. H. Read, and G. C. King, *J. Phys. B* **16**, 4039 (1983).

8. S. J. Buckman and D. S. Newman, *J. Phys. B* **20**, L711 (1987).

9. A. R. P. Rau, *Phys. Rev. A* **4**, 207 (1971).

10. D. R. Bates, A. Fundaminsky, H. S. W. Massey, and J. W. Leech, *Trans. Roy. Soc.* (London) **243**, 93 (1950).

11. J. Macek, *J. Phys. B* **1**, 831 (1968).

12. C. D. Lin, *Phys. Rev. A* **26**, 2305 (1982).

13. C. D. Lin, *Adv. Atom. Mol. Phys.* **22**, 77 (1986).

14. C. H. Greene and A. R. P. Rau, *Phys. Rev. Lett.* **48**, 533 (1982).

15. C. H. Greene and A. R. P. Rau, *J. Phys. B* **16**, 99 (1983).

16. M. R. H. Rudge and M. J. Seaton, *Proc. Roy. Soc. A* **283**, 262 (1965).

17. P. Selles, J. Mazeau, and A. Heutz, *J. Phys. B* **20**, 5183 (1987).

18. G. Breit, *Phys. Rev.* **35**, 569 (1930).

19. P. M. Morse and H. Feshbach, *Methods of Theoretical Physics*, McGraw-Hill, New York, 1953, Vol. 2, Chapter 12.

20. A. K. Bhatia and A. Temkin, *Rev. Mod. Phys.* **36**, 1050 (1964).

21. B. Holmberg, *Kgl. Fysiograf. Sällskap. Lund. Förh.* **26**, 135 (1956).

22. H. Diehl, S. Flügge, U. Schrödinger, A. Völkel, and A. Weiguny, *Z. Physik* **162**, 1 (1961).

23. S. Cvejanovic and F. H. Read, *J. Phys. B* **7**, 1841 (1974).

24. A. Temkin, *Phys. Rev. Lett.* **49**, 365 (1982).

25. S. Geltman, *J. Phys. B* **16**, L525 (1983).

26. P. G. Burke and A. J. Taylor, *Proc. Roy. Soc.* **287**, 105 (1965).

27. P. Marchand and J. Cardinal, *Can. J. Phys.* **57**, 1624 (1969).

28. P. Selles, J. Mazeau, and A. Huetz, *J. Phys. B* **20**, 5195 (1987).

29. J. B., Donahue, P. A. M. Gram, M. V. Hynes, R. W. Hamm, C. A. Frost, H. C. Bryant, K. B. Butterfield, D. A. Clark, and W. W. Smith, *Phys. Rev. Lett.* **48**, 1538 (1982).

30. P. Fournier-Lagarde, J. Mazeau, and A. Huetz, *J. Phys. B* **17**, L591 (1984).

31. I. Vinkalns and M. Gailitis, *Proc. 5th Int. Conf. Physics of Electtonic and Atomic Collisions*, Nauka, Lennigrad, 1967, Abstracts, pp. 648–650.

32. H. T. H. Piaggio, *Differential Equations*, Bell, London, 1958.

33. J. M. Feagin, *J. Phys. B* **28**, 1495 (1995).

34. A. Erdelyi, *Higher Transcendental Functions*, Vol. 1, McGraw-Hill, New York, 1953.

35. M. Abramowitz and I. A. Stegun, *Handbook of Mathematical Functions*, Dover, New York, 1970.

36. A. R. P. Rau, *J. Phys. B* **9**, L283 (1976).

37. T. A. Roth, *Phys. Rev. A* **5**, 476 (1972).

38. H. Klar and W. Schlecht, *J. Phys. B* **9**, 1699 (1976).

39. H. Goldstein, *Classical Mechanics*, Addison-Wesley, Cambridge, MA, 1950.

40. R. K. Peterkop, *Theory of Ionisation of Atoms by Electron Impact*, Colorado Univ. Press, Boulder, 1977.

41. F. B. M. Copeland and D. S. F. Crothers, *J. Phys. B* **27**, 2039 (1994).

42. C. Pan and A. F. Starace, *Phys. Rev. A* **45**, 4588 (1992).

43. F. Pichou, A. Heutz, G. Joyez, and M. Landau, *J. Phys. B* **11**, 3683 (1978).

44. J. F. McCann and D. S. F. Crothers, *J. Phys. B* **19**, L399 (1986).

45. D. S. F. Crothers and D. J. Lennon, *J. Phys. B* **21**, L409 (1988).

46. D. R. J. Carruthers and D. S. F. Crothers, *J. Phys. B* **24**, L199 (1991).

47. D. R. J. Carruthers, and D. S. F. Crothers, *Z. Phys. D.* **23**, 365 (1992).

48. U. Fano and C. D. Lin, *Atomic Physics*, Plenum Press, New York, 1975, Vol. 4, p. 47, especially p. 66ff.

49. U. Fano, *Phys. Today* **29**, 32 (1976).

50. A. R. Edmonds, *J. Physique* **34**(C4), 71 (1970).

51. A. F. Starace, *J. Phys. B* **6**, 585 (1973).

52. W. R. S. Garton and F. S. Temkin, *A Strophys. J.* **158**, 839 (1969).

53. U. Fano, *Colloq. Int. CNRS* **273**, 127 (1977).

54. D. W. O. Heddle, R. G. W. Kessing, and A. Parkin, *Proc. Roy. Soc. London Ser. A* **352**, 419 (1977).

55. J. N. H. Brunt, G. C. King, and F. H. Read, *J. Phys. B* **10**, 433 (1977).

56. A. R. P. Rau, *J. Phys. B* **16**, L699 (1983).

57. A. R. P. Rau, *Nucl. Instrum. Meth. B* **56**, 200 (1991).

58. G. J. Schulz, *Rev. Mod. Phys.* **45**, 378 (1973).

59. A. M. Loughan and D. S. F. Crothers, *Phys. Rev. Lett.* **79**, 4966 (1997).

60. A. M. Loughan and D. S. F. Crothers, *J. Phys. B* **31**, 2153 (1998).

61. D. S. F. Crothers and A. M. Loughan, *Phil. Trans. Roy. Soc. Lond. A* (in press).

62. R. Gans, *Ann. Phys. Lpz.* **47**, 709 (1915).

63. H. Jeffreys, *Prod. Lond. Math. Soc.* **23**, 428 (1923).

64. C. D. Lin and S. Watanabe, *Phys. Rev. A* **35**, 4499 (1987).

65. Y. Komninos, M. Chrysos, and C. A. Nicolaides, *J. Phys. B* **20**, L791 (1987).

66. J. M. Rost and J. S. Briggs, *J. Phys. B* **21**, L233 (1988).

67. W. C. Fon, K. A. Berrington, P. G. Burke, and A. E. Kingston, *J. Phys. B* **22**, 3939 (1989).

68. J. M. Rost and J. S. Briggs, *J. Phys. B* **22**, 3587 (1989).

69. Y. K. Ho, *Phys. Rev. A* **34**, 4402–4404 (1986).

70. L. C. G. Freitas, K. A. Berrington, P. G. Burke, A. Hibbert, H. Fukuda, N. Koyama, and M. Matsuzaw, *J. Phys. B* **20**, 2959–2973 (1987).

71. S. J. Buckman and C. W. Clark, *Rev. Mod. Phys.* **66**, 539 (1994).

72. U. Fano, *Phys. Rev.* **124**, 1866 (1961).

72. G. J. Schulz, *Phys. Rev. Lett.* **10**, 104 (1963).

73. R. G. W. Kessing, 1977 *Proc. Roy. Soc. Lond. Ser. A* **352**, 429 (1977).

75. D. Andrick, *J. Phys. B* **12**, L175 (1979).

76. G. J. Schulz, *Phys. Rev.* **112**, 150 (1958).

77. G. J. Schulz and R. E. Fox, *Phys. Rev.* **106**, 1179 (1957).

78. J. F. William, *J. Phys. B* **21**, 2107 (1988).

79. R. E. Kennerly, R. J. van Brunt, and A. C. Gallagher, *Phys. Rev. A* **23**, 2430 (1981).

80. D. W. O. Heddle, in *Electron and Photon Interactions with Atoms*, H. Kleinpoppen and M. R. C. McDowell, eds., Plenum press, New York, 1976, p. 671.

81. R. K. Nesbet, *J. Phys. B* **11**, L21 (1978).

82. R. P. Madden and K. Codling, *Phys. Rev. Lett.* **10**, 516 (1963).

83. F. H. Read, *J. Phys. B* **10**, 449 (1977).

84. J. M. Feagin and J. Macek, *J. Phys. B* **17**, L245 (1984).

85. J. Macek, and J. M. Feagin, *J. Phys. B* **18**, 2161 (1985).

86. A. Ronveauax, *Heun Differential Equations*, Oxford Science Pulbilations, 1997.

87. R. E. Langer, *Phys. Rev.* **51**, 669 (1937).

88. A. L. Robinson, *Science* **220**, 702 (1983).

89. W. Gasser, *Phys. Rev. B* **47**, 1324 (1992).

90. R. B. Laughlin, *Phys. Rev. Lett.* **50**, 1395 (1983).

91. P. V. Grujic and N. S. Simonovic, *J. Phys. B* **28**, 1159–1171 (1995).

92. Y. K. Ho and J. Calloway, *Phys. Rev. A* **34**, 130 (1986).

93. J. Muller, X. Yang, and J. Burgdorfer, *Phys. Rev. A* **49**, 2470 (1994).

94. D. Oza, *Phys. Rev. A* **33**, 824–837 (1986).

95. M. Domke, K. Schulz, G. Remmers, and G. Kaindl, *Phys. Rev. A* **53**, 1424 (1996).

96. M. Zubek, G. C. King, P. M. Rutter, and F. H. Read, *J. Phys. B* **22**, 3411 (1989).

97. S. J. Brotton, S. Cvejanovic, F. J. Currel, N. J. Bowring, and F. H. Read, *Phys. Rev. A* **55**, 318 (1997).

98. P. J. Hicks and J. Cromer, *J. Phys. B* **8**, 1866 (1975).

99. P. L. Altick, *J. Phys. B* **18**, 1841 (1985).

100. T. Rösel, J. Roder, L. Frost, K. Jung, and H. Erhardt, *Phys. Rev. A* **46**, 2539 (1992).

101. F. B. M. Copeland, and D. S. F. Crothers, *J. Phys. B* **28**, L763 (1995).

102. T. Jones, D. H. Madison, and M. K. Srivastava, *J. Phys. B* **25**, 1899 (1992).

103. X. Zhang, C. T. Whelan, and J. Walters, *J. Phys. B* **23**, L173 (1990).

104. C. Le Sech, *Chem. Phys. Lett.* **200**, 369 (1992).

105. J. R. Freidman, X. Guo, M. S. Lubell, and M. R. Frankel, *Phys. Rev. A* **46**, 652 (1992).

UNIFIED THEORY OF PHOTOCHEMICAL CHARGE SEPARATION

A. I. BURSHTEIN

The Weizmann Institute of Science, Rehovot, Israel

CONTENTS

Advances in Chemical Physics, Volume 114, Edited by I. Prigogine and Stuart A. Rice.
ISBN 0-471-39267-7   © 2000 John Wiley & Sons, Inc.

I. INTRODUCTION

Charge separation after light excitation of biological or chemical systems is the most important stage of natural and artificial photosynthesis. By these means solar energy is conserved and life originates. The secret of high efficiency of reaction centers is the fast and irreversible separation of ions, which are then involved in reduction and oxidation reactions on the periphery of the macromolecule [1]. The biological way of making these reactions possible is realized by means of the very special architecture of the reaction center. Primary ionization is followed by a sequence of electron-transfer steps, each of which brings an electron to a more distant and lower energy acceptor. Finally, the counterions are separated far enough apart to prevent or postpone their recombination for enough time until the photosynthesis is completed.

 The same goal can be achieved due to the photoexcitation of liquid solutions of electron donors (D) and acceptors (A). Primary ionization occurs during their encounters in solution, and the charged products can be separated well by diffusion, if they escape geminate recombination by backward electron transfer to either an excited or ground state. To involve the free ions in subsequent reactions of photosynthesis or in electric current, one has to optimize the charge separation quantum yield (that is a fraction of survived ions). This may be done by the proper choice of reactants (their redox potential and interactions) and solutions that differ by their viscosities and polarities. The photochemical ionization and charge separation in liquid solutions is an alternative way to conserve and utilize the light energy.

Although many systems were studied experimentally during the late twentieth century, very limited success was made because of a lack of fundamental electron-transfer theory applicable to mobile reactants. In the case of solid state [2] or intramolecular transfer [3,4] (including charge separation in biological systems [4,5]), the problem is simpler because the distance between the reactants is fixed. In liquid solutions the situation is different in principle, because the transfer rate is randomly modulated by encounter diffusion of the reaction partners. As a result, there is a violation of the fundamental "free-energy-gap law" discovered by R. Marcus [6]. The classical approach of Smoluchowski suitable for contact and irreversible reactions does not give a proper description of distant and/or reversible reactions of electron transfer [7]. The rate description of geminate recombination (the so-called exponential model employed in many experimental works [8]) is absolutely inappropriate [9].

The essential weaknesses of the model approaches were overcome only in the late 1990s. The distance dependence of the electron-transfer rate has been revised and shown to be noncontact and nonmonotonous (bell-shaped) for highly exergonic reactions in both polar [10] and nonpolar [11] solutions. The rest of the weaknesses of the conventional models of diffusion-assisted transfer reactions were then overcome by an original "unified theory" of bimolecular ionization followed by geminate recombination [12,13]. The diffusional control of geminate recombination was established [14], and the free-energy-gap (FEG) law for charge separation was essentially corrected [15,16]. The kinetics of ion accumulation and geminate recombination to the ground state or to neutral radicals was investigated and fitted to experimental data [17]. An essential role of the spin states of an ion pair and of their interconversion as a preliminary condition for subsequent recombination was disclosed [18]. The singlet–triplet conversion is sensitive to an external magnetic-field that affects the charge-separation quantum yield [19]. The magnetic-field effects have been shown to be a real fingerprint of the distant creation of ions in the course of bimolecular ionization [20]. They can be also detected through the fluorescence of exciplexes that can be formed before or after generation of solvent separated ions [21,22].

The outline of this chapter is as follows. In Section II the free energy and distance dependence of the transfer rate is analyzed and related to widely used contact, rectangular, and other reaction layer models. In Section III the FEG law is formulated and the reported results of its experimental verification in solid and liquid solutions give rise to the problems studied later. Section IV presents the competing approaches to diffusion-assisted bimolecular ionization: contact theory suggested by Smoluchowski and differential encounter theory (DET), which accounts for remote electron transfer and discriminates between contact and noncontact reactions in normal (N) and inverted (I) Marcus regions. DET provides us with a more accurate estimate of the rate and effective radius of diffusion-controlled ionization, as well as with the space distribution of its products, their quantum yield, and the FEG dependence of the latter. In Section V the geminate recombination of ions is described by means of a primitive "exponential model," phenomenologic "contact approximation," and the most appropriate theory of remote electron transfer, exponential or rectangular. All of them differ in the predicted FEG dependencies of the charge separation quantum yield and assume that the initial interion distance r_0 is either contact (in the exponential model) or left free as a fitting parameter (in other theories). This disadvantage has been overcome in Section VI within the original unified theory (UT), which operates with true initial conditions for geminate recombination, created by precursor bimolecular ionization. Having no fitting parameter at all, we studied the shape of the FEG law for recombination and concluded that it can be qualitatively distorted by diffusion, especially if ions start from inside of the recombination layer. The kinetics of ion accumulation and recombination is also studied here and continued in Section VII. There the theory is applied to ion transformation to neutral radicals via proton transfer in nonpolar solvents, straight recombination to the ground state in polar solvents, and competing recombination to the excited triplet state of reactants after preliminary spin conversion in ion pairs. In Section VIII the role of spin conversion in creating magnetic-field effects in the charge separation quantum yield is studied using the simple model of stochastic spin transitions as well as a realistic Hamiltonian description of spin evolution. In Section IX we turn to the most complex reactions composed of a few sequential and parallel processes, including spin conversion and exciplex formation from either primary excited reactants, or from the ions coming in contact after creation. In contact approximation the analytic study of the whole process becomes possible, including magnetic-field effects in quantum yields of charge separation and exciplex fluorescence. In conclusion, we focus attention on the most serious limitations of the unified theory and how they can be eliminated using the more general integral encounter theory, applied to a number of problems since the late 1990s.

II. POSITION-DEPENDENT TRANSFER RATES

The main input parameter of the theory is the rate of electron transfer at distance r between reactants, $W(r)$. This rate, taken from the microscopic quantum-mechanical theory, depends not only on the interparticle distance but also on a few other parameters peculiar to the particular energetic scheme of transfer.

A. Single-Channel Transfer Assisted Only by Classical Modes

The simplest model of a generally accepted energetic scheme consists of two parabolic energy levels $U_1(q)$ and $U_2(q)$, representing the initial and final states of an electron in donor and acceptor molecules, linearly coupled to a classical harmonic mode. This is either internal low frequency mode of the reactants, or a collective mode of their polar surroundings. The total reorganization energy of the classical modes λ determines the horizontal splitting of the parabolas, while the free energy of the transition, ΔG (from minimum to minimum), determines their vertical splitting (Fig. 1). The horizontal splitting is a measure of electron–phonon interaction, which is usually assumed to be strong enough. Under such a condition electron transfer can occur only in the vicinity of the intersection point of two levels, where their splitting is negligible and transition is actually resonant with respect to the total energy of the system. The height of this point, with respect to the bottom of the initial (left) well, determines the activation energy U for the electron transfer to the right (product) well. The flux of electrons through the intersection point is proportional to the equilibrium

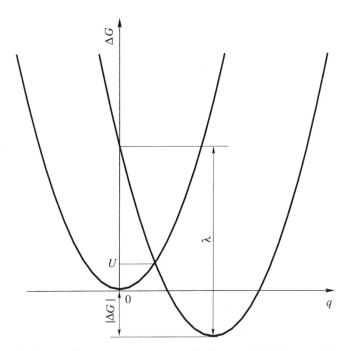

Figure 1. Energy diagram for electron transfer between initial (left) and final (right) parabolic levels.

density of states at this point, which is responsible for an Arrhenius factor in the general rate expression

$$W(r) = W_0(r, \gamma) \exp\left(-\frac{U(r)}{k_B T}\right) \qquad (2.1)$$

In line with the point intersection of energy levels of equal curvature, the activation energy is quadratic function of ΔG:

$$U = \frac{(\Delta G + \lambda)^2}{4\lambda} \qquad (2.2)$$

According to the classification of Marcus [23], the electron transfer is considered a "normal reaction" until $-\Delta G < \lambda$ and the activation energy decreases with increasing exergonicity of the process. At higher exergonicity, $-\Delta G > \lambda$, qualified as the *inverted region*, the intersection point appears at the opposite branch of the initial well and the activation energy starts to increase.

At fixed r the quadratic dependence of $\ln W$ on ΔG was confirmed experimentally and used as the basis for classification of the transfer regions as *normal* and *inverted*. They are separated by an activationless case $(-\Delta G = \lambda)$, where the cross-point lays at the bottom of the initial well and the reaction is the fastest. We will show that this famous bell-shaped dependence $\lg W(\Delta G)$, known as the *free-energy-gap law* (FEG law), may be essentially distorted in the liquid phase, when the encounter diffusion of reactants becomes a limiting stage.

The preexponent, generally speaking, depends on the rate of resonant tunneling V and the friction along the reaction coordinate, γ, which may be of either mechanical or dielectric origin. For weak and isotropic exchange interaction, the tunneling rate $V = V_0 \exp[-(r - \sigma)/L]$ is maximal at contact distance σ, but decreases drastically with reactant separation r. At large distances it is usually so weak, that the rate of transfer is given by the simplest ("golden rule") perturbation formula in which the pre-exponent does not depend on γ [24,25]:

$$W_0 = \frac{\sqrt{\pi} V^2}{\sqrt{\lambda k_B T}} \qquad (2.3)$$

With a decrease in interparticle distance, the tunneling magnifies and the perturbation theory becomes invalid. When the tunneling is too fast, the electron transfer is limited by the delivery of the reacting system to the intersection point. This is either diffusion in the energy space ($\sim \gamma$) up to the activation level, or diffusion in the reaction space ($\sim 1/\gamma$) to the intersection point. In fact, the preexponent is saturated with V and its upper limit is different at small, intermediate, and high friction levels. Numerous sophisticated theories were proposed to extend the theory to a strong interaction: the energy diffusion, transition state, and Landau–Zener approaches. The results were summarized in a few publications, separately for the normal [26] and inverted [27] regions. The boundaries between all the theories were established and depicted at the plane (V, γ), together with all the expressions for W_0 valid outside the perturbation theory limits. They establish the upper bounds for the electron transfer rates. For weak and remote intermolecular electron transfer, these limits can be hardly attained and therefore will be ignored in what follows.

The sharp distance dependence of $W_0(r)$ from Eq. (2.3) makes reasonable a popular *contact approximation* when the electron transfer is thought to be possible only at contact. It is usually introduced by *radiative* boundary conditions with corresponding rate constants of either ionization or recombination:

$$k_0 = \int W_I(r)d^3r, \quad k_c = \int W_R(r)d^3r \tag{2.4}$$

In contact approximation the peculiarities of $W_I(r)$ and $W_R(r)$ dependencies do not play a noticeable role. Such an approximation was used in the pioneering work of Smoluchowskii on binary ionization [28] and that of Hong and Noolandi for geminate ion recombination [29].

Although useful, the contact approximation is not nearly universally true and especially unfounded in the inverted region. From Eqs. (2.1)–(2.3) one can see that:

$$W(r) = V_0^2 \exp\left(-\frac{2(r-\sigma)}{L}\right)\frac{\sqrt{\pi}}{\sqrt{\lambda T}}\exp\left(-\frac{(\Delta G + \lambda)^2}{4\lambda T}\right) \tag{2.5}$$

From this point on we took $k_B = 1$. The sharp exponential decrease of the tunneling rate competes in this formula with the space dependence of the Arrhenius factor following from the pronounced r dependencies of both ΔG and λ.

In nonpolar solvents there is usually at least one low-frequency (classical) mode assisting electron transfer, whose reorganization energy $\lambda_i = $ const. However, in polar solvents there is an additional reorganization energy of the polar surroundings that depends on ion-pair separation r. In general, the total reorganization energy λ consists of two parts:

$$\lambda(r) = \lambda_i + \lambda_c\left(2 - \frac{\sigma}{r}\right) \tag{2.6}$$

The contact value of the polar term

$$\lambda_c = \left(\frac{1}{\epsilon_0} - \frac{1}{\epsilon}\right)\frac{e^2}{\sigma} \qquad (2.7)$$

is half as much as at infinite separation of reactants (ϵ and ϵ_0 are static and optical dielectric permitivities).

Consider, for instance, the forward electron transfer from excited donor to acceptor of electron, $[D^* \ldots A] \Rightarrow [D^+ \ldots A^-]$, and recombination of the generated ion pair to the ground state by backward electron transfer, $[D^+ \ldots A^-] \Rightarrow [D \ldots A]$. The exergonicity of ionization decreases with interparticle distance, because the energy of the Coulomb attraction between ions falls off:

$$\Delta G_I = \Delta G_i + T\left(\frac{r_c}{\sigma} - \frac{r_c}{r}\right) \qquad (2.8)$$

where ΔG_i is the free energy of ionization in the contact pair and $r_c = e^2/T$ is the Onsager radius of the Coulomb attraction. For the same reason, the exergonicity of recombination to the ground state increases (the free energy of backward transfer becomes more negative) with ion separation:

$$\Delta G_R = \Delta G_r - T\left(\frac{r_c}{\sigma} - \frac{r_c}{r}\right) \qquad (2.9)$$

where ΔG_r is the contact value of the recombination free energy. The extremum of the Arrhenius factor for either of the processes should be found from the condition

$$\Delta G(R_0) + \lambda(R_0) = 0 \qquad (2.10)$$

At $r = R_0$ the activation energy turns to 0 and the Arrhenius factor reaches its maximum value. On one side of the maximum the transfer is normal, on the other side it is inverted and both of them are activated, while in the middle the transfer is activationless.

If the point of maximum, R_0, is shifted far away from contact, then the ascending branch of the Arrhenius factor is predominant over $W_0(r)$ at short distances. Therefore, the electron transfer at contact is suppressed and the total rate $W(r)$ becomes a nonmonotonic, bell-shaped curve (Fig. 2). This fact was first established in Ref. 10 for polar media and in Ref. 11 for nonpolar solvents. We will thereafter classify this phenomenon as *noncontact* electron transfer and distinguish it from the conventional *contact transfer* with monotonic and sharply decreasing $W(r)$. The contact

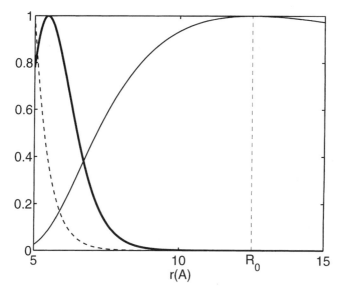

Figure 2. Arrhenius factor (thin line), exchange interaction $V(r)/V(0)$ (dashed line)and resulting rate of electron transfer $W(r)/W_{max}$ (thick line) as functions of interparticle distance $r \geq \sigma = 5\,\mathring{A}$, in a highly polar solution. (From Ref. 30.)

transfer occurs at shorter $R_0 \leq \sigma$ when the exponential drop of $W(r)$ is not qualitatively changed but only slightly accelerated.

From this analysis we see the importance of the position of the activationless point, R_0. Let us first consider the *nonpolar solvent*, where according to Eqs. (2.6)–(2.7):

$$\lambda \equiv \lambda_i = \text{const} \quad \text{at} \quad \epsilon = \epsilon_0 \tag{2.11}$$

By substitution of Eqs. (2.11) and (2.8) into Eq. (2.10), we obtain the equation on R_0, the point of activationless ionization:

$$\Delta G_i + T\left(\frac{r_c}{\sigma} - \frac{r_c}{R_0}\right) = \lambda_i \tag{2.12}$$

From Eqs. (2.11) and (2.9) the similar equation for the point of activationless recombination, R_0', can be obtained

$$\Delta G_r - T\left(\frac{r_c}{\sigma} - \frac{r_c}{R_0'}\right) = \lambda_i \tag{2.13}$$

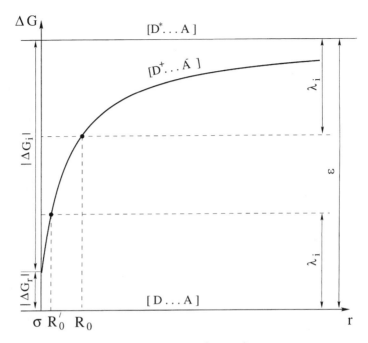

Figure 3. Energy level of excited reactants, $[D^*\ldots A]$, ion pair in the Coulomb well, $[D^+\ldots A^-]$, and the ground state, $[D\ldots A]$, in nonpolar solvent. The excitation energy $\varepsilon \gg 2\lambda_i$. The activationless points are shown by circles on the ion-pair energy curve.

As can be seen from Fig. 3, the ionization free energy is $\Delta G_I = E_{[D^+\ldots A^-]} - E_{[D^*\ldots A]}$, whereas the free energy of recombination to the ground state is $\Delta G_R = E_{[D\ldots A]} - E_{[D^+\ldots A^-]}$. The former becomes equal to λ_i at a larger distance than the latter, so that $R_0 > R_0'$. When the exergonicity of ionization is as high as shown in Fig. 3, the activationless point for ionization is shifted so far away that the forward electron transfer should be noncontact. Quite the opposite; the point of activationless recombination is too close to contact. At higher exergonicity of recombination it does not exist at all; there is only a descending branch of the Arrhenius factor at $r > \sigma$ and the recombination is contact for sure.

If the energy of photoexcitation $\varepsilon = E_{D^*} - E_D$ is fixed, the exergonicity of the forward and backward transfers are rigidly tied:

$$\Delta G_I + \Delta G_R = \Delta G_i + \Delta G_r = \varepsilon \qquad (2.14)$$

By substitution of one acceptor for another $(A \to A')$ one can change the exergonicity of both reactions but in opposite directions. In fact, this is the

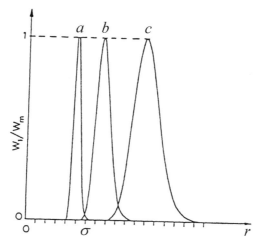

Figure 4. Spatial dispersion of ionization rate in nonpolar solvent ($\epsilon = \epsilon_0 = 2$) at ΔG_i:
(a) 0; (b) $3\lambda_i$; (c) $6\lambda_i$ ($W_m^a : W_m^b : W_m^c = 1 : 2 \times 10^{-3} : 6 \times 10^{-10}$). (From Ref. 11.)

usual way of studying the free-energy dependence of electron transfer. Using a variety of different acceptors, one can push the curve $[D^+ \ldots A^-]$ up or down. In a wide series of similar acceptors the free energy of transfer can be significantly varied, so that ionization changes from contact to noncontact (Fig. 4). The rate of resonant ionization is monotonous if not exponential, but at $|\Delta G_i| > \lambda_i$ it takes the bell shape and moves out of contact. As this takes place, the recombination remains contact until $|\Delta G_i|$ is less than $\varepsilon - \lambda_i$.

In polar solvents the situation is different. The equation for the activationless point is the same for ionization and recombination:

$$\Delta G_{i,r} = \lambda_i + \lambda_c \left(2 - \frac{\sigma}{R_0} \right) \tag{2.15}$$

The higher the exergonicity of either ionization or recombination, the larger the shift of the rate maximum. In comparison with nonpolar solvents, the widths of the remote reaction layers are larger, but all the rest is qualitatively the same (Fig. 5). The efficiency of electron transfer rapidly decreases with the radius of the reaction layer, due to a drop in the tunneling rate at long distances. The higher the exergonicity, the more remote and weaker is electron transfer in polar solutions. This conclusion is valid for both forward and backward electron transfer: each of them may be either contact or remote depending on the corresponding free energy. According to relation

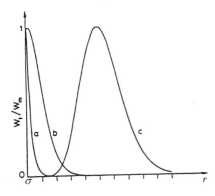

Figure 5. Spatial dispersion of ionization rate in polar solvent at ΔG_i : (a) 0; (b) $1.5\lambda(\sigma)$; (c) $3\lambda(\sigma)$ $[\lambda(\sigma) = \lambda_i + \lambda_c = 1.3$ eV; $W_m^a : W_m^b ; W_m^c = 9 \times 10^{-5} : 1 : 6 \times 10^{-12}]$. (From Ref. 11.)

(2.14), when ionization is quasiresonant, then recombination is highly exergonic and vice versa. Therefore, in polar solvents either ionization is contact while recombination is noncontact, or quite the reverse.

The relative positions of reaction layers play a crucial role in the unified theory of forward and backward electron transfer (Section VI). If ionization occurs outside the recombination layer, the quantum yield of charge separation is large, but it should be much less if the ionization and recombination layers interchange the places. In the latter case, the separating ions must cross the outer recombination layer, where most of them can be discharged. Although the difference in starting positions of ions is only a few angstroms, it cannot be ignored because the characteristic length of tunneling is even smaller, $L \sim 0.5 \div 1.0$ Å. The situation with charge separation is similar to that of the tunneling scanning microscope that makes possible macroscopic registration of separate atoms. The extraordinary sensitivity of electron tunneling to the space separation of reactants, considerably extends the systems response to a variation in the inter-reactant distance.

B. Multichannel Electron Transfer

Not only the classical motion along the reaction coordinate can assist the electron transfer, but as well the high-frequency vibrations of reactants. If there is only a single essentially quantum mode in the system (with $\hbar\omega \gg T$), then one should take into account a lot of additional reaction channels resulting from the intersection of vibronic levels of reactants and reaction products (Fig. 6). In the classical theory of nonadiabatic electron transfer, assisted simultaneously by classical surroundings and a single quantum mode ω, the position-dependent rates of ionization and

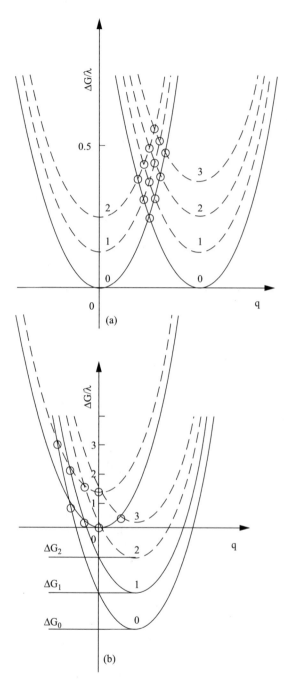

Figure 6. Multichannel transfer from the reactant (left) to a product (right) well in the normal (*a*) and inverted (*b*) regions (from Ref. 30). The dashed lines indicate the vibronic levels and ∘—their cross-points where resonant electron transfer occurs.

recombination are well defined [31,32]:

$$W_I(r) = V_I^2 \sum_{-\infty}^{\infty} P(n)\sqrt{\frac{\pi}{\lambda T}} \exp\left[-\frac{(\Delta G_I + \lambda + \hbar\omega n)^2}{4\lambda T}\right] \quad (2.16)$$

$$W_R(r) = V_R^2 \sum_{-\infty}^{\infty} P(n)\sqrt{\frac{\pi}{\lambda T}} \exp\left[-\frac{(\Delta G_R + \lambda + \hbar\omega n)^2}{4\lambda T}\right] \quad (2.17)$$

where $\Delta G_I + \hbar\omega n$ and $\Delta G_R + \hbar\omega n$ are the free energies for ionization and recombination through the nth vibronic channel:

$$P(n) = \begin{cases} \dfrac{e^{-S}S^n}{n!}, & n \geq 0 \\[2mm] \exp\left(\dfrac{\hbar\omega n}{T}\right)P(|n|), & n < 0 \end{cases} \quad (2.18)$$

The Franck–Condon parameter $S = \lambda_q/\hbar\omega$ is related to a quantum-mode reorganization energy λ_q.

As usual, the distance dependence of the multichannel rates is controlled mainly by the sharply decreasing matrix elements

$$V_I(r) = V_I^0 e^{-(r-\sigma)/L} \quad (2.19)$$

$$V_R(r) = V_R^0 e^{-(r-\sigma)/l} \quad (2.20)$$

where L and l are the space decrements of the exchange interaction for the forward and backward electron transfer. Even at contact, the exchange interaction is assumed so weak that the perturbation theory remains valid. In other words, the preexponential factors in Eqs. (2.16) and (2.17) are smaller than the rate of approaching the cross-points by a low-frequency oscillator carrying the electron. Otherwise, the saturation of electron transfer at short distances is inevitable [33,34] as well as the interference of reaction channels [35]. The higher the number of the activationless channel, the smaller the corresponding Franck–Condon factor $P(n)$ at $S \ll 1$, and the saturation of electron transfer is scarcely possible.

In addition to the r-dependent exchange interaction, the distance-dependent energies, $\Delta G(r)$ and $\lambda(r)$, contribute to the space dispersion of the multichannel rates. This is of no less importance than for single-channel reactions. Again, one should discriminate between nonpolar and highly polar solutions. In ultimate cases $\epsilon = \epsilon_0$ in the former situation and $\epsilon = \infty, r_c = 0$ in the latter. Correspondingly, either $\lambda = \lambda_i = \text{const}$ (in nonpolar solvents), or $\Delta G_I, \Delta G_R = \text{const}$ (in superpolar solvents).

Let us start from a nonpolar solvent and consider the electron transfer between singlet and triplet excited states of the reactants via an ion pair. Unlike the excitation energy, the singlet–triplet splitting ε is rather small and even smaller than the depth of the Coulomb well, Tr_c/σ. Setting, for instance, $\mathcal{E} = 2\lambda_i$, we can render the single-channel electron transfer activationless at the same point R_0, for both the ionization of singlet and recombination to triplet (Fig. 7). Now let us see what is changed when both these processes are multichannel reactions. The vibronic states of the ion pair have the same energy curves as its ground state, but shifted up by $\hbar\omega n$. Therefore, their activationless points, (circles), are shifted to the left from R_0. For similar reasons the activationless points for recombination to the triplet vibronic states are shifted toward the right (crosses). Because of our choice of $\mathcal{E} = 2\lambda_i$, the rate of a single-channel transfer ($S = 0$) is the same for ionization and recombination (if $V_I = V_R$). It is represented by the highest bell-shaped curve in Fig. 8. In the case of multichannel transfer ($S \neq 0$) this curve is split into a number of others, representing the transfer to the excited vibronic states. For recombination all of them are shifted rightward and damped by decreasing tunneling factor. As a result, the recombination layer remains at the same place, having become less efficient. In contrast to this, the vibronic components of ionization are shifted toward the contact where tunneling is much faster. As a result, the total ionization becomes quasicontact and much stronger.

In highly polar solvents the situation is much simpler and more universal. For either ionization or recombination the transfer to higher vibronic states becomes activationless closer to the contact. This is clearly seen in Fig. 9.

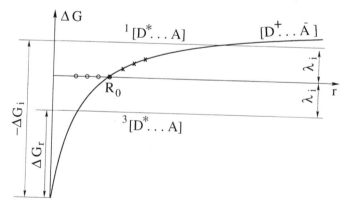

Figure 7. Energy level of excited singlet, $^1[D^*\ldots A]$, ion pair in the Coulomb well, $[D^+\ldots A^-]$, and the triplet state, $^3[D^*\cdots A]$, in nonpolar solvent. The excitation energy $\mathcal{E} = 2\lambda_i$. The activationless points for ionization to higher vibronic states of the ion pair are shown by circles and by crosses for recombination to higher vibronic states of triplets.

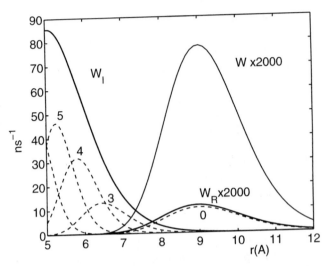

Figure 8. Distance-dependent rates of multichannel ionization (W_I) and recombination (W_R) at $\mathcal{E} = -\Delta G_I - \Delta G_R = 2\lambda_i = 0.5\,\text{eV}$ in a nonpolar solution: $\epsilon = \epsilon_0$ (from Ref. 30). In this condition, the rate W for a single-channel reaction ($S = 0$) is the same for both ionization and recombination (thin line). For multichannel transfer these rates, W_I and W_R are different (thick lines). The vibronic components of these rates are shown by dashed lines and nominated by their numbers $n\,[\sigma = 5\,\text{Å}\quad T = 300\,\text{K}, (V_I^0)^2\sqrt{\pi}/T = (V_R^0)^2\sqrt{\pi}/T = 10^4\,\text{ns}^{-1}, -\Delta G_i = 1.16\,\text{eV}]$.

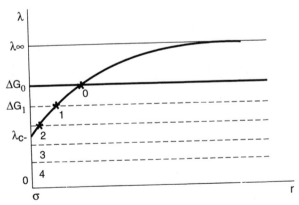

Figure 9. The reorganization energy $\lambda(r)$ in a highly polar solution ($\epsilon = \infty$) and the free energies of transfer to the ground state (thick line), or to vibronic states of the products (dashed lines). Activationless transitions to these states occur at the points marked by asterisks.

The transfer to excited products is less exergonic and the smaller $-\Delta G_n$, the closer to the contact is the intersection point, $-\Delta G_n = \lambda$. For the highest vibronic states there is no activationless points at $r > \sigma$ and the electron transfer to them is essentially contact. This is clearly seen in Fig. 10, where the total rates of the forward and backward transfer are split into vibronic components. For low n, the rate has a bell shape, but for higher n it becomes monotonic, quasiexponential. Therefore, the multichannel reaction zone is always closer to contact than that of a single-channel reaction. The corresponding rate of transfer may be either bell-shaped or quasiexponential depending on the extent of the exergonicity of the single-channel reaction (Fig. 10).

C. Model Rates

Noncontact electron transfer in condensed matter is now a widely recognized phenomenon. The transformation of the exponential ionization rate to the bell-shaped curve at $-\Delta G_I > 2\,\text{eV}$ was reported in Ref. 36. Taking $\varepsilon \gg 2\lambda_c$ as in Ref. 37, it is easy to get the noncontact distance dependence for both ionization and recombination rates. If we restrict our consideration to only polar solvents, then the general conclusion stands—the higher exergonicity of any transfer the more noncontact and weaker it is, as shown schematicly in Fig. 11.

Since the real shape of the transfer rate is rather complex, a few simple approximations were proposed to model them. If electron transfer is monotonous, sharply decreasing function of the distance, then by neglecting the space dispersion of such interactions, one can substitute them by δ-functions located at the contact distance:

$$W_I(r) = \frac{k_0 \delta(r - \sigma)}{4\pi\sigma^2}, \qquad W_R(r) = \frac{k_c \delta(r - \sigma)}{4\pi\sigma^2} \qquad (2.21)$$

This is in fact the contact approximation used in the Smoluchowskii approach to bimolecular reactions in solution [28] and applied to geminate recombination by Hong and Noolandi [29]. This approximation is a widespread method based on a few phenomenologic parameters (e.g., k_0 and k_c) and applied to a variety of contact reactions in solutions. For atom or proton transfer it is quite suitable.

However, for electron or energy transfer this approximation is too rough. A little bit better is an exponential model of the transfer rate that is in essence contact, but accounts for a finite width of the reaction layer, L:

$$W(r) = W_c e^{-2(r-\sigma)/L} \qquad (2.22)$$

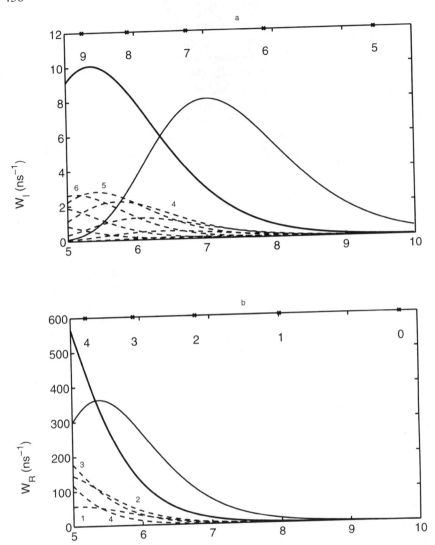

Figure 10. The rates of ionization (at $-\Delta G_i = 2\lambda_c$) (*a*) and recombination (at $-\Delta G_r = 1.5\lambda_c$) (*b*) in polar solvent ($r_c = 7\text{Å}$, $\lambda_i = 0$, $\lambda_c = 1.4\,\text{eV}$, $\sigma = 5\,\text{Å}$, $T = 300\,K$, $\hbar\omega/T = 6$) [30]. The single-channel reactions ($S = 0$) are exhibited by thin lines, and the thick lines are used for multichannel reactions ($S = 2$). The latter are decomposed in contributions related to different channels (dashed lines) nominated by numbers of vibronic states n (the corresponding activationless points are marked by asterisks placed at the top).

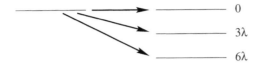

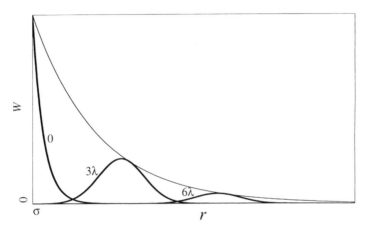

Figure 11. Transition from contact to remote transfer in polar solvent with increasing contact reorganization energy $\lambda = \lambda_i + \lambda_c$. Above: energy scheme of electron transfer at different exergonicity values. Below: r dependence of corresponding rates (thick lines) in comparison with an exponential decrease of the normalized tunneling rate (thin line).

This results from an "exchange interaction" [38] responsible for the resonant transfer of triplet excitation, but is also well applicable to electron transfer in normal Marcus region [7,39].

However, this is not the case if the electron transfer is not contact, as in the case of inverted region. One has to either use the real $W(r)$ dependence, or model it by similar bell-shaped functions. The best one proposed in Ref. 40 has a symmetric bell shape (Fig. 12):

$$W(r) = \frac{W_0}{\cosh^2\left(\dfrac{r - R}{\Delta}\right)} \tag{2.23}$$

This model was shown to be well suited for an analytic treatment of noncontact, diffusion-accelerated bimolecular reactions. Even simpler, the rectangular model of a reaction layer (Fig. 12) was often used instead:

$$W(r) = \begin{cases} 0 & \sigma < r \le R - \Delta \\ W_0 & R - \Delta < r < R + \Delta \\ 0 & R + \Delta \le r \end{cases} \tag{2.24}$$

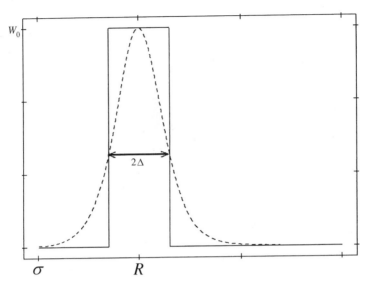

Figure 12. Bell-shaped [Eq. (2.23)] and rectangular [Eq. (2.24)] models of the position dependent transfer rate $W(r)$.

This model was also used a few times to calculate nonstationary ionization in binary encounters at contact $(R - \Delta = \sigma)$ [41] and in the remote reaction layer $(R - \Delta > \sigma)$ [42]. It has also been applied to remote geminate recombination of an ion pair, to calculate analytically the charge separation quantum yield at different initial separations of ions [43]. Of course, the main parameters of the model rates, their width Δ and location R, should be related to the reorganization and free energies of the system, responsible for these parameters. The height, W_0, is also related to them as well as to the rate of tunneling.

III. FREE-ENERGY GAP (FEG) LAW

The fundamental property of the transfer rate follows from Eqs. (2.1) and (2.2):

$$\ln\left(\frac{W}{W_0}\right) = -\frac{(\Delta G + \lambda)^2}{4\lambda T} \tag{3.1}$$

This has come to be known as the *free-energy-gap* (FEG) law, described earlier. If the lefthand side of this relationship is measured and the free energy ΔG is varied in a wide row of different acceptors, then the parabolic dependence (3.1) is expected to be found with a maximum in the activationless point ($\Delta G = -\lambda$). One branch of such a parabola is related to the normal transfer region; the opposite, to the inverted one. The latter was a surprising prediction of the Marcus theory of electron transfer; in the inverted region the rate of transfer decreases with increasing exergonicity.

A. Multichannel Intramolecular Transfer

The FEG law is represented by a fully symmetric parabolic curve for only the single-channel reaction (Fig. 13). If one of the quantum modes also assists the electron transfer, then each channel has the corresponding parabola drawn down and shifted to higher free energies. The lowering of the electron transfer is due to the Franck–Condon factor $P(n)$, which is less, the higher the degree of excitation of the vibronic level. The total free energy dependence becomes slightly asymmetric (Fig. 13) but holds qualitatively the same shape.

The FEG law was checked and confirmed experimentally in Ref. 44 but only for intramolecular electron transfer when the donor–acceptor separation r is fixed and what is measured is really $W(r)$ (Fig. 14). The results were reviewed and discussed in Ref. 45 and other articles published in the same special issue.

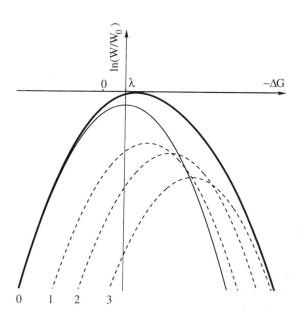

Figure 13. The free-energy-gap (FEG) law for single-channel (thin line) and multichannel (thick line) electron transfer, decomposed into vibronic components (dashed lines).

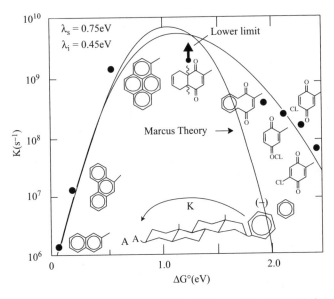

Figure 14. Intramolecular rate constants in 2-methyloxacyclopentane solution at 296 K, as a function of the free-energy change. The electron transfer occurs from biphenylanions to the eight different acceptor moieties (shown adjacent to the data points), in eight bifunctional molecules of the general structure shown in the center. (From Ref. 44.)

B. FEG Law for Ionization and Subsequent Recombination to the Ground State

At any fixed distance between the donor and acceptor of an electron there is a definite relationship between the energies of the excited reactants, the ion pair, and the ground state. In line with Eq. (2.14) and Fig. 3, the sum of the free energies of ionization and recombination to the ground state is equal everywhere to the energy of the excited reactants, ε. Where ionization is quasiresonant, the recombination is highly exergonic and vice versa. In Fig. 3, one limit gives way to another at shorter distances.

When the distance is fixed the same can be done by changing the reaction partner. As we saw, the energy of the ion state can be varied this way, over a wide range, between excited and ground states, changing simultaneously the free energies of ionization and recombination. In such a case the FEG laws for the forward and backward electron transfer are correlated and their relationship depends on the ratio $\varepsilon/2\lambda$. As can be seen in Fig. 15, there are two limiting cases of (A) small ($\varepsilon < 2\lambda$) and (C) large ($\varepsilon > 2\lambda$) excitation energies, separated by the border case (B) where $\varepsilon = 2\lambda$. Both ionization

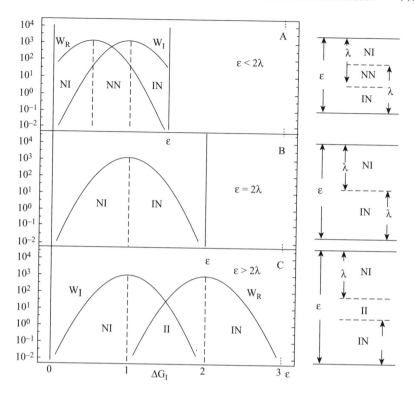

Figure 15.   Dependence of $\ln W_I$ and $\ln W_R$ on ΔG_I in the range between 0 and ε for small (A), intermediate (B), and large (C) ε as compared to 2λ. The corresponding energy diagrams show the separation of normal regions for ionization (recombination) within λ strips from the top (bottom), from the inverted regions (outside these strips).

and recombination can occur in either the normal (N) or inverted (I) regions, depending on where is the energy level of the ion pair, regarding the energies of the excited and ground states. If it is very close to the excited level, then ionization is in the normal while recombination is in the inverted region (NI). If the energy of the ion pair is much closer to the ground state then vice versa, the ionization is inverted while recombination is a normal transfer (IN). Case B is exhausted by only these two opposite situations. However, in case A both transfers can be simultaneously normal (NN) when the ionized state is approximately in the middle. In the same situation, but in case C they are simultaneously in the inverted region (II).

This classification is very important for recognizing the situation and making the right choice of appropriate approximations. For instance, the energy situation at contact in highly polar solvents determines the space dependence of the transfer rates. Both of them are contact only in region NN found in case A $[\varepsilon < 2\lambda(\sigma)]$ at $\Delta G_i \approx \Delta G_r \approx \varepsilon/2$. This is the most favorable situation for the application of contact or exponential approximations, (Eq. 2.21) or (Eq. 2.22). On the contrary, in case C $[\varepsilon > 2\lambda(\sigma)]$ at least one of the processes occurs in the inverted region and is therefore noncontact, described by the alternative models, Eq. (2.23) or (2.24). In what follows we will mainly concentrate on case B $[\varepsilon = 2\lambda(\sigma)]$ when either the forward transfer is contact and the backward transfer noncontact, or vice versa. For this particular case the curves representing the FEG laws for single channel reactions at fixed distance coincide (Fig. 15, case B). For the multichannel reactions they are slightly different. The asymmetry of the ionization curve shown in Fig. 13 arises at its right branch, while the same asymmetry of the recombination curve should appear at the opposite left branch, where recombination is inverted.

Even more important is the space dispersion of what is measured in liquid solution. Whatever it is, the effect is averaged over the distance between the reactants modulated by their encounter diffusion. The result does not always simply relate to any $W(r)$, as we shall see below.

C. FEG Law for Kinetically Controlled Binary Ionization

The irreversible ionization of the excited donor or acceptor of an electron in liquid solution can be represented by the simplest kinetic scheme:

$$D^* + A \rightarrow [D^+ \ldots A^-] \quad \text{or} \quad D + A^* \rightarrow [D^+ \ldots A^-] \qquad (3.2)$$

If the concentration of nonexcited reactants, c, is rather small this is a second-order transfer reaction that can be under kinetic or diffusional control, depending on how fast is the electron transfer in comparison with encounter diffusion.

At slow ionization and fast diffusion the forward electron transfer is expected to be under kinetic control and its rate should be $k_0 c$, where the kinetic rate constant k_0 defined in Eq. (2.4) is diffusion-independent. The real dependence of this constant on the contact free energy ΔG_i may be approximately estimated, provided $W_I(r)$ is a sharp quasiexponential function, as in Eq. (2.22) [14,15]:

$$k_0(\Delta G_i) \approx W_I(\sigma)v \qquad (3.3)$$

where the reaction volume $v \approx 2\pi(\sigma^2 L + \sigma L^2 + L^3/2)$ does not depend on ΔG_i. In Fig. 16 this approximation is compared with exact calculation of k_0 from Eq. (2.4) for a highly polar solvent. It was expected to work well for contact ionization, in the normal region, but much worse in the inverted

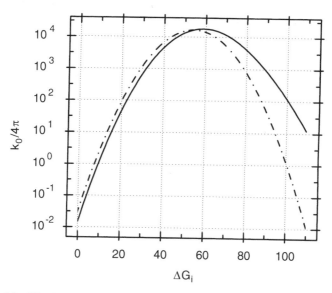

Figure 16. Kinetic rate constant for ionization: solid line—exact calculation from Eq. (2.4); dashed–dotted line—approximate (contact) estimate, Eq. (3.3). [$\lambda(\sigma) = \lambda_c = 55$ T, $W_0(\sigma) = 10^3 \text{ ns}^{-1}, \sigma = 5 \text{ Å}, r_c = 7 \text{ Å}, L = 1 \text{ Å}$].

region, where $W_I(r)$ is not contact. Indeed, the normal branches of both curves almost coincide while the real inverted branch is lifted up, rendering the whole curve asymmetric. The effect is similar to that produced by the multichannel transfer (see Fig. 13). Taking into account either of these effects makes no qualitative changes in the quasiparabolic shape of the FEG law.

In contrast to this conclusion, the very first experimental study of the FEG law in liquid solution discovered that the rate constant of binary ionization does not depend on the free energy in the inverted region (Fig. 17a).* This discovery by Rehm and Weller [6] put into doubt the whole Marcus theory and initiated a lot of experimental examinations and theoretical revisions of the FEG law. There is no need to review them because the most reliable explanation of the Rehm-Weller paradox given by Marcus and Siders [46] lies in the fact that the fastest transfer (at the top of FEG curve) is controlled

---

*Note, that in the original presentation of the data shown in Fig. 17a the authors used $\Delta G_i \equiv \Delta G_{23}$ for the abscissa instead of $-\Delta G_i$ as became common later. Therefore, the inverted region for ionization in Fig. 17a is to the left unlike in our other figures, where it is to the right.

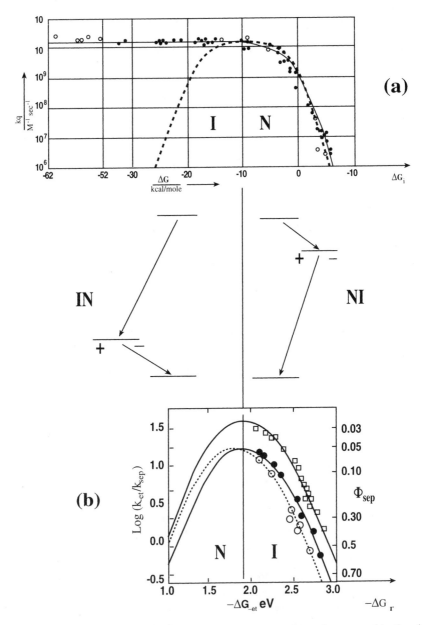

Figure 17. FEG law for (*a*) bimolecular ionization and (*b*) geminate recombination; in between the typical energy schemes for IN and NI transfer at contact. (*a*) Fluorescence rate constant k_q as a function of the free energy of ionization for different reactants in a polar solvent [6]. Dashed line represents the expected FEG law arising from the Marcus theory. (*b*) The backward transfer rate k_{-et} or charge separation quantum yield (righthand axis) versus recombination exergonicity $-\Delta G_{-et} \equiv \Delta G_r$ for different ion pairs fitted to the corresponding Marcus parabolas [47].

by encounter diffusion rather than by the reaction itself. What is measured in this region is not the kinetic but the diffusional rate constant, which is almost independent of the free energy. We will return to this explanation in the next section.

D. FEG Law for Geminate Recombination to the Ground State

The kinetic rate constant for binary recombination has a similar FEG dependence as that for ionization, especially in polar solutions [15]. A different situation arises with geminate recombination following binary ionization:

$$[D^+ \ldots A^-] \to [D \ldots A]$$

In this case the reactants are not uniformly distributed in space at the beginning but appear at a definite distance $r_0 \geq \sigma$. Starting from this distance they can come closer and react or be separated and survive. The fraction of survived ions $\varphi(r_0)$ depends essentially on the initial separation. This is the charge separation quantum yield that can be studied experimentally as a function of the free energy of recombination. To examine the FEG law with these data, there should be a theory that relates somehow the quantum yield to the rate of recombination.

The simplest theory of this kind, known as an *exponential model*, presumes that:

1. Ions are born in a sphere of radius σ and also recombine there with a uniform rate of backward electron transfer k_{-et},
2. Ions escape the sphere and become free (never return back) with a permanent rate

$$k_{sep} = \frac{3 r_c \tilde{D}}{\sigma^3 [e^{r_c/\sigma} - 1]} \qquad (3.4)$$

We have already seen that both ionization and recombination are remote reactions, occurring at $r \geq \sigma$ and never inside the eigenvolume of the reactants. However, the first assumption is a plausible approximation in a special case NN, when the forward and backward transfers are contact. The second assumption is much worse. Although k_{sep} is by definition the inverse time of diffusional escape from the Coulomb well (or the sphere of birth), it cannot be considered as a first-order rate of the process because the charge separation proceeds nonexponentially. Using such a rate in kinetic equations, one actually assumes that the charge separation occurs by a single jump from contact to infinite distance.

Against all odds, these kinetic equations are the formal basis of the exponential model. They were written for the survival probabilities of ions in the contact sphere, m_c, and fully separated ions, m_∞:

$$\frac{dm_c}{dt} = -k_{-et}m_c - k_{sep}m_c \tag{3.5a}$$

$$\frac{dm_\infty}{dt} = k_{sep}m_c \tag{3.5b}$$

For instantaneous photoionization the initial conditions are $m_c(0) = 1$ and $m_\infty = 0$.

The total survival probability

$$\Omega = m_c + m_\infty \tag{3.6}$$

can be found from the general solution of Eqs. (3.5). This probability decreases exponentially with time approaching the charge separation quantum yield φ_0. The same relaxation normalized to a fraction of recombined ions, $1 - \varphi_0$, can be represented as follows:

$$R(t) = \frac{\Omega - \varphi_0}{1 - \varphi_0} = e^{-t/t_e} \tag{3.7}$$

where

$$\frac{1}{t_e} = k_{-et} + k_{sep} \tag{3.8}$$

Although the exponentiality of this function owes its name to the model, it cannot be justified by a rigorous treatment of geminate recombination (see Section V).

A more complex situation arises with the charge separation quantum yield. In the exponential model this is

$$\varphi_0 = \frac{1}{1 + k_{-et}/k_{sep}} = \frac{1}{1 + z/\tilde{D}} \tag{3.9}$$

where

$$z = \frac{k_{-et}\sigma^3}{3r_c}\left(e^{r_c/\sigma} - 1\right) \tag{3.10}$$

In most cases the only measured quantity is φ_0, which is actually z, provided the coefficient $\tilde{D}$ of the encounter diffusion of ions is known. It is precisely z that is studied as a function of the recombination free energy. Another issue is whether $z/\tilde{D}$ can be considered as a ratio k_{-et}/k_{sep}, because neither of this rates exists in reality. Despite this fact, the FEG law inherent to z was attributed to the expected free-energy dependence of k_{-et} and widely studied experimentally. The typical data are shown in Fig. 17b for three different ion pairs. All of them lie in the inverted region and strongly support the predictions of the Marcus theory, namely, that recombination is slower at higher exergonicity. The opposite normal region of recombination is rarely available, because at low recombination free energy the thermal transitions from the ground to ion state can be hardly precluded. Nonetheless, in some systems both the ascending and descending branches of the FEG curve were obtained (Fig. 18). However, the FEG "hump" in recombination is often narrower then in ionization [49]. This may be a manifestation of the fact that the origin of the fingerlike feature in Fig. 18 may be somewhat different from what was expected assuming $k_{-et} \propto W_R(\sigma)$.

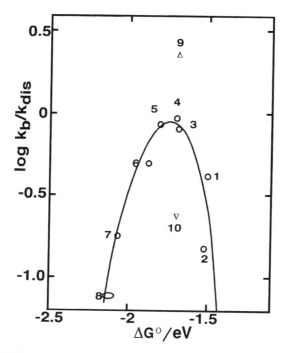

Figure 18. FEG law for recombination of ions formed in the quenching of triplet excited ruthenium(II) compounds by 10 different aromatic amines ($k_b \equiv k_{-et}, k_{dis} \equiv k_{sep}, \delta G^\circ \equiv \Delta G_r$) [48].

In Section VI we will take account for the fact that in general both ionization and recombination are not contact. The photogenerated ions initially appear either inside or outside the recombination layer and enter it by diffusion. Therefore the geminate recombination may be as well controlled by diffusion as bimolecular ionization while the exponential model implies that the geminate reaction is under kinetic control. Diffusional distortion of the conventional FEG law may be responsible for a more steep "hump" that changes the shape with solvent viscosity.

IV. DIFFERENTIAL ENCOUNTER THEORY FOR BINARY IONIZATION

Generally speaking, the bimolecular electron transfer of type (3.2) can be either kinetic or diffusional reaction. In the latter case the "rate constant" is not a constant at all but a time-dependent coefficient. There are a few fundamental approaches to diffusional reactions that provide the information about this dependence. They are considered below.

A. Contact Approximation

In classical works of Smoluchowski and others [28,50–54] the time-dependent reaction "constant" is given by the diffusional flux of reactants that enter the reaction sphere of contact radius or react in the thin layer of volume v adjacent to contact. For bimolecular ionization of neutral reactants this is

$$k_I(t) = 4\pi D r^2 \frac{\partial n}{\partial r}\bigg|_{r=\sigma} = k_0 n(\sigma, t) \qquad (4.1)$$

where the contact kinetic constant k_0 is given by Eq. (3.3). The last equality in (4.1) is actually the "radiation" boundary condition to a pair distribution function of the reactants, $n(r, t)$, which obeys the diffusional equation in free space provided there is no potential interaction between reactants:

$$\dot{n} = \frac{D}{r^2}\frac{\partial}{\partial r}r^2\frac{\partial n}{\partial r} \qquad (4.2)$$

This equation should be solved with the initial condition $n(r, 0) = 1$, if the excited reactants are randomly and homogeneously distributed around the partner in the beginning. Using the solution of (4.2) in (4.1), Collins and Kimball obtained the following result [51]:

$$k_I(t) = k_i\left(1 + \frac{k_0}{k_D}e^x \text{erfc}\sqrt{x}\right) \qquad (4.3)$$

where $k_D = 4\pi\sigma D$ is the diffusional rate constant, $x = (1 + k_0/k_D)^2 Dt/\sigma^2$ and

$$k_i = \lim_{t\to\infty} k_I(t) = \frac{k_0 k_D}{k_0 + k_D} \qquad (4.4)$$

is the stationary rate constant. More general results that account for potential interaction between reactants are also available [7,51,55]. The overall view of time evolution of $k_I(t)$ is shown in Fig. 19.

At fast electron transfer $(k_0 \gg k_D)$ the ionization is controlled by diffusion and

$$k_I(t) = \begin{cases} k_0 & \text{at} \quad t = 0 \\ 4\pi\sigma D \left(1 + \dfrac{\sigma}{\sqrt{\pi Dt}}\right) & \text{at} \quad t \to \infty \end{cases} \qquad (4.5)$$

Only in the opposite limit, under kinetic control $(k_0 \ll k_D)$, when $k_I(t) \equiv k_0$, one may expect to be able to reproduce the FEG law inherent to k_0 from the experimentally measured values of k_I. This is the case deeply in the normal and inverted regions, but not in between where k_0 reaches the maximum value that can be higher than k_D. In such a case the transfer is controlled by diffusion at the top of the FEG curve.

To describe the entire free-energy dependence of the stationary rate constant k_i, one should use the general expression (4.4), which differs

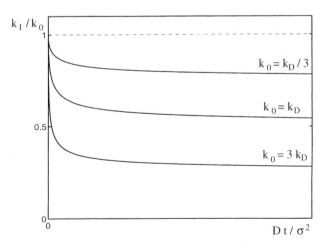

Figure 19. Time dependence of ionization rate constant, $k_I(t)$, in contact approximation at different ratios k_0/k_D. Dashed line represents kinetic control limit $(D \to \infty)$.

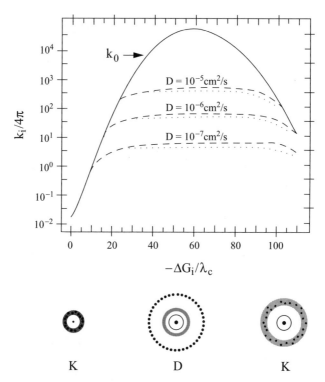

Figure 20.   The FEG law in case B of Fig. 15 ($\varepsilon = 2\lambda_c$) for the kinetic rate constant (solid line) in comparison to the free-energy dependence of the stationary rate constants k_i (dashed lines) and their contact estimates (dotted lines), at different diffusion rates. Bottom: the ionization layers (shadowed) and initial charge distributions (dots) in the normal and inverted regions, under kinetic control (K) and in the diffusional regime (D) [15]. [$W_0(\sigma) = 10^3$ ns^{-1}, $\lambda_c = 55$ T, $r_c = 7$ Å, $\sigma = 5$ Å .

essentially from k_0. The main distinction lies in the fact that the top of the parabolic FEG curve is cut by a diffusional plateau that is lower and wider the slower diffusion (dashed lines in Fig. 20). Such a flat "roof" of $k_i(\Delta G_i)$ dependence can be associated with a plateau obtained in the Rehm and Weller experiments (Fig. 17a). This opportunity mentioned in their work was first analyzed quantitatively by Marcus and Siders [46] using an exponential model instead of contact approximation. This is a better choice but only for the normal region because in the inverted region the ionization is neither contact nor exponential. Fortunately, the general encounter theory removes all limitations on the shape of $W_I(r)$. By means of this theory the FEG law can be revised at least numerically, using the realistic ionization rate that is nonmonotonous in the inverted region.

B. Differential Encounter Theory

The limitations of contact approximation were overcome only in 1963/64, when a few groups in Russia independently, but intuitively, derived a set of equations for remote energy transfer accelerated by encounter diffusion [56]. The results were noticed and more consistently rederived in Ref. 57 and later on in Ref. 58. Originally only the population of the excited state was considered, which was quenched with the rate $W(r)$. In Ref. 59 the theory was further extended for a density matrix and Hamiltonian formalism and termed *encounter theory*. To be more specific, we should designate it as "differential" encounter theory (DET) because the main kinetic equations remain differential and are consistent with the rate concept. DET is solely a version of the more fundamental *integral encounter theory* (IET) derived later from the first principles [60].

As applied to chemistry, the differential encounter theory accounts for the time dependence of the ionization rate constant $k_I(t)$, employing the position-dependent transfer rate of arbitrary shape, $W_I(r)$. The formal equations of chemical kinetics are the same for the contact and differential encounter theory:

$$\dot{N} = -k_I(t)cN - \frac{N}{\tau_D} \tag{4.6a}$$

$$\dot{P} = k_I(t)cN \tag{4.6b}$$

where $N(t)$ is the survival probability for excitations generated at $t = 0$, τ_D is their lifetime in the absence of quenchers, and $P(t)$ is the probability of product (charge) accumulation in either of reactions (3.2). Immediately after the δ-pulse excitation there are no ions but 100% excitations, so that $N(0) = 1, P(0) = 0$. At permanent excitation with a rate I, there is a stationary concentration of excitations $I \int_0^\infty N(t)dt$, so that the relative quantum yield of fluorescence quenching is [61]

$$\eta = \int_0^\infty \frac{N(t)dt}{\tau_D} = \frac{1}{1 + c\kappa\tau_D} \tag{4.7}$$

In the last expression η is presented as the Stern–Volmer law, where $\kappa(c)$ is the corresponding stationary quenching constant. The concentration dependence of the Stern–Volmer constant is characteristic of any theoretical method used for quantum yield calculation and differential encounter theory is the best in this respect [62].

The difference with contact approximation appears from the definition of the rate constant

$$k_I = \int W_I(r)n(r,t)d^3r \qquad (4.8)$$

expressed in terms of a real $W_I(r)$ directly and indirectly, via the pair distribution $n(r,t)$, which obeys the equation including $W_I(r)$:

$$\dot{n} = -W_I(r)n(r,t) + \frac{D}{r^2}\frac{\partial}{\partial r}r^2\frac{\partial n}{\partial r} \qquad (4.9)$$

If there is no exciplex formation or other contact reactions, this equation should be solved with reflecting boundary condition

$$4\pi Dr^2\frac{\partial n}{\partial r}\bigg|_{r=\sigma} = 0 \qquad (4.10)$$

and the same initial condition $n(r,0) = 1$, assuming the random and uniform distribution of reactants at the beginning.

In binary approximation, the relative motion of them is specified by the encounter diffusion coefficient $D = D_D + D_A$, where the diffusion coefficients of the donors and acceptors, D_D and D_A, are hidden. In nonbinary theories they appear separately in the higher-order concentration corrections. In general, one should distinguish between the so-called trapping problem when acceptors are immobile ($D_A = 0$) and only excited donors (or excitons) move and approach acceptors to where the energy is transferred. Alternatively, energy donors are immobile ($D_D = 0$) and become the targets for moving acceptors. This is the *target problem*, which has an exact solution if point acceptors are moving independently [63,64]. An important fact is that this solution coincides with that given by Eqs. (4.6)–(4.10) with $D \equiv D_A$. Hence, the differential encounter theory is not binary but exact in all orders in c for this particular case (target problem) [62].

Qualitatively the general solution of the problem is similar to the contact one. The ionization rate constant $k_I(t)$ falls off with time from its initial kinetic value $k_I(0) = k_0$ defined in Eq. (2.4) to the stationary rate constant $k_i = k_I(\infty)$, approached at $t \gg R_Q^2/D$. The latter can be expressed via the stationary ion-pair distribution, n_s:

$$k_i = \int W_I(r)n_s(r)d^3r = 4\pi R_Q D \qquad (4.11)$$

The stationary equation for n_s, which follows from Eq. (4.9), takes the form

$$W_I n_s = D \frac{D}{r^2} \frac{\partial}{\partial r} r^2 \frac{\partial n_s}{\partial r} \qquad (4.12)$$

and should be solved with the same boundary condition (4.10). The last equality in Eq. (4.11) is actually a definition of the effective reaction radius R_Q that can be smaller and larger than σ. In the former case the reaction is kinetic and $k_i = k_0$, while in the latter it is controlled by diffusion and $R_Q \equiv R_s$ increases monotonously with the solvent viscosity (D^{-1}).

The kinetics of the excitation decay, $N(t)$, is comprised of the initial static quenching, which develops nonexponentially with time, and the long time diffusional asymptotics, which is finally exponential and proceeds with the rate ck_i [65]. The static quenching is essentially a multiparticle process that starts with the rate ck_0, but then slows down. This retardation is common to any particular transfer rate. Looking for instance at the exponential model (2.22), we have [65,66]

$$N(t) = e^{-t/\tau_D} \times \begin{cases} \exp(-ck_0 t) & \text{at} \quad W_c t \ll 1 \\ \exp\left[-\frac{4\pi}{3} c \left(\sigma + \frac{L}{2} \ln(W_c t) \right)^3 \right] & \text{at} \quad W_c t \gg 1 \end{cases} \qquad (4.13)$$

In contrast to static quenching, diffusional quenching is a binary process as its contact analog. It also proceeds with retardation, but finally becomes exponential:

$$N(t) = e^{-t/\tau_D - 4\pi R_Q D c t} \qquad (4.14)$$

For binary reactions in solution, the diffusional dependence of the effective radius, $R_Q(D)$, plays the same role as the energy dependence of the reaction cross section for gas-phase kinetics. The effective radius can be obtained from the ratio $k_i/4\pi D$, if the rate constant and diffusion coefficient are measured simultaneously. By changing solvents or their compositions, one can vary viscosity and study $R_Q(D)$ over a wide range of reactant mobilities. On the other hand, one can obtain the same dependence by means of encounter theory for any particular model of $W_I(r)$. For instance, in contact approximation

$$R_Q = \sigma \frac{k_0}{k_0 + 4\pi\sigma D}$$

A similar dependence, although more complex, was found for a few other models considered below. Fitting these results to the experimental data, one can discriminate between the models and specify through the best fit the important microscopic parameters of electron transfer.

An example of such a procedure was given in Ref. 67. The quenching of pheophytin a fluorescence, by toluquinone in different solvents, was analyzed under pulsed and stationary photoexcitation. The results were compared with the theoretically calculated reaction radius [59] for the exponential model of the ionization rate (2.22):

$$R_Q = \sigma + \frac{L}{2}\left[\ln(\gamma^2\beta_m) + 2\theta\left(\beta_m, \frac{2\sigma}{L}\right)\right] \tag{4.15}$$

where

$$\theta(x,y) = \frac{K_0(2\sqrt{x}) - y\sqrt{x}K_1(2\sqrt{x})}{I_0(2\sqrt{x}) + y\sqrt{x}I_1(2\sqrt{x})}, \qquad \beta_m = \frac{W_c L^2}{4D}, \gamma = e^C \approx 1.781$$

where C is the Eiler constant and $K(x)$ and $I(x)$ are the modified Bessel functions. The best fit to the experimental data shown in Fig. 21 was obtained at $W_c = 1.8 \times 10^{10}\,\text{s}^{-1}, \sigma = 4\,\text{Å}$ and $L = 5.4\,\text{Å}$. The agreement was even better for the diffusional dependence of the Stern–Volmer constant, which is reached at $W_c = 4 \times 10^{10}\,\text{s}^{-1}$ and $L = 4\,\text{Å}$. Unfortunately, even a smaller L seems to be too large for electron tunneling. This may indicate that the transfer is noncontact as in the inverted region. Therefore, the same experimental data were approached with the bell-shaped $W_I(r)$ from Eq. (2.23). The real shape of the ionization rate was shown to be almost the same at $L = 1.5\,\text{Å}$ in a solvent of moderate polarity $[\epsilon = 9.85, \Delta G_I(\infty) = 0.5\,\text{eV}]$. However, the model is preferable because it allows the exact solution of the problem obtained in Ref. 40:

$$R_Q = R + \Delta\left\{C + \xi(s+1) + \frac{\Delta Q_s(\alpha) + \sigma\sqrt{1-\alpha^2}Q_s^1(\alpha)}{\Delta P_s(\alpha) + \sigma\sqrt{1-\alpha^2}P_s^1(\alpha)}\right\} \tag{4.16}$$

where

$$\alpha = \tanh\left(\frac{\sigma - R}{\Delta}\right), \qquad s = \frac{1}{2}\left[\sqrt{\frac{1 - 4W_0\delta^2}{D}} - 1\right]$$

where $\xi(x)$ is the logarithmic derivative of the Γ function, and $P_s^M(\alpha)$ and $Q_s^M(\alpha)$ are the attached Legendre functions. Assuming $\sigma = 10\,\text{Å}, \Delta = 2\,\text{Å}$, a reasonable agreement with the experimental results was reached, which is

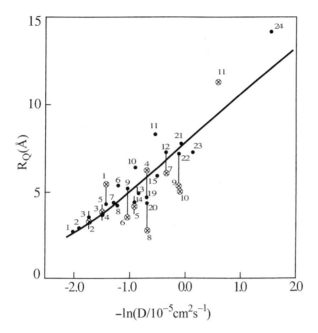

Figure 21. Effective radii of pheophytin a fluorescence quenching by toluquinone in solvents of different viscosity values. The solid line is the theoretical prediction of the exponential model. Crossed points represent data from the pulse experiments; full (solid) points show selected data from the stationary experiments. Thin vertical lines link points corresponding to the same solvents, but obtained with the different methods mentioned [67].

better the farther the maximum of $W_I(r)$ is from the contact (Fig. 22). The agreement with the rectangular model of the ionization layer was much worse at large R_Q [40], although the analytic solution for this model is also available. [41,68,69].

There are also some approximate but nonmodel methods for calculating the reaction radius at arbitrary $W_I(r)$. One of them, "closure approximation" [70], accepts the general contact result (4.4), but generalizes the definitions of the kinetic and diffusional constants, k_0 and k_D. The former is defined as in Eq. (2.4); the latter is given by the following expression:

$$\frac{k_0^2}{k_D} = \int\int W_I(r)G(r,r')W_I(r')d^3r\,d^3r' ,$$

where $G(r,r')$ is the Green function of the diffusional equation (4.2) with reflecting boundary condition. Unfortunately, this generalization is not much better than the original contact approximation. It is reasonable at relatively

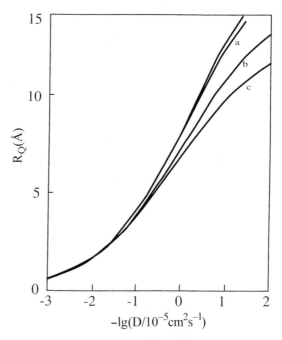

Figure 22. Variation of R_Q with D at different positions of the bell-like ionization layer at $\Delta = 2\,\text{Å}$: (a) $R = 16\,\text{Å}$, $W_0 = 1.4\,\text{ns}^{-1}$; (b) $R = 13\,\text{Å}$, $W_0 = 2.16\,\text{ns}^{-1}$; (c) $R = 11\,\text{Å}$, $W_0 = 3.0\,\text{ns}^{-1}$. The upper curve represents the best interpolation of the experimental data shown in Fig. 21.

fast diffusion when R_Q is less than or close to σ (or R), but not in the opposite limit when the reaction radius is essentially larger and the reaction is strongly controlled by diffusion. In this limit one can reduce the general expressions (4.15) and (4.16) to a well-known logarithmic diffusional dependence of the reaction radius [40,50,59,71]:

$$R_Q \to \sigma + \left(\frac{L}{2}\right) \ln\left(\frac{\gamma^2 W_c L^2}{4D}\right) = R_s \quad \text{for the exponential rate} \quad (4.17a)$$

$$R_Q \to R + \Delta\left(\ln\frac{\overline{W_0 \Delta^2}}{D} + C\right) = R_s \quad \text{for the bell-shaped rate.} \quad (4.17b)$$

This dependence is well approximated by the solution to the following equation:

$$W_I(R_s)\tau_c = 1 \quad (4.18)$$

where τ_c is the time of diffusional crossing of the reaction layer adjacent to the sphere of radius R_Q. This time is $L^2/4D$ for the exponential rate and Δ^2/D for the bell-like rate, which is also exponential at large distances. An exhaustive analytical study of the upper and lower bounds of the effective reaction radius at any transfer rate was presented in Ref. 72.

Equation (4.18) reveals the physical nature of the reaction radius R_s of the diffusional reaction. This is the radius of a "black sphere" around a nonexcited reactant. The excitations reaching the border of this sphere from the outside are quenched there in a thin reaction layer and do not enter. These events are sometimes associated with the reactive contacts between reactants at the distance R_s, which is larger than σ and depends on D. Then the approximate description of nonstationary quenching can be obtained from the contact model, through the substitution R_s for σ in Eq. (4.5). Such a modified formula was used in Ref. 73 to fit the experimental data and find $R_s(D)$ and k_0 by this means. Changing the solvent viscosity by pressure, the authors varied R_Q in the range 8–12 Å and also came to the conclusion that R_s "is significantly larger than the sum of the molecular radii of the reactants." Hence, one should expect that the charged products of diffusional ionization are generated far away from contact, at a distance $r \approx R_s > \sigma$. Below we will prove this statement.

C. Distribution of Charged Products

The total number of ionized products, $P(t)$, gradually accumulated during the transfer reaction can be related to a pair distribution of ions, $m(r,t)$:

$$P(t) = c \int m(r,t)d^3r \qquad (4.19)$$

By substituting this equation and Eq. (4.8) into Eq. (4.6b), one can easily obtain the following differential equation [12]:

$$\dot{m}(r,t) = W_I(r)n(r,t)N(t) \qquad (4.20)$$

with initial condition $m(r,0) = 0$. The total number of ions generated at distance r from a counterion, in a dr thick spherical layer, by time t, is $m(r,t)dr$. Here

$$m(r,t) = W_I(r) \int_0^t n(r,t')N(t')dt' \qquad (4.21)$$

where $n(r,t)$ and $N(t)$ are solutions to Eqs. (4.9) and (4.6a). After the completion of ionization (or decay) of the excited states, this distribution takes the final form:

$$m_0(r) = m(r, \infty) = W_I(r) \int_0^\infty n(r, t')N(t')dt' \qquad (4.22)$$

Its normalized shape is

$$f_0(r) = \frac{m_0(r)}{\int m_0(r)d^3r} \qquad (4.23)$$

This distribution would be saved if ions do not move and recombine. Although this is not the case in reality, the distribution does not become less important, because it establishes the initial conditions for the subsequent charge recombination (backward electron transfer).

The pair distributions of this kind were first studied in Refs. 12 and 14. The important conclusion was that they coincide in shape with $W_I(r)$ at fast diffusion, for kinetic ionization, but in the limit of slow diffusion, which controls ionization, the distributions of ions have a bell shape with the maximum shifted out of contact, to $r \approx R_s$.

Let us demonstrate this statement by the example of exponential $W_I(r)$, defined by Eq. (2.22). Solving numerically the DET equations with this particular $W_I(r)$ and $\tau_D = \infty$, three typical distributions were obtained in Ref. 14 (Fig. 23). In the case of kinetically controlled ionization, there is an exponential distribution of charged products (region K) reproducing the shape of the ionization rate. As was expected, the products of diffusional ionization are located near the sphere of reaction radius R_S and form the bell-shaped distribution shifted out of contact. If the reactants are immobile, the remote electron transfer is static (S) and the distribution of products is the widest and monotonous. Static distributions occur in radiation chemistry as a result of illumination or irradiation of cooled solid solutions [74,75], but this is of minor interest in the context of the present chapter.

To clarify the role of excitation decay, let us trace the time evolution of the normalized distribution $f(r, t) = m(r, t)/\int m(r, t)d^3r$, from start to finish at $\tau_D = \infty$ (Fig. 24). In the beginning the ionization is uniform as for a kinetic reaction, since $n(r, t) \approx 1$. Therefore, initially the ion distribution is quasiexponential, like $W_I(r)$, even at slow diffusion. After the nearest excitations have been quenched and a dip is formed at contact in the distribution $n(r, t)$, further ionization occurs only at the border of the reaction sphere R_s. It is in this range of interior distances that the final distribution reaches its maximum. However, the excitation decay interrupts the process at time $t \approx \tau_D$ and fixes the corresponding intermediate distribution, which is closer to contact the shorter is τ_D. A comprehensive

f_0

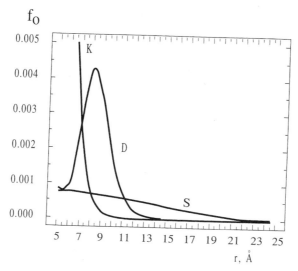

Figure 23. Normalized distribution over initial charge separations under (K) kinetic control ($D = \infty$), (D) diffusional control ($D = 10^{-6}\,\mathrm{cm^2/s}$) and (S) static ionization ($D = 0$). The rest of the parameters are $W_c = 10^3\,\mathrm{ns^{-1}}$, $L = 0.75\,\text{Å}$, $\sigma = 5\,\text{Å}$, $c = 0.1\,\mathrm{M}$, $\tau_D = \infty$ [14].

$f(r,t),$

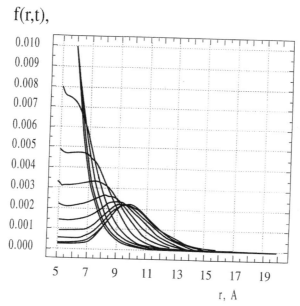

Figure 24. Normalized distribution of charged products at different times, from 0 to ∞ at $\tau_D = \infty$. The remaining parameters are $W_c = 10^3\,\mathrm{ns^{-1}}, L = 1.0\,\text{Å}$, $\sigma = 5\,\text{Å}$, $c = 0.03\,\mathrm{M}$, $D = 10^{-6}\,\mathrm{cm^2/s}$.

m_0

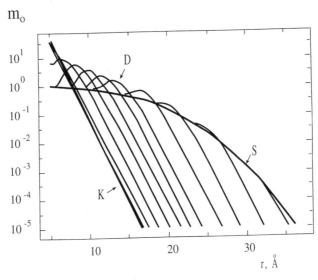

Figure 25. Initial charge distributions for $D = \infty, 10^{-3}, 10^{-4}-10^{-9}, 10^{-11}, 10^{-13}, 10^{-16}$, $10^{-20}, 0$ cm^2/s at the same parameters as in Fig. 23. Symbols K, D, and S denote kinetic, diffusional and static ionization [14].

comparison of the initial charge distributions at long (100-ns) and short (1-ns) lifetimes was given in Fig. 5 of Ref. 16, for nonmodel $W_I(r)$ in the normal, activationless, and inverted regions at different D.

For the exponential model, the gradual transformation of the final distribution $m_0(r)$ with decreasing D is shown in Fig. 25 [14]. The displacement of the maximum with D should approximately obey $R_s(D)$ dependence given in (4.17a). The more accurate characteristic of the average charge separation is $\bar{R}$ defined via dispersion of interion distances:

$$\bar{R}^2 = \int_\sigma^\infty r^2 f_0(r) d^3 r \qquad (4.24)$$

The normalized charge distributions $f_0(r)$, obtained from Fig. 25 in Ref. 14 were used to calculate the corresponding values of $\bar{R}$ depicted by stars in Fig. 26. As was expected, $\bar{R}$ changes between $R_{min} \approx \sigma + L$ in the kinetic control limit ($D \to \infty$) and R_{max} that is attained in the opposite, static limit ($D \to 0$). Between these limits shown by horizontal lines in Fig. 26 the ionization is controlled by diffusion. In this region ("D") $\bar{R}$ practically coincides with the positions of the maxima that are given by R_s. To the right of it, in region K the ionization is kinetic controlled and effective radius R_Q

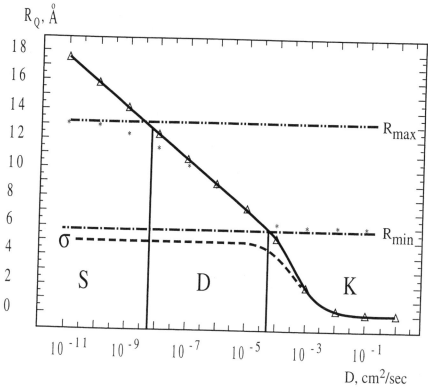

Figure 26. The mean charge separation $\bar{R}$ (asterisks) in comparison to the effective radius of binary ionization R_Q, shown by a solid line connecting triangles, and effective radius of contact ionization, shown by a dashed line. The horizontal lines indicate the minimal and maximal charge separation; S, D, and K respectively denote the regions of static, diffusional, and kinetic control [14].

is much less than σ. This radius, shown by the solid line connecting triangles, was calculated from Eq. (4.12). With a similar radius of contact model $R_c = \sigma[k_0/(k_0 + k_D)]$, shown by the dashed line, $R_Q(D)$ coincides only in the kinetic region and is essentially larger in region D, as it should be. In the static region S, $R_Q(D)$ is even larger than $\bar{R}$ but this is an artifact. In this region the majority of ions are generated in the course of initial, nonstationary (static) ionization described by Eq. (4.13). The minor contribution of final binary ionization (4.14) is negligible, so that R_Q does not determine the real charge separation in the quasi-static region.

The main conclusion is that the acceleration of either the excitation decay or diffusion moves the distant distribution of ions, peculiar to diffusional control, backward to contact until it takes the shape of $W_I(r)$.

Recently Murata and Tachiya measured the nonexponential transient effect in fluorescence quenching and compared it with $N(t)$ calculated numerically by means of encounter theory [76]. Using the nonmodel $W_I(r)$ from Eq. (2.5), they reached the best fit at $L \approx 2$ Å and used the data obtained for calculating the corresponding initial charge distributions. It was confirmed that at highly exergonic ionization $(-\Delta G_I = 1-2\,\text{eV})$ these distributions have maxima shifted $1-3$ Å out of contact, as it should be under diffusional control. The position of the distributions, or more precisely $\bar{R}$, indicate the average starting distance between ions when geminate recombination begins.

D. Quantum Yield of Ionization and Fluorescence

Here the quantum yield of ionization is understood as the total amount of ions produced in the primary act of electron transfer, $\psi = P(\infty)$, provided that initially $N(0) = 1, P(0) = 0$. According to Eqs. (4.19), (4.22), and (4.8), this is

$$\psi = c \int m_0(r) d^3r = c \int d^3r W_I(r) \int_0^\infty n(r, t') N(t') dt' = \int_0^\infty c k_I(t') N(t') dt'$$

$$(4.25)$$

By inserting here the solution for Eq. (4.6a)

$$N(t) = e^{-t/\tau_D - c \int_0^t k_I(t') dt'} \tag{4.26}$$

we obtain after integration

$$\psi = 1 - \int_0^\infty \frac{N(t) dt}{\tau_D} = 1 - \eta \tag{4.27}$$

In an important particular case that we often apply

$$\psi = 1, \qquad \eta = 0 \quad (\text{at} \quad \tau_D = \infty) \tag{4.28}$$

More generally, the ionization quantum yield is expressed via the Stern–Volmer constant κ, as follows from Eqs. (4.27) and (4.7):

$$\psi = \frac{c\kappa\tau_D}{1 + c\kappa\tau_D} \tag{4.29}$$

In the lowest-order concentration expansion of $\kappa(c)$, the Stern–Volmer constant does not depend on c but is a function of τ_D [62]. For instance, when ionization occurs at contact with the kinetic rate constant k_0 we have [16,77]

$$\kappa = \frac{k_0 k_D}{k_D + \dfrac{k_0}{1 + \sqrt{\tau_d/\tau_D}}} = \frac{k_0}{1 + \dfrac{\tilde{\tau}_d}{1 + \sqrt{\tilde{\tau}_d/\tilde{\tau}_D}}} \tag{4.30}$$

where $k_0 = 4\pi \int W_I(r) r^2 dr = k_I(0)$ whereas $k_D = 4\pi\sigma D$ is the diffusional rate constant within the contact approximation, and $\tau_d = \sigma^2/D$ is an encounter time that is proportional to the solvent viscosity. Dimensionless times $\tilde{\tau}_d$, $\tilde{\tau}_D$ are in units of $\tau_0 = 4\pi\sigma^3/k_0$.

Comparing Eqs. (4.30) and (4.3), we see that

$$\kappa \to k_i \quad \text{at} \quad \tau_D \gg \tau_d \tag{4.31}$$

Thus, at rather long decay times, the information available from the stationary fluorescence via κ or from kinetic study via k_i is the same. However, it is not the case when $\tau_D \ll \tau_d$. The former parameter accumulates all the information about the initial nonstationary development of ionization; the latter does not. Because initially the ionization is faster, κ/k_i is larger the slower the diffusion of reactants. This was demonstrated in Figs. 3 and 4 of Ref. 16. It was also proved in this work that contact approximation for κ, Eq. (4.30), is reasonable for long decay times, but less so for much shorter decay times, at least under diffusional control.

As was stated, the ionization quantum yield ψ gives the total amount of primary born ions. If they are reactive, part of them disappear during the geminate reaction, but the rest are separated and become free ions. Their share, known as *charge separation quantum yield*, $\bar{\varphi}$ will be discussed in the next section.

E. FEG Law for Bimolecular Ionization

The dashed lines in Fig. 20 represent the free-energy dependence of the stationary rate constant of ionization, k_i, calculated numerically from Eqs. (4.11) and (4.12) with the general ionization rate (2.5). Using an exponential model for the same rate as in Marcus and Siders [46], one can properly reproduce only the left half of this figure, because only in the normal region the ionization rate is approximately exponential. The closure approximation used in Ref. [36] covers the whole range of free energies, but for relatively fast diffusion, when R_Q does not exceed essentially $\sigma + L$. This approximation could be appropriate for only the solid curve shown in Fig. 20. However, in the ln (logarithmic) plot the difference between the exact results and the worst (contact) approximation (dotted curves) is not pronounced too much, even in the inverted region. The main conclusion is the same; if the maximum kinetic rate is too fast, the top of the FEG curve is cut by the diffusional plateau, which is lower and wider the slower the diffusion. This conclusion remains valid even in the more general case when ionization is a multichannel process, as in Eq. (2.16). Although the top of kinetic FEG curve becomes more wide and flat, it is cut similarly by the diffusional plateau (Fig. 27).

Is this a real explanation of the origin of the Rehm–Weller plateau (Fig. 17a)? Very strong arguments supporting this idea were presented in Ref. 49. The authors studied the time-dependent rate of ionization, $k_I(t)$, from the very beginning, when $k_I(0) = k_0$, and up to the end, when $k_I \to k_i$. They demonstrated that the experimental free-energy dependence of k_0 obeys the conventional FEG law, while $k_i(\Delta G_i)$ is actually cut by the diffusional plateau. In fact, the authors analyzed not only the initial and final curves, $k_I(0, \Delta G_i)$ and $k_I(\infty, \Delta G_i)$, but also a number of the curves corresponding

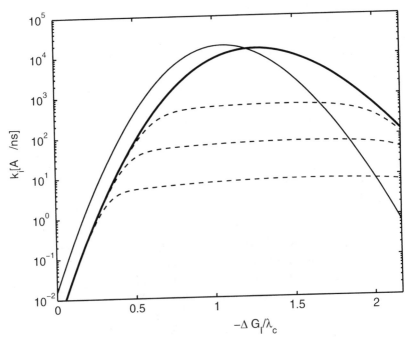

Figure 27. The rate constant k_0 of kinetic multichannel ionization at $S = 2$ (thick line) in comparison with that of a single-channel ionization shown in Fig. 20 (thin line). Dashed lines represent the stationary rate constants k_i of the multichannel ionization at $c = 0.1$ M and different $D = 10^{-5}, 10^{-6}, 10^{-7}$ cm^2/s ($\lambda_c = 1.4$ eV, $\lambda_i = 0, \sigma = 5$ Å, $(V_I^0)^2\sqrt{\pi}/T = 10^4$ ns^{-1}, $\epsilon = 80$) [30].

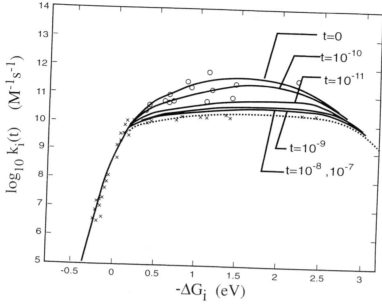

Figure 28. The FEG laws for ionization rate constants $k^{cat}(t) \equiv k_I(t)$ obtained by numerically solving the encounter theory equations (solid curves) except for $t = 0$, which is the kinetic rate constant k_0. Dashed line represents the stationary rate constant calculated with closure approximation in Ref. 36. The hollow circles (o) represent the experimental data obtained in Ref. 78; crosses (tickmarks) (×) represent the original data of Rehm and Weller [6].

to intermediate times, $k_I(t, \Delta G_i)$ (Fig. 28). There is no difference between all the curves deep in the normal or inverted regions, where ionization is kinetic and $k_I \equiv k_0 = $ const. Only at the top of the curve, where ionization is too fast, is ionization under diffusional control and does it have nonstationary development with time.

This is firm evidence of a diffusional reaction at the top of the FEG curves. Usually the diffusional plateau is so long that the inverted region is not accessible in the limited range of ΔG_i. The first evidence to the contrary has emerged—the plateau is driven out of the FEG curve, if the maximal k_0 is less or comparable with k_D. Such a system was first discovered and studied experimentally in Ref. 79. This is the photoinduced bimolecular reaction of a homologous series of Ru (II) diimines with cytochrome (cyt) c in its oxidized and reduced forms. The rate of bimolecular electron transfer was observed to drop off in the inverted region for either oxidation or reduction reactions (Fig. 29). The simplest contact estimate of the stationary rate constant, (4.4), fits the experimental data well, provided k_0 is calculated

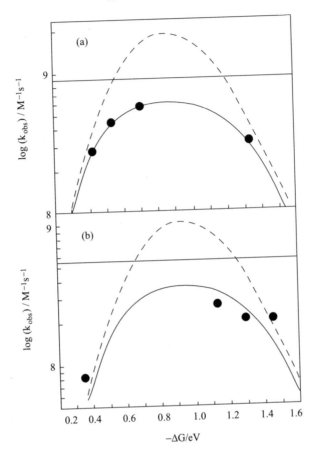

Figure 29. Semilog plots of the stationary rate constant k_{obs} (solid curves) versus the free energy of electron transfer between Ru(II) diimine complexes and (a) Fe(II) cytochrome c and (b) Fe(III) cytochrome c. The multichannel kinetic rate constant k_0 (dashed curve) and diffusional rate constant k_D (horizontal line) were calculated taking into account the electrostatic repulsion between reactants at $L = 3.33\,\text{Å}$, $\sigma = 7.4\,\text{Å}$, $V_0 = 200\,\text{cm}^{-1}$ [79].

theoretically from Eqs. (2.4) and (2.16) and electrostatic repulsion between reactants is taken into account. Because of this factor and the suspiciously large closest approach distance (23 Å), the maximal kinetic rate constant in this system does not really exceed the diffusional rate constant too much. In fact, $k_i \approx k_0$ everywhere and the classical FEG law is not essentially distorted by diffusion. This conclusion can and must be verified experimentally by inspecting diffusional dependence of the results at least at the top of FEG curve.

As stated above, if ionization is kinetic everywhere the distribution of the reaction products has the same shape as $W_I(r)$. The difference between distributions in the normal and inverted regions is that the distribution, as well as the reaction layer, is contact in the former and remote in the latter. This is always true for very small and very large exergonicities, even though in between the ionization is controlled by diffusion. As shown in Fig. 20, the radius of the ionization layer increases monotonously with $|\Delta G_i|$, but ions are predominantly generated in the middle of this layer only under kinetic control (region K). Under diffusional control (region D) they are mainly concentrated in a spherical layer of much larger radius R_s. Therefore, the mean charge separation increases nonmonotonously with $|\Delta G_i|$ (Fig. 30), but essentially exceeds σ almost everywhere, except in the narrow region of the lowest exergonicity.

With this result, it is very surprising that the exponential model assuming quite the opposite (contact generation of ion pairs), succeeded in the theoretical description of the FEG law for recombination. Let us compare again what is known for bimolecular ionization and geminate recombination. In Fig. 31 the Rehm–Weller results for k_i and the corresponding kinetic curve, $k_0(\Delta G_i)$, are compared with the FEG law for the recombination rate k_{-et}. There are two more surprises that raise the following questions:

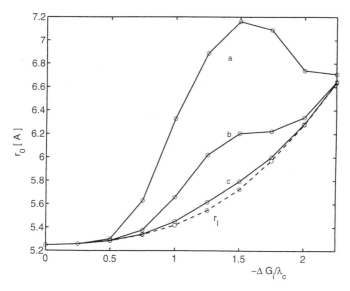

Figure 30.   Mean separation of charges in ion pairs produced by a multichannel ionization $(S = 2)$, $r_0 \equiv \bar{R}$, in comparison with the mean radius of the ionization layer r_I (dashed line) at $c = 0.1M$ and different diffusion coefficients: (a) $D = 10^{-6}\,\mathrm{cm}^2/\mathrm{s}$, (b) $D = 10^{-5}\,\mathrm{cm}^2/\mathrm{s}$, (c) $D = 10^{-4}\,\mathrm{cm}^2/\mathrm{s}$ [30].

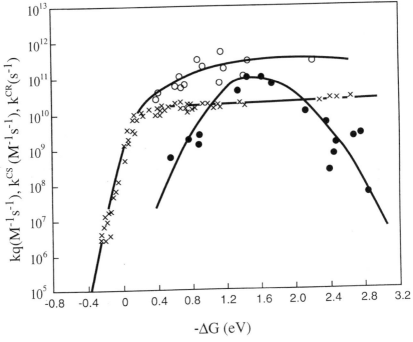

Figure 31. Experimental data for the free-energy dependence of the stationary rate constant of bimolecular ionization, $k_q \equiv k_i$ ($\times$), obtained by Rehm and Weller [6], initial (kinetic) rate constant $k^{CS} \equiv k_0$ ($\circ$), obtained in Ref. 78, and *charge recombination rate* $k^{CR} \equiv k_{-et}$ of the geminate ion pairs ($\bullet$), obtained by Mataga et al. [80,81]. The smooth curves are drawn arbitrarily to visualize the energy-gap law.

- Why does diffusion that distorts the ionization rate in the activationless region, pushing it down, not perturb the recombination rate to such an extent?
- If this is the true FEG law for recombination, why is the corresponding "hump" much narrower than that for ionization, even though the width of both should be determined by the same λ_c?

We will try to answer these questions in the next two sections.

V. CONTACT AND REMOTE GEMINATE RECOMBINATION

Sometimes the ion pair recombination/separation may be considered as a separate problem. For instance, the radical ion pairs (RIPs) generated by irradiation of a cooled sample do not begin to move and react until the sample is warmed up. In such a case one could study the geminate recombination of RIP when it starts after heating, provided the ion-pair distribution at this moment, $m_0(r)$, is known. The same is true for the isothermal conditions, if the RIP generation due to pulse excitation is a much faster process than the following geminate recombination. Starting from the moment when the accumulation of ions stops, one may look for only a monotonous decrease of their number due to geminate recombination. The main features of this process can be elucidated, assuming that all pairs were initially separated by distance r_0. In this case their distribution

$$m_0(r) = \frac{\delta(r - r_0)}{4\pi r^2} \tag{5.1}$$

used as the initial condition for the equation for geminate recombination, provides us with a corresponding Green function. With any other initial conditions the recombination is described by the convolution of this Green function with arbitrary $m_0(r)$, provided there is a significant time separation between the accumulation and recombination stages. In the present section we accept this limitation but will overcome it in the next one.

A.   Contact Approximation

In the normal region the recombination occurs in a narrow reaction layer adjacent to the contact sphere. In contact approximation the width of this layer is said to be zero and the real recombination rate $W_R(r)$ is substituted by the corresponding expression from Eq. (2.21) expressed via k_c. This rate constant enters the boundary condition similar to (4.1), but for the ion-pair distribution function $m(r, t)$:

$$4\pi \tilde{D} r^2 \left(\frac{\partial m}{\partial r} + \frac{r_c}{r^2} m \right) \bigg|_{r=\sigma} = k_c m(\sigma, t) \tag{5.2}$$

Since ions are charged, the diffusional operator in the lefthand side (LHS) of the equation accounts for the Coulomb attraction between them and makes an allowance for a different value of the diffusion coefficient $\tilde{D}$, which is not necessarily equal to D. The same distinction exists between the free diffusion equation for neutral reactants, (4.2), and that for an ion pair:

$$\dot{m} = \frac{\tilde{D}}{r^2} \frac{\partial}{\partial r} r^2 \left(\frac{\partial m}{\partial r} + \frac{r_c}{r^2} m \right). \tag{5.3}$$

The set of equations (3.5) of the exponential model is replaced in contact approximation by a single equation (5.3) that accounts for all interior distances, r, and not just contact and infinite. Therefore, the total number of ions survived by time t is

$$\Omega(t) = \int m(r,t)d^3r \qquad (5.4)$$

instead of sum $m_c + m_\infty$. This quantity was an object of numerous investigations. Here we follow the pioneering article of Hong and Noolandi [29] and more recent work [9] that contain a valuable analytic solution to the problem.

The solution of Eq. (5.3) can be expressed via the Green function $G_0(r,r',t)$ that obeys the similar equation

$$\dot{G_0} = \frac{\tilde{D}}{r^2}\frac{\partial}{\partial r}r^2\left(\frac{\partial G_0}{\partial r} + \frac{r_c}{r^2}G_0\right) \qquad (5.5)$$

and the initial condition

$$G_0(r,0) = \frac{\delta(r - r')}{4\pi r^2} \qquad (5.6)$$

but reflecting the boundary condition

$$\left.\left(\frac{\partial G_0}{\partial r} + \frac{r_c}{r^2}G_0\right)\right|_{r=\sigma} = 0 \qquad (5.7)$$

If one uses the initial condition (5.1), then, in fact, $m(r,t)$ is the Green function of Eq. (5.3) with the radiation boundary condition, $G(r,r_0,t)$, which can be expressed through $G_0(r,r',t)$ as follows [9]:

$$m(r,t) \equiv G(r,r_0,t) = G_0(r,r_0,t) - k_c\int_0^t G_0(r,\sigma,t-\tau)G(\sigma,r_0,\tau)d\tau \qquad (5.8)$$

Making the Laplace transformation of this equality, one has

$$\tilde{G}(r,r_0,s) = \tilde{G}_0(r,r_0,s) - k_c\tilde{G}_0(r,\sigma,s)\tilde{G}(\sigma,r_0,s) \qquad (5.9)$$

Solving this equation at $r = \sigma$ for $G(\sigma,r_0,s)$ and using the result in the righthand side (RHS) of Eq. (5.9), we obtain

$$\tilde{G}(r,r_0,s) = \tilde{G}_0(r,r_0,s) - \frac{k_c\tilde{G}_0(r,\sigma,s)\tilde{G}_0(\sigma,r_0,s)}{1 + k_c\tilde{G}_0(\sigma,\sigma,s)} \qquad (5.10)$$

After integration over r, taking into account the normalization condition $\int \tilde{G}(r, r', s)d^3r = 1/s$, we finally get

$$\tilde{\Omega}(r_0, s) = \int \tilde{G}(r, r_0, s)d^3r = \frac{1 + k_c\left[\tilde{G}_0(\sigma, \sigma, s) - \tilde{G}_0(\sigma, r_0, s)\right]}{s\left[1 + k_c\tilde{G}_0(\sigma, \sigma, s)\right]} \tag{5.11}$$

Thus in contact approximation the general solution of the problem is expressed via the simplest Green function of free diffusion in the ion pair, $\tilde{G}_0(r, r', s)$.

B. Charge Separation Quantum Yield in Contact Approximation

A quantum yield of ions that escaped geminate recombination is simply

$$\varphi(r_0) = \Omega(r_0, t = \infty) \tag{5.12}$$

This can be easily calculated using Eq. (5.11):

$$\varphi(r_0) = \lim_{s \to 0} s\tilde{\Omega}(r_0, s) = \frac{1 + k_c\left[\tilde{G}_0(\sigma, \sigma, 0) - \tilde{G}_0(\sigma, r_0, 0)\right]}{1 + k_c\tilde{G}_0(\sigma, \sigma, 0)} \tag{5.13}$$

Taking into account a well-known expression for the Green function at $s = 0$

$$\tilde{G}_0(r, r', 0) = \begin{cases} \dfrac{1}{4\pi r_c \tilde{D}}\exp\left(\dfrac{r_c}{r}\right)\left[1 - \exp\left(-\dfrac{r_c}{r'}\right)\right] & \text{at}\quad r < r' \\[2ex] \dfrac{1}{4\pi r_c \tilde{D}}\left[\exp\left(\dfrac{r_c}{r}\right) - 1\right] & \text{at}\quad r > r' \end{cases} \tag{5.14}$$

one can deduce from Eq. (5.13) the final result [29]:

$$\varphi(r_0) = \frac{1 + (k_c/4\pi r_c\tilde{D})e^{r_c/\sigma}\left[e^{-r_c/r_0} - e^{-r_c/\sigma}\right]}{1 + (k_c/4\pi r_c\tilde{D})[e^{r_c/\sigma} - 1]} = \frac{1}{1 + Z/\tilde{D}} \tag{5.15}$$

This quantum yield is parametrically dependent on the initial charge separation r_0. The larger r_0 the more ions escape geminate recombination.

If not only recombination is contact, but ionization is contact as well, then $r_0 = \sigma$, as in the exponential model, and instead of the general expression (5.15), we have

$$\varphi(\sigma) = \frac{1}{1 + (k_c/4\pi r_c\tilde{D})[e^{r_c/\sigma} - 1]} = \frac{1}{1 + z/\tilde{D}} \tag{5.16}$$

There is nothing surprising about the fact that in this particular case the result (5.16) can be identified with Eq. (3.9) of the exponential model, if one sets

$$k_c = k_{-et} \frac{4\pi}{3} \sigma^3 \qquad (5.17)$$

From this relationship we can elucidate the physical meaning of the phenomenologic parameter k_{-et}. If the width of the reaction layer $(L/2) \ll \sigma$, then $k_c \approx W_R(\sigma) 2\pi\sigma^2 L$, as in Eq. (3.3), and

$$k_{-et} = \frac{3L}{2\sigma} W_R(\sigma) \qquad (5.18)$$

The geometric factor in this relationship is due to a different shape of the reaction zone in the contact and exponential models. This is the spherical layer of width L in the former and the entire sphere of radius σ in the latter. The geometric factor, which is the ratio of these two values, accounts for an objective difference between models, but does not affect the free-energy dependence, which is the same for k_{-et} and k_c as well as for $W_R(\sigma)$.

Comparing expressions (5.15) and (5.16), we see the principal difference that arises when ions do not start from a contact distance. Capital Z is a function of the diffusion coefficient $\tilde{D}$ while z introduced in Eq. (3.10) is diffusion independent. The relationship between them is depicted by the following equation [82]:

$$Z = \frac{qz}{1 + (1-q)z/\tilde{D}} \qquad (5.19)$$

where

$$q = \frac{1 - \exp(-r_c/r_0)}{1 - \exp(-r_c/\sigma)}, \qquad z = \frac{k_c}{4\pi r_c}\left(e^{r_c/\sigma} - 1\right) \qquad (5.20)$$

In general, $Z(\tilde{D})$ increases monotonously, but levels off approaching the plateau qz, which is lower the larger the initial separation of ions r_0 (Fig. 32). This is the reason for subdividing geminate reactions into the same classes as bimolecular reactions: diffusional and kinetic. The former are accelerated by diffusion but the latter do not depend on it:

$$Z = \begin{cases} \dfrac{q}{1-q}\tilde{D} & \text{diffusional geminate recombination,} \quad \tilde{D} \ll (1-q)z \\[2mm] qz & \text{kinetic geminate recombination,} \quad \tilde{D} \gg (1-q)z \end{cases}$$

$$(5.21)$$

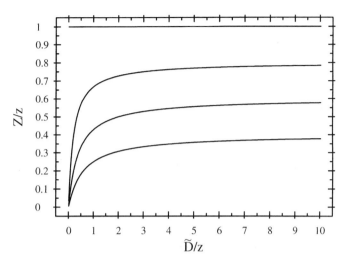

Figure 32. Diffusional dependence of Z at different starting distances between ions corresponding to $q = 0.4$, 0.6, 0.8, and 1.0. The latter corresponds to $r_0 = \sigma$ when the exponential model result is reproduced (horizontal line $Z = z$).

If geminate recombination is kinetically controlled, then $Z = qz$ reproduces the bell-shaped curve of the FEG law peculiar to z, as well as for $W_R(\Delta G_r)$. But if the top of this curve is too high, it can be cut by the diffusional plateau $q\tilde{D}/(1 - q)$ in exactly the same manner as in Fig. 20 or 27. However, such a distortion of the FEG law is not universal because the contact approximation is not valid at any free energies, but only in the normal region where recombination is really contact. If this is not the case the situation is altered, as will be shown later in this section.

The geminate recombination is actually controlled by diffusion, if initial separation of ions is so large that their transport from there to contact takes more time than the reaction itself. The exponential model excludes such a situation from the very beginning, assuming that ions are born in the same place where they recombine. What is the situation in reality? The answer to this question can be obtained experimentally from the diffusional (viscosity) dependence of Z, which is constant only under kinetic control. Initially we knew only a single experimental investigation of this kind made in Ref. 83 (Fig. 33). These data were obtained for similarly charged ions that are unlikely to be created at contact because of the Coulomb repulsion between them. The results confirm that the geminate recombination in this case is certainly not kinetic. This provides the straightforward evidence that the diffusional stage of geminate recombination should be seriously taken into account.

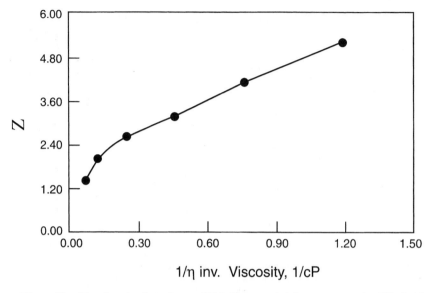

1/η inv. Viscosity, 1/cP

Figure 33. The viscosity dependence of Z (arbitrary units) that represents the diffusional acceleration of geminate recombination. Points are experimental data for recombination of Ru(bpy)$^{2+}$ with MV^{2+} from Ref. 83.

On the other hand, the $Z(\tilde{D})$ dependence is not linear, as it is expected to be in the diffusional limit according to Eq. (5.21). This is an indication that some important factor such as the spin state of the ion pair has not been taken into account. This particular pair is born in the triplet state and should undergo the spin conversion to a singlet state from where the recombination is possible. We will return to this problem in Section VIII.

C. Kinetics of Contact Geminate Recombination

It has been known for a long time that the survival probability of ions does not fit the exponential law (3.7), which gave the name to the corresponding model. At least the long time behavior of $\Omega(t)$ in contact approximation obeys the power law obtained in Ref. 29:

$$\Omega(t) = \varphi(r_0) \left[1 + \frac{e^{r_c/\sigma}}{4\pi\tilde{D}r_c/k_c - 1} \sqrt{\frac{r_c^2}{\pi\tilde{D}t}} \right] \quad \text{at} \quad t \gg \frac{\rho^2}{\tilde{D}} \quad (5.22)$$

where ρ is the greatest from σ, r_0, r_c. This is the universal result valid for kinetic and diffusional geminate recombination in polar and nonpolar solvents.

In polar solvents ($r_c \rightarrow 0$) the charge separation quantum yield (5.15) can be presented as follows:

$$\varphi = \frac{\varphi_0 + \delta}{1 + \delta} \tag{5.23}$$

where $\varphi_0 = (1 + k_c/k_D)^{-1}$, $k_D = 4\pi\sigma\tilde{D}$, and $\delta = r_0/\sigma$. The general expression for relaxation kinetics (from $t = 0$ up to ∞) can be expressed through the same parameters [9]:

$$R(t) = \frac{\Omega(t) - \varphi}{1 - \varphi} = \text{erf}\left\{\frac{r_0 - \sigma}{2\sqrt{\tilde{D}t}}\right\} + \exp\left\{\frac{\delta}{\varphi_0} + \frac{\tilde{D}t}{\sigma^2\varphi_0^2}\right\} \times \text{erfc}\left\{\frac{\sqrt{\tilde{D}t}}{\sigma\varphi_0} + \frac{\sigma\delta}{2\sqrt{\tilde{D}t}}\right\} \tag{5.24}$$

This complex kinetics is far from exponential. Even for contact start $r_0 = \sigma$, suggested by exponential model, it differs qualitatively from the primitive exponential law (3.7):

$$R(t) = e^{t/t_0}\text{erfc}\sqrt{\frac{t}{t_0}} = \begin{cases} 1 - \sqrt{4t/\pi t_0} & \text{at} \quad t \ll t_0 \\ \sqrt{t_0/\pi t} & \text{at} \quad t \gg t_0 \end{cases} \tag{5.25}$$

where the relaxation time of this process

$$t_0 = \tau_0\varphi_0^2 \tag{5.26}$$

is proportional to the separation time of the inert particles $\tau_0 = \sigma^2/\tilde{D}$.

There is also a remarkable difference in the diffusional dependence of t_0 and its exponential model analog, t_e. They are compared in Table I, which discriminates between two situations, a and b. Situation a, both models lead to almost the same results; in situation b the results are qualitatively different. In case a, diffusion is much faster than the reaction and the

TABLE I
Two Limits of Contact Geminate Recombination

	Contact Approximation		Conditions	Exponential Model
a	$\dfrac{1}{t_0} = \dfrac{1}{\tau_0} = \dfrac{\tilde{D}}{\sigma^2}$		$\dfrac{\varphi_0 \approx 1}{k_D \gg k_c}$	$1/t_e = k_{\text{sep}} = 3\dfrac{\tilde{D}}{\sigma^2}$
b	$\dfrac{1}{t_0} = \dfrac{1}{\tau_0}\left(\dfrac{k_c}{k_D}\right)^2 = \dfrac{k_c^2}{16\pi^2\sigma^4\tilde{D}}$		$\dfrac{\varphi_0 \ll 1}{k_D \ll k_c}$	$1/t_e = k_{-et} = \dfrac{3k_c}{4\pi\sigma^3}$

recombination is interrupted by diffusional separation of the contact born ion pair. In our earlier works [9,84] we associated this process with diffusional recombination on the basis that its effective rate $1/t_0$ or k_{sep} is proportional to the diffusion coefficient. For similar reasons (dependence of the rate on k_c or k_{-et}) the alternative case b was attributed to kinetic recombination. However, this reasoning is not appropriate for such an identification. As was indicated in Eq. (5.21), geminate recombination can be diffusional if the ions generated far from the contact move slowly approaching the reaction zone. If they are born in contact, the recombination should be considered as kinetic in both cases, a and b. The difference between the cases demonstrated by Table I is of another origin.

Although φ_0 or, better still, Z, characterize the efficiency of geminate recombination, the relaxation times are mainly characteristics of charge separation. In case a such a time is a purely kinematic parameter, $1/k_{sep}$. The time increases monotonously with decreasing $\tilde{D}$, but becomes too long in region b, where recombination competes favorably with separation. The fast reaction prevents the separation of the majority of ions and thus terminates the process at a much shorter time. However, the estimate of this time in the exponential and contact models is essentially different. In the exponential model this is an inverse rate of static recombination, k_{-et}^{-1}, which does not depend on diffusion. To the contrary, in the contact model the same time is linear in $\tilde{D}$ and tends to 0 when diffusion stops. The explanation of this surprising result given in Ref. 84 will be given later.

A nonexponential behavior is peculiar not only for highly polar solutions but also for solvents of moderate polarity. The latter, studied numerically in Ref. 9, was shown to be also well approximated with Eqs. (5.25) and (5.26), provided τ_0 is replaced by the encounter time τ_c accounting for the Coulomb well:

$$\tau_c = \frac{r_c^2}{\tilde{D}[1 - \exp(-r_c/\sigma)]^2} \qquad (5.27)$$

Comparison of the results shown in Fig. 34 proves that the charge separation kinetics is far from exponential, especially at the very end. The analytical treatment of geminate recombination in solutions showed that at lower polarity the main kinetics is closer to exponential decay, but the universal character of the $1/\sqrt{t}$ asymptotics of charge separation was confirmed [85]. The characteristic time of nonexponential relaxation can be introduced by the relation

$$R(t_e) = \frac{1}{e}$$

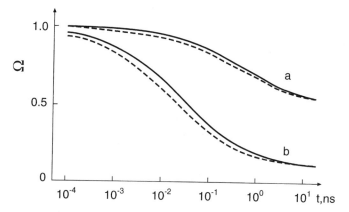

Figure 34. The survival probability $\Omega(t)$ for ions starting from contact at (*a*) $\varphi_0 = 0.5$ and (*b*) $\varphi_0 = 0.1$, provided $\tilde{D} = 10^{-5}\,\mathrm{cm^2/s}$ and $\sigma = 7\,\text{Å}$. Solid lines represent numeric solutions for solvents of moderate polarity ($r_c = 2\sigma$); dashed lines offer analytic solution (5.25) for highly polar solvents ($r_c = 0$) [9].

The diffusional dependence of this time remains nonmonotonous in a solution of lower polarity, although it becomes flatter at the maximum shifted to a smaller $\tilde{D}$ (Fig. 35).

Within the exponential model there is a simple linear relationship between t_e and φ_0, which follows from Eqs. (3.8) and (3.9):

$$t_e = \frac{1 - \varphi_0}{k_{-et}} \equiv \frac{3k_c}{4\pi\sigma^3}[1 - \varphi_0] \tag{5.28}$$

The best way to examine the model is to compare this relation with what follows from contact approximation for $t_e(\varphi_0)$ dependence, in the alternative cases of contact and noncontact creation of ion pairs. For contact born pairs the parabolic dependence arises from Eq. (5.26) instead of the linear one [9]:

$$t_e = \alpha t_0 = \frac{4\pi\sigma^3\alpha}{k_c}\,\varphi_0(1 - \varphi_0) \tag{5.29}$$

The numeric factor $\alpha = 1.565$ appears to be due to a nonexponentiality of the relaxation law (5.25). To find t_e at any $\tilde{D}$ for noncontact starts, one should use the general expression (5.24) or its exact equivalent from Ref. 84:

$$R(t) = 1 + e^{-\delta^2/4\tau}\left[\mathrm{er}\left(\frac{\delta}{2\sqrt{\tau}} + \frac{\sqrt{\tau}}{\varphi_0}\right) - \mathrm{er}\left(\frac{\delta}{2\sqrt{\tau}}\right)\right]$$

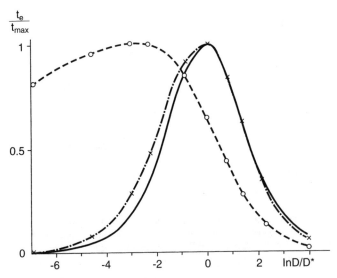

Figure 35. Relaxation time dependence on the diffusion coefficient for $r_c = 0$, $t_{max} = 0.196$ ns (solid line); $r_c = 14$ Å, $t_{max} = 1.265$ ns (dashed–dotted line); $r_c = 70$ Å, $t_{max} = 227$ ns (dashed line). Everywhere $D^* = k_c[\exp(r_c/\sigma) - 1]/4\pi r_c = 10^{-5}$ cm^2/s and the relaxation time is related to its maximum value t_{max} [9].

where $\tau = \tilde{D}t/\sigma^2$, $\mathrm{er}(x) = \exp(x^2)\mathrm{erfc}(x)$, and $\mathrm{erfc}(x)$ is a complementary error function.

In the limits $0 < \tilde{D} < \infty$ the quantum yield φ_0 changes in the finite interval, from 0 to 1, and is therefore preferable as an argument of t_e. The family of the curves $t_e(\varphi_0)$ shown in Fig. 36 demonstrates a striking difference to the same result of the exponential model. The latter is represented by a straight line, whereas in contact approximation we obtain at contact start the inverse parabola, with a maximum at $\varphi_0 = 0.5$ and t_e turns to zero as $\varphi_0 \to 0$. On the contrary, at any finite initial separation $t_e \to \infty$ in this limit.

D. Remote Geminate Recombination

Although at moderate and large φ_0 the contact approximation provides a reasonable approximation of the real $\tilde{t}_e(\varphi_0)$ dependence, it does not work in the limit $\tilde{D} \to 0$, where $\varphi_0 \approx 0$. If there is no diffusion, the ions recombine at the same distance r_0, where they were created with the rate $W_R(r_0) = 1/t_e(0)$. The corresponding values of $\tilde{t}_e(0)$ are shown by the horizontal cutoffs in Fig. 36, and the real curves should approach these values as $\varphi_0 \to 0$. Nothing like this can happen in contact approximation.

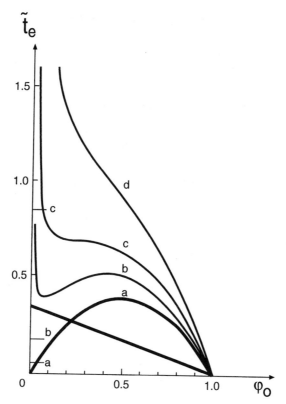

Figure 36. Variation of the reduced relaxation time $\tilde{t}_e = t_e k_c / 4\pi\sigma^3$ as a function of φ_0 in the exponential model (straight line) and in contact approximation starting from the contact (a) $\delta = 0$ and outside it: (b) $\delta = 0.05$, (c) $\delta = 0.1$, (d) $\delta = 0.2$. Horizontal cutoffs show the corresponding times of static recombination at $L/2\sigma = 0.03$ (for curve d this time equals 23.6) [9].

Assuming that $W_R(r) = 0$ at any finite separation of ions, it implies that they are stable at $\tilde{D} = 0$ so that $t_e(0) = \infty$.

To overcome such a drawback of contact approximation and remove all the limitations on ion diffusion, one should turn to the theory of remote geminate recombination. The latter was actually developed in Ref. 84. Assuming no reaction at contact, we used the reflecting boundary condition instead of Eq. (5.2):

$$\frac{\partial e^{-r_c/r} m(r,t)}{\partial r}\bigg|_{r=\sigma} = 0 \qquad (5.30)$$

but included the recombination term into Eq. (5.3):

$$\dot{m} = \frac{\tilde{D}}{r^2} \frac{\partial}{\partial r} r^2 e^{r_c/r} \frac{\partial}{\partial r} e^{-r_c/r} m - W_R(r)m \qquad (5.31)$$

The share of pairs that survived by instant t is

$$\Omega(t|r_0) = \int m(r, t; r_0, 0) d^3 r \qquad (5.32)$$

where r_0 is the initial separation of ions. It obeys the following equation conjugated to Eq. (5.31) [86,87]:

$$\frac{\partial}{\partial t} \Omega(t|r) = \frac{\tilde{D} e^{-r_c/r}}{r^2} \frac{\partial}{\partial r} r^2 e^{r_c/r} \frac{\partial}{\partial r} \Omega(t|r) - W_R(r)\Omega(t|r) \qquad (5.33)$$

with the following initial and boundary conditions:

$$\Omega(0|r) = 1, \qquad \frac{\partial}{\partial r} \Omega(t|r) = 0, \qquad \Omega(t|\infty) = 1 \qquad (5.34)$$

The quantum yield of the charge separation

$$\varphi(r_0) = \lim_{t \to \infty} \Omega(t|r_0) = \Omega(\infty|r_0) \qquad (5.35)$$

where $\Omega(\infty|r)$ may be found as a stationary solution of Eq. (5.33).

For a straightforward comparison of the contact and remote geminate recombination, let us concentrate on the simplest case of the highly polar solution ($r_c = 0$) and the exponential model of the transfer rate (2.22). Then the analytic expression for the quantum yield can be obtained from Eq. (5.33) [84,88]:

$$\varphi(r_0) = \frac{2\lambda}{1 + \delta} \left\{ K_0(\alpha) - I_0(\alpha) \frac{K_0(\alpha_0) - \sqrt{x/\lambda} K_1(\alpha_0)}{I_0(\alpha_0) - \sqrt{x/\lambda} I_1(\alpha_0)} \right\} \qquad (5.36)$$

where $K_0, I_0, K_1,$ and I_1 are modified Bessel functions, [89]

$$\alpha^2 = 4\lambda x e^{-\delta/\lambda}, \qquad \delta = \frac{r_0}{\sigma} - 1$$

$$\alpha_0 = \alpha|_{\delta \to 0}, \qquad \lambda = \frac{L}{2\sigma}$$

The most important parameter

$$x = \frac{\sigma L W_0}{2\tilde{D}} = \frac{2\pi\sigma^2 L W_0}{4\pi\sigma\tilde{D}} = \frac{k_c}{k_D} \qquad (5.37)$$

discriminates between the fast diffusion limit ($x \ll 1$), where $\varphi \approx 1$, and the slow diffusion limit ($x \gg 1$), where $\varphi \ll 1$. In the extreme case

$$\alpha^2 = \frac{L^2 W_R(r_0)}{\tilde{D}} \gg 1 \qquad \left(\text{or} \quad x \gg \frac{e^{\delta/\lambda}}{4\lambda} \gg 1 \right)$$

the recombination becomes quasistatic and can be accomplished earlier than when particles leave their initial positions. Therefore in the quasistatic limit

$$\varphi(r_0) \approx \frac{\lambda}{1+\delta} \sqrt{\frac{2\pi}{\alpha}} e^{-\alpha}$$

turns to zero when $\tilde{D} \to 0$.

In contact approximation ($\lambda \to 0$ at $x = \text{const}$) this limit is lacking. In this approximation Eq. (5.36) reduces to Eq. (5.23). The latter is valid for both small and large x but cannot be extended to a quasistatic limit. The difference between contact and remote geminate recombination makes itself evident by comparing the charge separation quantum yields as functions of diffusion coefficient or, better still, of $\varphi_0 = (1 + k_c/k_D)^{-1}$, the quantum yield in the exponential model. The results are presented in Fig. 37. In contact approximation the charge separation quantum yield (5.23) is linear in φ_0 and does not turn to 0 at $\varphi_0 = \tilde{D} = 0$, provided the ions did not start from contact ($r_0 \neq \sigma$). This is nonsense, which is removed by a more appropriate approach that accounts for remote electron transfer in the course of charge separation.

With the exponential model there is even a more serious problem: the lack of diffusion-controlled geminate recombination. It keeps finite the rate of recombination near contact, but assumes that ions are born also at contact and need no time to reach the recombination zone. In fact, this is rarely the case. Most often $r_0 > \sigma$, and using the theory of remote transfer, one can easily distinguish between kinetic, diffusional, and static geminate recombination. The extended variant of such a theory will be compared in detail with an exponential model and fit to the same experimental data at the end of Section VIII. In particular, the space distributions will be obtained of discharged products of geminate recombination. In spinless theory, it can

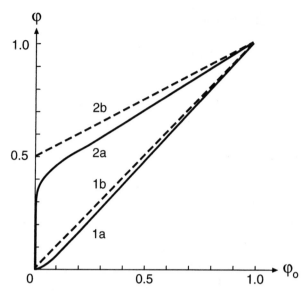

Figure 37. The dependence of the quantum yield $\varphi(r_0)$ on the exponential model quantum yield φ_0. Solid lines indicate remote electron transfer ($\lambda = 0.1$); dashed lines, contact approximation ($\lambda = 0$). Curves 1 account for contact start ($\delta = 0$); curves 2, for start outside the reaction layer ($\delta = 1$) [84].

also be found from Eq. (5.31) with the initial condition (5.1):

$$\chi(r, r_0) = W_R(r) \int_0^\infty m(t)dt = W_R(r)\tilde{G}(r, r_0, 0) \qquad (5.38)$$

A number of such distributions, calculated taking into account spin conversion and shown in Fig. 74 (later), differ in the same way as the distribution of products of bimolecular ionization, shown in Figs. 23 and 25. They reproduce the shape of $W_R(r)$ in the fast diffusion (kinetic control) limit, take the bell-shaped profile shifted from contact under diffusional control and do not change initial shape (5.1) in the static case, when $\tilde{D} = 0$. The total quantum yield of discharged ions

$$\chi(r_0) = \int \chi(r, r_0)d^3r = -\int_0^\infty \int \dot{m}d^3r\,dt = -\Omega(t|r_0)|_{t=0}^{t=\infty} = 1 - \varphi(r_0) \qquad (5.39)$$

Only in the case of contact start ($r_0 = \sigma$) can it be properly estimated with exponential model as $k_c/(k_c + k_D)$.

The kinetics of geminate recombination in contact approximation was presented by Eq. (5.24). Its relaxation time is a nonmonotonous function of $\tilde{D}$. In case of contact start, it becomes an inverted parabola in coordinates used in Fig. 36. The ascending branch of this curve qualitatively contradicts the opposite dependence obtained within the "exponential model." This is because the latter completely ignores the possibility of repeated contacts. Their number $\bar{\nu}$ increases with time monotonously and initially $\bar{\nu} \propto \sqrt{4\tilde{D}t}/L$. At each contact ions recombine with a small probability $z_0^2 = 4\lambda x$, so that the survival probability after $\bar{\nu}$ contacts is

$$\Omega = e^{-z_0^2 \bar{\nu}(t)} = \exp\left(\sqrt{\frac{4k_c^2 t}{\pi \sigma^3 k_D}}\right)$$

The characteristic time of its decay $t_0 \propto \sigma^3 k_D / k_c^2 \approx \tau_0 (k_D/k_c)^2$ as shown in Table I (case b).

Although in general the contact approximation is more appropriate than the exponential model, it is not valid in the quasistatic limit. To extend consideration to this region, one should employ the theory of remote electron transfer using numeric methods for the solution of Eq. (5.33). This was actually done in Ref. 84 with the DCR-3 program written by Krissinel and Shokhirev. A detailed description of the algorithm and the accuracy control method was given in Ref. 90. The results for the relaxation time t_e are presented in Fig. 38, where the true charge separation yield φ was used as an argument instead of its exponential model analog φ_0. Thus, both the abscissa and ordinate are measurable quantities, and one curve differs from another only by the initial separation of ions, r_0. As was expected, no problem arises in the static limit, when $\tilde{D} = \varphi = 0$. All $t_e(0)$ are finite and coincide with the inverse $W_R(r_0)$. All the rest is qualitatively the same as in contact approximation (compare with Fig. 36).

In principle, Figure 38 allows one to find out what is r_0, provided both quantities, t_e and φ, were measured. Unfortunately, not all ions start from the same initial separation. Usually they are distributed over the distance with a normalized probability $f_0(r_0)$. It is given in Eq. (4.23) with the assumption that ions were generated by binary photoionization. In fact, one should average the charge separation quantum yield $\varphi(r_0)$ over such a distribution to get a final result comparable with the real experimental data. We will prove this statement in Section VI.

E. Rectangular Model of Recombination Layer

The exponential model of the reaction zone profile given by Eq. (2.22) is an appropriate approximation for remote recombination only for the normal

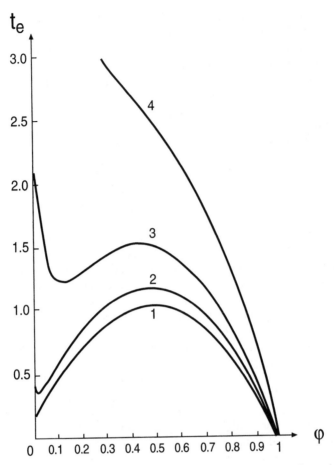

Figure 38. Dependence of decay time t_e on actual quantum yield φ for various initial separations: (1) $\delta = 0$; (2) $\delta = 0.05$; (3) $\delta = 0.1$; (4) $\delta = 0.2$ [84].

Marcus region. To extend the theory to higher exergonicity (to the inverted region), one can use the simpler, rectangular model of the reaction layer given in Eq. (2.24). In the normal region the layer is adjacent to the contact sphere ($\sigma = R - \Delta$) and the results are similar to those obtained with the exponential model. However, in the inverted region the layer is moved away and ions can be initially created either (*a*) inside or (*b*) outside the inner border of the reaction layer (Fig. 39). The results in (*a*) are qualitatively different from those in (*b*).

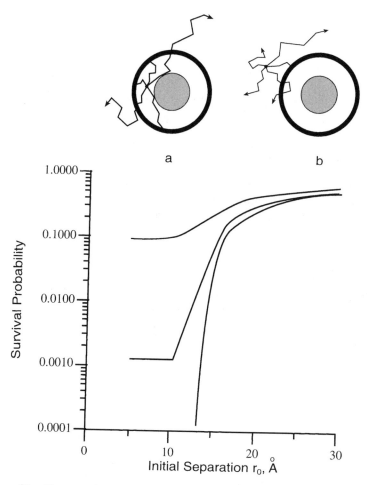

Figure 39.   The separation quantum yield (survival probability at $t = \infty$) as a function of the initial distance between the ions for $\tilde{D} = 10^{-5}\,\mathrm{cm^2/s}$ and three recombination rates (from top to bottom), $W_0 = 10, 100, 1000\,\mathrm{ns^{-1}}$ ($\sigma = 5\,\text{Å}$, $R = 12.5\,\text{Å}$, $\Delta = 2.5\,\text{Å}$ $r_1 = 10\,\text{Å}$, $L = 5\,\text{Å}$). Top: (*a*) the start from inside the recombination layer related to the left, horizontal branches of the curves; (*b*) the outside start related to the right branches approaching the maximum $\varphi = 1$ [43].

Let us concentrate on the quantum yield of charge separation from the starting distance r_0, $\varphi(r_0) = \Omega(\infty|r_0)$. It obeys the stationary variant of Eq. (5.33):

$$\frac{\tilde{D}e^{-r_c/r_0}}{r_0^2}\frac{\partial}{\partial r_0}r_0^2 e^{r_c/r_0}\frac{\partial}{\partial r_0}\varphi(r_0) = W_R(r_0)\varphi(r_0) \tag{5.40}$$

with the following boundary conditions:

$$\frac{\partial \varphi(r_0)}{\partial r_0}\bigg|_{\sigma} = 0, \qquad \varphi(\infty) = 1 \tag{5.41}$$

For highly polar solutions ($r_c = 0$) Eq. (5.40) becomes

$$\frac{\tilde{D}}{r_0^2}\frac{\partial}{\partial r_0} r_0^2 \frac{\partial}{\partial r_0} \varphi(r_0) = W_R(r_0)\varphi(r_0) \tag{5.42}$$

Its solution for rectangular $W_R(r)$ obtained in Ref. 43 has a different form when the starts are from interior of the sphere restricted by reaction layer, from inside of the layer itself and from outside:

$$\varphi(r_0) = \begin{cases} \dfrac{2}{C} & \sigma < r_0 < r_1 \\[2ex] \dfrac{(qr_1 + 1)e^{q(r_0 - r_1)} + (qr_1 - 1)e^{-q(r_0 - r_1)}}{Cqr_0} & r_1 < r_0 < r_2 \\[2ex] 1 - \dfrac{(r_1 + L)}{r_0} + \dfrac{(qr_1 + 1)e^{qL} + (qr_1 - 1)e^{-qL}}{Cqr_0} & r_2 < r_0 < \infty \end{cases}$$

$$\tag{5.43}$$

where $r_1 = R - \Delta$, $r_2 = R + \Delta$, $r_2 - r_1 = 2\Delta = L$, and

$$C = (qr_1 + 1)e^{qL} - (qr_1 - 1)e^{-qL}, \qquad q = \sqrt{\frac{W_0}{\tilde{D}}} \tag{5.44}$$

At $qL \gg 1$ these expressions reduce to those obtained in Ref. 91.

As can be seen, the quantum yield does not depend on the initial separation, as long as it is less than the inner radius of the recombination layer, r_1. The recombination layer screens the ions started from inside. The separation quantum yield ("survival probability" of ions) is smaller the faster the recombination but, sharply increases when the starting point is shifted outside (Fig. 39). When diffusion becomes slower, the quantum yield reduces especially for the shortest starts. With an increase in the residence time in the recombination layer, $\tau_e = RL/\tilde{D}$, the layer becomes non-transparent for particles started from inside (Fig. 40).

The quantum yield of the particles with an inner start

$$\varphi(r_0 < r_1) = \frac{1}{\cosh(qL) + qr_1 \sinh(qL)} \tag{5.45}$$

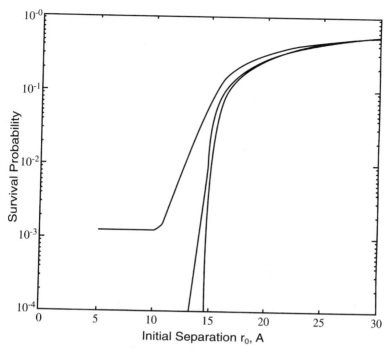

Figure 40. The separation quantum yield (survival probability at $t = \infty$) as a function of the initial distance between the ions for $W_0 = 100 \text{ ns}^{-1}$ and three diffusion coefficients (from top to bottom): $\tilde{D} = 10^{-5}, 10^{-6}, 10^{-7} \text{ cm}^2/\text{s}$. The remaining parameters are the same as in Fig. 39 [43].

turns to 0 in the slow diffusion limit ($q \to \infty$). In the opposite, fast diffusion limit, it approaches the plateau, independent on r_0:

$$\lim_{q \to 0} \varphi(r_0 < r_1) = \frac{1}{1 + W_0 R L / \tilde{D}} \qquad (5.46)$$

This is an upper limit reached at

$$qL = \sqrt{\frac{W_0 L^2}{\tilde{D}}} \ll 1 \qquad (5.47)$$

Under this condition the recombination during a single diffusional crossing of the reaction layer ($\tau_c = L^2/\tilde{D}$) is small. However, the total effect, presented by the term $W_0 R L \tilde{D} = (W_0 \tau_c)(\tau_e/\tau_c) = (qL)^2 (R/L)$ in Eq. (5.46), may be larger than 1 if the number of crossings, $\tau_e/\tau_c = R/L$, is even larger.

The limit (5.46) coincides with what was predicted by the exponential model, Eq. (3.9), at $r_c \to 0$, if one sets

$$z = W_0 L R, \tag{5.48}$$

that is, $k_{-et} = W_0 = k_c/v$, where the volume of the thin reaction layer is $v \approx 4\pi R^2 L$ and

$$k_c = \int W_R(r) d^3 r \approx 4\pi R^2 L W_0$$

This constant, which remains the main characteristic of the reactivity, even of noncontact layers, should not be mixed up with a kinetic rate constant for bimolecular recombination:

$$k_K = \int W_R(r) \exp\left(\frac{r_c}{r}\right) d^3 r \approx k_c \exp\left(\frac{r_c}{R}\right)$$

Presenting the general result (5.45) in the same way as Eq. (5.15) we should set

$$Z = z \frac{\cosh(qL) + qr_1 \sinh(qL) - 1}{q^2 L R} \tag{5.49}$$

This quantity exceeds significantly z in the slow diffusion (high-viscosity) region, but approaches this value in the fast diffusion (low viscosity) limit (Fig. 41):

$$Z_0 = \lim_{q \to 0} Z = z \tag{5.50}$$

Quite the opposite behavior is peculiar for ions created outside the recombination layer. Their separation quantum yield can be written as

$$\varphi(r_0 \ge r_2) = 1 - \frac{r_2}{r_0} + \frac{qr_1 \cosh(qL) + \sinh(qL)}{qr_0 \cosh(qL) + q^2 r_0 r_1 \sinh(qL)} \tag{5.51}$$

If condition (5.47) is met and the recombination layer is narrow ($L \ll R$), this result can be simplified to the following:

$$\varphi(r_0 \ge r_2) \approx 1 - \frac{r_2}{r_0} \frac{q^2 R L}{1 + q^2 R L} = 1 - \frac{r_2}{r_0} \frac{k_c}{k_c + k_D} \tag{5.52}$$

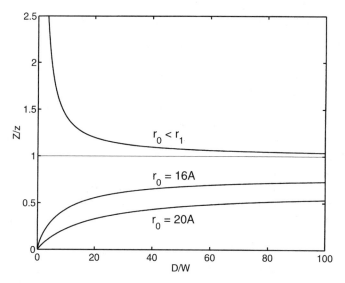

Figure 41. Deviations from the exponential model result ($Z \equiv z$) at slow diffusion of ions starting from inside the recombination layer ($r_0 < r_1 = 10\,\text{Å}$) and the same for two starts from outside: from $r_0 = r_2 = 16\,\text{Å}$ and $r_0 = 20\,\text{Å}$ at $W_0 = 10\,\text{ns}^{-1}$.

where $k_D = 4\pi R\tilde{D}$ is the diffusional rate constant for bimolecular recombination. An essential parameter $q^2 RL = k_c/k_D \lesssim 1$ is similar to that in contact approximation. By equating expression (5.52) to $1/[1 + Z/\tilde{D}]$, we obtain

$$Z = \frac{r_2}{r_0} \frac{z}{1 + \left(1 - \dfrac{r_2}{r_0}\right) q^2 RL} \qquad (5.53)$$

which is always smaller than z even in the fast diffusion limit when

$$Z_0 = \lim_{q \to 0} Z = \frac{r_2}{r_0} z < z \qquad (5.54)$$

This is actually a "quasikinetic limit" when the deviation from the exponential model is due only to noncontact creation of ions ($r_0 > r_1$). A more essential difference is seen in the opposite limit of slow diffusion, which controls geminate recombination. There Z increases linearly with $\tilde{D}$ approaching the quasikinetic limit from below, as well as in contact approximation (compare Figs. 41 and 32).

Thus, the diffusional (viscosity) dependence of Z is qualitatively opposite when the ions start from inside and outside the recombination layer (Fig. 41).

F. Smooth Recombination Layer and Real Initial Distributions

Instead of employing the simplified models of the reaction layer, one can use the real $W_R(r)$ dependence at fixed ΔG_r and the true initial distribution of ions $f_0(r)$ at different ionization free energies, ΔG_i. Changing the latter from 0, up to at least $2\lambda_c$, one can extend the mean initial separation of ions defined in Eq. (4.24) and shift ions far away from contact, even outside the remote recombination layer. In some hypothetical situations the energy of excited reactants (in a row of similar species) can increase without changing the energy of the charge-transfer state of the same species (Fig. 42). The latter is assumed to be as high as necessary for the inverted region marked "I." Then the recombination layer, although not rectangular, is shifted away from contact, while the averaged initial separation $\bar{R} \approx r_0$ may be either closer to or farther from it, depending on ΔG_i. An example of such a smooth layer for a single-channel ionization in the inverted region is shown in Fig. 43 together with a few initial distributions of ions at different ΔG_i. The larger $|\Delta G_i|$, the wider becomes the distribution that finally acquires the bell shape with a maximum shifted out of the recombination layer. The average

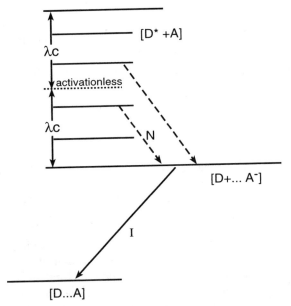

Figure 42. Energy levels and electron transitions at a fixed free energy of recombination, $\Delta G_r = \text{const}$, but arbitrary ionization free energy. The scheme shows forward electron transitions from excited states of different donors (dashed arrows), followed by the common backward transition to a ground state (solid arrow).

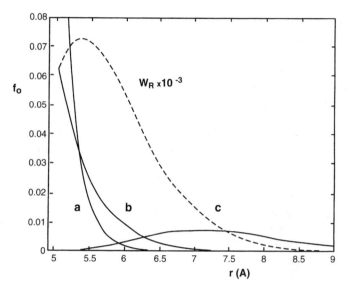

Figure 43. Initial distributions of ions for (a) $|\Delta G_i| = 0$, (b) $|\Delta G_i| = \lambda_c$, (c) $|\Delta G_i| = 2\lambda_c$ in comparison with the recombination rate $W_R(r)$ (dashed line). The other parameters are $w_i = 1.3 \times 10^3 \, \text{ns}^{-1}, \lambda_c/T = 55, T = 300 \, \text{K}, |\Delta G_r| = 1.5\lambda_c, \sigma = 5 \, \text{Å}$ [43].

initial separation of ions increases accordingly, being less at small $|\Delta G_i|$, but finally larger than the mean recombination distance.

Therefore, one should expect that the charge separation quantum yield should increase with $|\Delta G_i|$ as with r_0. If there were a family of donors having different energies of excited state D^*, but the same energy of the electron transfer state in a pair $[D^+...A^-]$ (Fig. 42), then it might be proved experimentally that the ionization free energy significantly affects the separation quantum yield. Alternatively, one can manipulate with a solvent viscosity, changing the initial charge separation and the width of their recombination zone. The acceleration of encounter diffusion strongly affects the charge separation quantum yield and related magnetic field effect (see Section VIII.H). The exponential model does not provide either of these effects in principle, because all initial and final distances except contact are excluded from the beginning.

To make a real comparison with the experimental data, one should average the partial quantum yield $\varphi(r_0)$, calculated for a given $W_R(r)$, with the normalized initial charge distribution (4.23) obtained by means of encounter theory:

$$\bar{\varphi} = \int_{\sigma}^{\infty} \varphi(r_0) f_0(r_0) 4\pi r_0^2 dr_0 \qquad (5.55)$$

Alternatively, the same can be done using numerical methods developed to trace the photoionization kinetics $P(t)$ from the beginning up to a full separation of charges [15]. As will be proved in the next section, $P(\infty) = \bar{\varphi}$ at $\tau_D = \infty$. The results shown in Fig. 44 were obtained in such a way using a program that generates simultaneously the initial distributions, recombination kinetics and charge separation quantum yield, accounting properly for Coulomb attraction if necessary [43]. We considered water where Onsager's radius is rather small (7 Å) and does not play a significant role. As was expected, the averaged separation quantum yield is beginning to sharply increase as soon as $|\Delta G_i|$ becomes larger than $|\Delta G_r|$. Under this condition, the ionization radius exceeds the recombination one and ions are generated outside the reaction layer.

The close resemblance of the curves $\varphi(r_0)$ and $\bar{\varphi}(\Delta G_i)$ seen in Figs. 39 and 44 is possible if the mean initial separation of ions $\bar{R} \equiv r_0$ is a monotonous function of ΔG_i. This is the case when the ionization is almost everywhere under kinetic control, so that $r_0 \approx r_I$ increases monotonously with ΔG_i (curve c in Fig. 30). However, one can see from the same figure that at slower diffusion, when ionization in the activationless region is diffusion controlled, a loop appears in the $r_0(\Delta G_i)$ dependence. Under such conditions a similar loop arises also in the $\bar{\varphi}(\Delta G_i)$ dependence obtained in Ref. 30 (Fig. 45). Since both effects have the same origin, the relationship

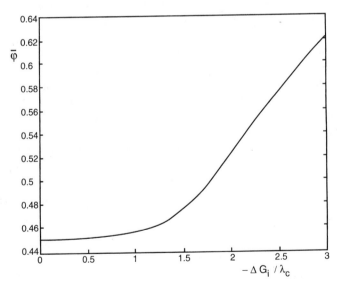

Figure 44. The averaged separation quantum yield $\bar{\varphi}$ as a function of the ionization free energy $|\Delta G_i|$ at $w_r = w_i = 1.3 \times 10^3 \text{ ns}^{-1}$ and $r_c = 7 \text{ Å}$ (the other parameters are the same as in Fig. 43) [43].

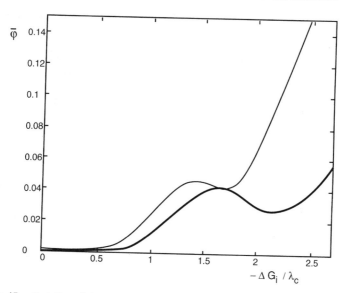

Figure 45. Variation of the separation quantum yield $\bar{\varphi}$ with an increase in the ionization exergonicity for a single-channel ionization, $S = 0$ (thick line), and for a multichannel ionization, $S = 2$ (thin line), at $D = \tilde{D} = 10^{-6}\,cm^2/s, c = 0.1\,M, r_c = 7\,\text{Å}$. The mean radius of a single-channel recombination zone is 6.08 Å [30].

between $\bar{\varphi}$ and $\bar{r}_0$ shown in Fig. 46 was expected to remain monotonous. In fact, there is no full identity in $\bar{\varphi}$ at the same $\bar{r}_0$ because the distributions of starting distances are different, even if the averaged values coincide.

In conclusion, it should be noted that to change ΔG_i keeping ΔG_r fixed (as in Fig. 42) is scarcely possible experimentally. What has really been done many times is a simultaneous changing of both ionization and recombination free energies, where their sum ε is fixed (Fig. 17). This case is more complex and less universal. A number of different situations classified in Fig. 15 (A, B, C) should be discriminated. Analyzing in the next section how the FEG dependence is affected by encounter diffusion of reactants and ions, we will concentrate mainly on situation B ($\varepsilon = 2\lambda_c$) illustrated in Fig. 47. However, finally we will show how Z changes with ΔG_i in all cases at slow diffusion and different lifetimes of the excitation.

VI.   PHOTOIONIZATION FOLLOWED BY GEMINATE RECOMBINATION

A.   Ion Accumulation without Recombination

As an introduction to this section, let us consider a hypothetical situation in which photogenerated ions are mobile but cannot recombine ($k_c = W_R = 0$).

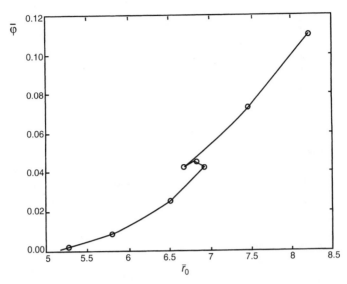

Figure 46. The relationship between the averaged charge separation quantum yield $\bar{\varphi}$ and the mean initial separation of ions $\bar{r}_0$, for a single-channel reaction (the remaining parameters are the same as in Fig. 45).

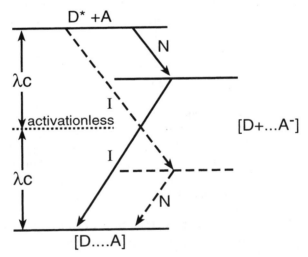

Figure 47. A scheme of energy levels and electron transitions at fixed excitation energy $\varepsilon = 2\lambda_c$, but for different positions of charge-transfer state. The forward and backward electron transitions corresponding to the IN case are depicted by dashed arrows; those related to the NI case are shown by solid arrows.

Then their distribution changes with time according to the kinetic equation

$$\dot{m}(r,t) = W_I(r)n(r,t)N(t) + \frac{\tilde{D}}{r^2}\frac{\partial}{\partial r}r^2\left(\frac{\partial m}{\partial r} + \frac{r_c}{r^2}m\right) \qquad (6.1)$$

which is a synthesis of Eqs. (4.20) and (5.3). Initially there are no ions

$$m(r,0) = 0 \qquad (6.2)$$

but they are generated from excitations, whose distribution around acceptors, $n(r,t)$, and the total amount, $N(t)$, are determined by differential encounter theory [Eqs. (4.6a) and (4.8)–(4.10)]. As the ions do not recombine, the boundary condition (5.2) becomes

$$4\pi\tilde{D}r^2\left(\frac{\partial m}{\partial r} + \frac{r_c}{r^2}m\right)\Bigg|_{r=\sigma} = 0 \qquad (6.3)$$

In this case the diffusional redistribution following the photogeneration of ions does not change the total amount of them, $P(t)$, which still obeys Eqs. (4.19) and (4.6b):

$$P(t) = c\int m(r,t)d^3r, \quad \text{and} \quad \dot{P} = k_I(t)cN \qquad (6.4)$$

Although the pair distribution $m(r,t)$ broadens with time and finally turns to 0, the total number of mobile ions monotonously increases with time, as it did in the case of immobile ions.

B. Photogeneration of Ions Followed by Their Recombination

If there is geminate recombination of ions with the rate $W_R \neq 0$, the results are qualitatively different. First of all, the kinetic equation for the pair distribution function should be composed of Eqs. (4.20) and (5.31), which accounts for remote recombination in line with the diffusion of ions:

$$\dot{m}(r,t) = W_I(r)n(r,t)N(t) + \frac{\tilde{D}}{r^2}\frac{\partial}{\partial r}r^2 e^{r_c/r}\frac{\partial}{\partial r}e^{-r_c/r}m - W_R(r)m \qquad (6.5)$$

Unlike Eqs. (5.31) and (6.1), this equation includes both bimolecular ionization and geminate recombination. Therefore, the initial rise of the ion concentration may give way to a partial concentration decrease after ionization has been finished. The initial and boundary conditions remain the same as in Eqs. (6.2) and (6.3) or (5.30), but only one of the equalities (6.4) still holds:

$$P(t) = c \int m(r,t)d^3r, \quad \text{and} \quad \dot{P} \neq k_I(t)cN \qquad (6.6)$$

This is a very important change. The general definition (4.19) cannot change, but the total ion density P does not fit the differential rate equation any more. Suggesting Eq. (6.5) in 1992 [12], the present author actually made a break with the rate concept entirely. The same equation was proposed simultaneously and independently by Dorfman and Fayer [13]. The combination of differential encounter theory with Eqs. (6.5) and (6.6) constitutes the so-called unified theory (UT) of photoionization followed by geminate recombination. Much later it becomes clear that this theory is a particular case of a more general integral encounter theory (IET) and may be deduced from it when ionization is an irreversible process [92]. More recently IET has been strongly developed and extended to a number of new applications, including multiple energy transfer and biexcitonic (nonlinear) photoconductivity. Since this should be a subject of a separate review, we will only briefly note the main achievements of IET in Section X.

The general solution to Eq. (6.5) can be expressed via the Green function of Eq. (5.31), $G(r, r', t - t')$:

$$m(r,t) = \int_0^t dt' \int d^3r' G(r,r',t-t')W_I(r')n(r',t')N(t') \qquad (6.7)$$

while the total density of ions is related in the same way to the survival probability $\Omega(t|r')$:

$$P(t) = c \int W_I(r')d^3r' \int_0^t \Omega(t-t'|r')n(r',t')N(t')dt' \qquad (6.8)$$

The survival probability obeys Eq. (5.33) with the corresponding initial and boundary conditions given in Eq. (5.34). As $t \to \infty$, it turns to the charge separation quantum yield $\varphi(r) = \Omega(\infty|r)$, as well as when $P(\infty)$ turns to the free-ion quantum yield, ϕ. According to Eq. (6.8)

$$\phi = P(\infty) = c \int W_I(r')\varphi(r')d^3r' \int_0^\infty n(r',t')N(t')dt' = \psi\bar{\varphi} \qquad (6.9)$$

where ψ was introduced in Eq. (4.25) and $\bar{\varphi}$ is exactly the same averaged quantum yield of charge separation that was introduced in Eq. (5.55). As expected, at $\tau_D = \infty$ Eq. (6.9) reduces to an identity $\phi \equiv \bar{\phi}$, because in this case the quantum yield of primary ionization is 100% ($\psi = 1$).

C. Kinetics of Charge Accumulation and Recombination

The shape of the kinetic curves $P(t)$ strongly depends on how fast the primary ionization is compared to the geminate recombination. The stationary rate of the former is $1/\tau_D + ck_i = 1/t_i$ while the characteristic time of the latter is t_e, roughly estimated in Eq. (5.29) for contact born pairs. In the slow ionization (fast recombination) limit

$$t_e \ll t_i \qquad (6.10)$$

If this inequality is reversed, the ionization is fast, whereas recombination is slow. These two opposite limits should be considered separately.

In the case of slow ionization, setting $\Omega \approx \varphi$, one can get the following from Eq. (6.8):

$$P(t) \approx \phi - c \int W_I(r')\varphi(r')d^3r' \int_t^\infty n(r',t')N(t')dt' \qquad (6.11)$$

At $t \gg R_Q^2/D$ the ionization reaches its stationary regime where $n(r,\infty) = n_s(r)$ and $k_I(\infty) = k_i$ defined in Eq. (4.11). Using these values in Eq. (6.11), we can use it to calculate the following asymptotics of the process:

$$P(t) \approx \phi - \psi_s\bar{\phi}e^{-(1/\tau_D + ck_i)t} \quad \text{at} \quad t \gg \frac{R_Q^2}{D} \qquad (6.12)$$

where the quantum yields of ionization and charge separation are

$$\psi_s = \frac{ck_i\tau_D}{1 + ck_i\tau_D} \quad \text{and} \quad \bar{\phi} = \int \varphi(r)f_s(r)d^3r \qquad (6.13)$$

The latter is averaged over the normalized stationary distribution of the reactants

$$f_s(r) = \frac{W_I n_s(r)}{\int W_I n_s(r)d^3r}$$

According to Eq. (6.12), at slow ionization $P(t)$ grows monotonously approaching the plateau $P(\infty) = \phi$ from below, as it did in the absence of recombination (Fig. 48). However, in the latter case $\phi \equiv \psi$, while now $\phi = \psi\bar{\phi} < \psi$ because part of the ions, $1 - \bar{\phi}$, do not escape geminate recombination.

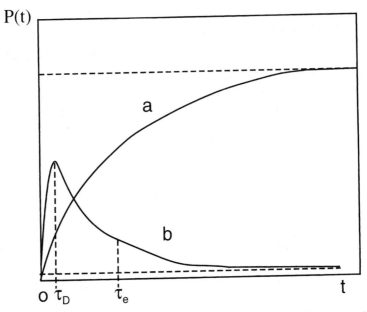

Figure 48. Schematic representation of recombination kinetics for slow (*a*) and fast (*b*) ionization. Dashed horizontal lines indicate the escape probabilities φ (free-ion quantum yields).

Qualitatively different behavior is peculiar to the opposite case of fast ionization. Usually it is fast as the result of abrupt decay of the excited state, which interrupt the ionization at its earliest stage. At this stage it proceeds exponentially with a rate constant $k_I(0) = k_0 = \int W(r)d^3r$ as in Eq. (4.13). Since the decay of $N(t')$ is much faster then the recombination given by $\Omega(t - t'|r)$, the latter can be factored out of the integral at time t. Then Eq. (6.8) transforms to the following:

$$P(t) = c \int W_I(r')\Omega(t|r')d^3r' \int_0^t n(r',t')N(t')dt' \qquad (6.14)$$

At the earliest stage of ionization

$$n(r,t) = 1 \quad \text{and} \quad N(t) = e^{-t/\tau_D - ck_0 t} \qquad (6.15)$$

Using these approximations in Eq. (6.14), we obtain

$$P(t) \approx \psi_0 \int \Omega(t|r')f_0(r')d^3r'\left[1 - e^{-t/\tau_D - ck_0 t}\right] \qquad (6.16)$$

where the ionization quantum yield and the distribution of ions are

$$\psi_0 = \frac{ck_0\tau_D}{1 + ck_0\tau_D} \quad \text{and} \quad f_o(r) = \frac{W_I(r)}{\int W_I(r)d^3r} \tag{6.17}$$

The latter distribution is a particular case of the more general, Eq. (4.23), and can be deduced from it by using an approximation (6.15) in Eq. (4.22).

According to Eq. (6.16) the time dependence $P(t)$ is not monotonous in the case of fast ionization. It consists of a short ascending branch determined by the expression in brackets, followed by a much longer descending tail, determined by the averaged quantity $\Omega(t|r)$. The former branch describes the fast accumulation of ions; the latter represents their subsequent geminate recombination (Fig. 48). Only in the limit of instantaneous ionization (at $\tau_D \rightarrow 0$), this complex kinetics degenerates into a single descending branch, which we studied in the previous section.

To clarify the interrelationship between the real and instantaneous ionization, we studied in Ref. 14 how the initial conditions for the latter should be chosen to reach the maximal coincidence of the descending branches of both. If ionization is neither fast nor slow, one has to solve numerically Eq. (6.5) together with the supplementary Eqs. (4.6a) and (4.8)–(4.10). We used the numeric procedure based on the expanded DCR program [93] that was successfully employed earlier [9], as well as in the previous section. From the simultaneously calculated initial distribution, we found $\bar{R}$ to be a rough estimate of the initial distance between ions, which may be identified with r_0, if contact approximation is used. This comparison is shown in Fig. 49 for the case when $\psi = 1$ ($\tau_D = \infty$). Under this condition, the difference between $P_{\max}$ and 1 testifies to how far we are from the fast diffusion limit when $P_{\max} \approx 1$. We used the exponential model (2.22) for both ionization and recombination, which is possible in the NN region (see Fig. 15). As can be seen from Fig. 49, the maximum is rather high and only slightly exceeds the free ion quantum yield ϕ. This is always the case when recombination is kinetically-controlled, specifically, $k_c/k_D \ll 1$, with k_c defined in Eq. (2.4).

Although there is only a descending branch in the kinetics of recombination after instantaneous ionization, it finally approaches well the long tail of the photoionization curve, provided the choice of starting distance was properly made. The closer r_0 is to $\bar{R}$, the better is the coincidence of the long-time asymptotics of charge separation and the quantum yields of free ions.

Using contact approximation for recombination, we reached the best fit to our data at $r_0 = \bar{R} = 7.3\,\text{Å}$, which is essentially larger than the contact distance $\sigma = 5\,\text{Å}$. The exponential model is completely inadequate.

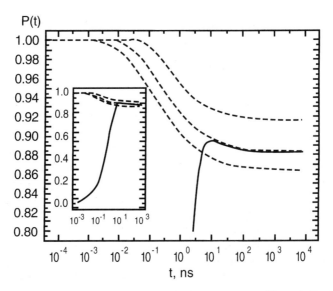

Figure 49. Kinetics of photoionization (solid lines) at $\tau_D = \infty (\phi \equiv \bar\phi)$, $\bar{R} = 7.3\,\text{Å}$ and $k_c/k_D = 0.2[(W_R(\sigma) = 3.96\,\text{ns}^{-1})]$, $\tilde{D} = D = 10^{-5}\,\text{cm}^2/\text{s}$, $c = 0.1\,\text{M}$, $W_I(\sigma) = 10^3\,\text{ns}^{-1}$, $L = 0.75$, $\sigma = 5\,\text{Å}$, $r_c = 0$. The dashed lines depict the kinetics of recombination from the initial distance $r_0 = 6; 7.3; 10\,\text{Å}$ (from bottom to top) [14].

However, the advantage of contact approximation is also lost as soon as recombination becomes diffusion-controlled. To make it clear we compare in Fig. 50 the properly averaged charge separation quantum yield $\bar\phi$ (solid line) with the partial quantum yield $\varphi(r_0)$ obtained in the contact approximation for a number of initial charge separations r_0 (dashed lines). All calculations were made for polar solutions ($r_c = 0$) and exponential $W_R(r)$. As seen from the inset, at small $x = k_c/k_D$ both $\bar\phi$ and $\varphi(r_0)$ decrease similarly, provided r_0 is properly chosen. However, at high x (slower diffusion) the significant discrepancy appears that becomes qualitative as $x \to \infty$. This is the static limit where the contact approximation leads to an unphysical result seen as plateau $\varphi(x) \to \varphi_\infty$. Contrary to these expectation, for any distant recombination, as well as for exponential one, the $\lim_{x\to\infty} \varphi = 0$. Hence, for slow diffusion (viscous solutions) there is no alternative to the theory of remote electron transfer.

D. FEG Law for Geminate Recombination

Only now can we approach the key problem of the free-energy dependence of the charge separation quantum yield in either the kinetic ($k_c/k_D \ll 1$) or diffusional limit ($k_c/k_D \gg 1$). As seen from Fig. 50, in the former case (fast

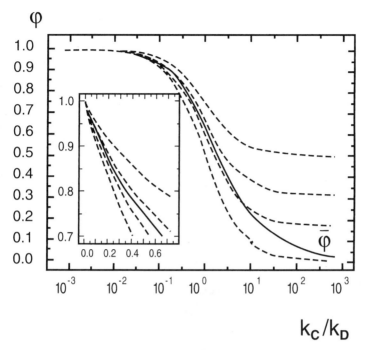

Figure 50.   Space-averaged charge separation quantum yield (solid line) and that obtained with the contact approximation for $r_0 = 5, 6, 7.3, 10$ Å   (dashed lines from bottom to top) as functions of $k_c/k_D \propto 1/\tilde{D}$. The remaining parameters are the same as in Fig. 49 [14].

diffusion) $\bar{\varphi}$ only slightly deviates from 1 $(1 - \bar{\varphi} \ll 1)$, while in the latter case (slow diffusion) the yield is really small $(\bar{\varphi} \ll 1)$. The results of numeric calculations for different rate of diffusion are shown in Fig. 51 [15]. Unfortunately, in such a presentation of the data nothing is clear, except for the fact that the whole curve is shifting up with an acceleration of diffusion and finally should degenerate in a horizontal line $\bar{\varphi} = 1$. This is a trivial reflection of the competition between Z and $\tilde{D}$ in Eq. (5.15), which is won by diffusion as $\tilde{D} \to \infty$. To understand how the recombination itself depends on diffusion one should turn to a family of $Z(\Delta G_r)$ curves obtained from the same data (Fig. 52) [15,82]. They can be easily compared with a similar curve in the kinetic controlled limit $(\tilde{D} \to \infty)$, as well as for the exponential model prediction made for z in Eq. (3.10). The latter should reproduce the free-energy dependence of k_{-et}, which according to the contact estimate (5.18) duplicates the FEG dependence of $W_R(\sigma)$ given by Eq. (2.5) or (2.17). As for kinetic recombination, it should be analyzed here before an interpretation of the results at slower diffusion is attempted.

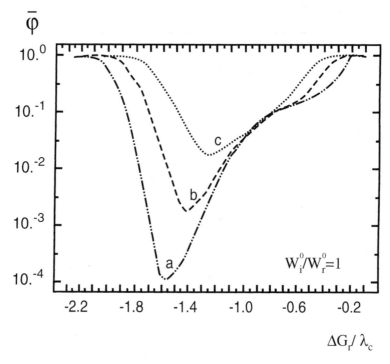

Figure 51. Quantum yield of charge separation, $\bar{\varphi}$, against recombination free energy at contact ΔG_r for different encounter diffusion coefficients of ions: (a) $\tilde{D} = 10^{-7}\,\mathrm{cm^2/s}$, (b) $\tilde{D} = 10^{-6}\,\mathrm{cm^2/s}$, (c) $\tilde{D} = 10^{-5}\,\mathrm{cm^2/s}$ [15].

It has already been shown that Z is a measure of the recombination rate in the conventional presentation of

$$\bar{\varphi} = \int_{\sigma}^{\infty} \varphi(r_0) f_0(r_0) 4\pi r_0^2 dr_0 = \frac{1}{1 + Z/\tilde{D}} \qquad (6.18)$$

According to Eqs. (5.32) and (5.35), we obtain

$$\varphi(r_0) = \int m(r, \infty; r_0, 0) d^3r \qquad (6.19)$$

where $m(r, t; r_0, 0)$ is the Green function of Eq. (5.31).

The geminate recombination is under kinetic control when W_R is relatively small. Hence we can solve Eq. (5.31) with a perturbation theory

Z

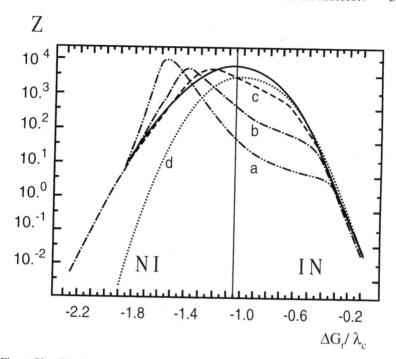

Figure 52.   The free-energy dependence of Z at the same $W_0 = 10^{12}$ s^{-1} for ionization and recombination, $\lambda_c = 55$ T and (*a*) $\tilde{D} = 10^{-7}$ cm^2/s, (*b*) $\tilde{D} = 10^{-6}$ cm^2/s, (*c*) $\tilde{D} = 10^{-5}$ cm^2/s, in comparison with the kinetic limit (solid line) and exponential model expectation (dotted line *d*) [15].

linear in W_R. The solution can be expressed via the Green function G_0 of Eq. (5.5):

$$m(r, t; r_0, 0) = G_0(r, r_0, t) - \int d^3 r' W_R(r') \int_0^t G_0(r, r', t - t') G_0(r', r_0, t') dt'$$

$$(6.20)$$

Inserting this result into Eq. (6.19), we obtain after integration over r:

$$\varphi(r_0) = 1 - \int W_R(r') \tilde{G}_0(r', r_0, 0) d^3 r' \qquad (6.21)$$

where $\tilde{G}_0(r', r_0, s)$ is the Laplace transformation of the Green function. In contact approximation $\tilde{G}_0(r', r_0, 0)$ is given by Eq. (5.14).

Substituting Eq. (6.21) into Eq. (6.18), we find that for geminate recombination under kinetic control

$$\bar{\varphi} \approx 1 - \int d^3 r' W_R(r') \int \tilde{G}_0(r', r_0, 0) f_0(r_0) d^3 r_0 = 1 - \frac{Z}{\tilde{D}} \tag{6.22}$$

where we kept only linear term expanding $\bar{\varphi}$ in $Z/\tilde{D}$, which is a small correction in kinetic limit. As can be seen from Eq. (5.5), $G_0(r, r_0, t) \equiv p(r, r_0, \tilde{D}t)$ is actually a universal function of dimensionless time $\tau = \tilde{D}t$ and

$$\tilde{p}(r', r_0, 0) = \int_0^\infty p(r', r_0, \tau) d\tau = \tilde{D} \tilde{G}_0(r', r_0, 0) \tag{6.23}$$

As follows from Eqs. (6.22) and (6.23)

$$Z = \int d^3 r' W_R(r') \int \tilde{p}_0(r', r_0, 0) f_0(r_0) d^3 r_0 \equiv Z_0 \tag{6.24}$$

does not depend on diffusion of ions as it should be in the kinetic limit.

Such an important parameter as Z_0 should now be compared with its analog qz, obtained for the same limit in Eq. (5.21) with contact approximation. In this approximation we have, from Eqs. (2.21) and (5.14):

$$W_R(r') = \frac{k_c}{4\pi\sigma^2} \delta(r' - \sigma) \quad \text{and}$$

$$\tilde{p}(\sigma, r_0, 0) = \frac{1}{4\pi r_c} \exp\left(\frac{r_c}{\sigma}\right) \left[1 - \exp\left(-\frac{r_c}{r_0}\right)\right]$$

Using these results in Eq. (6.24) we reduce it to the following:

$$Z_0 = \frac{k_c}{4\pi r_c} \exp\left(\frac{r_c}{\sigma}\right) \int \left[1 - \exp\left(-\frac{r_c}{r_0}\right)\right] f_0(r_0) d^3 r_0 = z \int q(r_0) f_0(r_0) d^3 r_0 \tag{6.25}$$

where q and z were defined in Eq. (5.20). If there were an initial distribution with the shape (5.1) then Z_0 would be identical to $q(r_0)z$ as was expected. In fact, Z_0 is equal to $\bar{q}z$, where $\bar{q}$ is a mean quantity. It is averaged over the real initial distribution $f_0(r)$, created by the forward (bimolecular) electron transfer and therefore depends on diffusion of the reactant. In highly polar solvents (as $r_c \to 0$)

$$\bar{q} = \int \left(\frac{\sigma}{r_0}\right) f_0(r_0) d^3 r_0 = \left\langle \frac{\sigma}{r_0} \right\rangle \tag{6.26}$$

A similar result can be obtained when the recombination layer is not contact but rectangular and initial separation is larger than its external radius r_2. In the latter case $q = r_2/r_0$ as has been shown in Eq. (5.54). This case, as well as *contact approximation*, should be identified with the IN regions of electron transfer (inverted for ionization and normal for recombination).

In the NI regions (normal for ionization, inverted for recombination), the result is a bit different: $q = 1$. It was obtained in Eq. (5.50) with the rectangular model of the remote recombination layer, assuming that initially ions were born near the contact. Here we have to approve this result by inserting into the general formula (6.24) the rectangular $W(r)$ from Eq. (2.24) and the Green function for the inner start from Eq. (5.14):

$$\tilde{p}_0(r', r_0, 0) = \frac{1}{4\pi r_c} \left[\exp\left(\frac{r_c}{r'}\right) - 1 \right]$$

By these means we obtain for a narrow recombination layer

$$Z_0 = \frac{W_0}{4\pi r_c} \int_{R-\Delta}^{R+\Delta} \left[\exp\left(\frac{r_c}{r}\right) - 1 \right] 4\pi r^2 dr \approx \frac{k_c}{4\pi r_c} \left[\exp\left(\frac{r_c}{R}\right) - 1 \right] = z \quad (6.27)$$

where $k_c = W_0 4\pi R^2 L$ and $L = 2\Delta$. In highly polar solvents $z = W_0 RL$ as in Eq. (5.48).

The most general expression for Z_0, valid in all regions and for arbitrary $W_R(r)$, is the following:

$$Z_0 = \int Z(r) f_0(r) d^3 r \qquad (6.28)$$

where

$$Z(r) = \frac{1 - e^{-r_c/r}}{r_c} \int_\sigma^r e^{r_c/R} W_R(R) R^2 dR$$
$$+ \int_r^\infty \frac{e^{r_c/R} - 1}{r_c} W_R(R) R^2 dR \qquad (6.29)$$

This equation was obtained in Ref. 82 from the lowest order perturbation theory solution of Eq. (5.40). Roughly speaking, Z_0 is proportional to a space-averaged value of $W_R(r)$ while in the exponential model $z \propto W_R(\sigma)$. This is the main reason why they differ so much in Fig. 52. In its right part, within the IN region, the difference is mainly numeric, as between the contact and exponential models, because there the recombination layer is really adjacent to the contact. The descending branches of both curves can be brought to full coincidence by the appropriate choice of exponential model parameters. This model leads to a perfect parabolic shape of the FEG law depicted in Fig. 52. Therefore, on the left side the discrepancy is very large between the model and real FEG dependence, which is not as symmetric. On this side, in the NI region, the remote recombination layer is far from contact and the average rate there is higher than at contact. Therefore, the real ascending branch of $Z_0(\Delta G_r)$ is shifted up relative to that of the exponential model.

Now we can understand why the behavior of $Z(\Delta G_r)$ is also so different to the right and to the left, when it is controlled by diffusion. According to the result shown in Fig. 41, in the former case (IN region) Z increases with diffusion approaching Z_0 from below, while in the latter case (NI region) it decreases approaching Z_0 from above. Therefore, at rather slow diffusion the quasiparabolic shape peculiar to the kinetic curve $Z_0(\Delta G_r)$ is strongly distorted and a fingerlike ledge appears at the top. This "finger" is much narrower and shifted to the left from the maximum of the kinetic curve. A similar narrow pike appears at the top of the $Z(\Delta G_r)$ curve, instead of a smooth and wide maximum of $Z_0(\Delta G_r)$, when one deals with multichannel electron transfer [30]. Those peaks in $Z(\Delta G_r)$ can be easily taken for a true FEG law peculiar only for $Z_0(\Delta G_r)$. All the more that experimentally only the left (ascending) part of the curve is often available including rarely the small region near the maximum (to avoid thermal ionization, which is not negligible at lower exergonicity).

Hence, the striking difference between the shape and the width of the FEG curves for ionization and recombination demonstrated in Fig. 31 may be attributed to a peak that represents the false FEG law for the backward electron transfer. As shown in Fig. 53, the latter is narrower than the real FEG curves for either recombination or ionization. The same is true regarding the data shown in Fig. 17 (note that the abscissa axis is reversed there in regard to later figures). The very small values of $\bar{\varphi} \equiv \Phi_{sep} < 0.1$ at the top of the curves in Fig 17b is clear evidence for diffusional control of geminate recombination. If there were a true FEG law obtained from kinetically controlled geminate recombination, $\bar{\varphi}$ would be much closer to 1. When $\bar{\varphi} \ll 1$ as in [94], this is a straight indication of diffusion-controlled recombination to which the exponential model is inapplicable. Parabolic interpolation of similar data found in [95] leads only to imitation of the free-energy-gap (FEG) law. This statement could be easily inspected by altering the viscosity of solution. The location of the false maximum and the whole shape of the curve should be very sensitive to diffusion when $\bar{\varphi} \ll 1 (k_{-et}/k_{sep} \gg 1)$. Unfortunately, this has not been done yet.

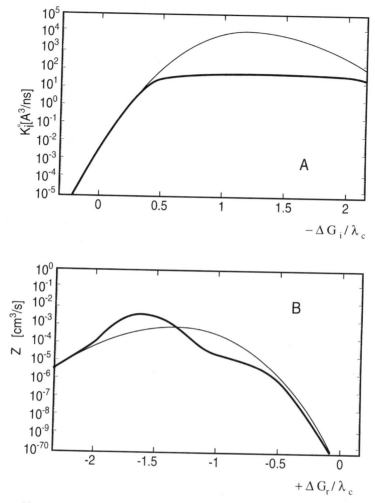

Figure 53. Free-energy law for the multichannel ionization (A) and recombination (B) for situation b (see Fig. 15) $\varepsilon = 2\lambda_c$ at $S = \lambda_q/\hbar\omega = 2$ and $D = \tilde{D} = 10^{-6}\,\mathrm{cm}^2/\mathrm{s}$. (The results were obtained with data taken from Ref. 30.)

The shape and position of a narrow feature in $Z(\Delta G)$ dependence should also be sensitive to a few other factors: the reorganization energy (related to the energy of excitation) and the lifetime of the excitation. The latter affects the shape of the charge distribution in (6.18) as well as diffusion. This expectation is illustrated by Fig. 54 for three typical cases related to A, B, and C energy diagrams shown in Fig. 15. The location of the feature depends on where is the interchange between the situations a and b shown in Fig. 39.

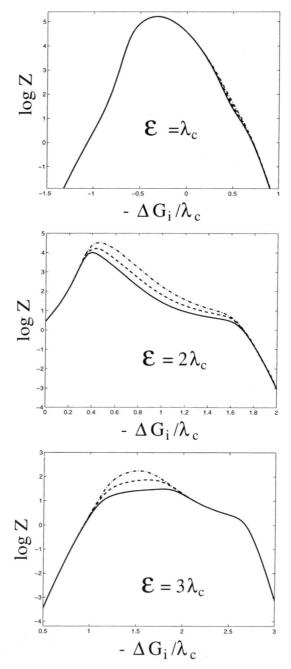

Figure 54. The free-energy dependence of Z for $\varepsilon = \lambda_c$, $\varepsilon = 2\lambda_c$, and $\varepsilon = 3\lambda_c$ at $\tau_D = 100$ ns (solid line), $\tau_D = 1$ ns (dashed line), and $\tau_D = 0.1$ (dot–dashed line). Diffusion coefficient $D = 10^{-7}$ cm^2/s [16].

In case A it takes place approximately at the border between the NN and IN regions; in case B, a little bit to the left of the NI/IN border; and in case C, in the II region. Therefore, it moves from the right to left with increasing excitation energy ε. In case A, it locates at the descending branch and is insignificant because the recombination there is kinetic and weakly depends on diffusion and lifetime. In other cases the peak is well expressed and enhanced with the increasing decay rate of excitation.

Hence, the answers to the questions raised in the conclusion of Section IV are the following:

- Diffusion does perturb the FEG law for recombination when the latter is under diffusion control. This happens when diffusion is rather slow compared with the true recombination rate, provided either ionization or recombination is not contact.

- Free-energy dependence that is too narrow for the recombination parameter Z, compared to the stationary ionization rate constant, may be attributed to a feature that appears under diffusion control and can imitate the true FEG law for recombination.

To verify these conclusions, one should change or at least control the diffusion of reactants and ions to prove that the backward electron transfer (BET) is really very fast and therefore controlled by diffusion.

On the other hand, the BET may be slowed down by different means to approach the true kinetic limit. For instance, the ion pair generated from triplet excitation is also in a triplet state, and its recombination into the ground state is strongly prohibited by spin-selection rules. When such a system was experimentally studied [96] the free-energy-gap law was reported to be verified within the framework of the exponential model. However, the real situation is much more complex because the theory should be generalized to take into account the spin evolution of the ion pair due to superfine interaction or an external magnetic field. The latter was actually shown to affect significantly the quantum yield of free ions and other reaction products. The magnetic-field effects are worthy of special attention, and we will address them in the following sections.

VII. SEPARATION OF SINGLET-BORN RIPs

A. Extended Exponential Model

Some effort was made to bring the exponential model closer to reality. The real-time evolution of ion concentration after δ-pulse excitation is generally composed from ascending and descending branches, while the original exponential model is represented by the latter only (Fig. 49). However, this is a minor disadvantage that has been easily removed by incorporation of the rate analogue of Eq. (4.6a) to the set (3.5) [16,81]:

$$\dot{N} = -k_i cN - \frac{N}{\tau_D} \tag{7.1a}$$

$$\frac{dm_c}{dt} = -k_{-et}m_c - k_{sep}m_c + k_i cN \tag{7.1b}$$

$$\frac{dm_\infty}{dt} = k_{sep}m_c \tag{7.1c}$$

where k_i is the stationary rate of ionization introduced in Eq. (4.11). From the exact solution of these equations we can get the kinetics of ion accumulation and recombination, which consists of both branches:

$$P(t) = k_i c t_i \left[k_{sep} t_e + \frac{k_{-et} t_e e^{-t/t_e} - (1 - k_{sep} t_i) e^{-t/t_i}}{1 - t_i/t_e} \right] \tag{7.2}$$

where $1/t_i = k_i c + 1/\tau_D$.

Now the exponential model prediction can be directly compared with the experimental data or the analogous result of the unified theory. The latter, given by Eq. (6.8), can be expressed via the time-dependent distribution of photogenerated ions, $m(r,t)$, introduced in Eq. (4.21):

$$P(t) = c \int m(r,t) d^3r + c \int d^3r \int_0^t \dot{\Omega}(t'|r) m(t-t', r) dt' \tag{7.3}$$

If there is no recombination ($\Omega \equiv 1$), Eq. (7.3) reduces to Eq. (4.19) and $P \to \psi$ at $t \to \infty$. If recombination does occur, then $m(r,t) \to m_0(r)$ at $t \gg t_i$, $\Omega(t|r) \to \varphi(r)$ at $t \gg t_e$ and $P \to \phi = \psi\bar{\varphi}$ as in Eq. (6.9). In this limit the final distribution of ions $m_0(r)$ determines both ψ and $\bar{\varphi}$ as well as the whole result, $P(\infty)$. However, the total kinetics (7.1) does not universally depend on m_0 but is determined by the intermediate ion distribution $m(r,t)$.

The original exponential model as well as its generalization given above are not sensitive to such details because both admit only contact creation of ions, at $r = \sigma$. However, this restriction can be removed if one uses the exponential model only for an approximate estimate of Ω in the general equation (7.3). This was done in Ref. 17 by a simple extension of Eq. (3.7) for the arbitrary starting distances:

$$\Omega(t|r) = \varphi(r) + [1 - \varphi(r)]e^{-t/t_e} \tag{7.4}$$

where $\varphi(r)$ is given by Eq. (5.15). Using $\varphi(r)$ instead of $\varphi(\sigma)$ in Eq. (7.4), we provide for a right limit $\Omega(r, \infty) = \varphi(r)$. However, this limit is never approached exponentially [9,85,97]. The long-time asymptotics of the recombination kinetics is quasiexponential only in nonpolar solvents where the Coulomb well is rather deep. The escape from it looks like a single jump out with a rate k_{sep}, but Shushin's assumption [97] that ions first fall at the bottom of the well and then start to separate from the very contact was not confirmed. The majority of separated ions escaped recombination because they never attended the contact region but moved away from where they were created.

Even better agreement with numerical simulations of geminate recombination kinetics can be reached if one corrects the ion separation rate as suggested in Ref. 85:

$$\frac{1}{t_e} = [k_{-et} + k_{\text{sep}}]\gamma\left(\frac{\sigma}{r_c}\right) \tag{7.5}$$

The correction factor is

$$\gamma(x) = \frac{4x^3[\cosh(1/x) - 1] - 2x}{Ei(1/x) + E_1(1/x) - F(x)}$$

where

$$F(x) = \exp(1/x)[x + x^2 + 2x^3] + \exp(-1/x)[x - x^2 + 2x^3] - 4x^3 - 6x$$

where $Ei(y)$ and $E_1(y)$ are the integral exponent functions [89]. This factor noticeably exceeds 1 at $x < 0.1$ and tends to $\frac{1}{3}$ at $x > 1$ (Fig. 55). In between it can be presented as $\gamma = \beta x/3$ where β values were tabulated [85] within the interval $0.05 > x > 1$. Since in polar solvents, at $x > 1$, Eq. (3.3) does not hold at all, the validity region of our generalized exponential model is confined to small values of x, where t_e should be corrected. But even without such a correction this generalization of the exponential model is better than the previous one, at least in the NI region, where distant creation of ions is essential. Nevertheless, it remains contact with respect to recombination and roughly simplifies the long-time asymptotics of this process assumed to be exponential in Eq. (7.4). We will start from inspection of this point by analyzing the real kinetics of ion and radical accumulation in nonpolar solvents.

B. Photogeneration of Free Ions and Radicals

The evacuation of ions from the deep Coulomb well is hardly possible, but the separation of discharged particles is not a problem. Of course, the neutral products of geminate recombination, $[D \ldots A]$, are of no interest, but

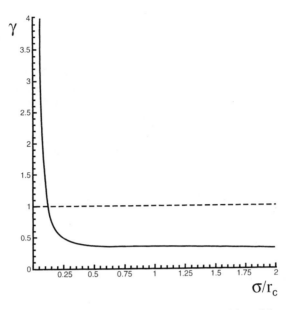

Figure 55. Correction factor for the generalized exponential model rates, $\gamma(r)$, as a function of the Coulomb well size, r_c [17].

sometimes the recombination proceeds via proton transfer from cation to anion producing the free radicals $\dot{D}$ and $\dot{A}$, which are observable with the FT-EPR (Fourier transform–electron paramagnetic resonance) technique used in other studies [98,99]. The full reaction scheme of irreversible free radical photogeneration is

$$D^* + A \longrightarrow {}^1[D^+ \dots A^-] \to D^+ + A^- \tag{7.6a}$$

$$\downarrow$$

$${}^1[\dot{D} \dots \dot{A}] \to \dot{D} + \dot{A} \tag{7.6b}$$

To describe the production of both ions and radicals, we now need two corresponding pair distribution functions, instead of the single one used in Eq. (6.5). These are $m(r,t)$ for ions, as before, and $g(r,t)$ for radicals. The corresponding kinetic equations for both are the following:

$$\frac{\partial}{\partial t} m = -W_R(r)m + \tilde{D}\frac{1}{r^2}\frac{\partial}{\partial r}r^2 e^{r_c/r}\frac{\partial}{\partial r}e^{-r_c/r}m + W_I(r)nN(t) \tag{7.7a}$$

$$\frac{\partial}{\partial t} g = W_R(r)m + \bar{D}\frac{1}{r^2}\frac{\partial}{\partial r}r^2 e^{-U_R/T}\frac{\partial}{\partial r}e^{U_R/T}g \tag{7.7b}$$

where $\tilde{D}$ and $\bar{D}$ are encounter diffusion coefficients of ion radicals and neutral radicals, $-r_c/r$ is the electrostatic potential of the Coulomb well, and $U_R(r)$ is the interradical interaction. The initial and boundary conditions are the same for both distributions:

$$m(r,0) = g(r,0) = 0, \qquad \frac{\partial}{\partial r} e^{-r_c/r} m(r,t) \bigg|_{r=\sigma} = \frac{\partial}{\partial r} e^{U_R/T} g(r,t) \bigg|_{r=\sigma} = 0$$

$$(7.8)$$

The total number of ions and neutral products of their recombination are

$$P = c \int_\sigma^\infty m(r,t) 4\pi r^2 dr, \qquad R = c \int_\sigma^\infty g(r,t) 4\pi r^2 dr \qquad (7.9)$$

As the result of a particle conservation law

$$N + \int_0^t \frac{N(t')dt'}{\tau_D} + P + R = 1 \qquad (7.10)$$

the knowledge of two quantities, $N(t)$ and $P(t)$, is sufficient to obtain the third, the total number of radicals $R(t)$. However, only the distantly separated free radicals are observable with the FT-EPR technique used in Refs. 98 and 99. Those that are too close cannot be detected because of a dominant exchange coupling, resulting in splitting or broadening of the EPR spectra to such an extent that the nominal $\pi/2$ microwave pulse is ineffective in creating any transverse magnetization. Therefore, the EPR registration was "successful only after complete separation of the radicals," out of the "invisible cage" of radius R_0. Hence, the measured number of the radicals

$$R_{\text{exp}} = \int_{R_0}^\infty g(r,t) d^3 r \qquad (7.11)$$

may be essentially less than P when $R_0 \gg \sigma$. This number can not be found from the conservation law (7.10), but only from Eq. (7.11), through the straightforward integration of $g(r,t)$ found as a solution to Eqs. (7.7). The same distribution should be used for estimation of the width of the superfine lines of the EPR spectra. The distance between radicals determines the magnitude of the magnetic dipole–dipole interaction responsible for broadening of the superfine structure. Because of a space expansion of the pair distribution function $g(r,t)$, this contribution to the linewidth decreases with time improving the resolution of the ESR spectra.

To illustrate the main features of ion and radical generation and separation, let us refer to the simplest case of kinetic ionization when approximation (6.15) holds at all times. Using this approximation in Eq. (6.8), we can reduce it to the following:

$$P(t) = ck_0 \int_0^t \bar{\Omega}(t - \tau)e^{-\tau/t_i}d\tau \qquad (7.12)$$

Here $1/t_i = ck_0 + 1/\tau_D$ and

$$\bar{\Omega}(t) = \bar{\varphi} + [1 - \bar{\varphi}]e^{-t/t_e} \qquad (7.13)$$

proceeds exponentially in time as in Eq. (7.4), but the separation quantum yield is averaged over the normalized initial distribution of ions, $f_0(r)$. As shown in Eq. (6.17), the latter is identical in shape with $W_I(r)$ when ionization is kinetic. After integration we obtain from Eqs. (7.12) the following interpolation formula for the survival probability of charged particles:

$$P(t) = \psi\left[\bar{\varphi} + \frac{(1 - \bar{\varphi})e^{-t/t_e} + (1 - \bar{\varphi}t_i/t_e)e^{-t/t_i}}{1 - t_i/t_e}\right] \qquad (7.14)$$

where $\psi = k_0 c t_i = ck_0\tau_D/[1 + ck_0\tau_D]$ is the photoionization quantum yield in the case of kinetic electron transfer ($k_i = k_0$). If we replace here $\bar{\varphi}$ by its exponential model estimate

$$\varphi(\sigma) = k_{\text{sep}}t_e = \frac{k_{\text{sep}}}{k_{-et} + k_{\text{sep}}}$$

then Eq. (7.14) immediately reduces to Eq. (7.2), applied to kinetic ionization.

Now we are ready to compare the results of the unified theory with those obtained with the exponential model. In Fig. 56 the exciton quenching kinetics, $N(t) = \exp(-t/t_i)$, is shown together with the accumulation kinetics of ion radicals and neutral radicals, $P(t)$ and $R(t)$. As always, there can be two alternative cases of slow ($t_i \gg t_e$) and fast ionization ($t_i \ll t_e$). In the former case the concentration of radicals adiabatically follows excitation decay, so that both of them increase monotonously with the same rate $1/t_i$:

$$P(t) \approx \psi\bar{\varphi}\left[1 - e^{-t/t_i}\right], \qquad R(t) \approx \psi(1 - \bar{\varphi})\left[1 - e^{-t/t_i}\right] \qquad (7.15)$$

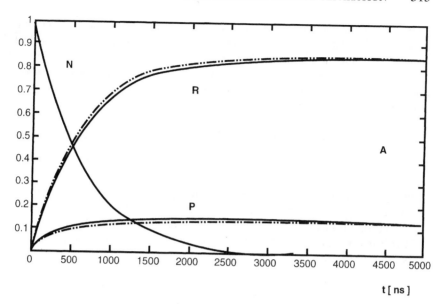

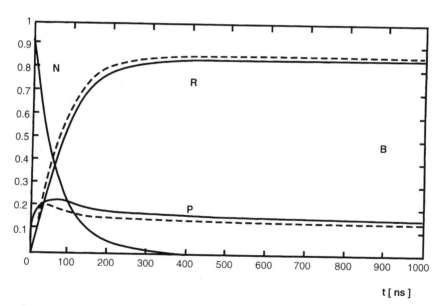

Figure 56. Kinetics of exciton quenching $N(t)$ and accumulation of ions $P(t)$ and radicals $R(t)$ at slow (A) and fast (B) kinetic ionization in normal region ($|\Delta G_i| = 0.9\lambda = 42$ T): (A) $W_0\sqrt{\lambda/T} = 1$ ns^{-1}, (B) 100 ns^{-1}. The remaining parameters are $c = 0.1$ M, $W_r = 10$ ns^{-1}, $D = \tilde{D} = \bar{D} = 10^{-5}$ cm^2/s, $r_c = 50$ Å, $\sigma = 5$ Å, $L = 1$ Å, $l = 0.03$ Å [17]. Dashed curves are an exponential approximation of the accumulation kinetics.

This rate is not sensitive to either recombination of charges or their Coulomb attraction. The latter plays the dominant role in the alternative case of fast ionization when the recombination and separation of charges occur much later then their creation. During the short ionization stage $P(t)$ quickly reaches its maximum and only then slowly approaches from above the final value $P(\infty) = \psi\bar{\varphi}$:

$$P(t) \approx \psi\left[e^{-t/t_e} - e^{-t/t_i} + \bar{\varphi}(1 - e^{-t/t_e})\right], \qquad R(t) \approx \psi(1 - \bar{\varphi})\left[1 - e^{-t/t_e}\right]$$

(7.16)

Although these model expressions reproduce qualitatively the time behavior of ion and radical concentrations, they deviate significantly from the exact results at long times when approaching the quantum yields. The asymptotic behavior is not well reproduced because it is not exponential. According to Refs. 85 and 97, the latter follows the power law:

$$P(t) - P(\infty) \propto \frac{1}{\sqrt{t}}$$

(7.17)

Within the available accuracy of numeric calculations, we confirm this expectation (see Fig. 57), which constitutes the main difference between the exact theory and the exponential model descriptions of reaction kinetics.

Unfortunately, these results cannot be applied to accumulation kinetics of free radicals generated in the photoreduction of anthraquinone by triethylamine in alcoholic solutions that was studied in Refs. 98 and 99. The ionization time reported there is $t_i = 0.8$ ns, while the buildup time constant of radical accumulation varies from 60 to 560 ns.

Moreover, in alcohols studied by Dinse et al. [98,99], the polarity is so high that these solvents can be scarcely considered as nonpolar. The increase in solvent polarity makes the Coulomb well smaller and facilitates evacuation of radicals. Since Coulomb stabilization does not suffice to provide significant delay in formation of radicals, their accumulation must end in a few nanoseconds but not last about 100 ns as it did, in fact. This is actually an indication that the real binding potential has much deeper well than that originating from the Coulomb attraction.

Dinse et al. fitted their data to a mysterious well having a depth of about 5–8 T, and the minimum shifted far away from contact (28–46 Å). More reasonable modeling of the binding potential has been proposed [97] to explain the same discrepancy between predicted buildup times and those observed experimentally [100]. This modeling was based on a statistical theory of mean force potential that oscillates at short distances and has a deep well near the contact [101]. This is an effect of the short-range liquid

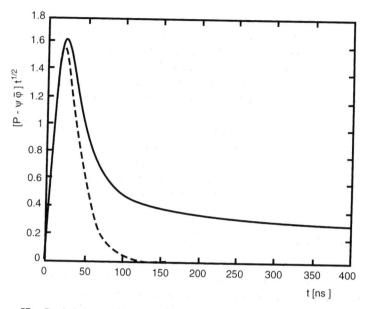

Figure 57. Deviations of $P(t)$ from its long-time limit $P(\infty) = \psi\bar{\varphi}$. Unlike exponential approximation (dashed line), the real kinetics (solid line) approaches the asymptotic result, which is a constant value in used anamorphosis [17].

structure that was described in other terms in Refs. [102–104]. Any complex formation should also produce a similar effect. In particular, the exciplex formation can hinder the separation of ions, and they are accumulated with the exciplex dissociation time $1/k_{diss}$, which may be much longer than t_e. The radical accumulation can be controlled by either exciplex dissociation or proton transfer, if the latter is still slower. Anyhow, the role of the exciplexes should be separately investigated, and we will devote Section IX to exciplexes.

C. Ion Accumulation Followed by Recombination to the Ground State

Quite the opposite problem arises from the application of the exponential model to another system studied [81]. We address the reaction between excited perylene (Per), which serves as acceptor of an electron and N,N-dimethyl-o-toluidine (o-DMT) as a donor:

$$D + {}^{1}A^{*} \longrightarrow {}^{1}[D^{+} \ldots A^{-}] \begin{array}{l} \nearrow D^{+} + A^{-} \\ \\ \searrow D \ldots A \end{array} \qquad (7.18)$$

This system was experimentally investigated among many others, by Mataga et al. [81]. The observed kinetics of ion separation consists of fast ionization followed by slower but sharp recombination. An excellent fit to this kinetics obtained with the exponential model in Mataga et al. [81] is a real surprise. Since the solution is of high polarity, one would more expect the power-law diffusional asymptotics of charge separation instead of the exponential one.

Since the quasiexponential evacuation from the well is likely when it is much deeper, we were in doubt as to whether the regular Coulomb law, $-r_c/r$, accounts appropriately for the shape and depths of the real well [18]. If the interion attraction $U_R(r)$ is actually different, one should use it instead of the Coulomb interaction in the generalized equation (6.5):

$$\dot{m}(r,t) = W_I(r)n(r,t)N(t) + \frac{\tilde{D}}{r^2}\frac{\partial}{\partial r}r^2 e^{-U_R/T}\frac{\partial}{\partial r}e^{U_R/T}m - W_R(r)m \quad (7.19)$$

Sometimes $U_R(r)$ is assumed to be a model potential that accounts not only for electrostatic attraction but also for the molecular structure of the solvent at short interion distances [103,104]. We also estimated the resulting structural corrections to either reactant or product distributions, but did not find them significant. These structural corrections result in a quasiharmonic behavior of the pair distributions near the contact, as if there were a few small wells in $U_R(r)$ at short distances. These peculiarities, which were taken into consideration, do not change the reaction kinetics qualitatively and may be ignored in our case.

On the contrary, the correction to the Coulomb well may be of more importance, making it deeper at the very contact, if the effect of nonlocal screening is taken into account. As has been known for a long-time, this can be done by taking into consideration the space dispersion of the dielectric constant. We took the effective dielectric constant from the simplest model accounting for the absence of screening at short distances [105]:

$$\epsilon(r) = \frac{\epsilon}{1 + (\epsilon/\epsilon_0 - 1)\gamma \exp(-r/\Lambda)} = \frac{\epsilon}{\Gamma(r)} \quad (7.20)$$

where $\epsilon_0 = 2$ is the optical dielectric constant, ϵ is its static value in the continuum and Λ is a fitting parameter. The correction for the excluded volume of finite-size particles, $\gamma = 2(\Lambda^2/\sigma^2)(\cosh(\sigma/\Lambda) - 1)$. As a result, the Coulomb potential has the following r dependence:

$$U_R(r) = \frac{r_c}{r}\Gamma(r) \quad (7.21)$$

This potential provides us with a sharp feature near the very contact that looks like an additional narrow and rather deep well. It originated from the same Coulomb attraction, but was not screened by intermediate solvent molecules. The depth of this well can be adjusted by the independent parameter Λ.

The space dispersion of ϵ also affects the ionization and recombination rates, changing the r dependence of ΔG and λ. The r dependence of the outer-sphere reorganization energy calculated for the potential (7.21) is given by the following formula [105]:

$$\lambda(r) = \lambda_c \left[2\Phi\left(\frac{\sigma}{\Lambda}\right) - \frac{\sigma}{r}\left(1 - \gamma e^{-r/\Lambda}\right) \right] \tag{7.22}$$

where $\Phi(x) = 1 - x^{-1}(1 - \exp(-x))$. If $\Lambda = 0$, then Eq. (7.22) reduces to the conventional Marcus result (2.6) with $\lambda_i = 0, \lambda_c = T(r_0 - r_c)/\sigma$ and $r_0 = e^2/(\epsilon_0 T)$ is determined by the optical dielectric constant, $\epsilon_0 = 2$. After a similar correction of the ionization free energy, its dependence on r looks as follows [18]:

$$\Delta G_I(r) = \Delta G_i + T\frac{r_c}{\sigma}\left[1 - \frac{\sigma}{r} + \gamma\left(\frac{\epsilon}{\epsilon_0} - 1\right)\left(e^{-\sigma/\Lambda} - \frac{\sigma}{r}e^{-r/\Lambda}\right)\right] \tag{7.23}$$

As before the free energies of ionization and recombination are bounded by the relationship

$$-\Delta G_I - \Delta G_R = \epsilon_0$$

where ϵ_0 is the energy of excitation.

Mataga et al. [81] performed the experiment in acetonitrile solution, $\epsilon \simeq 40$, so that $r_c = 14 \text{ Å}$ at room temperature. Perylene (Per) served as an electron acceptor in reaction (7.18) was known to have the following parameters:

$$\sigma = 7 \text{ Å}, \qquad L_{I,R} = 1 \text{ Å}, \qquad c = 0.4 \text{ M}, \qquad -\Delta G_r = 2.44 \text{ eV}$$

For the level structure and excitation lifetime of perylene, we borrowed data from Ref. 106:

$$\epsilon_0 = 2.82 \text{ eV}, \qquad \tau_A = 7 \text{ ns}, \qquad \hbar\omega = 43.5 \text{ eV}$$

The phenomenologic reaction rates reported by Mataga et al. [81] are

$$k_{-et} = 0.68 \text{ ns}^{-1} \quad \text{and} \quad k_{sep} = 0.6 \text{ ns}^{-1}$$

To reach the best correspondence of the exponential model to experiment, values must be in accordance with the encounter diffusion coefficient $\tilde{D}$ and the rate constant for recombination, k_c, defined in Eq. (2.4). Such a correspondence was established in Eqs. (3.4) and (5.17). Using the aforementioned data for k_{sep} and k_{-et} we can deduce the following from these equations:

$$\tilde{D} = 3.1 \cdot 10^{-6} \text{cm}^2/\text{s} \quad \text{and} \quad k_c = 977 \, \mathring{A}^3/\text{ns} = 5.9 \cdot 10^8 \, \text{L}/(\text{Ms})$$

The stationary rate constant of ionization can be estimated from the phenomenologic ionization rate $K_q = ck_i$ measured [81]. It was found in these data that $k_i = 1.19 \cdot 10^4 \, \mathring{A}^3/\text{ns} = 7.2 \times 10^9 \, \text{L}/(\text{Ms})$. For the total excitation decay rate $1/t_i = ck_i + 1/\tau_A$, we obtained the value of 3ns^{-1} that was extracted from the plot of the reaction kinetics presented in Ref. 81.

It is a pity that a lot of necessary input data, such as diffusion coefficient, ionization rate, acceptor lifetime, and even its concentration c, were not reported in this work. Therefore, we had to find them indirectly or in other sources. On the other hand, for this very reason we used the aforementioned estimates as only a starting point for the fitting procedure and did not restrict ourselves by fixing these values. It should be stressed, nevertheless, that only approaching kinetic data with ab initio estimated rates, can one realize that there is a serious problem concerning interpretation of experimental results.

The most serious problem arises when the single-channel rate (2.5) is applied to both ionization and recombination for calculating the kinetics of ion accumulation. With the parameters listed above, the latter appears to be monotonic and much slower than the real one that goes through the maximum and was well approximated by the exponential model (solid line in Fig. 58). In the case of a single-channel electron transfer ($S = 0$), due to a very large free energy gap $|\Delta G_r|$, the recombination rate in this system is completely suppressed by the Arrhenius factor in Eq. (2.5). Since there is practically no recombination, $\Omega \approx 1$ and $\dot{\Omega} \approx 0$, the last term in Eq. (7.3) turns to 0 and accumulation of ions becomes monotonic (dashed line in Fig. 58).

To get the pronounced maximum in the accumulation kinetics within the unified theory, we have to accelerate the charge recombination. To speed up this process, one can try to increase the exchange interaction, but it is already about the upper limit for tunneling rates, $V_{I,R}^0 \sim 10^3 \text{ns}^{-1}$. The only alternative way to reach the same goal is to diminish the free-energy gap for recombination. This can be done assuming that the backward electron transfer (BET) is a multichannel process assisted by the vibronic modes of the neutral products (see Section II.B and Fig. 59a). Since $|\Delta G_r| \gg \lambda_c$, the single-channel recombination is strongly inverted and the maximum of the $W_R(r)$ distribution is shifted far from the contact (Fig. 60). If a number of

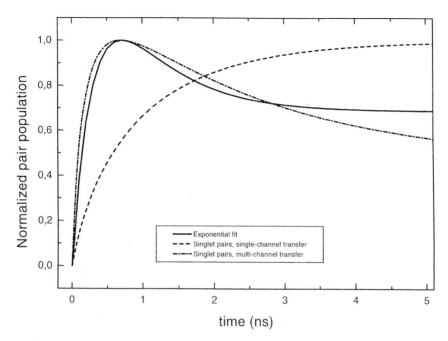

Figure 58. Time development of the normalized population of ion pairs. The solid line is the best fit to the experimental data in Ref. 81 within the exponential model. The dashed line is the ion accumulation calculated with the unified theory, but accounting only for the one-channel forward and backward transfers between singlet states. The dash–dotted line is the same, but for the multichannel recombination accounting for the vibronic state of the neutral products [18].

vibronic states are involved in the BET in polar solvents, the maximum moves closer to contact and its height increases correspondingly. In Fig. 60 we plot the ionization and recombination rates for the different number of vibronic states involved. The larger S is, the higher the states are involved in electron transfer and the stronger becomes the recombination of ions. However, the ionization rate decreases in parallel, if at the same time the vibronic states of the ions are taken into account. These results are different from those depicted in Fig. 10. There the ionization was initially even more "inverted" than recombination, while here it is almost normal from the beginning (typical NI case). Therefore the ionization rate, $W_I(r)$, decreases simultaneously with an increase in recombination rate, $W_R(r)$, when the number of assisting quantum states grows. As a result the ionization becomes even slower than it was in the case of a single-channel transfer, and the problem only intensifies.

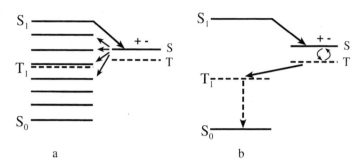

Figure 59. Schemes of ion-pair recombination via the multichannel backward electron transfer to the ground state (*a*) and a single-channel recombination to triplet excitation of the neutral product (*b*).

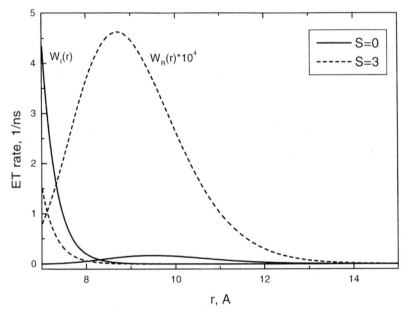

Figure 60. Ionization rates $W_I(r)$ (monotonous curves) and recombination rates $W_R(r) \times 10^4$ (bell-shaped curves) at different electron–phonon interactions (Frank–Condon factors for assisting vibrational mode, S); $\epsilon(r) = $ const [18].

To find some exit from such a difficult situation, we made an artificial assumption that the recombination is a multichannel transfer while ionization remains a single channel reaction and maintains the fast accumulation of ions. Huge asymmetry between forward and backward ET arises then, which does not seem to be physically reasonable. Only for the sake of completeness, we illustrate the aforementioned situation in Fig. 58. For this illustration we have taken the spatial dispersion $\epsilon(r)$ into account only in the potential (7.21) that enters the diffusional equation (7.19), but used $\epsilon = \text{const}$ for calculation of $\lambda(r)$ and $\Delta G(r)$. The deep electrostatic well near the contact significantly affects the relative motion of the charged products. However, when ET rates are calculated, the $\epsilon(r)$ dependence affects mainly the magnitude of $W_{I,R}(r)$ that can be easily compensated for by varying the fitting parameters, $V_{I,R}^0$. Those used in Eqs. (2.16) and (2.17) for calculating kinetics (dash–dotted curve shown in Fig. 58) were:

$$V_I^0 = 7500\,\text{ns}^{-1}, \qquad V_R^0 = 3400\,\text{ns}^{-1}$$

$$D = \tilde{D} = 3.1 \times 10^{-6}\,\text{cm}^2/\text{s}$$

$$\Lambda = 1.4\,\text{Å}, \quad \hbar\omega = 1.7\,\text{T}$$

$$\text{In recombination}: \quad \lambda_q = 24.1\text{T} \quad (S = 14.2)$$

$$\text{In ionization}: \quad \lambda_q = 0 \quad\quad (S = 0)$$

With such a large number of fitting parameters, the set reported above is not unique, and practically the same fits can be obtained with some parameters varied. However, we must stress that the fit cannot be improved. Fortunately the theory is not flexible at all. The falloff after the maximum cannot be made sharper by any means because the curve has a distinct, slowly decaying diffusional tail as predicted in Eq. (7.17).

D.   Recombination via Ion Pair Undergoing Singlet–Triplet Conversion

Although we made ionization fast and got a maximum in the $P(t)$ dependence, interpolation of the experimental data is still not satisfactory. Moreover, for this partial success we paid too high a price, suggesting different mechanisms for forward and backward electron transfers. Fortunately, there is a unique way to eliminate such an asymmetry, if we take into consideration the spin states of the ion pair. The forward electron transfer from the excited singlet state produces the ion pair in its singlet state as well. However, the spin conversion from the singlet to triplet state of the ion pair opens up the way to geminate recombination from the latter to the excited triplet state of the neutral particles (Fig. 59b):

$$D + {}^1A^* \longrightarrow {}^1[D^+\ldots A^-] \to D^+ + A^- \qquad (7.24a)$$

$$\downarrow\uparrow$$

$$D + {}^3A^* \longleftarrow {}^3[D^+\ldots A^-] \to D^+ + A^- \qquad (7.24b)$$

This reaction way should manifest itself by a pronounced magnetic-field dependence of the ion recombination and separation [107–109]. In fact, in Ref. 110 a magnetic-field effect was observed for one of the systems studied in the work under consideration [81] (pyrene $+N,N$-dimethylaniline in acetonitrile solution). This had been the straightforward experimental indication that spin effects should be taken into account in these systems. Surprisingly, they were ignored until the late 1990s.

To investigate the BET to the excited triplet state of perylene, we start with the assumption that the spin conversion is very fast. In this case, the distribution of the ion pairs between the singlet and triplet states is at equilibrium at any time and the ratio of their populations, P_T and P_S, reflects the threefold degeneracy of the triplet state, in the absence of a magnetic field,

$$P_T = 3P_S \qquad (7.25)$$

As has been proved a single-channel recombination from the singlet ion pair to the ground state is so slow that it could be neglected in the scheme (7.24). Therefore, to account for BET from only triplet states of the ion pair, we need only substitute $\tilde{W}_R(r) = (3/4)W_R(r)$ for $W_R(r)$ in the conventional unified theory. The latter is the ET rate calculated for the free-energy gap between the ion pair and molecular triplet states ΔG_r^t, whose modulus is now smaller: $|\Delta G_r^t| = 1.32\,\mathrm{T} \ll |\Delta G_r|$.

Having included all the effects of nonlocal screening, we were able to obtain more physically reasonable fitting parameters:

$$V_I^0 = 145\,\mathrm{ns}^{-1}, \qquad \tilde{V}_R^0 = 135\,\mathrm{ns}^{-1}, \qquad \Lambda = 1.5\,\text{Å} \qquad (7.26)$$

Although the theory became more consistent, one can see from Fig. 61 that the fit does not improve anymore, compared to that obtained previously. The extension of this theory to multichannel electron transfer does not make it fit better, but results solely in an insignificant change of the numerical values of the parameters.

A too-smooth decrease of the ion population after passing the maximum is incompatible with a much sharper and faster approach to the asymptotic value obtained with the exponential model, not to mention a too slow recombination tail, which is characteristic of any consistent diffusional theory. The only possibility to improve the fit is to consider a *two*-component system, with one component rapidly decaying, and the other one gradually growing, at the same time preventing the full disappearance of ions at earlier times and compensating for the slow diffusional decay of the tail, remaining from the fast component.

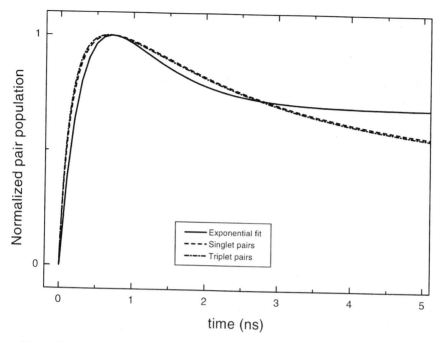

Figure 61. Normalized population of ion pairs. Solid line — best analytic presentation of the experimental data from Ref. 81; dashed line — best fit for backward electron transfer from the singlet ion pair to the ground state (the same as in Fig. 58); dot–dashed line — best fit for the backward electron transfer through the triplet ion pair to the excited triplet of the neutral product [18].

This idea was first proposed in Ref. 107, where the one component was attributed to exciplex (contact ion pair) formation while the other to free ion accumulation. The authors of Ref. 107 claimed that only in this case were they able to simulate the fast fall off following the maximum in the total ion population. To make this model treatable, the authors of Ref. 107 assumed that geminate ion recombination is instantaneous in comparison with other times. This assumption does not appear to be justified for the present system. If there is no exciplex formation as shown in Ref. 111, then only delayed geminate recombination of the triplets, postponed by the spin conversion from singlets, can explain the results obtained. Instead of exciplexes and free ions, the two components under our consideration are ion pairs in singlet and triplet states correspondingly, whose total number is the sum of the sharp and smooth curves developed in the timescale of geminate events.

Using the simplest method of accounting for the spin conversion exploited in Refs. 112–116, we must introduce the rates of forward and backward stochastic spin transitions between singlet and triplet states:

$$k_{ST} = 3k_{TS} \qquad (7.27)$$

Their ratio accounts for the relative singlet and triplet states multiplicity noted in Eq. (7.25). Now we can write the kinetic equations for the state populations, accounting for pumping the singlet and recombination from the triplet along with spin transitions between them:

$$\frac{\partial}{\partial t} m_S(r,t) = \frac{\tilde{D}}{r^2} \frac{\partial}{\partial r} r^2 e^{-U_R/T} \frac{\partial}{\partial r} e^{U_R/T} m_S(r,t) + W_I(r)n(r,t)N(t)$$

$$- k_{ST} m_S(r,t) + k_{TS} m_T(r,t) \qquad (7.28a)$$

$$\frac{\partial}{\partial t} m_T(r,t) = \frac{\tilde{D}}{r^2} \frac{\partial}{\partial r} r^2 e^{-U_R/T} \frac{\partial}{\partial r} e^{U_R/T} m_T(r,t) - W_R(r)m_T(r,t)$$

$$+ k_{ST} m_S(r,t) - k_{TS} m_T(r,t) \qquad (7.28b)$$

In Fig. 62 we plot the best fit of $P(t)$ to the experiment (which is rather good finally), together with constituting the fit total populations of singlet and triplet ion pairs, $P_S = \int m_S d^3 r$ and $P_T = \int m_T d^3 r$. The following parameters were taken for a single-channel electron transfer: $V_I^0 = 182\,\text{ns}^{-1}$, $V_R^0 = 390\,\text{ns}^{-1}, D = \tilde{D} = 4 \times 10^{-5}\,\text{cm}^2/\text{s}, \Lambda = 2.75\,\text{Å}, k_{TS} = 0.8\,\text{ns}^{-1}$. These parameters look physically reasonable, although we have too many of them to make a unique choice. Besides, refining the theory by better accounting for the solvent effects and more realistic (Hamiltonian) dynamics of the spin conversion will have an impact on the values of the parameters. It would also be better to operate with more diluted systems then that studied in Ref. 81 at excessively high concentrations ($c \sim 0.3$–$0.5M/L$).

The main result, however, should remain untouched. Only coexistence of the singlet ion pair, subjected to a fast spin conversion and the triplet one accumulated through this channel, allows us to obtain a very good fit to the experimental data. The sharp decrease in total amount of ions actually reflects an exponential in the time spin conversion, which does not affect much longer diffusional tails of both ion fractions.

It is common knowledge that the photoionization and geminate recombination of ion radicals affect the spin polarization of the reaction products and constitute numerous magnetic field effects [108,109,117,118]. Here we see the feedback: the kinetics of charge separation itself depends on the spin dynamics during geminate recombination. This is an important conclusion that places in doubt all the results obtained with full ignorance of the spin states of the radical products. They remain valid only in the limit of fast spin conversion during the encounter, which is unlikely to be the case.

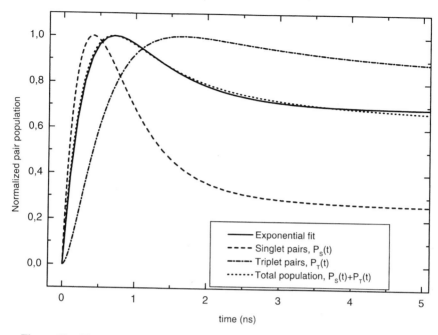

Figure 62. Normalized population of ion pairs at spin conversion control: our fit (dotted line) to the experimental data from Ref. 81 (solid line). Dashed line — simulated population of singlet ion pairs; dot–dashed line — the same for the triplet ion pairs [18].

Otherwise, the spin dynamics can affect not only the kinematics of the charge separation but also the separation quantum yield.

In particular, the yield should be different in the zero magnetic field that we have considered and in the very large magnetic field. In the latter case the transitions to highly split T_{+1} and T_{-1} sublevels, induced by superfine interaction, are switched off and these components of the triplet state do not participate in the process. Since only transitions between the singlet and T_0 sublevel still occur, their rate in both direction should be the same:

$$k_{ST} = k_{TS} \qquad (7.29)$$

Comparing this relationship to Eq. (7.27), we see that the pumping of the triplet is weaker and its population smaller. As a result, more ions can escape recombination and the total charge separation yield should be higher. Keeping all parameters the same, we repeated the calculations but substituted Eq. (7.29) for Eq. (7.27). The results, presented in Fig. 63, confirmed our expectations—the magnetic-field effect (MFE) in the charge separation

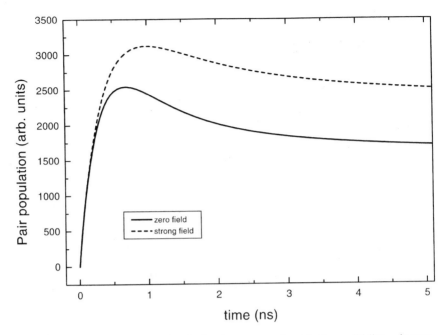

Figure 63. Magnetic field-effect in the charge separation kinetics: solid line — in zero field (the same as dotted one on Fig. 62); dashed line — in high magnetic field [18].

is above 40%. We will obtain similar effects in the next two sections in other systems.

VIII. SEPARATION OF TRIPLET ION PAIR

The discovery of magnetic field effects in liquid-phase chemistry marked the beginning of the active EPR and NMR study of photochemical radical reactions, which has become an independent line of investigations known as "spin chemistry" [108,109]. Using very sophisticated methods of magnetic spectroscopy to solve the spin part of the problem, most if not all of the authors simplified as much as possible the kinematic aspect of radical ion pair (RIP) evolution. This part was usually considered in the framework of the notorious "exponential model." As was shown above, the model attributes the Poisson distribution to the lifetimes of the RIPs, ignoring their initial distribution in interparticle distances and the diffusional kinematics of the motion [120].

Nonetheless, the exponential model widely accepted in photochemistry was useful at the beginning. Making optical registration of excited reactants and their charged products, researchers were able to inspect the model's predictions regarding the FEG law in the fluorescence and quantum yield of separated ions, ignoring their spin states. In the previous section we proved that this is impossible if the spin conversion in RIPs controls the charge recombination and separation. This is even more correct when ions are produced from the triplet state of the reactants, so that RIP appears also in a triplet (Fig. 64):

$$^3D^* + A \Longrightarrow {}^3[D^+ \ldots A^-] \qquad (8.1)$$

As usual, the ion pair may be separated by diffusion, if it escapes geminate recombination to the ground state. The latter is allowed from the singlet state, but forbidden from the triplet one. Therefore, the triplet–singlet conversion in RIPs is a necessary step that makes recombination possible and leads to an essential decrease in the quantum yield of separated (survived) ions. The process proceeds according to the kinetic scheme proposed in Ref. 83 for the reaction of photoexcited Ru-trisbipyridine (D) with methylviologen as an electron acceptor (A) (Fig. 65):

$$D^+ + A^- \leftarrow {}^3[D^+ \ldots A^-] \rightleftharpoons {}^1 [D^+ \cdots A^-] {\Large \langle}{\begin{array}{l} \nearrow D^+ + A^- \\ \searrow {}^1[D \ldots A] \end{array}} \qquad (8.2)$$

Changing the rate of spin conversion, one can change the ratio of triplet and singlet products of the reaction, as well as the share of survived ions and those recombined during the geminate stage.

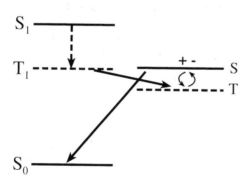

Figure 64. Scheme of triplet ionization and subsequent triplet ion-pair recombination to the ground state following its conversion to a singlet state.

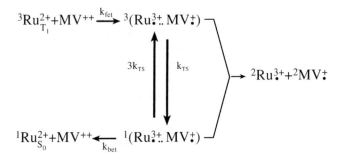

Figure 65. Reaction scheme for the forward and backward electron transfer (with contact rates $k_{fet} \equiv ck_i$ and $k_{bet} \equiv k_{-et}$ correspondingly) assisted by the spin conversion with a rate $k_{TS} \equiv k_s$ that suppresses the ion-pair separation in either the triplet or singlet state [19].

A. Spin Conversion Rate

The mechanism of spin conversion in a chosen system is not the superfine interaction, but the individual spin relaxation in both radical ions assumed to proceed with the same longitudinal and trasverse relaxation times T_1 and T_2. We will concentrate on the simplest example of the Δg mechanism of spin conversion in an ion pair with moderate magnetic fields.

In general, the spin evolution is described by the kinetic equation for the pair density matrix $\hat{\rho}$ of RIP:

$$\dot{\hat{\rho}}(t) = -i[\mathbf{H}, \hat{\rho}] + \hat{\mathbf{R}}\hat{\rho} \qquad (8.3)$$

where $\mathbf{H}$ is the spin Hamiltonian and $\hat{\mathbf{R}}$ is the relaxation superoperator, accounting for the population and phase relaxation in the collective states basis of the ion pair ($|T_+\rangle, |T_-\rangle, |T_0\rangle$ and $|S\rangle$).

The spin Hamiltonian in such a basis takes the form

$$\mathbf{H} = \frac{1}{2}\begin{pmatrix} \omega_+ & 0 & 0 & 0 \\ 0 & -\omega_+ & 0 & 0 \\ 0 & 0 & 0 & \omega_- \\ 0 & 0 & \omega_- & 0 \end{pmatrix} \qquad (8.4)$$

where

$$\omega_\pm = \frac{(g_+ \pm g_-)\beta}{\hbar}H \qquad (8.5)$$

β is the Bohr magneton, g_+ and g_- are g factors of positive and negative ions and H is the magnetic field. The relaxation superoperator in the Liouville space that has the rank 16×16 can be taken from Ref. 121. It should and will be used later to account for an arbitrary strong magnetic field.

Fortunately, for moderate magnetic fields, when

$$\omega_- \ll \frac{1}{T_2} \tag{8.6}$$

the quasistationary solution for off-diagonal elements may be obtained by setting

$$\dot{\rho}_{ij} = 0 \quad \text{for} \quad i \neq j \tag{8.7}$$

in the general kinetic equations (8.3). Using this solution in the rest of the equations, one can transform them into the set of rate equations for only the diagonal elements of the density matrix:

$$\dot{\rho}_{SS} = \left(\frac{\omega_-^2 T_2}{4} + \frac{1}{T_2} - \frac{1}{2T_1} \right) \rho_{T_0 T_0} - \left(\frac{\omega_-^2 T_2}{4} + \frac{1}{T_2} + \frac{1}{2T_1} \right) \rho_{SS}$$
$$+ \frac{\rho_{T_+ T_+} + \rho_{T_- T_-}}{2T_1} \tag{8.8a}$$

$$\dot{\rho}_{T_+ T_+} = \frac{\rho_{SS} + \rho_{T_0 T_0}}{2T_1} - \frac{\rho_{T_+ T_+}}{T_1} \tag{8.8b}$$

$$\dot{\rho}_{T_0 T_0} = \left(\frac{\omega_-^2 T_2}{4} + \frac{1}{T_2} - \frac{1}{2T_1} \right) \rho_{SS} - \left(\frac{\omega_-^2 T_2}{4} + \frac{1}{T_2} + \frac{1}{2T_1} \right) \rho_{T_0 T_0}$$
$$+ \frac{\rho_{T_+ T_+} + \rho_{T_- T_-}}{2T_1} \tag{8.8c}$$

$$\dot{\rho}_{T_- T_-} = \frac{\rho_{SS} + \rho_{T_0 T_0}}{2T_1} - \frac{\rho_{T_- T_-}}{T_1} \tag{8.8d}$$

In moderate magnetic fields we can assume that

$$T_2 = T_1 = T \tag{8.9}$$

and both of them are field-independent, as in the case of the rotational relaxation mechanism at $g_\pm \beta H \ll \hbar / \tau_\theta$, where τ_θ is the orientational relaxation time [119]. For equal relaxation times, the general set of equations (8.8) reduces to only two equations for the total populations of the triplet state, $m_T = \rho_{T_+ T_+} + \rho_{T_0 T_0} + \rho_{T_- T_-}$, and that of singlet state, $m_S \equiv \rho_{SS}$:

$$\dot{m}_S = -3k_s m_S + k_s m_T \tag{8.10a}$$
$$\dot{m}_T = 3k_s m_S - k_s m_T \tag{8.10b}$$

with a microscopically defined spin conversion rate

$$k_s = \left[\frac{1}{2T} + \frac{\omega_-^2 T}{12} \right] = k_{TS} = \frac{k_{ST}}{3} \tag{8.11}$$

simply related to two others used in Eqs. (7.28). According to the inequality (8.6), the second term in Eq. (8.11) is a small correction sensitive to the magnetic field. Hence, k_s has an order of magnitude of the inverse relaxation time and quadratic dependence on the magnetic field, though it changes only slightly within the stochastic (rate) description of the spin conversion obtained in Eq. (8.10).

B. Triplet Ionization Followed by Stochastic Spin Conversion

At first the charge separation was studied in systems where recombination is not spin-forbidden, because the reaction proceeds from the excited singlet via singlet RIP to the ground state. Therefore, in the original *unified theory* we also operated with a single pair distribution function of ions, $m(r, t)$, which describes their accumulation, resulting from ionization, and recombination proceeding with a rate $W_R(r)$. However, in the previous section we were compelled to discriminate between distributions of singlet and triplet RIPs, m_S and m_T, because the former was pumped by bimolecular ionization, while recombination is possible only from the latter. Here the situation is quite the opposite: the triplet RIPs are products of ionization, while only the singlet RIPs are allowed to recombine. Taking into account the spin conversion with the position independent rate k_s, we can rewrite a system of differential equations (7.28) for the present case:

$$\frac{\partial}{\partial t} m_S(r, t) = \frac{\tilde{D}}{r^2} \frac{\partial}{\partial r} r^2 e^{r_c/r} \frac{\partial}{\partial r} e^{-r_c/r} m_S(r, t) - W_R(r) m_T(r, t)$$
$$- 3k_s m_S(r, t) + k_s m_T(r, t) \tag{8.12a}$$

$$\frac{\partial}{\partial t} m_T(r, t) = \frac{\tilde{D}}{r^2} \frac{\partial}{\partial r} r^2 e^{r_c/r} \frac{\partial}{\partial r} e^{-r_c/r} m_T(r, t) + W_I(r) n(r, t) N(t)$$
$$+ 3k_s m_S(r, t) - k_s m_T(r, t) \tag{8.12b}$$

From now on we can ignore the r dependence of ϵ, noted in (7.20), because it did not prove to be very essential. The initial and boundary conditions for both distributions are the same:

$$m_T(r, 0) = m_S(r, 0) = 0, \qquad \frac{\partial}{\partial r} e^{-r_c/r} m_T(r, t) \Big|_{r=\sigma} = \frac{\partial}{\partial r} e^{-r_c/r} m_S(r, t) \Big|_{r=\sigma} = 0 \tag{8.13}$$

The total survival probability of the ionized products is

$$P(t) = c \int_\sigma^\infty [m_T(r,t) + m_S(r,t)] 4\pi r^2 dr \qquad (8.14)$$

and $P(\infty) = \phi$ as always.

The solution to Eqs. (8.12) may be obtained in the following form:

$$\begin{pmatrix} m_T(r,t) \\ m_S(r,t) \end{pmatrix} = \int_0^t dt' \int d^3r' \begin{pmatrix} p_T(r,r',t-t') & p_S'(r,r',t') \\ p_S(r,r',t-t') & p_T'(r,r',t') \end{pmatrix} \begin{pmatrix} W_I(r')n(r',t')N(t') \\ 0 \end{pmatrix} \qquad (8.15)$$

We need only p_T and p_S, which are the Green functions for triplet and singlet RIPs originating from an initially excited triplet state, because another pair, p_T' and p_S', are the similar functions but for RIPs produced from a singlet excitation. From Eqs. (8.14) and (8.15) one can easily obtain

$$P(t) = c \int d^3r \int d^3r' W_I(r') \int_0^t p(r,r',t-t')n(r',t')N(t')dt' \qquad (8.16)$$

Here the total charge distribution in RIP

$$p(r,r',t-t') = p_T(r,r',t-t') + p_S(r,r',t-t') \qquad (8.17)$$

is composed of triplet and singlet components that yield the homogeneous differential equations

$$\frac{\partial}{\partial t} p_T = -k_s p_T + 3k_s p_S + \tilde{D} \frac{1}{r^2} \frac{\partial}{\partial r} r^2 e^{r_c/r} \frac{\partial}{\partial r} e^{-r_c/r} p_T \qquad (8.18a)$$

$$\frac{\partial}{\partial t} p_S = k_s p_T - 3k_s p_S + \tilde{D} \frac{1}{r^2} \frac{\partial}{\partial r} r^2 e^{r_c/r} \frac{\partial}{\partial r} e^{-r_c/r} p_S - W_R(r) p_S \qquad (8.18b)$$

with the same reflecting boundary conditions as in Eq. (8.13), but with another initial conditions peculiar to Green functions:

$$p_T(r, r_0, 0) = \frac{\delta(r - r_0)}{4\pi r^2}, \qquad p_S(0) = 0 \qquad (8.19)$$

The free ion quantum yield $\phi = P(\infty)$ resulting from Eqs. (8.16) and (4.22)

$$\phi = c \int d^3r \int d^3r_0 p(r, r_0, \infty) m_0(r_0) = \psi \bar{\phi} \qquad (8.20)$$

is again a product of the total ionization quantum yield ψ, given by Eq. (4.25) and the charge separation quantum yield

$$\varphi(r_0) = \int_\sigma^\infty p(r, r_0, \infty) 4\pi r^2 dr \tag{8.21}$$

averaged over a normalized initial distribution of ions (4.23):

$$\bar{\varphi} = \int \varphi(r_0) f_0(r_0) d^3 r_0 \tag{8.22}$$

C. Contact Recombination

The main difficulties encountered in solving the set of equations (8.18), results from the complex space dependence of the recombination rate $W_R(r)$. However, at low exergonicity (in the normal region) the recombination can be considered as contact, so that its rate given in Eq. (2.21) is expressed through the single parameter k_c. Using this parameter one can omit the last term in Eq. (8.18b), but account for recombination through the radiation boundary condition different from that given in Eq. (8.13):

$$\left(\frac{\partial p_T}{\partial r} + \frac{r_c}{r^2} p_T \right) \Bigg|_{r=\sigma} = 0, \qquad 4\pi \tilde{D} \sigma^2 \left(\frac{\partial p_S}{\partial r} + \frac{r_c}{r^2} p_S \right) \Bigg|_{r=\sigma} = k_c p_S(\sigma, t) \tag{8.23}$$

In this approximation the rigorous solution for $\varphi(r_0)$ obtained in Ref. 19 was expressed through the Green function G_0, defined by Eqs. (5.5)–(5.7):

$$\varphi(r_0) = 1 - \frac{(1/4) k_c [\tilde{G}_0(\sigma, r_0, 0) - \tilde{G}_0(\sigma, r_0, 4k_s)]}{1 + (3/4) k_c \tilde{G}_0(\sigma, \sigma, 4k_s) + (1/4) k_c \tilde{G}_0(\sigma, \sigma, 0)} \tag{8.24}$$

Let us first correct our previous results that were obtained when we did not discriminate between spin states. This can be done only in the limit of fast conversion ($k_s \to \infty$), when the latter does not control the rate of the whole process. When k_s is large enough

$$\tilde{G}_0(\sigma, r_0, 4k_s) \approx \frac{G_0(\sigma, r_0, 0)}{4k_s} \quad \text{and} \quad \tilde{G}_0(\sigma, \sigma, 4k_s) \approx \frac{G_0(\sigma, \sigma, 0)}{4k_s}$$

may be neglected as small corrections and Eq. (8.24) reduces to the following:

$$\varphi(r_0) = 1 - \frac{(k_c/4)\tilde{G}_0(\sigma, r_0, 0)}{1 + (k_c/4)\tilde{G}_0(\sigma, \sigma, 0)} = 1 - R_0(r_0) \qquad (8.25)$$

where the recombination quantum yield.

$$R_0(r_0, \alpha) = \frac{\alpha\tilde{G}_0(\sigma, r_0, 0)}{1 + \alpha\tilde{G}_0(\sigma, \sigma, 0)} \qquad (8.26)$$

is the same as in Eq. (5.13), but with $\alpha = \frac{k_c}{4}$ instead of k_c.

Equation (8.25) establishes the lower limit for the charge separation quantum yield, which is almost the same as in the spinless theory (Section V.B), except that an effective reaction constant in Eq. (8.24), $k_c/4$, is one-fourth as much as in Eq. (5.13), because of the equilibrium weight of the only reactive (singlet) state. With this reservation one may rewrite the quantum yield as in Eq. (5.15) and confirm Eqs. (5.19) and (5.20), although again $k_c/4$ must be substituted for k_c:

$$z = \frac{(k_c/4)\left[e^{r_c/\sigma} - 1\right]}{4\pi r_c} \rightarrow \frac{k_c}{16\pi\sigma} \quad \text{at} \quad r_c \rightarrow 0 \qquad (8.27)$$

Hence, we confirm the classification on kinetic and diffusional geminate recombination made in Eq. (5.21), as well as the rough explanation of a particular $Z(\tilde{D})$ dependence (Fig. 33), related to diffusional control. However, now we have an additional factor that can control the same process. This is the spin conversion that was expected in Section V to be able to essentially modify the results if it becomes too slow. By analyzing this case as an alternative to the fast spin conversion limit, we will show that spin conversion may be responsible for a nonlinear $Z(\tilde{D})$ dependence shown in Fig. 33.

D. Spin Control in Highly Polar Solutions

To simplify the analysis (which still remains rather complex), we will restrict ourselves to a highly polar solution, setting $r_c = 0$. Then we can proceed further with the analytic calculations because the important Green function for this particular case is known [see Eq. (4.11) in Ref. 122]:

$$\tilde{G}_0(\sigma, r_0, s) = \frac{1}{4\pi r_0 \tilde{D}} \frac{\exp\left[-(r_0 - \sigma)\sqrt{s/\tilde{D}}\right]}{1 + \sigma\sqrt{s/\tilde{D}}} \qquad (8.28)$$

By substituting Eq. (8.28) into Eq. (8.24), we obtain the following for highly polar solutions:

$$\varphi(r_0) = 1 - \frac{(k_c/4)[1 - \exp\{-y\}/(1+x)]}{4\pi r_0 \tilde{D} + (k_c/4)(r_0/\sigma)[1 + 3/(1+x)]} \qquad (8.29)$$

Here

$$y = \frac{r_0 - \sigma}{\sigma}x \quad \text{and} \quad x = \sqrt{\frac{4k_s\sigma^2}{\tilde{D}}} = \sqrt{\kappa \frac{k_c}{k_{\tilde{D}}}} \qquad (8.30)$$

where $k_{\tilde{D}} = 4\pi\sigma\tilde{D}$ and

$$\kappa = \frac{16\pi\sigma^3 k_s}{k_c} = 12\frac{k_s}{k_{-et}} \qquad (8.31)$$

Parameter y is a measure of the spin conversion during delivery time from the initial distance to the contact, $(r_0 - \sigma)^2/\tilde{D}$, while another parameter, x, measures the same, but during encounter time, $\sigma^2/\tilde{D}$. In the limiting case of ultrafast spin conversion, both these parameters turn to ∞ and Eq. (8.29) reduces to the spin-independent result, equivalent to Eq. (8.25):

$$\varphi(r_0) = 1 - \frac{(k_c/4)}{4\pi r_0 \tilde{D} + (k_c/4)(r_0/\sigma)} \quad \text{at} \quad y, x \gg 1 \qquad (8.32)$$

At any finite rate of spin conversion, the latter controls the recombination but in a different way, depending on whether it is fast ($\kappa \gg 1$) or slow ($\kappa \ll 1$). One should also discriminate between

- Distant starts, when $r_0 > 2\sigma$ and hence $y > x$
- Close starts, when $\sigma < r_0 < 2\sigma$ and hence $y < x$

All the differences are seen pretty well in Fig. 66. The spin conversion affects the results when y and/or x become small. Therefore, at large κ only the very right part in Fig. 66a is affected, at $k_{\tilde{D}}/k_c > [(r_0 - \sigma)/\sigma]^2\kappa$, where $y < 1$. There all descending branches of the curves shown on Fig. 66A stick together. This happens over the right edge of the kinetic plateau and the later, the larger is the starting distance. Only then the recombination comes under conversion control and slows down with diffusion. The nature of this effect has a common origin for any starting conditions and will be explained later.

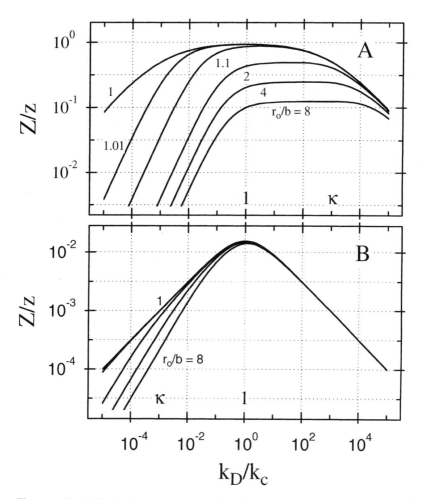

Figure 66. Diffusional dependence of Z in contact approximation with $k_c = \int W_R(r)d^3r = 628 \, \text{Å}^3/\text{ns}$, $\tilde{D} = 5 \times 10^{-5} \, \text{cm}^2/\text{s}$. Calculated from Eq. (8.29) for different initial separations (from bottom to top: $r_0/\sigma = 8, 4, 2, 1.1, 1.01, 1$) at (a) fast spin conversion, ($\kappa = 10^3$, $\sigma = 5\text{Å}$) and (b) slow spin conversion ($\kappa = 10^{-3}$) [19].

The other parts of the curves is easy to recognize: they consist of ascending (diffusional) branches and kinetic plateaux as in Fig. 32. According to Eq. (5.21) the closer the starting point, the higher the plateau and the sharper the ascent to it. For a highly polar solvent the height of the plateau should be in accordance with Eq. (6.26):

$$\frac{Z_0}{z} = q = \frac{\sigma}{r_0} \tag{8.33}$$

As $r_0 \to \sigma$ only the kinetic plateau remains, which represents the exponential model, but even this plateau gives way to a descending branch when diffusion becomes so fast that interrupts the encounters before the spin conversion is completed.

A quite different situation arises when the spin conversion is slow ($\kappa \ll 1$). There is no plateau and not only descending, but also ascending branches of the curves are affected by the spin conversion (Fig. 66B). However, this effect is so different for distant and close starts, that it should be described separately.

E.   Distant Starts

Distant starts are compatible with contact recombination in the IN regions (see Figs. 43, 47, and 10). Keeping r_0 much larger than σ, we can see how the situation changes by slowing down the spin conversion rate k_s. The plateau shortens and finally disappears, while the ascending branch undergoes a break in its slope, which is lower the smaller is κ (Fig. 67). As was mentioned earlier, $y > 1$ for distant starts, although x can be either larger or smaller than 1 provided $r_0 \gg \sigma$. In the latter case we can set $y \gg 1$, to get from Eq. (8.29) the simplified expression of main interest,

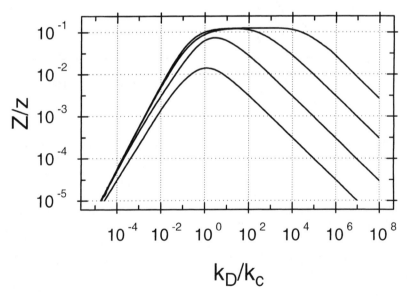

Figure 67. The same as in Fig. 66 but for fixed $r_0 = 8\sigma$ and different $\kappa = 10^{-3}$, $0.1, 10, 10^3$ (from bottom to top).

losing only the descending branch at the largest $k_{\tilde{D}}/k_c$:

$$\varphi(r_0) = 1 - \frac{k_c}{4k_{\tilde{D}}}\frac{\sigma}{r_0}\frac{1+x}{1+k_c/k_{\tilde{D}}+x(1+k_c/4k_{\tilde{D}})} = \begin{cases} 1 - R_0\left(r_0,\dfrac{k_c}{4}\right) & \text{at} \quad x \gg 1 \\ 1 - \dfrac{1}{4}R_0(r_0,k_c) & \text{at} \quad x \ll 1 \end{cases}$$

(8.34)

For the fast spin conversion ($x \gg 1$) we again confirm the results that can be obtained from Eqs. (8.25) and (8.27) at $r_c = 0$. The recombination is proceeding as in the spinless theory, but its rate is 4 times less, because it can proceed only from a singlet state. However, when the conversion is slower ($x \ll 1 \ll y$), the equilibration of spin states during the encounter is unattainable, so that only one out from four RIPs delivered to contact can react, but with a full rate available for the singlet state. Therefore, the quantum yield of recombination is four times smaller then in the case when the spin states were neglected.

The location of the point $x = 1$, where $k_{\tilde{D}}/k_c = \kappa$, plays an important role. When κ is large, it is positioned in the limits of the kinetic plateau. There it plays no role at all because $Z \equiv Z_0$ has the same value (8.33) to the left and to the right of the point. However, at smaller κ the point moves left and down to the ascending branch where $k_c \ll k_{\tilde{D}}$ and recombination is diffusion-controlled. The branch becomes divided into two parts, that can be deduced from Eq. (8.34) [19]:

$$Z(r_0) = \frac{\sigma\tilde{D}}{r_0\dfrac{4+\sqrt{4k_s\sigma^2/\tilde{D}}}{1+\sqrt{4k_s\sigma^2/\tilde{D}}} - \sigma} = \begin{cases} \dfrac{\sigma}{r_0-\sigma}\tilde{D} & \text{at} \quad \dfrac{k_{\tilde{D}}}{k_c} \ll \kappa \\ \dfrac{\sigma}{4r_0-\sigma}\tilde{D} & \text{at} \quad \dfrac{k_{\tilde{D}}}{k_c} \gg \kappa \end{cases}$$

(8.35)

The lower part of the branch is sharper and does not differ from that given by a spinless theory in Eq. (5.21). Only at $k_{\tilde{D}}/k_c = \kappa$ all curves start to deviate from this standard behavior and their upper part get a smaller slope. Thus the spin-effect in diffusional recombination comes to light as the bend of a quasilinear $Z(k_{\tilde{D}})$ dependence near $k_{\tilde{D}}/k_r = \kappa$. This bending point lays the lower the slower is the spin conversion. Hence the ascending branches of the curves are composed from two parts: the lower, with a higher slope, which indicates the region of regular diffusion control, and the upper one, with a slower slope, where in addition the spin control hinders the recombination.

Though very useful for understanding the general features of the phenomenon, this picture can not be inspected experimentally in a full range of $\tilde{D}$ and k_s variation but only in a very narrow strip of both diffusion and spin-conversion rates. Going from the log/log plot to the usual coordinates we can see in this restricted region the bunch of selected curves with close and reliable Z, but with better resolution (Fig. 68). Initially, when only diffusion controls the recombination, all curves merge into one. However, at faster diffusion the spin conversion is not completed during encounter time

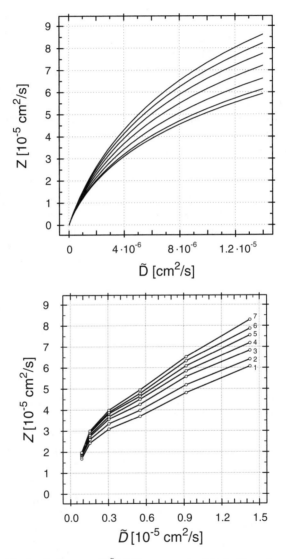

Figure 68. The band of curves $Z(\tilde{D})$ falling approximately in the region experimentally studied in [121, 123]: theoretical prediction (at the top) and experimental findings (at the bottom).

and the slower conversion the less effective is the recombination through the singlet state of ion pair. Therefore Z is lower than that in spin-less theory and dispersion of the curves is more pronounced at this stage. The general $Z(\tilde{D})$ behavior looks much more similar to what was presented in Fig. 33 as experimental findings. More recent experimental investigations of the effect verified qualitatively the whole picture shown in Fig. 68 [123]. The non-linearity of $Z(\tilde{D})$ curves is beyond experimental error and clearly shows the importance of singlet-triplet transitions in RIPs.

If this is the case, then an essential decrease in the ion separation quantum yield $\bar{\varphi}(k_s)$ with k_s should be expected in the transition region. This is what we really see in Fig. 69. Since the rate of the spin conversion (8.11) is quadratic in ω_- as well as in H, the charge separation quantum yield must decrease with the magnetic field H, as has been obtained experimentally [121]. For the quantitative description of the magnetic-field effect (MFE), one should know not only the $k_s(H)$ dependence but also the distribution of initial interion distances $f_0(r)$ as well. The latter should be used to average $\varphi(r_0)$ in a general way recommended in Eq. (8.22). This procedure was done and analyzed in an original work [19]. It was shown to affect the results much more essentially when ionization is not kinetic but diffusional. In the latter case, the mean starting distance r_e is really large and

$$\bar{\varphi} = 1 - \left\langle \frac{\sigma}{4r_0} \right\rangle \frac{1+x}{1+x/4} = \varphi(r_e) \tag{8.36}$$

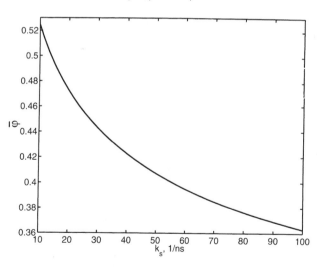

Figure 69. The drop in the averaged charge separation quantum yield with the rate of spin conversion, accelerated by a magnetic field. Parameters $r_e = 15\,\text{Å}$, $\sigma = 12\,\text{Å}$, while $\tilde{D} = 5 \times 10^{-5}\,\text{cm}^2/\text{s}$ [19].

where $r_e = \langle 1/r_0 \rangle^{-1} = \left[\int_\sigma^\infty f_0(r_0) 4\pi r_0 \, dr_0 \right]^{-1}$. The magnetic-field effect in this case is the most pronounced [19].

F. The Closest Starts

Here we come to an alternative extreme, assuming $y = 0$, that is, $r_0 = \sigma$ as was suggested by the exponential model. Then Eq. (8.29) reduces to the following:

$$\varphi = \frac{1 + (k_c/k_{\tilde{D}}) + x}{1 + (k_c/k_{\tilde{D}}) + x[1 + (k_c/4k_{\tilde{D}})]} \tag{8.37}$$

Presenting this result in a usual way, it is easy to see that

$$Z = z \frac{\sqrt{k_c \kappa k_{\tilde{D}}}}{k_c + k_{\tilde{D}} + \sqrt{k_c \kappa k_{\tilde{D}}}} \tag{8.38}$$

Since this expression is invariant with respect to the permutation of $k_{\tilde{D}}$ and k_c, the diffusional dependence of Z plotted against $\ln(k_{\tilde{D}}/k_c)$ (Fig. 70) is symmetric at any κ.

When $\kappa \gg 1$, there are three different regions:

$$Z = \begin{cases} z\sqrt{\kappa \left(\dfrac{k_{\tilde{D}}}{k_c} \right)} = \dfrac{1}{2}\sqrt{k_s \sigma^2 \tilde{D}} & \text{at} \quad \sqrt{\dfrac{k_{\tilde{D}}}{k_c}} \ll 1/\kappa & (a) \\[3mm] z & \text{at} \quad 1/\kappa \ll \sqrt{\dfrac{k_{\tilde{D}}}{k_c}} \ll \kappa & (b) \\[3mm] z\sqrt{\kappa \left(\dfrac{k_c}{k_{\tilde{D}}} \right)} = \dfrac{k_c}{\pi}\sqrt{\dfrac{k_s}{\tilde{D}}} & \text{at} \quad \kappa \ll \sqrt{\dfrac{k_{\tilde{D}}}{k_c}} & (c). \end{cases} \tag{8.39}$$

In the intermediate diffusion region b we confirm the existence of the kinetic plateau $Z = z$ predicted by the exponential model. Here is no dependence on spin conversion, unlike in both side regions, a and c, where recombination increases with k_s. If the rate of conversion decreases, making $\kappa \ll 1$, the side regions are joined together expelling the kinetic plateau.

It is remarkable that diffusion accelerates the process in region a and decelerates it in region c. This effect may be understood if one takes into account that the efficiency of the spin conversion is determined by the product $k_s t_e$, where t_e is the survival time of RIPs, before ions either

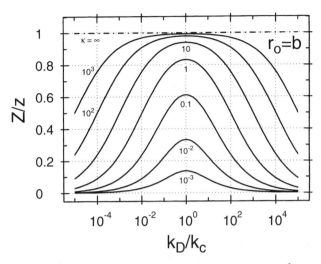

Figure 70. Diffusional dependence of Z for contact RIPs, $r_0 = \sigma = 5$ Å, at different rates of spin conversion characterized by parameter $\kappa = 10^{-3}, 10^{-2}, 0.1, 1, 10, 10^2, 10^3$ ($k_c = \int W_R(r)d^3r = 628$ Å^3/ns, $\tilde{D} = 5\times 10^{-5}$ cm^2/s) [19].

recombine or are separated. For the RIPs born in contact, this time obtained in Eq. (5.29) can be represented as follows:

$$t_e = 1.6\frac{4\pi\sigma^3 k_{\tilde{D}}}{\left(k_c + k_{\tilde{D}}\right)^2} \tag{8.40}$$

It clearly shows that t_e increases until $k_{\tilde{D}} < k_c$ and decreases at $k_{\tilde{D}} > k_c$. The results given in Eq. (8.39) for regions a and c may be presented by the common formula

$$Z = z\sqrt{2.5k_s t_e} \tag{8.41}$$

although in region a Z increases, whereas in region c it decreases with diffusion.* Since both the ascending and descending branches of $Z(\tilde{D})$ are spin-controlled, the quantum yield in regions a and c is rather sensitive to the variation of k_s but not as much in between, within the kinetic plateau of region b. The latter is wider the higher is κ. Thus we come to the conclusion that k_s has its greatest impact on the charge recombination if $\kappa < 1$, that

---

*The latter is also seen in Figs. 66 and 67 because the origin of the decending branch is the same for distant and close starts.

is, at

$$k_c > 16\pi\sigma^3 k_s, \qquad k_{-et} \gg k_s \qquad (8.42)$$

As was noted under such a condition, the kinetic region b disappears and the recombination is limited everywhere by the time of the spin conversion. In this case Eq. (8.41) holds at any speed of diffusion and reaches the maximum value

$$Z_{max} = z\sqrt{\frac{4\pi k_s\sigma^3}{k_c}} = \sqrt{\frac{k_s k_c \sigma}{64\pi}} \qquad \text{at} \qquad k_{\tilde{D}} = k_c$$

If condition (8.42) is met, there is no place for the exponential model anywhere.

The parabolic dependence (8.40) as well as the symmetrical result (8.38) are peculiar only for RIPs started from the closest approach distance $r_0 = \sigma$. At even minor but nonzero initial ion separation, the symmetry of $Z(\tilde{D})$ dependence is lost [19]. The ascending branches of the curves shown in Fig. 70 approach each other and even merge, while their descending branches are pushed apart. With a further increase of r_0, the whole picture gradually transforms to that shown in Fig. 67.

Additional changes were shown to occur at $k_{\tilde{D}}/k_c \ll 1$, due to lack of accuracy available in contact approximation [19]. For relatively fast diffusion the latter is well approved, but the difference with the theory of remote electron transfer applied to the same system increases and becomes qualitative as $\tilde{D} \to 0$, as we have already shown in Figs. 37 and 50. However, with this restriction the geminate recombination of charges in the IN regions, when they are initially separated, and in the NN regions (the most favorable case for the exponential model) can be described by contact approximation, at least approximately. It is much worse with the situations in the NI regions (the most widespread), as well as in the II regions, where ions can be initially inside the remote recombination layer. Contact approximation cannot be applied to the latter in principle. It is possible to use a rectangular model instead, but it may not be easier than to apply the general theory of remote transfer to the same phenomena.

In short, the diffusional dependence of the reaction rate and quantum yields is the most important source of information on bimolecular reaction mechanisms in liquids. It plays the same role in the chemical kinetics of condensed matter as the velocity (energy) dependence of cross sections does in gases. For instance, the experimental study of fast radical reactions in solutions turned attention of theorists to the chemical anisotropy of the contact reaction. Finally, it resulted in the development of the exhaustive theory of radical recombination assisted by translational and rotational motion of reactants [124]. Here we also profited from a limited number of works where rarely available experimental information on diffusional (viscosity) effects was presented. The magnetic field effects are of main interest when the reaction is spin-sensitive and controlled by conversion. Later in this section, we will give an example of a thorough experimental study of these effects accompanied by interpretation of the data using the most general theory of diffusion- and spin-assisted reactions of electron transfer.

G.   Dynamic Spin Conversion in a Two-Level System

We addressed the same reaction whose scheme was shown in Fig. 65. The experimental results of interest are reproduced in panel A of Fig. 71; in panel B, their representation has been adopted to indicate more clearly the viscosity dependence of the magnetic field effect (MFE) in the charge

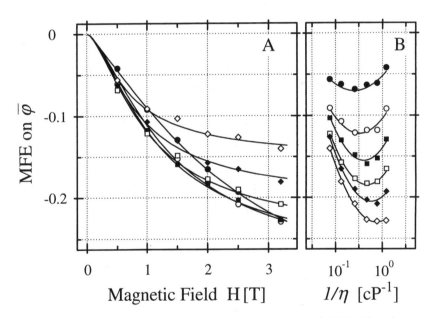

Figure 71. (A) Magnetic-field dependence of MFE on $\bar{\varphi}$, $M = [\bar{\varphi}(H)/\bar{\varphi}(0) - 1]$ versus H, observed for solvent mixtures (H_2O/ACN 1 : 1 with ethylene glycol) of different viscosities [in centipoise (cP)]: $\diamondsuit$ 13.40, $\blacklozenge$ 7.68, $\square$ 3.88, $\blacksquare$ 2.15, $\circ$ 1.29, $\bullet$ 0.83. The solid lines represent the best fits obtained on the basis of the exponential model. (Diagram adapted from Ref. 83.) (B) Data of Fig. 71, panel A, represented as the viscosity dependence of the MFE at various magnetic fields in tesla (T): $\diamondsuit$ 3.2, $\blacklozenge$ 2.5, $\square$ 2.0, $\blacksquare$ 1.5, $\circ$ 1.0, $\bullet$ 0.5. The solid lines represent second-order polynomial approximations drawn to aid the viewer [20].

separation quantum yield, which is

$$M = \frac{\bar{\varphi}(H) - \bar{\varphi}(0)}{\bar{\varphi}(0)} \qquad (8.43)$$

It has been proved [20] that the family of curves $M(\eta)$ at different magnetic fields can serve as a fingerprint of distance dependent electron transfer. To make it possible the theory should be essentially generalized.

Earlier in this section the unified theory was extended to the case of singlet and triplet states of RIPs with stochastic transitions between them characterized by the rate k_s. Such a description of spin conversion is reasonable only for low magnetic fields fitting condition (8.6). For the system under study, it must not exceed $5 \cdot 10^{-3}\,T$, although in fact the applied fields are much larger. Therefore, the dynamic description of the spin conversion should be substituted for a stochastic one. For this goal, the kinetic evolution of RIPs has to be described by means of a distance dependent spin density matrix $\hat{\rho}(r, t)$:

$$\frac{\partial}{\partial t}\hat{\rho}(t) = \frac{\tilde{D}}{r^2}\frac{\partial}{\partial r}r^2 e^{-U_R/T}\frac{\partial}{\partial r}e^{U_R/T}\hat{\rho} - i[\mathbf{H}, \hat{\rho}]_- + \hat{\mathbf{R}}\hat{\rho} - \frac{1}{2}[\mathbf{W}, \hat{\rho}]_+ + \hat{\rho}_0(t)$$

$$(8.44)$$

Apparently we included in Eq. (8.3), in line with the encounter diffusion of RIPs, their recombination and generation. The former is represented by operator $\mathbf{W}$ in the anticommutator $[x, y]_+ = xy + yx$, and the latter is given by the source term $\hat{\rho}_0(t)$.

From now we will restrict consideration to the simplest model of a RIP having only two spin levels, reactive singlet (S) and the single unreactive state (T). This can be envisaged as a restriction to S and T_0 as is often done with organic RIPs in the high-field limit, when other triplet states $T_\pm$ are so decoupled that either coherent or incoherent transitions between them become negligible. However, the RIPs under consideration have strong spin–orbit coupling, causing effective spin relaxation among all triplet sublevels, which is field-independent up to very high fields. Because of this $T_\pm$ states should, generally speaking, be taken into account. Although available for a numeric solution, the full four-level system is rather difficult to analyze. Here our aim is not to fit the particular experimental results, but rather to qualitatively study the viscosity effect, as has been done in Ref. 20.

In the two-state basis all the operators included in Eq. (8.44) have the simplest form:

$$\mathbf{H} = \frac{1}{2}\begin{pmatrix} 0 & \omega_- \\ \omega_- & 0 \end{pmatrix}, \quad \mathbf{W} = \begin{pmatrix} W_R(r) & 0 \\ 0 & 0 \end{pmatrix}, \quad \hat{\rho}_0 = \begin{pmatrix} cW_I(r)n(r,t)N(t) & 0 \\ 0 & 0 \end{pmatrix}$$

$$(8.45)$$

The relaxation superoperator $\hat{\mathbf{R}}$ can be obtained from its general form for the four-level system adduced in Ref. 125. The decoupling of $T_\pm$ sublevels is achieved by setting the longitudinal relaxation time to infinity. Then the elements of $\hat{\mathbf{R}}$, which remains nonzero, are the following:

$$\hat{R}_{ST_0,ST_0} = \hat{R}_{T_0S,T_0S} = 2\hat{R}_{SS,SS} = 2\hat{R}_{T_0T_0,T_0T_0} = -2\hat{R}_{SS,T_0T_0} = -2\hat{R}_{T_0T_0,SS} = -\frac{1}{T_2}$$

$$(8.46)$$

where T_2 is the transversal relaxation time assumed to be the same for both radicals in a pair. The initial and boundary conditions to Eq. (8.44) remain the same, as for the diagonal elements elements of $\hat{\rho}$ (level populations) in the case of stochastic conversion:

$$\hat{\rho}(r,0) = 0, \qquad \hat{\rho}(r \to \infty, t) = 0, \qquad \frac{\partial}{\partial r}e^{U_R/T}\hat{\rho}(r,t)\bigg|_{r=\sigma} = 0 \qquad (8.47)$$

Since we are interested here only in quantum yields that are expressed via

$$\hat{\sigma}(r) = \int_0^\infty \hat{\rho}(r,t)dt$$

it is reasonable to get an equation for this quantity by integrating Eq. (8.44) over time:

$$\frac{\tilde{D}}{r^2}\frac{\partial}{\partial r}r^2 e^{-U_R/T}\frac{\partial}{\partial r}e^{U_R/T}\hat{\sigma} - i[\mathbf{H},\hat{\sigma}]_- + \hat{\mathbf{R}}\hat{\sigma} - \frac{1}{2}[\mathbf{W},\hat{\sigma}]_+ = -\hat{\sigma}_0(r) \qquad (8.48)$$

where

$$\hat{\sigma}_0(r) = \int_0^\infty \hat{\rho}_0(r,t)dt = \begin{pmatrix} cm_0(r) & 0 \\ 0 & 0 \end{pmatrix} = \psi\begin{pmatrix} f_0(r) & 0 \\ 0 & 0 \end{pmatrix} \qquad (8.49)$$

where $\psi = c\int m_0 d^3r$ as in Eq. (4.25) and f_0 is a normalized initial distribution (4.23). Finally, the fraction of discharged reaction products $\psi - \phi$ can be found by integration of the dissipative term over r:

$$\psi - \phi = \frac{1}{2}\int Sp[\mathbf{W},\hat{\sigma}(r)]_+ d^3r = \psi\bar{\chi} \qquad (8.50)$$

where

$$\bar{\chi} = 1 - \bar{\phi} = \int d^3r \int \chi(r,r_0)f_0(r_0)d^3r_0 \qquad (8.51)$$

is the averaged quantum yield of recombination and $\chi(r, r_0)$ is the same, provided ions, started from r_0, recombine at r.

Except for some limiting cases, it is hardly possible to solve the set of these equations analytically. Krissinel et al. solved it numerically [20], employing a nonequidistant finite-difference scheme for the space coordinate r and implicit Euler scheme, with variably adjusted time steps, for joint integration of unified theory equations including Eqs. (4.6a) and (4.8)–(4.10) for $n(r, t)$ and $N(t)$. For simplicity, the position-dependent rates of forward and backward electron transfer were made the same:

$$W_I(r) = W_R(r) = W_c \exp\left[-\frac{2(r - \sigma)}{L}\right] \qquad (8.52)$$

as in Eq. (2.22). Details of the numerical procedure employed may be found elsewhere [14,15,93]. For the special case of RIPs generated and recombining in contact, there is an analytic solution to the problem obtained by Mints and Pukhov [126]. This is the only analytic solution available and it is in excellent agreement with the results of numeric calculation obtained for this particular case. This case is the most favorable for the exponential model, but we are mainly interested in quite the opposite situation, when RIPs are born not at contact but far from it, due to diffusion control of ionization. The results of our calculations compared with the exponential model predictions, are reported below.

H. Unified Theory versus Exponential Model

Besides assuming contact creation and recombination of RIPs, the exponential model employs the first-order rate constant k_{sep}, Eq. (3.4), to describe their separation, which is tantamount to ignoring interradical distance as a continuous degree of freedom. Mathematically the dynamics of the exponential model is represented by the same Eq. (8.44) if one substitutes $-k_{\text{sep}}\hat{\rho}$ for the diffusional term and uses the contact approximation (2.21) for $W_R(r)$. Then, using spin Hamiltonian **H** and relaxation operator $\hat{\mathbf{R}}$, specified in Eqs. (8.45) and (8.46), the solution of the exponential model equation can be obtained analytically. Such a solution was first provided by Lüders and Salikhov [127]. However, there are some sign inconsistencies in their kinetic equations and their final analytic result for the recombination probability is not in full accord with the analogous solution obtained in Ref. 20 with MAPLE, a mathematical computation program [128]. Here we provide the corrected analytic solution that was well verified numerically in Ref. 20:

$$\varphi(\omega_-) = k_{\text{sep}} \, Sp\hat{\sigma} = \frac{1}{1 + \dfrac{k_{-et}}{k_{\text{sep}}} Q(\omega_-)} \qquad (8.53)$$

where

$$Q(\omega_-) = \frac{\omega_-^2 T_2^2 + T_2(2k_{sep} + k_{-et}) + 4}{2\omega_-^2 T_2^2 + T_2[T_2(k_{sep} + k_{-et}) + 4](2k_{sep} + k_{-et}) + 2k_{-et}T_2 + 8}$$

(8.54)

is the factor describing the impact of spin control on the recombination in the framework of the exponential model. The case $Q = 1$ corresponds to the conventional spinless model result φ_0, in Eq. (3.9). In the limit of the high field $Q = \frac{1}{2}$, which is equivalent to the system with $k_{-et}/2$ substituted for k_{-et}: at fully equilibrated S and T_0 sublevels the state of RIPs subjected to recombination is half populated.

Panel A of Fig. 71 shows the observed MFE for the charge separation quantum yield in solutions whose viscosity variation ranges over more than a decade. The general characteristic of the MFE is a decrease in M with the magnetic field, with a curved shape tending to saturation. However, it is not yet reached at 3.2 T, the highest field applied in these experiments. The amplitude of the MFE curve is significantly affected by a change in solvent viscosity. Although an increase of viscosity tends to enhance the (negative) MFE at low fields, this tendency is reversed at high fields. The viscosity dependent behavior of the MFE shown in Fig. 71, panel B is represented differently, as $\bar{\varphi}$ versus $\log(1/\eta)$. Such a presentation gives a clearer picture, since curve crossing is avoided. What is borne out in this representation quite conspicuously is that for any field the MFE passes a minimum at intermediate viscosities. Concurrent with an increase in absolute value, the position of the minimum shifts to higher values of $1/\eta$ as the field is increased. In the highest field, the low-viscosity branch of the curve is cut off at the lowest viscosity experimentally explored.

The curves fitting the experimental points in panel A of Fig. 71 resulted from theoretical calculations, based on the exponential model [83]. These employed a spin formalism with the full 4×4 spin density matrix. The fits which reproduce the shape of the MFE curves very well, were used to determine specific values of T_2, k_{sep}, and k_{-et} for each solvent viscosity, whereupon it was found that T_2 is rather insensitive to η, k_{sep} varies approximately linearly with $1/\eta$ and k_{-et} also shows a pronounced covariance with $1/\eta$, albeit not a linear one. The first two findings were expected. Spin relaxation in ruthenium complexes is mainly determined by an intramolecular spin–orbit-coupling-based mechanism that is viscosity-independent, whereas the separation rate (3.4) is linear in $\tilde{D}$ and in high polarity solvents is simply

$$k_{sep} = \frac{3\tilde{D}}{\sigma^2}.$$

(8.55)

Unlike previous findings, the solvent viscosity dependence of k_{-et} is unexpected within perturbation theory of weak nonadiabatic electron transfer [23–27]. Therefore, it was argued [83] that the observed variation of the k_{-et} parameter with solvent viscosity might indicate the occurrence of a pronounced dynamic solvent effect (strong nonadiabatic electron transfer) [26,27]. This conclusion was later revised and disproved [20] using conventional, viscosity-independent rates of electron transfer (8.52) with the following parameters: $\sigma = 10 \text{ Å}, W_c = 10^3 \text{ ns}^{-1}$, and $L = 0.75 \text{ Å}$. The unified theory of electron transfer was used not for accurate fitting to the same data but rather to qualitatively describe and explain them in principle. For both approaches T_2 was fixed to 10 ps and k_{-et} was determined such that for $D = \tilde{D} = 10^{-5} \text{ cm}^2/\text{s}$ the exponential model yielded the same value of $\bar{\varphi}$ as the unified theory. The solvent was considered as highly polar, setting $U_R = 0$. Note that according to Eq. (8.5) and for the average value of 1, for $\Delta g = g_+ - g_-$ the mixing angular frequency was $\omega_- \approx 100 H \text{ rad/ns}$, where H is measured in tesla. The general behavior of $\bar{\varphi}(D)$ curves is qualitatively similar in both the exponential and unified theory: they monotonously increase but slower at a higher field.

The principal difference between the two approaches is seen only from MFE curves exhibited in Fig. 72. The results obtained with the exponential model (A) and unified theory (B) are compared, using as a measure of ion mobility k_{sep} from (8.55) in the former case and the diffusion coefficient (same for reactant and ions) in the latter case. Comparing this figure with Fig. 71 (B), it is clear that the exponential model cannot account for the characteristic minimum in the viscosity dependence of the MFE curves, if all parameters except k_{sep} are kept constant. On the contrary, the curves for the MFE obtained with the unified theory display a qualitatively different behavior with minima (i.e., maxima of the field-effect amplitude) at intermediate viscosities, shifting to lower viscosities with increasing mixing frequency ω_-. In fact, for the approximate range of parameters corresponding to the experimentally scanned range of D (cf. dashed box in panel B of Fig. 72), the calculated curves exhibit approximately the same behavior as the observed ones shown in panel B of Fig. 71. Thus the observed viscosity dependence of the MFE curves do indeed reflect specific features disclosed with the unified theory that are not present in the exponential model. In the following, we will demonstrate that these features come to the light due to a unified description of the forward and backward reactions considered as remote electron transfer.

From Eq. (8.51) we can deduce that

$$\bar{\varphi} = 1 - \int \chi(r_0) f_0(r_0) d^3 r_0 = \int \varphi(r_0) f_0(r_0) d^3 r_0 \qquad (8.56)$$

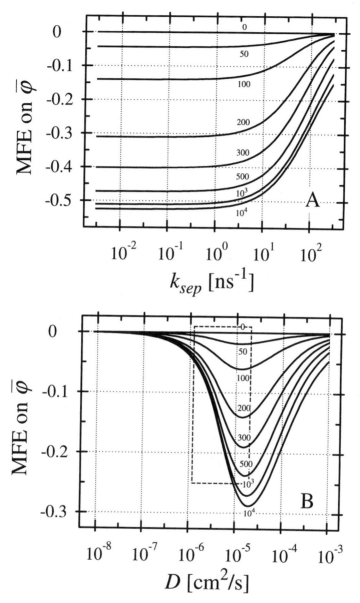

Figure 72. Viscosity dependence of the MFE at various magnetic fields measured in mixing frequency ω_- (the numbers assigned to the curves denote ω_- in rad/ns as the rate of $S - T_0$ conversion): (A) exponential model; (B) unified theory (the rectangle in dashed lines indicates the region approximately spanned by the experimental data in Fig 71, panel B). Calculation parameters: $\sigma = 10\,\text{Å}$, $\tau_D = 1210\,\text{ns}$, $c = 0.1\,\text{M}$, $W_c = 10^3\,\text{ns}^{-1}$, $L = 0.75\,\text{Å}$, $T_2 = 10\,\text{ps}$ [20].

where the quantities

$$\varphi(r_0) = 1 - \chi(r_0) = 1 - \int \chi(r, r_0)d^3r \quad \text{and} \quad \chi(r, r_0) = \frac{1}{2\psi} Sp[\mathbf{W}, \hat{\sigma}(r|r_0)]_+$$

$$(8.57)$$

as well as $\hat{\sigma}(r|r_0)$ are obtained using $f_0(r) = \delta(r - r_0)/4\pi r^2$ in Eq. (8.49). Equation (8.56), as well as its spinless precursor, Eq. (5.55), express the total charge separation quantum yield via the product of partial yield, from a given initial distance r_0, and the distribution of these distances $f_0(r_0)$ formed by ionization. Both these factors play an equally important role in the creation of studied phenomenon and should now be examined more intently.

Figure 73 shows the partial magnetic-field effect, $M(r_0) = [\varphi(r_0, H) - \varphi(r_0, 0)]/\varphi(r_0, 0)$ versus D, at different initial separations of radical ions. As

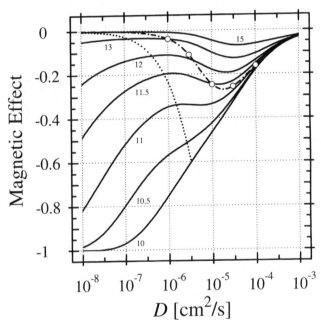

Figure 73. Partial MFE as a function of the diffusion coefficient at different initial separations (in Å) denoted by numbers on the curves. All the curves were obtained at the same parameters as in Fig. 72, but for a single $\omega_- = 10^3$ rad/ns. The same ω_- for the dashed–dotted line borrowed from Figure 72, panel B, where the distribution of initial distances was created by precursor ionization. The circles on this curve represent situations that will be further explored in Fig. 75. The dotted line connects points with a constant $\varphi(r_0)$ value of 0.01. In the region to the left of this curve the $\varphi(r_0)$ values are < 0.01; to the right they are higher. [20]

follows from the figure, for a given D, the MFE amplitude decreases with increasing initial separation. This behavior is, of course, expected since $\varphi(r_0, H)$ is larger when the RIP starts at a larger r_0 and the magnetic field can exhibit its effect only insofar as backward electron transfer can take place. Concerning the behavior of the individual MFE curves, the most striking feature is the occurrence of a loop with a minima clearly manifested at $r_0 > 10.6\,\text{Å}$ and still indicated as shoulders at $r_0 < 10.6\,\text{Å}$. The minima become shallower and move to higher D values, that is, lower viscosities, with increasing r_0. Phenomenologically, the whole curve can be divided into three parts that were attributed in Ref. 20 to

1. Kinetic geminate recombination (the upper part and the fastest diffusion)

2. Diffusional geminate recombination (the interior of the loop, moderate diffusion)

3. Static recombination (the lowest part and the slowest diffusion)

Passing from high to low D values, the slope of the MFE curve is positive, negative, positive. On the contrary, the slope of a similar curve obtained with the exponential model is permanently positive. This is because the model is just an interpolation between static and kinetic geminate recombination, missing the intermediate part essentially controlled by diffusion.

The subdivision of the real MFE into three regions is confirmed by the shape analysis of the distribution of recombination products, $\chi(r, r_0)$. It was made in Ref. 20 for a single curve from Fig. 73, related to a starting distance $11.5\,\text{Å}$, but for six selected D values (Fig. 74).

1. Cases $D = 10^{-4}$ and $3 \times 10^{-5}\text{cm}^2/\text{s}$ in Fig. 74. In these cases the length parameter $\Lambda = \sqrt{DT_2}$, scaling the average distance of diffusion that RIP undergoes during spin relaxation, is correspondingly 3.1 and $1.7\,\text{Å}$, so that $L, r_0 - \sigma < \Lambda$. Under these conditions diffusion may establish a rather uniform distribution of RIPs over the reaction zone before spin conversion is completed. Hence spin equilibration occurs uniformly over all the positions, and $\chi(r, r_0)$ should closely match the shape of $W_R(r)$, independently of r_0. In our example it must be exponential as in Eq. (8.52), which is really verified by calculations.

2. Cases $D = 10^{-5}, 3 \times 10^{-6}, 10^{-6}\text{cm}^2/\text{s}$ in Fig. 74, corresponding to the values of $\Lambda = 1.0, 0.5, 0.3\,\text{Å}$ which are comparable to L and equal to or smaller than $r_0 - \sigma$. Since the spin evolution occurs while RIPs are on the way to positions of higher reactivity, the deviation from exponential $\chi(r, r_0)$ becomes more and more pronounced as D decreases. As in the case of diffusion-controlled bimolecular ionization, there is actually a maximum in $\chi(r, r_0)$ located approximately at R_Q, if r_0 is

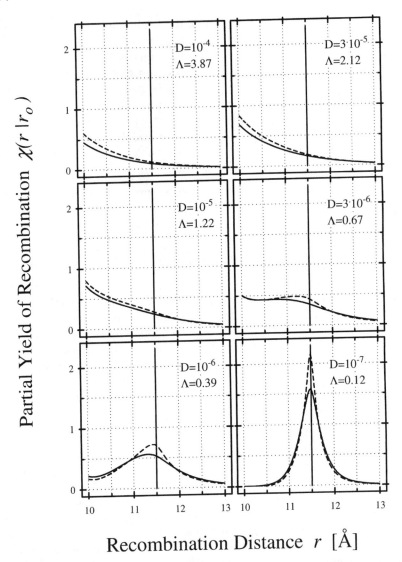

Figure 74. Distribution function $\chi(r, r_0)$ for RIPs discharged at distance r provided they were born at distance $r_0 = 11.5$ Å marked by the vertical bars. Solid lines — reaction in zero field; dashed lines — reaction in magnetic field corresponding to $\omega_- = 10^3$ rad/ns. The values of the diffusion constant and the corresponding relaxation length $\Lambda = \sqrt{DT_2}$ are given in each of diagram [20].

chosen sufficiently far from it. Since the characteristic condition for this regime is $\Lambda \leq r_0 - \sigma$, the cases $\Lambda = 1.7$ and $0.3\,\text{Å}$ were considered as borderline cases. In fact, for these values the particular MFE curve considered is close to its maximum and minimum, respectively.

3. Case $D = 10^7\,\text{cm}^2/\text{s}$ in Fig. 74, that is, $\Lambda \ll L$. The maximum of $\chi(r, r_0)$ is very close to starting distance r_0, because at so slow diffusion RIPs actually react where they were born.

In both the kinetic and static regimes the distribution $\chi(r, r_0)$ is not sensitive to diffusion, although for different reasons. This leads to a viscosity dependence of the MFE with a positive slope sign as expected from the exponential model. However, the intermediate diffusional regime where the sign is negative is lost in the model and therefore it fails to reproduce the non-monotonic behavior of the MFE versus D curves.

When comparing these curves for different r_0, we note that the minima and the maxima come closer as r_0 decreases. This is because as $r_0 \rightarrow \sigma$ the kinetic and static regimes expand towards each other and the range for the diffusional regime becomes narrower. Finally, the branch with negative slope disappears. This may also be interpreted as disappearance of the diffusion-controlled regime, because the closer to contact the RIPs are born, the less is the impact of the delivery process on the recombination.

In Fig. 73 the dashed–dotted line indicates the curve from Fig. 72 (B) related to the same mixing frequency $\omega_- = 10^3\,\text{rad/ns}$ as others, but representing the $M(D)$ dependence averaged over a distribution $f_0(r_0)$ created by bimolecular ionization. The latter factor was passive until now, but here it plays a crucial role in transforming the S-like individual curves in Fig. 73 to bell-like curves from Fig. 72 (B), obtained after space averaging. For any distribution over r_0, one may introduce an effective starting distance r_s, by assigning it that r_0 value that would yield the same MFE as resulting from real averaging. The particular values of r_s are indicated by the crossing-points of the dashed–dotted line with others (there $r_s = r_0$). Evidently, r_s increases with increasing viscosity as well as $\bar{R}$ defined in Eq. (4.24) and shown by stars in Fig. 26. This is the main reason for so drastic a difference in shapes between the $M(D)$ curves obtained with the exponential model and the unified theory (Fig. 72). The fast shortening of the starting distance with diffusion results in a steep decrease of $M(D)$, which is even sharper than within the loop and begins earlier.

This conclusion is strongly supported by the results of numeric calculations of $f_0(r_0)$ for the present system shown in Fig. 75. They are made for a number of situations denoted by open circles in Fig. 73. At slower diffusion, the average initial distance $r_s \approx \bar{R}$ is farther from contact and the distribution around it is wider. Such a transformation of initial

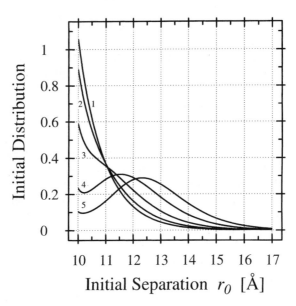

Figure 75. Initial distribution of RIPs, $\psi f_0(r_0)$. The pertinent values of the diffusion coefficient D, cm^2/s, are (1) 10^{-4}; (2) 3×10^{-5}; (3) 10^{-5}; (4) 3×10^{-6}; (5) 10^{-6}. The other parameters are the same as in Fig. 72 [20].

conditions was actually realized experimentally [83] by changing viscosity. The results that we have just discussed were initially attributed to the viscosity-dependent rate of backward electron transfer, k_{-et} and considered as an indication of a pronounced dynamic solvent effect that might be responsible for such a dependence. As was demonstrated, the unified theory can qualitatively account for the observed viscosity dependence of MFE without assuming an unusual viscosity dependence of the electron-transfer rate itself. However, we focused here only on principal aspects of the phenomena that must be studied more carefully in detail. In particular, we actually restricted ourselves to the NN case, assuming the exponential transfer rate (8.52) for both forward and backward transfer. As we saw from the investigation of the FEG law, some other situations (IN or NI) are not of less interest, but the peculiarities of MFE in these cases are open for further study.

IX. PHOTOGENERATION OF THE EXCIPLEXES

There are two points of view on the origin of the exciplexes produced by encounters of excited molecules with acceptors or donors of electrons in dilute solutions. The exciplexes [or contact ion pairs (CIPs)] can be either immediately prepared by the contact electron transfer between the reactants, or result from the aggregation of the solvent separated ion pair first generated by remote electron transfer. There was evidence in favor of the primary creation of the exciplexes [Scheme I in Eq. (9.1)] [129–134] as well as for their secondary origin (Scheme II) proposed in Refs. 135 and 136. In principle, the exciplexes, $^1[D^+A^-]$, and solvent separated ion pairs, $[D^+ \dots A^-]$, were generated in parallel [130,137,138] and the reversible reaction between them was finally assumed to be possible [139,140]. In such a general reaction scheme all reaction channels are actually included

$$
\begin{array}{ll}
& ^1[D^+ \dots A^-] \rightleftharpoons {}^3[D^+ \dots A^-] \qquad \text{Scheme II} \\
^1D^* + A \Big\langle \quad \uparrow\downarrow & \\
& ^1[D^+A^-] \qquad \qquad \qquad \qquad \text{Scheme I}
\end{array}
\tag{9.1}
$$

As usual, we will denominate the survival probability of excited reactant as $N_S = [^1D^*]$ and the concentrations of the neutral partner as $c = [A]$. We will also use the distributions of ion pairs in the singlet and triplet states, $m_S = {}^1[D^+ \dots A^-]$ and $m_T = {}^3[D^+ \dots A^-]$.

As has been mentioned, the reversible singlet–triplet conversion in the ion–radical pair firmly established experimentally long ago [141] was shown to be responsible for numerous magnetic-field effects, [108,109,119] The intrapair spin conversion leads to a further transformation of the ion–radical pair described by the following supplementary kinetic scheme:

$$
[DA] + h\nu_{fl} \longleftarrow {}^1[D^+A^-] \rightleftharpoons {}^1[D^+ \dots A^-] \Big\langle \begin{array}{l} D^+ + A^- \qquad (9.2a) \\ \\ \uparrow\downarrow \quad D \dots A \end{array}
$$

$$
D \dots A + h\nu_{ph} \longleftarrow {}^3D^* \dots A \longleftarrow {}^3[D^+ \dots A^-] \rightarrow D^+ + A^-. \tag{9.2b}
$$

We now have to introduce the exciplex density N_e and the density of the triplet state $N_T = [^3D^*]$. These quasi-particles are the products of the reaction as well as the free ions presented in concentration $N^+ = [D^+] = [A^-] = N^-$. Below we will propose a set of kinetic equations which establish the relationship between all these quantities, provided there are stochastic transitions from the singlet to triplet ion pair with the rate $3k_s$ and from the triplet to singlet pair with the rate k_s. We also accept the statement made in Ref. 142 that the spin conversion in the excited reactants does not compete with triplet formation from the ion pair and can be neglected.

A.   Basic Equations of the Unified Theory

The main equations for the exciton quenching remain practically the same as
in Section IV:

$$\dot{N}_S = -k(t)cN_S - \frac{1}{\tau_S}N_S \tag{9.3a}$$

$$k(t) = \int W_I^S(r)n(r,t)d^3r + k_e\,n(\sigma,t) \tag{9.3b}$$

$$\frac{\partial}{\partial t}n = -W_I^S(r)n + D\frac{1}{r^2}\frac{\partial}{\partial r}r^2\frac{\partial}{\partial r}n \tag{9.3c}$$

The main difference lies in a generalization of the quenching rate constant
$k(t)$, which accounts not only for a remote ionization with rate $W_I(r)$ but
also the contact creation of the exciplex with the rate constant k_e. Therefore,
the pair distribution should also satisfy not the reflecting but "radiative"
boundary condition:

$$4\pi Dr^2\frac{\partial}{\partial r}n(r,t)\bigg|_{r=\sigma} = k_e n(\sigma,t) \tag{9.4}$$

while initial conditions remain the same, $N_S(0) = 1, n(r,0) = 1$. The
contact creation of the exciplex competes with the remote generation of
solvent separated ions with the position dependent rate (2.5):

$$W_I^S(r) = V_I^2\sqrt{\frac{\pi}{\lambda T}}\exp\left[-\frac{(\Delta G_I + \lambda)^2}{4\lambda T}\right]. \tag{9.5}$$

Here $\Delta G_I(r)$ is the ionization free energy and $\lambda(r)$ is the reorganization
energy of a single classical (low-frequency) assisting mode.

The kinetic equations for the singlet and triplet ion pairs are

$$\dot{m}_S = k_s\,m_T - 3k_s\,m_S + \tilde{D}\frac{1}{r^2}\frac{\partial}{\partial r}r^2 e^{r_c/r}\frac{\partial}{\partial r}e^{-r_c/r}m_S - W_R^0 m_S$$
$$+ W_I^S n(r,t)N_S \tag{9.6a}$$

$$\dot{m}_T = -k_s\,m_T + 3k_s\,m_S + \tilde{D}\frac{1}{r^2}\frac{\partial}{\partial r}r^2 e^{r_c/r}\frac{\partial}{\partial r}e^{-r_c/r}m_T - W_R^T m_T \tag{9.6b}$$

where $W_R^T(r)$ is the rate of triplet pair recombination to the triplet state of the
products and $W_R^0(r)$ is the recombination of the singlet ion pair to the

groundstate. The difference between recombination rates

$$W_R^0(r) = V_0^2 \sqrt{\frac{\pi}{\lambda T}} \exp\left[-\frac{(\Delta G_R + \lambda)^2}{4\lambda T}\right] \quad \text{and}$$

$$W_R^T(r) = V_T^2 \sqrt{\frac{\pi}{\lambda T}} \exp\left[-\frac{(\delta G_R + \lambda)^2}{4\lambda T}\right] \tag{9.7}$$

is due mainly to a difference in the free energies of recombination to the ground state (ΔG_R) and to the triplet state of the neutral reactant (δG_R). They obey the relationship

$$\Delta G_R = \delta G_R + \varepsilon_T \tag{9.8}$$

where ε_T is the energy of the triplet excitation. Sometimes, the rate of ion-pair recombination to the ground state is much slower than that to the triplet state and may completely ignored.

To account for the reversible formation of exciplexes by collapse of a singlet pair at the closest distance of ion approach, we have to introduce the rate constant of such an event, k_a, as well as the unimolecular rate k_d of exciplex dissociation to a solvent separated singlet ion pair. Then the boundary conditions for Eqs. (9.6) take the form

$$\mathbf{j}m_S = 4\pi\sigma^2\tilde{D}\frac{\partial}{\partial r}e^{-r_c/r}m_S(r,t)\bigg|_{r=\sigma} = k_a m_S(\sigma,t) - \frac{k_d N_e(t)}{c} \tag{9.9a}$$

$$\mathbf{j}m_T = 4\pi\sigma^2\tilde{D}\frac{\partial}{\partial r}e^{-r_c/r}m_T(r,t)\bigg|_{r=\sigma} = 0 \tag{9.9b}$$

Here we presume that the triplet exciplexes do not exist because of instability.

The population of exciplexes obeys the equation

$$\dot{N}_e = ck_e n(\sigma,t)N_S(t) + c k_a m_S(\sigma,t) - \frac{1}{\tau_E}N_e \tag{9.10}$$

Here we account for both channels of contact creation of exciplexes, from neutral reactants and solvent separated ions, and for two channels of their decay, with the total rate $1/\tau_E = 1/\tau_{\text{exc}} + k_d$. In line with dissociation to an ion pair (with the rate k_d), the dissipation (via delayed fluorescence or radiationless transfer to the ground state) occurs with the rate $1/\tau_{\text{exc}}$. Another product of recombination (from the triplet ion pair) is the excited triplet of the donor molecule whose density obeys the equation

$$\dot{N}_T = c \int W_R^T(r) m_T(r,t) d^3r \, - \, \frac{1}{\tau_T} N_T \tag{9.11}$$

The decay rate of the triplet state $1/\tau_T$ is also composed from a spin-forbidden dissipation via the phosphorescence channel and the radiationless transfer to the groundstate. Experimentally the density of the triplet excitation can be determined by either phosphorescence or by triplet–triplet absorption of the probe light.

The standard initial conditions for all the products are as follows:

$$m_S(r,0) = 0, \quad m_T(r,0) = 0, \quad N_e(0) = 0, \quad N_T(0) = 0 \, .$$

One can solve at least numerically the whole set of equations and obtain the time behavior of any species, but very often only the quantum yields of luminescence or charge separation are of interest.

In particular, the quantum yield of the exciplex fluorescence is

$$\phi_e = \frac{1}{\tau_{fl}} \int_0^\infty N_e(t) dt = \eta_0 \, \bar{\varphi}_E \tag{9.12}$$

where τ_{fl} is the fluorescence lifetime and $\eta_0 = \tau_{exc}/\tau_{fl}$ is the quantum yield of fluorescence, provided the exciplex is stable. In fact, the exciplexes dissociate so that their quantum yield

$$\bar{\varphi}_E = \frac{1}{\tau_{exc}} \tilde{N}_e(0) \, , \tag{9.13}$$

where $\tilde{N}_e(s)$ is the Laplace transformation of $N_e(t)$.

Similar characteristics exist for the triplet phosphorescence yield as well:

$$\phi_T = \frac{1}{\tau_{ph}} \int_0^\infty m_T(t) dt = \eta_{ph} \, \bar{\varphi}_T \, , \tag{9.14}$$

where $1/\tau_{ph}$ is the phosphorescence rate and $\eta_{ph} = \tau_T/\tau_{ph}$ is the quantum yield of phosphorescence. The quantum yield of triplets

$$\bar{\varphi}_T = \frac{\tilde{N}_T(0)}{\tau_T} \tag{9.15}$$

can be alternatively obtained by the straightforward measurements of the triplet absorption.

The total number of ion radicals generated by bimolecular ionization is

$$P(t) = c \int [m_S(r,t) + m_T(r,t)]d^3r \qquad (9.16)$$

Since $N_S(0) = 1$, the quantum yield of free ions is

$$\phi = P(\infty) = \lim_{s \to 0} c \int [s\tilde{m}_S(r,s) + s\tilde{m}_T(r,s)]d^3r \qquad (9.17)$$

This yield, as well as others, depends crucially on the spin-conversion rate k_s.

The formal solutions of Eqs. (9.10) and (9.11) relates them to reactant and product distributions:

$$N_e(t) = c \int_0^t e^{-[(t-t')/\tau_E]}[k_e n(\sigma,t')N_S(t') + k_a m_S(\sigma,t')] dt' \qquad (9.18a)$$

$$N_T(t) = c \int_0^t e^{-[(t-t')/\tau_T]} \int W_R^T(r)m_T(r,t')d^3r \, dt' \qquad (9.18b)$$

On the other hand, the ion distributions are related to the exciplex density via the boundary condition (9.9a). To eliminate such a dependence, one has to substitute (9.18a) into Eq. (9.9a), transforming the latter into the following:

$$\mathbf{j}m_S = k_a m_S(\sigma,t) - k_d \int_0^t e^{-[(t-t')/\tau_E]}[k_e n(\sigma,t')N_S(t') + k_a m_S(\sigma,t')] dt' \qquad (9.19)$$

B. Specifying the Reaction Scheme: Scheme I

Setting $W_I^S(r) = 0$, we reduce the general reaction scheme to a particular case, which we shall refer to as scheme I. Then, the exciplexes are primarily produced only during the encounters of neutral reactants and their total number is given by the integral of the first term in Eq. (9.10):

$$P_e(t) = ck_e \int_0^t n(\sigma,t')N_S(t')dt' \qquad (9.20)$$

The quantum yield of such exciplexes is given by the expression

$$\psi_e = P_e(\infty) = ck_e\tilde{\theta}(0) \qquad (9.21)$$

where

$$\tilde{\theta}(s) = \int_0^\infty e^{-st'}\theta(t')dt' \quad \text{and} \quad \theta(t') = n(\sigma, t')N_S(t').$$

Since all exciplexes are generated at contact, their normalized space distribution is evidently

$$f_e(r) = \frac{\delta(r - \sigma)}{4\pi r^2} \tag{9.22}$$

Taking advantage of this knowledge, we can use a trick, attributing the exciplex production to a space distributed process with the rate proportional to $f_e(r)$. Excluding the first term under the integral in Eq. (9.19), we must multiply it by $f_e(r)$ and insert into Eq. (9.6) instead of the last term, which becomes zero. Then the modified equation for singlets takes another form but the boundary condition to it becomes homogeneous:

$$\dot{m}_S = k_s m_T - 3k_s m_S + \tilde{D}\frac{1}{r^2}\frac{\partial}{\partial r}r^2 e^{r_c/r}\frac{\partial}{\partial r}e^{-r_c/r}m_S - W_R^0 m_S$$

$$+ k_d k_e f_e(r)\int_0^t e^{-[(t-t')/\tau_E]}\theta(t')dt' \tag{9.23a}$$

$$\mathbf{j}m_S = k_a\left[m_S(\sigma, t) - k_d\int_0^t e^{-\frac{t-t'}{\tau_E}}m_S(\sigma, t')]\,dt'\right] \tag{9.23b}$$

The kinetic equation and the boundary condition for $m_T(r)$ remain the same. The Laplace transformations of the boundary conditions are quite usual

$$\mathbf{j}\tilde{m}_S = \tilde{K}(s)\tilde{m}_S(\sigma, s), \qquad \mathbf{j}\tilde{m}_T = 0 \tag{9.24}$$

but the effective recombination constant of singlet ions is

$$\tilde{K}(s) = k_a\left(1 - \frac{k_d}{s + k_d + 1/\tau_{\text{exc}}}\right) = k_a\frac{1 + s\tau_{\text{exc}}}{1 + (s + k_d)\tau_{\text{exc}}} \tag{9.25}$$

After such a reformulation of the problem, its general solution can be obtained in the following form:

$$m_S(r, t) = k_d k_e \int f_e(r')d^3r'\int_0^t p_{SS}(r, r', t - \tau)\,d\tau\int_0^\tau e^{-\frac{\tau-t'}{\tau_E}}\theta(t')dt' \tag{9.26a}$$

$$m_T(r, t) = k_d k_e \int f_e(r')d^3r'\int_0^t p_{ST}(r, r', t - \tau)\,d\tau\int_0^\tau e^{-\frac{\tau-t'}{\tau_E}}\theta(t')dt' \tag{9.26b}$$

Here $p_{SS}(r, r', t)$ and $p_{ST}(r, r', t)$ are the Green functions for ion pair distributions proceeding from singlet excitation: $p_{SS}(r, r', 0) = \delta(r - r')$, $p_{ST}(r, r', 0) = 0$. They obey the homogeneous equations (9.6) whose Laplace transformation may be presented as follows:

$$\frac{-\delta(r - r')}{4\pi r^2} + s\tilde{p}_{SS} = k_s \tilde{p}_{ST} - 3k_s \tilde{p}_{SS} + \mathbf{L}\tilde{p}_{SS} - W_R^0(r)\tilde{p}_{SS} \qquad (9.27a)$$

$$s\tilde{p}_{ST} = -k_s \tilde{p}_{ST} + 3k_s \tilde{p}_{SS} + \mathbf{L}\tilde{p}_{ST} - W_R^T(r)\tilde{p}_{ST} \qquad (9.27b)$$

where

$$\mathbf{L} = \tilde{D} \frac{1}{r^2} \frac{\partial}{\partial r} r^2 e^{r_c/r} \frac{\partial}{\partial r} e^{-r_c/r}$$

and the boundary conditions are the same as in Eq. (9.24).

C. Contact Recombination

In general the analytic solution of the problem is still not possible. One needs to employ the contact approximation for recombination to the triplet and ground states from the triplet and singlet ion pairs respectively. Such a simplification is reasonable if the backward electron transfer in both channels occurs in the normal transfer region when $W_R(r)$ is quasiexponential and differs from zero only at contact [11,30]. Otherwise one should assume that the backward charge transfer is carried out by a proton and therefore is contact transfer, even in the inverted region [17]. If the recombination of ion pairs is contact, the rate of the reactions can be characterized by the kinetic rate constants:

$$k_r^T = \int W_R^T(r) d^3 r \qquad \text{and} \qquad k_r^S = \int W_R^0(r) d^3 r \qquad (9.28)$$

The former describes triplet formation; the latter represents recombination to the ground state.

In the contact approximation, the electron transfer at $r > \sigma$ is ignored by setting all $W_R = 0$ in Eqs. (9.27):

$$\frac{-\delta(r - r')}{4\pi r^2} + s\tilde{p}_{SS} = k_s \tilde{p}_{ST} - 3k_s \tilde{p}_{SS} + \mathbf{L}\tilde{p}_{SS} \qquad (9.29a)$$

$$s\tilde{p}_{ST} = -k_s \tilde{p}_{ST} + 3k_s \tilde{p}_{SS} + \mathbf{L}\tilde{p}_{ST} \qquad (9.29b)$$

but the boundary conditions (9.24) must be changed to account for the recombination with the rate constants (9.28):

$$\mathbf{j}\tilde{p}_{SS} = [\tilde{K}(s) + k_r^S] \tilde{p}_{SS}(\sigma, r', s), \qquad \mathbf{j}\tilde{p}_{ST} = k_r^T \tilde{p}_{ST}(\sigma, r', s) \qquad (9.30)$$

Performing the Laplace transformation of Eqs. (9.26) and using the results in Eq. (9.17), we obtain the following:

$$\phi = k_d \tau_E \psi_e \varphi(\sigma)$$

The quantum yield of charge separation in contact approximation

$$\varphi(\sigma) = \lim_{s \to 0} s \int [\tilde{p}_{SS}(r, \sigma, s) + \tilde{p}_{ST}(r, \sigma, s)] d^3 r \qquad (9.31)$$

was expressed in Ref. 21 via the Green function of the free diffusion of ions (without any recombination), $\tilde{G}_0$, and $\tilde{p}_{ST}(\sigma, \sigma, 0)$:

$$\varphi(\sigma) = \frac{1 - (k_r^T - K - k_r^S)\tilde{p}_{ST}(\sigma, \sigma, 0)}{1 + (K + k_r^S)\tilde{G}_0(\sigma, \sigma, 0)} \qquad (9.32)$$

Here $K = \tilde{K}(0) = k_a/(1 + k_d \tau_{exc})$ is the rate constant of the reversible association of ions into the exciplex

$$\tilde{p}_{ST}(\sigma, \sigma, 0) = \frac{3k_s J(\sigma, 0)}{1 + 3k_s k_r^T J_0(\sigma, 0) + (K + k_r^S)[\tilde{G}_0(\sigma, \sigma, 0) - 3k_s J_0(\sigma, 0)]} \qquad (9.33)$$

and the auxiliary function

$$J(r', s) = \frac{\tilde{G}_0(\sigma, r', s) - \tilde{G}_0(\sigma, r', s + 4k_s)}{4k_s[1 + k_r^T \tilde{G}_0(\sigma, \sigma, s + 4k_s)]} \qquad (9.34)$$

The quantum yield of the exciplexes defined by Eq. (9.13) takes the following general form:

$$\bar{\varphi}_E = \psi_e \left[\frac{1}{1 + k_d \tau_{exc}} + \frac{k_d \tau_{exc}}{1 + k_d \tau_{exc}} \chi_E \right] \qquad (9.35)$$

where the quantum yield of the backward electron transfer to the exciplex state is

$$\chi_E = K \frac{\tilde{G}_0(\sigma, \sigma, 0) - \tilde{p}_{ST}(\sigma, \sigma, 0)[1 + k_r^T \tilde{G}_0(\sigma, \sigma, 0)]}{1 + (K + k_r^S)\tilde{G}_0(\sigma, \sigma, 0)} \qquad (9.36)$$

There is a simple physical interpretation of these results. The quantum yield of the exciplexes that have not dissociated after creation is represented by the first term in Eq. (9.35), but the exciplexes produced by ion association are taken into account by the second term.

The quantum yield of triplets obtained in Ref. 21

$$\bar{\varphi}_T = \psi_e \, k_d \tau_E \psi_T \tag{9.37}$$

is expressed via the quantum yield of the backward electron transfer to the triplet state:

$$\psi_T = k_r^T \tilde{p}_{ST}(\sigma, \sigma, 0) \tag{9.38}$$

Generally speaking, the total quantum yield of triplets (9.37) is a product of the quantum yields of sequential reactions of exciplex formation (ψ_e), dissociation of the exciplex to a solvent separated ion pair ($k_d \tau_E$) and the recombination of the triplet ion pair produced by spin conversion (ψ_T).

D.   Polar Solvents

All the quantum yields in Eqs. (9.32), (9.35), and (9.38) are expressed through $\tilde{p}_{ST}(\sigma, \sigma, 0)$, the auxiliary function $J(\sigma, 0)$, and finally via the Green function $\tilde{G}_0(\sigma, \sigma, s)$. In highly polar solvents ($r_c = 0$) the latter can be found from its well known general form (8.28):

$$\tilde{G}_0(\sigma, \sigma, s) = \cfrac{1}{4\pi\sigma\tilde{D}\left[1 + \sigma\sqrt{s/\tilde{D}}\right]} = \cfrac{1}{k_{\tilde{D}}\left[1 + \sigma\sqrt{s/\tilde{D}}\right]} \tag{9.39}$$

where $k_{\tilde{D}}$ is the rate constant of diffusional recombination, without any electrostatic interaction between ions. Under such conditions

$$3k_s J(\sigma, 0) = \frac{3}{4k_{\tilde{D}}} \frac{\sigma\sqrt{4k_s/\tilde{D}}}{Z} \tag{9.40}$$

$$\tilde{p}_{ST}(\sigma, \sigma, 0) = \cfrac{3\sigma\sqrt{4k_s/\tilde{D}}}{4(k_{\tilde{D}} + K + k_r^S)Z + 3(k_r^T - K - k_r^S)\sigma\sqrt{\dfrac{4k_s}{\tilde{D}}}} \tag{9.41}$$

where $Z = (1 + k_r^T/k_{\tilde{D}} + \sigma\sqrt{4k_s/\tilde{D}})$.

Inserting $\tilde{G}_0(\sigma,\sigma,0) = 1/k_{\tilde{D}}$ and $\tilde{p}_{ST}(\sigma,\sigma,0)$ from (9.41) into Eq. (9.39), we have

$$\varphi(\sigma) = \frac{1}{1 + k^*/k_{\tilde{D}}} , \qquad (9.42)$$

where the effective kinetic rate constant k^* is given by the expression

$$\frac{k^*}{K + k_r^S} = 1 + \frac{3(k_r^T - K - k_r^S)}{4(K + k_r^S)} \times \frac{\sigma\sqrt{4k_s/\tilde{D}}}{1 + k_r^T/k_{\tilde{D}} + \sigma\sqrt{4k_s/\tilde{D}}} = \mu \qquad (9.43)$$

If there is no spin conversion ($k_s = 0$), then $\mu = 1$ and the effective rate constant is a real rate constant for the recombination of the singlet ion pair to either the exciplex or ground state, $K + k_r^S$. The same is true for the special case when $k_r^T = K + k_r^S$, because the spin conversion does not change anything, if the rates of recombination from the singlet and triplet ion pairs are the same. If $k_r^T < K + k_r^S$, then the kinetic rate constant is smaller ($\mu < 1$) because the transformation of ion pairs from the singlet to triplet state slows down their recombination to the exciplex. In the extreme case of $k_r^T = 0$,

$$\varphi(\sigma) \to \varphi_0(\sigma) = \left[\frac{1 + (K + k_r^S)\mu_0}{k_{\tilde{D}}}\right]^{-1} , \qquad \text{where} \qquad \mu_0 = \frac{4 + \sigma\sqrt{4k_s/\tilde{D}}}{4 + 4\sigma\sqrt{4k_s/\tilde{D}}}$$

Since μ_0 decreases with the spin conversion rate from 1, as $k_s = 0$, to $\frac{1}{4}$ at $k_s \to \infty$, the corresponding kinetic constant $k^* = \mu_0(K + k_r^S)$ falls off correspondingly. In the opposite case of fast triplet recombination, $k_r^T > K + k_r^S$, the effect changes the sign: $\mu > 1$. In the limit $k_r^T \to \infty$, the charge separation quantum yield approaches the minimum

$$\varphi(\sigma) \to \frac{1}{1 + (K + k_r^S)/k_{\tilde{D}} + (3/4)\sigma\sqrt{4k_s/\tilde{D}}}$$

which is lower the faster the spin conversion. In Fig. 76, the change in the effective rate constant k^* with the spin conversion rate k_s is shown for the extreme cases of $k_r^T = 0$ (descending curve) and $k_r^T = \infty$ (ascending curve) and for the border case $k_r^T = K + k_r^S$ (horizontal line).

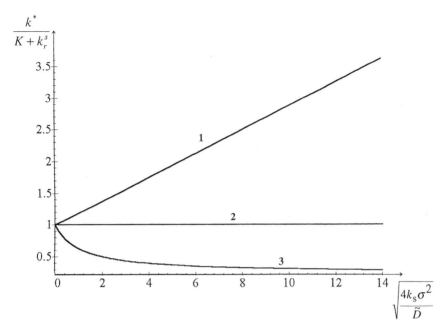

Figure 76. Effective rate constant k^* as a function of the spin conversion rate k_0 for (1) $k_r^T = \infty$, (2) $k_r^T = K + k_r^S$ and (3) $k_r^T = 0$ at $K + k_r^S = 4k_{\tilde{D}}$ [21].

Using Eq. (9.41) for highly polar solvents, the quantum yield of triplet recombination, Eq. (9.38), may be represented in the following form:

$$\psi_T = \frac{3k_r^T F(k_s)}{4(k_{\tilde{D}} + K + k_r^S) + 3(k_r^T - K - k_r^S)F(k_s)} \qquad (9.44)$$

where

$$F(k_s) = \frac{\sigma\sqrt{4k_s/\tilde{D}}}{1 + k_r^T/k_{\tilde{D}} + \sigma\sqrt{4k_s/\tilde{D}}} \qquad (9.45)$$

Fraction $F(k_s)$ monotonously increases with k_s, approaching 1 as $k_s \to \infty$. Near this limit the quantum yield takes the form

$$\psi_T \approx \frac{(3/4)k_r^T}{k_{\tilde{D}} + (3/4)k_r^T + (1/4)(K + k_r^S)} \qquad \text{at} \qquad k_{\tilde{D}}\sqrt{\frac{4k_s\sigma^2}{\tilde{D}}} \gg k_r^T \quad (9.46)$$

This is the ratio of the triplet recombination rate to the total rate of all channels: charge separation, and recombination to the triplet and singlet states. The recombination rate constants are taken with the equilibrium weights of the states $\left(\frac{3}{4}\ \text{and}\ \frac{1}{4}\right)$, because in this limit the equipartition distribution between spin sublevels is established. In another limit $(k_r^T \to \infty)$ it follows from Eq. (9.44) that

$$\psi_T \approx \frac{3\sigma\sqrt{4k_s/\tilde{D}}}{4[1 + (K + k_r^S)/k_{\tilde{D}}] + 3\sigma\sqrt{4k_s/\tilde{D}}} \quad \text{at} \quad k_{\tilde{D}}\sqrt{\frac{4k_s\sigma^2}{\tilde{D}}} \ll k_r^T$$

$$(9.47)$$

Obviously, the quantum yield of triplet recombination is zero at $k_r^T = 0$ but monotonously increases with the spin conversion rate at any $k_r^T > 0$, approaching the limit (9.46) as $k_s \to \infty$ (Fig. 77).

In polar solutions the quantum yield of the exciplex formation via backward electron transfer can be obtained from Eqs. (9.36) and (9.41):

$$\chi_E = K\frac{4 - 3F}{4(k_{\tilde{D}} + K + k_r^S) + 3(k_r^T - K - k_r^S)F} \qquad (9.48)$$

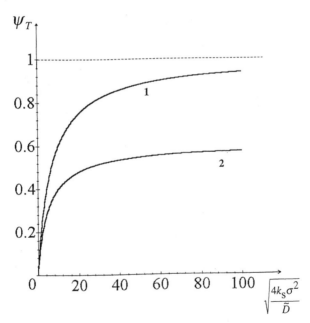

Figure 77. The quantum yield of triplet recombination ψ_T as a function of the spin conversion rate k_0 for (1) $k_r^T = \infty$, (2) $k_r^T = K + k_r^S$ at $K + k_r^S = 4k_{\tilde{D}}$ [21].

where F was introduced in Eq. (9.45). If there is no singlet to triplet conversion we obtain from here the trivial result

$$\chi_E = \frac{K}{k_{\bar{D}} + K + k_r^S} \tag{9.49}$$

which is the ratio of the rate of exciplex formation (K) to the sum of the rates of all competing channels of geminate charge recombination and separation. In the opposite limit of the ultrafast spin conversion $(k_s \to \infty)$ leading to the equipartition distribution between spin states, we have $F = 1$ and correspondingly

$$\chi_E = \frac{K/4}{k_{\bar{D}} + (K + k_r^S)/4 + (3/4)k_r^T} \tag{9.50}$$

This is also the ratio of the same rates including triplet pair recombination, but all of them are taken with their equilibrium weights. Hence, $\chi_E(k_s)$ decreases with the spin conversion rate from the value (9.49), up to the limit (9.50). The falloff is faster, with higher value's of k_r^T:

$$\chi_E = \begin{cases} \dfrac{K(4 + \sqrt{4k_s\sigma^2/\tilde{D}})}{4(k_{\bar{D}} + K + k_r^S) + (4k_{\bar{D}} + K + k_r^S)\sqrt{4k_s\sigma^2/\tilde{D}}} & \text{at} \quad k_r^T = 0 \\[3mm] \dfrac{K}{(k_{\bar{D}} + K + k_r^S) + (3/4)k_{\bar{D}}\sqrt{4k_s\sigma^2/\tilde{D}}} & \text{at} \quad k_r^T \to \infty \end{cases} \tag{9.51}$$

As can be seen from Eq. (9.35), qualitatively the same happens to $\bar{\varphi}_E$ whose decrease with k_s is sharper the higher is k_r^T (Fig. 78).

E.   Specifying the Reaction Scheme: Scheme II

In Scheme II, when the generation of exciplexes in encounters of neutral reactants is completely neglected, $k_e = 0$, and from Eqs. (9.6) the general solution for m_S and m_T can be easily obtained:

$$m_S(r, t) = \int_0^t \int p_{SS}(r, r', t - t') W_I^S(r')n(r', t')d^3r' \, N_S(t')dt' \tag{9.52a}$$

$$m_T(r, t) = \int_0^t \int p_{ST}(r, r', t - t') W_I^S(r')n(r', t')d^3r' \, N_S(t')dt' \tag{9.52b}$$

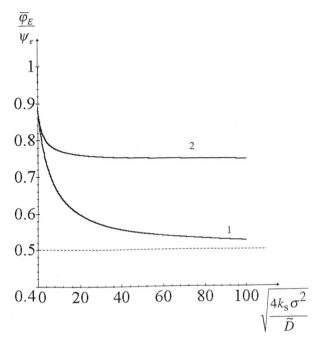

Figure 78. The quantum yield of the exciplexes for (1) $k_r^T = \infty$ and (2) $k_r^T = 0$ at $1/(1 + k_d\tau_{\text{exc}}) = \frac{1}{2}$ and $k_r^S = 0.1\,k_{\tilde{D}}, K + k_r^S = 4k_{\tilde{D}}$ [21].

These are the alternatives to Eqs. (9.26) derived for Scheme I, but the equations for the Green functions or their Laplace transformations, Eqs. (9.27), are the same, as well as the initial and boundary conditions for them.

Making the Laplace transformation of Eqs. (9.52) and using it in Eq. (9.17), we obtain the quantum yield of free ions in the conventional form (6.9):

$$\phi = \psi\bar{\phi}$$

where

$$\bar{\phi} = \int\int[\lim_{s\to 0} s\tilde{p}_{SS}(r, r', s) + \lim_{s\to 0} s\tilde{p}_{ST}(r, r', s)]f_0(r')d^3rd^3r' = \int \varphi(r')f_0(r')d^3r' \tag{9.53}$$

while

$$\psi = c\int m_0(r')d^3r' \quad \text{and} \quad m_0 = \int_0^\infty W_I^S(r')n(r', t')\,N_S(t')dt'$$

were introduced earlier, in Eqs. (4.25) and (4.22), as well as the normalized charge distribution (4.23): $f_0(r) = m_0(r)/\int m_0(r)d^3r$. The main difference between the charge separation quantum yields (9.31) and (9.53) comes from the difference between the initial distributions of ions in alternative schemes of their generation, $f_e(r)$ and $f_0(r)$. Whereas the former is contact [Eq. (9.22)], the latter is distributed and changes with diffusion (viscosity) and ionization free energy (see Section IV).

An exhaustive study of the role of spin interconversion in ion pairs was carried out in Ref. 19, assuming that recombination to the ground state does not occur ($k_r^S = 0$). Although we proved in Section VII.D that this channel can hardly compete with recombination to a triplet, one should not neglect its rate, but should include the latter into the total rate of singlet state recombination $K^{\ddagger} = K + k_r^S$. By this redefinition, we can generalize the previous consideration and obtain for contact recombination:

$$\varphi(r') = 1 - \frac{K^{\ddagger}\tilde{G}_0(\sigma, r', 0)}{1 + K^{\ddagger}\tilde{G}_0(\sigma, \sigma, 0)}$$
$$- \frac{3k_s(k_r^T - K^{\ddagger})[J(r', 0)(1 + K^{\ddagger}\tilde{G}_0(\sigma, \sigma, 0)) - J(\sigma, 0)K^{\ddagger}\tilde{G}_0(\sigma, r', 0)]}{(1 + K^{\ddagger}\tilde{G}_0(\sigma, \sigma, 0))[1 + 3k_s k_r^T J(\sigma, 0) + K^{\ddagger}(\tilde{G}_0(\sigma, \sigma, 0) - 3k_s J(\sigma, 0))]}$$

$$(9.54)$$

where $J(r', s)$ was expressed in Eq. (9.34) via the Green function $\tilde{G}_0(\sigma, r', s)$. For highly polar solvents ($r_c = 0$), it has a well known analytic expression (8.28). The latter was used in Ref. 19 for investigation of the complex dependence of the charge separation quantum yield (9.54) on all the parameters of the problem and initial separation of ions r'. As soon as the solutions for ψ and $f_0(r')$ on one hand and $\varphi(r')$ on the other hand are obtained, the yield calculation reduces to the quadrature given by Eq. (9.53).

Similar relationships exist for two other quantum yields as well. Making the Laplace transformation of Eq. (9.10), at $k_e = 0$, and inserting the result into definition (9.13), we obtain

$$\bar{\varphi}_E = cK\tilde{m}_S(\sigma, 0) = \psi \int \varphi_E(r') f_0(r') \, d^3r' \qquad (9.55)$$

where

$$\varphi_E(r) = K\tilde{p}_{SS}(\sigma, r, 0)] \qquad (9.56)$$

is a partial quantum yield of the exciplexes, equal to the ratio φ_e/η_0 studied in Ref. 19. Borrowing this quantity from there, we generalized it for $k_r^S \neq 0$:

$$\varphi_E(r') = K\left\{ \frac{\tilde{G}_0(\sigma, r', 0)}{1 + K^{\ddagger}\tilde{G}_0(\sigma, \sigma, 0)} - \tilde{p}_{ST}(\sigma, r', 0)\left[1 + \frac{(k_r^T - K^{\ddagger})\tilde{G}_0(\sigma, \sigma, 0)}{1 + K^{\ddagger}\tilde{G}_0(\sigma, \sigma, 0)}\right]\right\}$$

(9.57)

where

$$\tilde{p}_{ST}(\sigma, r', s) = 3k_s \frac{J(r', s)\left[1 + K^{\ddagger}\tilde{G}_0(\sigma, \sigma, s)\right] - J(\sigma, s)K^{\ddagger}\tilde{G}_0(\sigma, r', s)}{1 + 3k_s k_r^T(\sigma, s) + K^{\ddagger}\left[\tilde{G}_0(\sigma, \sigma, s) - 3k_s J(\sigma, s)\right]}$$

(9.58)

Similarly, from definition (9.15) and Eq. (9.52b), we get

$$\bar{\varphi}_T = c\int W_R^T(r)\,\tilde{m}_T(r, 0)d^3r$$

$$= \psi\int W_R^T(r)\int \tilde{p}_{ST}(r, r', 0)f_0(r')d^3r'\,d^3r \qquad (9.59)$$

In the contact approximation $W_R^T(r) = k_R^T\delta(r - \sigma)/4\pi r^2$, so that the last expression reduces to the following:

$$\bar{\varphi}_T = \psi k_R^T\int \tilde{p}_{ST}(\sigma, r', 0)f_0(r')d^3r' \qquad (9.60)$$

If ionization is also a contact one, then the averaging in Eqs. (9.53), (9.55), and (9.60) should be performed with $f_e(r')$ instead of $f_0(r')$. Otherwise, the ionization problem should be solved first and the results used for evaluating the integrals over the initial distributions. This is important if ionization is diffusional or even a kinetic one but in the inverted transfer region.

F.   Magnetic Field Effects in Scheme II

One should discriminate between some important particular cases following from Eqs. (9.54) and (9.57):

1. *No Triplet Production* ($k_r^T = 0$). The recombination of ion pairs is possible from only the singlet state (to either the exciplex or ground state). The charge separation quantum yield should increase with increasing the rate of the spin conversion to the triplet state.

2. *Spin-Independent Recombination* $(k_r^T = K^{\ddagger})$. The rate constant of recombination is the same for both singlet and triplet ion pairs. The quantum yields do not depend on spin conversion, as well as in the case $k_s = 0$.

3. *Mainly Triplet RIP Recombination* $(K^{\ddagger}/k_r^T \to 0)$. The triplet recombination is much faster than the singlet one. The quantum yield of charge separation decreases with increase of the singlet–triplet conversion rate constant.

Let us start from the last case when actually no exciplex formation occurs $(K^{\ddagger} = 0)$. Then $\varphi_E = 0$, while

$$\varphi(r_0) = \frac{4\left[1 + k_r^T \tilde{G}_0(\sigma, r_0, 4k_s)\right]}{4\left[1 + k_r^T \tilde{G}_0(\sigma, r_0, 4k_s)\right] + 3k_r^T\left[\tilde{G}_0(\sigma, r_0, 0) - \tilde{G}_0(\sigma, r_0, 4k_s)\right]}$$
(9.61)

If the ion pairs were born at contact in a polar solution, the result is even simpler:

$$\varphi(\sigma) = \frac{1 + \alpha + \dfrac{k_r^T}{k_{\tilde{D}}}}{1 + \alpha + \dfrac{k_r^T}{k_{\tilde{D}}}\left(1 + \dfrac{3}{4}\alpha\right)}$$
(9.62)

where $\alpha = \sqrt{4k_s\sigma^2/\tilde{D}}$ is the dimensionless rate of spin conversion. As was expected for case 3, the charge separation quantum yield shows the monotonous decrease with this parameter approaching the constant value

$$\varphi_{\infty} = \lim_{\alpha \to \infty} \varphi(\sigma) = \frac{1}{1 + 3k_r^T/4k_{\tilde{D}}}$$

In Fig. 79 the transition from $\varphi = 1$ to φ_{∞} is shown for contact born ion pairs. If there is no spin conversion $(\alpha \to 0)$ in line with the absence of singlet recombination, then all ions survive and $\varphi = 1$. The spin conversion brings into operation the recombination through the triplet channel. The higher the ratio $k_r^T/k_{\tilde{D}}$, the faster and deeper is the fall in the charge separation quantum yield. A similar decrease in this yield is seen for the noncontact ion pair (Fig. 80), but here the effect is smaller, the farther apart the ions are in the beginning. Although we have considered geminate recombination that is under diffusional control $(k_r^T \gg k_{\tilde{D}})$, no qualitative changes are expected in the kinetic control limit $(k_r^T \ll k_{\tilde{D}})$, except that the difference $1 - \varphi_{\infty}$ should be smaller.

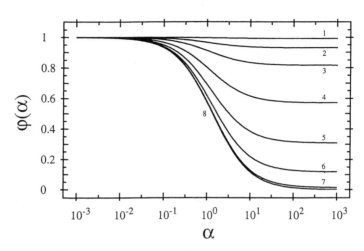

Figure 79. Spin conversion suppression of the separation of contact pairs that do not form exciplexes at $k_r/k_D =$ (1) 0.01; (2) 0.1; (3) 0.3; (4) 1; (5) 3; (6) 10; (7) 100; (8) $10^3 (\alpha = \sqrt{4 k_s \sigma^2 / \tilde{D}}$) [22].

Now let us turn to a more general situation when both recombination channels are active, from either the singlet or triplet states of RIPs. In Fig. 81 we plot the charge separation and fluorescence quantum yields for contact born pairs as functions of the spin conversion rate constant. At chosen ratio $K^{\ddagger}/k_{\tilde{D}} = 1$ and $k_s = 0$, we have $\varphi = \varphi_E = \frac{1}{2}$. In accordance with the

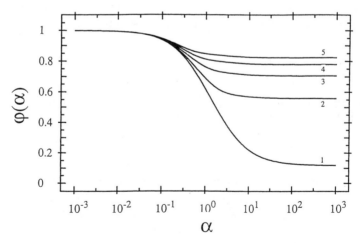

Figure 80. The same as in Fig. 79 but for ions initially separated by distance $r_0 = \sigma(1)$, $2\sigma(2), 3\sigma(3), 4\sigma(4), 5\sigma(5)$ at $k_r/k_D = 10$ [22].

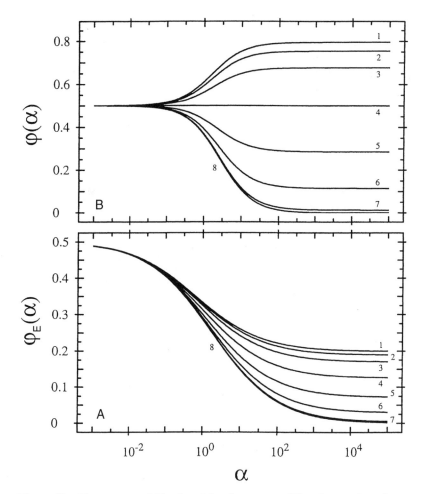

Figure 81. The quantum yields of exciplex fluorescence (A) and separation of contact born ions (B) as functions of the spin conversion rate $k_s(\alpha)$ at $K^{\ddagger}/k_{\bar{D}} = 1$ and $k_r^T/k_{\bar{D}} = (1)\,0.01;\ (2)\,0.1;\ (3)\,0.3;\ (4)\,1;\ (5)\,3;\ (6)\,10;\ (7)\,100;\ (8)\,10^3$ [22].

qualitative classification given above, the charge separation quantum yield does not depend on the spin conversion in case 2 (horizontal line 4), when the rate constants of recombination through both reaction channels are the same ($K^{\ddagger} = k_r^T$). If the singlet recombination is more efficient than the triplet one ($K^{\ddagger} > k_r^T$), the spin conversion increases the charge separation quantum yield approaching case 1. If triplet recombination dominates ($K^{\ddagger} < k_r^T$), the charge separation quantum yield decreases, approaching the limit $K^{\ddagger}/k_r^T = 0$, almost identical with curve 8 (case 3).

Contact generation of ion pairs is the most favorable condition for exciplex formation, but exciplex fluorescence is essentially depleted by spin conversion, opening competing channels for ion recombination and/or their separation. Even when triplet recombination is prohibited ($k_r^T = 0$), the decrease in fluorescence with the increasing spin conversion rate is minimal but nonzero. In this particular case one can obtain the following from Eq. (9.57):

$$\varphi_E(r_0) = K \frac{\tilde{G}_0(\sigma, r_0, 0) - \tilde{p}_{ST}(\sigma, r_0, 0)}{1 + K^{\ddagger}\tilde{G}_0(\sigma, \sigma, 0)} \qquad (9.63)$$

For ions created at the contact distance $r_0 = \sigma$, this expression allows for further simplification:

$$\varphi_E(r_0) = \frac{K}{K^{\ddagger} + k_{\tilde{D}}} \left\{ 1 - \frac{3}{4} \frac{\sigma\sqrt{4k_s/\tilde{D}}}{\left(1 + \sigma\sqrt{4k_s/\tilde{D}}\right)\left(1 + K^{\ddagger}/4k_{\tilde{D}}\right) + 3K^{\ddagger}/4k_{\tilde{D}}} \right\} \qquad (9.64)$$

This is the equation for the upper curve 1 from the family shown in Fig. 81. It describes the gradual decrease of φ_E, with k_s from 1/2 to 1/5 at $K^{\ddagger}/k_{\tilde{D}} = 1$.

There is an essential difference in the separation quantum yields of contact born ion pairs and those created out of contact ($r_0 > \sigma$). As shown in Fig. 82, in case 1 (when $K^{\ddagger} > k_r^T$), the separation quantum yield of the distant ion-pair changes with the rate constant of the spin conversion nonmonotonously—it increases at low k_s and decreases at high k_s. If the spin conversion is slow and can only partially depopulate the initial singlet state, during the pair lifetime $\sigma^2/\tilde{D}$, then the more ion pairs convert into the triplet state (which is practically stable), the higher are their chances to survive and be separated. However, if the fast spin conversion can equalize populations of both spin states before ions come in contact, then the backconversion from the triplet to singlet state stimulates the contact recombination to the exciplex, thus reducing the fraction of survived ions. This effect is most pronounced at $k_r^T = 0$. The curves 1 in Fig. 82 are the closest to this limit that can be easily obtained from the general Eq. (9.54) at $k_r^T = 0$:

$$\lim_{k_r^T \to 0} \varphi(r_0) = 1 - \frac{(\sigma/r_0)(1 + \alpha + 3\exp\{-\alpha(r_0/\sigma - 1)\})}{(1 + 4k_{\tilde{D}}/K^{\ddagger})(1 + \alpha) + 3} \qquad (9.65)$$

This curve has a maximum at finite α, at any $r > \sigma$.

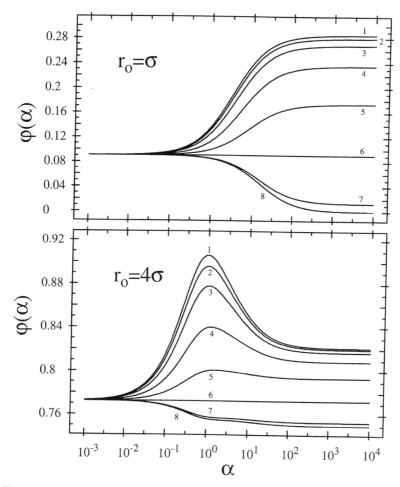

Figure 82. The charge separation quantum yield as a function of the spin conversion rate $\alpha = \sqrt{4k_s \sigma^2 / \bar{D}}$ for contact ($r_0 = \sigma$) and noncontact ($r_0 = 2\sigma$) ion pairs at $K^{\ddagger} = 10k_{\bar{D}}$ and $k_r^T / k_{\bar{D}} = (1)\ 0.01;\ (2)\ 0.1;\ (3)\ 0.3;\ (4)\ 1;\ (5)\ 3;\ (6)\ 10;\ (7)100;\ (8)10^3$ [22].

Assuming that k_s monotonously increases with magnetic field H as in Eq. (8.11), we may characterize the total effect of spin conversion by the relative difference in the quantum yield, at the infinitely fast conversion, $\varphi(r_0, \infty)$, and its spinless (or conversion-free) value $\varphi(r_0, 0)$:

$$M(r_0) = \frac{\varphi(r_0, \infty) - \varphi(r_0, 0)}{\varphi(r_0, 0)}, \qquad M_e(r_0) = \frac{\varphi_E(r_0, \infty) - \varphi_e(r_0, 0)}{\varphi_E(r_0, 0)}$$

$$(9.66)$$

These definitions are analogous to that given in Eq. (8.43), as $H \to \infty$. They are widely used for estimating the magnetic-field effect in charge separation as well as in exciplex fluorescence [140]. As can be seen from Fig. 82, parameter M does not necessarily coincide with a maximal spin conversion effect, but merely characterizes its scale.

Consider, for example, the simplest case of contact born ion pairs, when the general expression (9.54) reduces to the following:

$$\varphi(\sigma) = \varphi_0 \left[1 - \frac{A}{1 + A} \right] \tag{9.67}$$

where $\varphi_0 = [1 + K^{\ddagger}/k_{\bar{D}}]^{-1}$ and

$$A(k_s) = \frac{3(k_r^T - K^{\ddagger})\alpha\varphi_0}{4k_{\bar{D}}(1 + \alpha) + 4k_r^T}$$

From Eqs. (9.66) and (9.67), we obtain

$$M(\sigma) = -\frac{A(\infty)}{1 + A(\infty)} = \frac{3(K^{\ddagger} - k_r^T)}{K^{\ddagger} + 4k_{\bar{D}} + 3k_r^T} \tag{9.68}$$

In the absence of singlet recombination ($K^{\ddagger} = 0$), the spin conversion effect is negative and $M(\sigma)$ changes from 0 to -1 with increasing k_r^T (Fig. 83, top panel). The effect is 0 at $k_r = 0$ because there is no recombination at all and the quantum yield is 1 at any k_s. In contrast, at $k_r^T = \infty$ and $k_s = \infty$ all triplets immediately recombine, so that after instantaneous conversion there are no ions to separate: $\varphi(\sigma, \infty) = 0$ and $M = -1$. The exciplex formation causes M to increase up to the upper limit, which equals 3, because at spin conversion that is infinitely fast the quantum yield is 4 times higher than φ_0.

Similarly, the fluorescence quantum yield (9.57) in the case of contact born ions reduces to a much simpler expression:

$$\varphi_E(\sigma) = \varphi_e(\sigma)[1 - \tilde{p}_{ST}(\sigma, \sigma, 0)(k_{\bar{D}} + k_r^T)] \tag{9.69}$$

where $\varphi_e(\sigma) = K/(K^{\ddagger} + k_{\bar{D}})$ is the quantum yield in the absence of spin conversion and

$$\tilde{p}_{ST}(\sigma, \sigma, 0) = \frac{3\alpha}{4(K^{\ddagger} + k_{\bar{D}})(1 + k_r^T/k_{\bar{D}} + \alpha) + 3\kappa\alpha}$$

$$\longrightarrow \frac{3}{4k_{\bar{D}} + 3k_r^T + K^{\ddagger}} \quad \text{as} \quad \alpha \to \infty$$

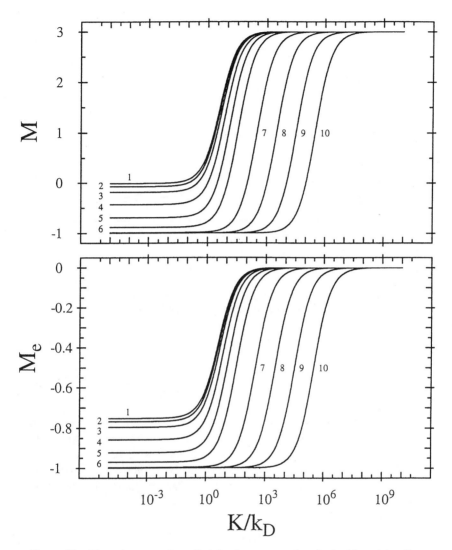

Figure 83. The spin conversion effect in charge separation (top) and exciplex fluorescence (bottom) for the contact born ion pair transforming to an exciplex with the rate K at $k_r^T/k_{\tilde{D}}$ =(1) 0.01; (2) 0.1; (3) 0.3; (4) 1; (5) 3; (6) 10; (7) 100; (8) 10^3; (9) 10^4; (10) 10^5 [22].

Using this result, we obtain for the spin conversion effect on the fluorescence quantum yield

$$M_{\mathrm{e}}(\sigma) = -\lim_{\alpha \to \infty} \tilde{p}_{ST}(\sigma, \sigma, 0)(k_{\tilde{D}} + k_r^T) = -\frac{3(k_{\tilde{D}} + k_r^T)}{4k_{\tilde{D}} + 3k_r^T + K^{\ddagger}} \qquad (9.70)$$

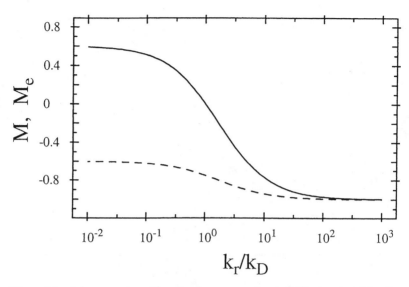

Figure 84. Spin conversion effects for charge separation (solid line) and exciplex fluorescence (dashed line), as functions of the triplet recombination rate at $K^{\ddagger}/k_{\bar{D}} = 1$ [22].

If no singlet recombination occurs ($K^{\ddagger} = 0$), this parameter varies from $-\frac{3}{4}$ in the absence of triplet recombination, to -1 at infinitely fast recombination. Singlet recombination causes this parameter to increase up to 0 (Fig. 83, bottom panel B).

Both M and M_e change rapidly from the minimum to the maximum values when the rate of singlet recombination becomes comparable to the separation rate measured by $k_{\bar{D}}$. In the middle of these S-like curves (at $K^{\ddagger} = k_{\bar{D}}$), the effect essentially depends on the triplet recombination rate (Fig. 84). This is especially the case with the sign alternating M, that turns to zero at $k_r^T = K^{\ddagger}$ as usual.

X. CONCLUSIONS

Here are the principal achievements of the present theory:

- The limitations of a popular contact approximation were overcome and the real microscopic rates of the forward and backward electron transfer were initially used.

- The primitive "exponential model" was discarded and diffusional control of geminate recombination was established at noncontact creation or recombination of ions.

- The "unified theory" eliminates an arbitrary choice of initial ion separation, links together the sequential stages of bimolecular ionization and geminate recombination, and for both of them predicts an essential distortion of the free-energy gap (FEG) law by encounter diffusion.

- The theory allows for further transformation of ions to neutral radicals or exciplexes, discrimination between the spin states of reactants and products, accounting for the spin and magnetic-field effects in charge accumulation and separation kinetics.

But even more important is clear realization of the fact that the unified theory makes a complete break with the traditional rate description of chemical reactions. It has been shown [92] that UT is simply a particular application to electron transfer of a much more general integral encounter theory (IET) developed earlier [143]. In application to forward and backward electron transfer, the rate concept is represented by the set of differential equations, (4.6a) and (6.4). As soon as UT takes into account the geminate recombination of ions, Eq. (6.4) is substituted by Eq. (6.6), which is apparently beyond the rate concept. Only Eq. (4.6a) for bimolecular ionization remains differential, while the geminate recombination cannot be presented this way any more.

In fact, the situation is even worse. Neither of the equations may be differential in the general case, but both of them should be integral. Breaking the traditional rate description of the process, the IET suggests the following set of integral equations instead of (4.6):

$$\dot{N} = -c \int_0^t R^*(\tau)N(t - \tau)d\tau - \frac{N}{\tau_D} \tag{10.1a}$$

$$\dot{P} = c \int_0^t R^\dagger(\tau)N(t - \tau)d\tau \tag{10.1b}$$

where two memory functions, $R^*(\tau)$ and $R^\dagger(\tau)$, are introduced, instead of the single rate constant $k_I(t)$. They are given in IET the unique definitions through the electron-transfer rates, $W_I(r)$ and $W_R(r)$, and corresponding pair distributions affected by transfer and diffusion. Equations (10.1) can be reduced to those of the unified theory in the particular case of irreversible ionization $(-\Delta G_i \gg T)$ [92]. However, neither resonant nor endergonic reactions $(\Delta G_i \geq 0)$ can be considered in this way. Therefore only the integral theory gives a proper description of the FEG law in the normal region when both the recombination channel (to the ground and excited singlet state) are taken into account [92]:

$$D + A \leftarrow D^* + A \; \rightleftharpoons \; [D^+ \ldots A^-] \begin{array}{l} \nearrow D^+ + A^- \\ \searrow [D \ldots A] \end{array} \qquad (10.2)$$

Because ions are stable products, their geminate recombination is always followed by bimolecular recombination in the volume:

$$[D \ldots A] \leftarrow [D^+ \ldots A^-] \rightarrow D^+ + A^- \rightarrow D + A \qquad (10.3)$$

This also can be taken into consideration only within IET formalism [144].

Apparently IET is a kind of "memory function formalism" widely used in other branches of physical kinetics, but it was first applied to chemistry. The IET seems to be the most fundamental, elegant, and powerful method that has been easily extended to energy-transfer phenomena, including singlet–triplet conversion, impurity and biexciton quenching, and triplet annihilation [144,145].

A very complex problem of the last kind, solved in 1998 [145], involves reactions of similar particles. This is delayed fluorescence from $^1A^*$ produced by a triplet–triplet annihilation after biexciton quenching and interconversion of singlet excitations [146]:

$$A + {}^1A^* \leftarrow {}^1A^* + {}^1A^* \rightarrow {}^1A^* + {}^3A^* \begin{array}{l} \nearrow A + {}^3A^* \rightarrow 2A \\ \searrow {}^3A^* + {}^3A^* \Longrightarrow A + {}^1A^* \end{array}$$

In both cases the anti-Smoluchowski behavior of the annihilation rate at short times is expected. It was really measured and given a reasonable explanation [146] that was confirmed and extended in Ref. [145] The biexcitonic reactions can also be responsible for bimolecular ionization in liquids and solids, which results in nonlinear photoconductivity of the systems. IET has been employed for the estimation of its stationary value [147], as well as for the kinetics of a photocurrent after δ pulse [148] and permanet excitation of arbitrary strength [149].

It is now widely recognized that the integral theory is the most fundamental and general approach to all reactions in solutions, assisted by the random walk of reactants and reaction products. A few alternative approaches of a similar kind were recently developed. Although they are partially compared in Refs. [62 and 150], a more complete comparison and analysis of the applications listed above should probably be left for another review. [151]

ACKNOWLEDGMENT

At the invitation of the Rockefeller Foundation, the author had an opportunity to spend several productive weeks during May–June 1999 to write the manuscript of this review, at the Villa Serbelloni in Bellagio, Italy. The quiet atmosphere and beautiful surroundings of the Villa certainly allowed concentration on the topic and strongly facilitated work on this review. Author is also grateful to Israel Science Foundation supported the majority of his original works cited in this chapter.

References

1. J. N. Onuchic and D. N. Beratan *Annu. Rev. Biophys. Biomol. Struct.* **21**, 349 (1992); C. C. Moser, and P. L. Dutton, *Biochimica Biophysica Acta* **1101**, 171 (1992).

2. J. R. Miller, *Science* **189**, 221 (1975); T. Guarr, M. Mc.Guire, S. Strauch, and G. McLendon *J. Am. Chem. Soc.* **105**, 616 (1983).

3. J. R. Miller, L. T. Calcaterra, and G. L. Closs *J. Am. Chem. Soc.* **106**, 3047 (1984); J. M. Warman, M. P. de Haas, M. N. Paddon-Row, E. Cotsaris, N. S. Hush, H. Oevering, and J. W. Verhoeven, *Nature* **320**, 615 (1986).

4. B. E. Bowler, A. L. Raphael, and H. B. Gray, *Long-Range Electron Transfer in Donor (Spacer) Acceptor Molecules and Proteins*, in Progress in Inorganic Chemistry: Bioinorganic Chemistry, Vol. 38, S. J. Lippard, ed., Wiley, New York, 1990, p. 259.

5. D. De Vault, *Quantum-Mechanical Tunneling in Biological Systems*, Camridge Univ. Press, Cambridge, UK, 1984.

6. D. Rehm and A. Weller, *Israel J. Chem.* **8**, 259 (1970).

7. D. E. Eads, B. G. Dismer, and G. R. Fleming, *J. Chem. Phys.* 1136 (1990).

8. R. J. Harrison, B. Pears, G. S. Beddard, J. A. Cowan, and J. K. M. Sanders *Chem. Phys.* **116**, 429 (1987); I. R. Gould, J. E. Mozer, D. Ege, and S. Farid *J. Am. Chem. Soc.* **110**, 1991 (1988); G. Grampp and G. Hetz, *Ber. Bunsenges. Phys. Chem.* **96**, 198 (1992).

9. A. I. Burshtein, A. A. Zharikov, N. V. Shokhirev, O. B. Spirina, and E. B. Krissinel, *J. Chem. Phys.* **95**, 8013 (1991).

10. B. S. Brunchwig, S. Ehrenson, and N. Sutin, *J. Am. Chem. Soc.* **106**, 6859 (1984).

11. A. I. Burshtein, P. A. Frantsuzov, and A. A. Zharikov, *Chem. Phys.* **155**, 91 (1991).

12. A. I. Burshtein, *Chem. Phys. Lett.* **194**, 247 (1992).

13. R. C. Dorfman and M. D. Fayer, *J. Chem. Phys.* **96**, 7410 (1992).

14. A. I. Burshtein, E. Krissinel, and M. S. Mikhelashvili, *J. Phys. Chem.* **98**, 7319 (1994).

15. A. I. Burshtein and E. Krissinel, *J. Phys. Chem.* **100**, 3005 (1996).

16. A. I. Burshtein and A. Yu. Sivachenko, *J. Photochem. Photobiol. A* **109**, 1 (1997).

17. A. I. Burshtein and P. A. Frantsuzov, *Chem. Phys. Lett.* **263**, 513 (1996).

18. A. I. Burshtein and A. Yu. Sivachenko, *Chem. Phys.* **235** 257 (1998).

19. A. I. Burshtein and E. Krissinel, *J. Phys. Chem. A* **102**, 816 (1998).

20. E. Krissinel, A. I. Burshtein, N. Lukzen, and U. Steiner, *Mol. Phys.* **96** 1083 (1999).

21. A. I. Burshtein, *Chem. Phys.* **247**, 275 (1999).

22. A. I. Burshtein and E. Krissinel, *J. Phys. Chem. A* **102**, 7541 (1998).

23. R. A. Marcus, *J. Chem. Phys.* **24**, 966 (1956); idem., **43**, 679 (1965).

24. V. G. Levich and R. R. Dogonadze, *Dokl. Akad. Nauk SSSR* **124**, 123 (1959); idem., **133**, 158 (1960).

25. A. M. Kuznetsov, *Charge Transfer in Physics, Chemistry and Biology,* Gordon & Breach, 1995.

26. Yu. Georgievskii and A. I. Burshtein, *J. Chem. Phys.* **101**, 10858 (1995).

27. Yu. Georgievskii, A. I. Burshtein, and B. Chernobrod, *J. Chem. Phys.* **105**, 3108 (1996).

28. M. V. Smoluchowski, *Z. Phys. Chem.* **92**, 129 (1917).

29. K. M. Hong and J. Noolandi, *J. Chem. Phys.* **68** (1978).

30. A. I. Burshtein and P. A. Frantsuzov, *Chem. Phys.* **212**, 137 (1996).

31. S. Efrima and M. Bixon, *Chem. Phys. Lett.* **25**, 34 (1974).

32. J. Jortner and M. Bixon, *J. Chem. Phys.* **88**, 167 (1988).

33. L. D. Zusman, *Chem. Phys.* **49**, 49 (1980).

34. A. I. Burshtein and B. I. Yakobson, *Chem. Phys.* **49**, 385 (1980).

35. A. I. Burshtein, P. A. Frantsuzov, and A. A. Zharikov, *J. Chem. Phys.* **96**, 4261 (1992).

36. M. Tachiya and S. Murata, *J. Phys. Chem.* **96**, 8441 (1992).

37. A. Yoshimori, K. Watanabe, and T. Kakitani, *Chem. Phys.* **201** 35 (1995).

38. D. L. Dexter, *J. Chem. Phys.* **21**, 836 (1953).

39. L. Song, S. F. Swallen, R. C. Dorfman, K. Weidemaier, and M. D. Fayer, *J. Phys. Chem.* **97**, 1374 (1993).

40. A. I. Burshtein and P. A. Frantsuzov, *J. Luminesc.* **51**, 215 (1992).

41. A. I. Burshtein and A. G. Kofman, *Opt. Spectrosc.* **40**, 175, (1976).

42. P. A. Frantsuzov, N. V. Shokhirev, and A. A. Zharikov, *Chem. Phys.* **236**, 30 (1995).

43. A. I. Burshtein and N. V. Shokhirev, *J. Phys. Chem. A* **101**, 25 (1997).

44. J. R. Miller, L. T. Calcaterra, and G. L. Closs, *J. Am. Chem. Soc.* **106**, 3047 (1984).

45. H. D. Roth, *J. Phys. Chem.* **97**, 13037 (1993).

46. R. A. Marcus and P. Siders, *J. Phys. Chem.* **86**, 622 (1982).

47. I. R. Gould and S. Farid, *J. Phys. Chem.* **97**, 13067 (1993).

48. T. Ohno, A. Yoshimura, and N. Mataga, *J. Phys. Chem.* **90**, 3295 (1986).

49. T. Kakitani, N. Matsuda, A. Yoshimori, and N. Mataga, *Prog. React. Kinet.* **20**, 347 (1995).

50. S. A. Rice, *Diffusion—Limited Reactions* in *Comprehensive Chemical Kinetics,* Elsiever, 1985, Vol. 25, p. 404.

51. F. C. Collins and G. E. Kimball, *J. Colloid Sci.* **4**, 425 (1949).

52. T. R. Waite, *J. Chem. Phys.* **28**, 103 (1958).

53. R. M. Noyes, in *Progress of Reaction Kinetics,* Ed. G. Porter, ed., Pergamon Press, 1961, Vol. 1.

54. U. M. Gosele, *Prog. React. Kinet.* **13**, 63 (1984).

55. M. Tachiya, *Radiat. Phys. Chem.* **21** 167 (1983)

56. N. N. Tunitskii and Kh. S. Bagdasar'yan, *Opt. Spectr.* **15**, 303 (1963); S. F. Kilin, M. S. Mikhelashvili, and I. M. Rozman, *Opt. Spectr.* **16**, 576 (1964); I. I. Vasil'ev, B. P. Kirsanov, and V. A. Krongaus, *Kinet. Kataliz* **5**, 792 (1964).

57. I. Z. Steinberg and E. Katchalsky, *J. Chem. Phys.* **48**, 2404 (1968).

58. G. Wilemski and M. Fixman, *J. Chem. Phys.* **58**, 4009 (1973).

59. A. B. Doktorov and A. I. Burshtein, *Sov. Phys. JETP* **41**, 671 (1975).

60. V. P. Sakun, *Physica A* **80**, 128 (1975); A. B. Doktorov, idem., *A* **90**, 109 (1978).

61. V. M. Agranovich and M. D. Galanin, *Electron Excitation Energy Transfer in Condensed Matter*, North-Holland, Amsterdam, 1982, A. I. Burshtein Theor. and Exp. Chem. **1**, 365 (1965).

62. A. I. Burshtein, I. V. Gopich, and P. A. Frantsuzov, *Chem. Phys. Lett.* **289**, 60 (1998).

63. M. Yokoto and M. Tonimoto, *J. Phys. Soc. Jpn.* **22**, 779 (1967).

64. K. Allinger and A. Blumen, *J. Chem. Phys.* **72**, 4608 (1980).

65. A. I. Burshtein, *Sov. Phys. Usp.* **27**, 579 (1984).

66. M. Inokuti, and F. Hirayama, *J. Chem. Phys.* **43**, 1978 (1965).

67. A. I. Burshtein, E. I. Kapinus, I. Yu. Kucherova, and V. A. Morozov, *J. Luminese.* **43**, 291 (1989).

68. A. Szabo, *J. Chem. Phys.* **93**, 6929 (1989).

69. P. A. Frantsuzov, N. V. Shokhirev, and A. A. Zharikov, *Chem. Phys. Lett.* **236**, 30 (1995).

70. G. Wilemski and M. Fixman, *J. Chem. Phys.* **60**, 866 (1974).

71. M. J. Philling and S. A. Rice, *J. Chem. Soc. Farad. Trans.* 2, **71**, 1563 (1975); Yu. A. Berlin, *Dokl. Akad. Nauk SSSR* **223**, 625 (1975).

72. A. A. Zharikov and N. V. Shokhirev, *Chem. Phys. Lett.* **190**, 423 (1992).

73. A. D. Scully, T. Takeda, M. Okamoto, and S. Hirayama, *Chem. Phys. Lett.* **228**, 32 (1994).

74. D. F. Thomas, J. J. Hopfield, and W. M. Augustyniak, *Phys. Rev.* **140**(1A), 202 (1965).

75. K. I. Zamaraev and R. F. Khairutdinov, *Russ. Chem. Rev.* **47** 518 (1978).

76. S. Murata and M. Tachiya, *J. Phys. Chem.* **100**, 4064 (1996); *J. Chim. Phys.* **93**, 1577 (1996).

77. A. I. Burshtein and Yu. I. Naberuhin, *Preprint 130*, Institute of Theoretical Physics, Ukrainean Academy of Science, 1974.

78. S. Nishikava, T. Asahi, T. Okada, N. Mataga, and T. Kakitani, *Chem. Phys. Lett.* **185**, 237 (1991).

79. C. Turro, J. M. Zaleski, Y. M. Karabatos, and D. G. Nocera, *J. Am. Chem. Soc.* **118** 6060 (1996).

80. N. Mataga, J. Kanda, and T. Okada, *J. Phys. Chem.* **90**, 3880 (1986).

81. N. Mataga, T. Asahi, J. Kanda, T. Okada, and T. Kakitani, *Chem. Phys.* **127**, 249 (1988).

82. A. I. Burshtein, *J. Chem. Phys.* **103**, 7927 (1995).

83. H.-J. Wolff, D. Bürßner, and U. Steiner, *Pure Appl. Chem.* **67**(1), 167 (1995).

84. A. I. Burshtein, A. A. Zharikov, and N. V. Shokhirev, *J. Chem. Phys.* **96** 1951 (1992).

85. A. A. Zharikov and N. V. Shokhirev, *Chem. Phys. Lett.* **186**, 253 (1991).

86. N. N. Korst, *Theor. Math. Phys.* **6**, 196 (1971).

87. M. Tachiya, *J. Chem. Phys.* **71**, 1276 (1979).

88. B. Sippp and R. Voltz, *J. Chem. Phys.* **79**, 434 (1983).

89. M. Abramowitz and I. A. Stegun, eds., *Handbook of Mathematical Functions* Dover, New York, 1965.

90. E. B. Krissinel, N. V. Shokhirev , and K. M. Salikhov, *Chem. Phys.* **137**, 207 (1989).

91. A. Yoshimori, K. Watanabe, and T. Kakitani, *Chem. Phys.* **201**, 35 (1995).

92. A. I. Burshtein and P. A. Frantsuzov, *J. Chem. Phys.* **106**, 3948, (1997).

93. E. B. Krissinel and N. V. Shokhirev, *Differential Approximation of Spin-Controlled* and *Anisotropic Diffusional Kinetics*, Siberian Academy Scientific Council 'Mathematical Methods in Chemistry', **1989**, *Preprint 30* (Russian); *Diffusion-Controlled Reactions*, Vol. 22, Krissinel' and Shokhirev, Inc., 1990, *DCR User's Manual*, 11-20-1990.

94. H. Gan, U. Leinhos, I. R. Gould, and D. G. Whitten, *J. Phys. Chem.* **99**, 3566 (1995).

95. I. R. Gould and S. Farid, *J. Phys. Chem.* **97**, 3567 (1993).

96. P. P. Levin, P. F. Pluznikov, and V. A. Kuzmin, *Chem. Phys. Lett.* **147**, 283 (1988).

97. A. I. Shushin, *Chem. Phys. Lett.* **118**, 197 (1985); idem., **162**, 409 (1989).

98. D. Beckert, M. Plüschau, and K. P. Dinse, *J. Phys. Chem.* **96**, 3193 (1992).

99. M. Plüschau, G. Kroll, K. P. Dinse, and D. Beckert, *J. Phys. Chem.* **96**, 8820 (1992).

100. H.-J. Werner, H. Staerk, and Albert Weller, *J. Chem. Phys.* **68**, 2419 (1978).

101. D. J. C. Chan, D. J. Mitchell, and B. W. Ninham, *J. Chem. Phys.* **70**, 2946 (1979).

102. B. I. Yakobson and A. I. Burshtein, *Int. J. Chem. Kinet.* *XII* 261 (1980).

103. S. F. Swallen, K. Weidemaier, H. L. Tavernier, and M. D. Fayer, *J. Phys. Chem.* **100**, 8106 (1996).

104. S. F. Swallen, K. Weidemaier, and M. D. Fayer, *J. Phys. Chem.* **104**, 2976 (1996).

105. A. A. Kornyshev and J. Ulstrup, *Chem. Phys. Lett.* **126**, 74 (1986).

106. Y. H. Meyer and P. Plaza, *Chem. Phys.* **200**, 235 (1995).

107. K. Schulten, H. Staerk, A. Weller, H.-J. Werner, and B. Nickel, *Z. Physikalische Chemie Neue Folge* **101**, 371 (1976).

108. K. M. Salikhov, Yu. N. Molin, R. Z. Sagdeev, and A. L. Buchachenko, *Spin Polarization and Magnetic Effects in Radical Reactions*, Budapest, Akad. Kiado, 1984.

109. U. E. Steiner and Th. Ulrich, *Chem. Rev.* **89**, 51 (1989).

110. A. Weller, H. Staerk, and R. Treichel, *Faraday Disc. Chem. Soc.* **78**, 217 (1984).

111. N. Mataga, Yu Kanda, T. Asahi, H. Miyasaka, T. Okada, and T. Kakitani, *Chem. Phys.* **127**, 239 (1988).

112. M. Tomkiewicz and M. Cocivera, *Chem. Phys. Lett.* **8**, 595 (1971).

113. W. Bube, R. Haberkorn, and M. E. Michel-Beyerle, *J. Am. Chem. Soc.* **100**, 5993 (1978).

114. L. Sterna, D. Ronis, S. Wolfe, and A. Pines, *J. Chem. Phys.* **73**, 5493 (1980).

115. H. Hayashi and S. Nagakura, *Bull. Chem. Soc. Jpn.* **57**, 322 (1984).

116. A. A. Zharikov and N. V. Shokhirev, *Z. Phys. Chem. Bd.* **99**, 2643 (1992).

117. Z. Schulten and K. Schulten, *J. Chem. Phys.* **66**, 4616 (1977).

118. R. Haberkorn, *Chem. Phys.* **26**, 35 (1977).

119. A. L. Buchachenko, R. Z. Sagdeev, and K. M. Salikhov, *Magnetic and Spin Effects in Chemical Reactions*, Nauka, Novosibirsk, 1978.

120. A. B. Doktorov, O. A. Anisimov, A. I. Burshtein, and Yu. N. Molin, *Chem. Phys.* **71**, 1 (1982); M. Tadjikov and Yu. N. Molin, *Chem. Phys. Lett.* **233**, 444 (1995).

121. D. Bürßner, H-J. Wolff, and U. Steiner, *Z. Phys. Chem. Bd.* **182**, 297 (1993), *Angewandte Chemie Int. Ed. Engl.* **33**, 1772 (1994).

122. A. A. Zharikov and A. I. Burshtein, *J. Chem. Phys.* **93**, 5573 (1990).

123. A. I. Burshtein, E. B. Krissinel, U. E. Steiner, *Mol. Phys.* (in press).

124. A. I. Burshtein, I. V. Khudiakov, and B. I. Yakobson, *Progress in Reaction Kinetics*, Pergamon Press, New York, 1984, Vol. 13, p. 221.

125. U. E. Steiner and D. Bürßner, *Z. Phys. Chem. Neue Folge* **196**, 159 (1990).

126. R. G. Mints and A. A. Pukhov, *Chem. Phys.* **87**, 467 (1984).

127. L. Lüders and K. Salikhov, *Chem. Phys.* **117**, 113 (1987).

128. K. M. Heel, M. Hansen, and K. Pickard, *The Maple V Learning Guide for Release 5*, Springer-Verlag, Berlin, 1977; M. Abell and J. Braselton, *The Maple V Handbook*, Academic Press, 1994.

129. A. Weller, *Z. Physikalische Chemie Neue Folge* **130**, 129 (1982).

130. A. Weller, H. Staerk, and R. Treichel, *Faraday Disc. Chem.* **78**, 271 (1984).

131. Th. Förster, in *The Exciplex*, M. Gordon and W. R. Ware, Academic Press, New York, 1975 p. 1.

132. K. Schulten, H. Staerk, A. Weller, H.-J. Werner, and B. Nickel, *Z. Physikalische Chemie Neue Folge* **101**, 371 (1976).

133. K. Kikuchi, Y. Takanashi, M. Hoshi, T. Niva, T. Katagiri, and T. Miyashi, *J. Phys. Chem.* **95**, 2378 (1991).

134. B. R. Arnold, D. Noukakis, S. Farid, J. I. Goodman, and I. R. Gould, *J. Am. Chem. Soc.* **117**, 4399 (1995).

135. Y. Tanimoto, K. Hasegawa, N. Okada, M. Itoh, K. Iwai, K. Sujioka, F. Takemura, R. Nakagaki, and S. Nagakura, *J. Phys. Chem.* **93**, 3587 (1989).

136. N. Kh. Petrov, V. N. Borisenko, M. V. Alfimov, T. Fiebig, and H. Staerk, *J. Phys. Chem.* **100**, 6368 (1996).

137. K. Kikuchi, Y. Takanashi, T. Katagiri, T. Niva, and T. Miyashi, *Chem. Phys. Lett.* **180**, 403 (1991).

138. I. R. Gould, R. H. Young, L. J. Mueller, and S. Farid, *J. Am. Chem. Soc.* **116**, 8176 (1994).

139. D. N. Nath and M. Chowdhuru, *Chem. Phys. Lett.* **109**, 13 (1984).

140. T. Fiebig, W. Kühnle, and H. Staerk, *J. Fluoresc.*, **7**(Suppl.), 29S (1997).

141. N. Mataga and M. Ottolenghi. in *Molecular Association*, R. Foster, ed., Academic Press, 1979, Vol. 2, p. 1.

142. N. Orbach and M. Ottolenghi in *The Exciplex*, M. Gordon and W. R. Ware, eds., Academic Press, New York, 1975, p. 75.

143. N. N. Lukzen, A. B. Doktorov, and A. I. Burshtein, *Chem. Phys.* 289 (1986); A. I. Burshtein and N. N. Lukzen, *J. Chem. Phys.* **103**, 9631 (1995); idem., **105**, 9588 (1996).

144. A. I. Burshtein and P. A. Frantsuzov, *J. Chem. Phys.* **107**, 2872 (1997).

145. A. I. Burshtein and P. A. Frantsuzov, *J. Luminesc.* **78**, 32 (1998).

146. B. Nickel, H. E. Wilhelm, and A. A. Ruth, *Chem. Phys.* **188**, 267 (1994).

147. P. A. Frantsuzov and A. I. Burshtein, *J. Chem. Phys.* **109**, 5957 (1998).

148. E. B. Krissinel, O. A. Igoshin, and A. I. Burshtein, *Chem. Phys.* **247**, 261 (1999).

149. O. A. Igoshin and A. I. Burshtein JCP 112, N24 (2000).

150. I. V. Gopich and A. I. Burshtein, *J. Chem. Phys.* **109** 2833 (1998).

151. A. I. Burshtein, Proceedings of Trombay Symposium on Radiation and Photochemistry (TSRP 2000). Bhabha Atomic Research Center, Mumbai, Part II, 545–556 (2000).

AUTHOR INDEX

Friedrich, B., 149(71), *190*
Friesner, R. A., 290(47), 302(56,58), *309–310*
Frishman, E., 125(32), 129(32), *188, 191*
Fröhlich, J., 15(25), *119*
Frost, C. A., 328(29), 377(29), *416*
Frost, L., 390(100), 396–398(100), 405(100), *418*
Fukuda, H., 366(70), 374(70), 388(70), 397(70), *417*
Fundaminsky, A., 313(10), *416*
Fuss, W., 253(188), *261*

Gailitis, M, 329(31), *416*
Galanin, M. D., 451(61), *585*
Galbraith, H. W., 76(54), *120*
Gallagher, A. C., 373(79), *417*
Galperin, M. Yu, 172(117), *191*
Gambogi, J. E., 195(31), 202(89,91), 205(31,89), 206(105), 215(105), 217(91,105), 226(89), 229(89), 242(89), *256, 258–259*
Gan, H., *586*
Gandhi, S. R., 138(59), *189*
Gans, R., 362(62), *417*
Garton, W. R. S., 356(52), *417*
Gasser, W., 386(89), *418*
Gaubatz, U., 138(52), 155(52), *189*
Gauss, J., 99(62), *120*
Gauthier, J. M., 124(13), 159(101), *188, 191*
Gazdy, B., 289(43), *309*
Geers, A., 196(39), 202(39), *257*
Gelbart, W. M., 195(8–9), 206(102), 217(102), *256, 258*
Geltman, S., 326–327(25), *416*
George, T. F., 159(87), *190*
Georges, R., 206(100), *258*
Georgievskii, Yu., 424(26–27), 550(26–27), *584*
Gerber, G., 201(75), 249(75), *258*
Gerger, R. B., 218(142), *260*
Gerstein, J. I., 159(86), *190*
Gilles, J. M. F., 76(53), *120*
Girard, B., 248(173), 230(173), *260*
Go, J., 195(36), 206(106), 215(106), 217(129), *257, 259*
Goldstein, H., 347(39), *416*
Golub, I. I., 130(41), *189*
Goodman, J. I., 557(134), *587*
Gopich, I. V., 452(62), 462(62), 548(150), 582(62), *585, 587*
Gordon, R. J., 201(76), 249(76), *258*
Gosele, U. M., 448(54), *584*

Goswami, D., 138(59), *189*
Gotlieb, D., 292(49), *309*
Gould, I. R., 421(8), 444(17), 506(95), 557(134), 581(138), *583–584, 586–587*
Gould, P. L., 145(60), *190*
Graf, G-M., 15(25), *119*
Gram, P. A. M., 328(29), 377(29), *416*
Grampp, G., 421(8), *583*
Gray, H. B., 421(4), *583*
Gray, S. K., 218(140), *260*
Greene, C. H., 313(14–15), 339(14–15), 345(15), 354(14), 359(14–15), 379(14), 381(15), 390(14), *416*
Grischkowski, D. G., 138(44), *189*
Grobe, R., 124(24–25), *188*
Gruebele, M., 195(17–18,37), 196(17,45–46,50,55), 198(61–62), 199(17,64–65), 200(17,50,55,68), 201(45,50,55,81), 204(37,50), 205(37,68), 207(37,55), 209(18,64), 211(17–18,64), 212(18,64,122), 213(17–18,64–65), 214(18,64), 215(17–18,64), 217(45,127), 218(18,37,50,64,122,140), 220(55), 224–225(55), 226(17,50,68), 229(17,37), 230(18,37,55), 231(17,45,55), 232(55,64), 233(64), 237(18,37,45,50,122,140), 238(122), 239(37,64), 240(46,68,122), 241(18,37,122), 242(17,64), 243(17,37,45,55), 244(17,50,55), 245(50,55,68), 249(37,50), 251(50,55), 252(185), 253(191), *256–261*
Grujic, P. V., 388(91*), 418*
Gruner, D., 206(101), *258*
Guarr, T., 421(2), *583*
Gudzenko, L. I., 159(89), *190*
Guisti-Suzor, A., 124(20), 149(66), 159(99), *188, 190–191*
Gunslaus, I. C., 201(79), *258*
Guo, X., 415(105), *418*
Gurvich, L. V., 159(89), *190*
Guyot-Sionnest, P., 194(2), *256*
Gwinn, W. D., 267(4), 269(4), 274(4), *308*

Haberkorn, R., 525(113), 526(118), *586*
Haberland, H., 149(69), *190*
Hagedorn, G., 5(8–9), *119*
Hagen, S. J., 201(80), *258*
Hahn, E. L., 138(48), *189*
Hall, M. A., 252(184), *261*
Halonen, L., 226(157), 228(157), *260*
Hamilton, C. E., 195(26), *256*
Hamilton, I. P., 269(25), 279(25), *309*

SUBJECT INDEX